THE NASA MARS CONFERENCE

AAS PRESIDENT
  Captain David C. Honhart

Office of the Chief of Naval Research

VICE PRESIDENT - PUBLICATIONS
  Diana P. Hoyt

Virtual Enterprises

SERIES EDITOR
  Dr. Horace Jacobs

Univelt, Incorporated

SERIES ASSISTANT EDITOR
  Robert H. Jacobs

Univelt, Incorporated

VOLUME EDITOR
  Duke B. Reiber

Media Consultant

COVER PHOTO: In February, 1977, Viking Lander 1 (VL-1) spent three days slowly digging the deep hole visible to the right of the meteorology boom's elbow. The science team was interested in acquiring a sample of deeper material to compare with a variety of earlier samples taken essentially from the surface around the lander, particularly to see if the "anomalously high sulfur content" would be consistent. At this point, VL-1 had been in operation for nearly seven months, and changes taking place in the martian environment were in evidence. Shadows were much lighter (less contrast) and the sky was brighter, indicating that the amount of dust in the atmosphere had increased significantly and was scattering light more efficiently. Because of the amount of surface sampling activity already conducted, a veneer of reddish martian dust had accumulated on exposed surfaces of the lander, as can be seen on the RTG wind cover (left) and the top of a leg support (center). The large boulder (left-center) is about 25 meters from the lander and was extensively imaged throughout VL-1's mission life. It was known to the Viking flight team as "Big Joe." [JPL/P-18641]

The NASA
Mars Conference:

· Our Current Knowledge and Understanding of Mars ·
· The Present and Future Unmanned Exploration of Mars ·
· Issues and Options for the Manned Exploration of Mars ·

Sponsored by
National Aeronautics and Space Administration
Space Science and Applications

Co-Sponsored by
American Astronomical Society

AN AMERICAN *Astronautical* SOCIETY PUBLICATION

THE NASA MARS CONFERENCE

Edited by
Duke B. Reiber

Volume 71
SCIENCE AND TECHNOLOGY SERIES
A Supplement to Advances in the Astronautical Sciences

Proceedings of the NASA Mars Conference held
July 21-23, 1986, at the National Academy of
Sciences, Washington, D.C.

Published for the National Aeronautics and Space Administration
and the American Astronautical Society
by Univelt, Incorporated, P.O. Box 28130, San Diego, California 92128

First Printing 1988

ISSN 0278-4017
ISBN 0-87703-293-9 (Hard Cover)
ISBN 0-87703-294-7 (Soft Cover)

Published for the National Aeronautics and Space Administration
and the American Astronautical Society
by Univelt, Inc., P.O. Box 28130, San Diego, California 92128

Printed and Bound in the U.S.A.

NASA MARS CONFERENCE

CONFERENCE PLANNING & COORDINATION

Sponsoring Representation . . . . Dr. Burton L. Edelson
Associate Administrator, Space Science & Applications
National Aeronautics and Space Administration

General Chairman . . . . Dr. Geoffrey A. Briggs
Director, Solar System Exploration Division
National Aeronautics and Space Administration

ORGANIZING COMMITTEE

Organizing Chairman . . . . Mr. James S. Martin, Jr
Aerospace Consultant

Chairman, Session 1 . . . . Dr. Gerald A. Soffen
Associate Director, Space & Earth Sciences, NASA-GSFC

Chairman, Session 2 . . . . Mr. James R. French, Jr
American Rocket Company

Chairman, Session 3 (am) . . . . Dr. Michael Duke
Solar System Exploration, NASA-JSC

Chairwoman, Session 3 (pm) . . . . Dr. Carol Stoker
National Research Council Fellow, NASA-ARC

Conference Coordinator . . . . Mr. A. Gary Price
Head, Office of External Affairs, NASA-LaRC

Administrative Coordinator . . . . Ms. Lu Agee
JL Associates

EDITORIAL COORDINATION

Editor, Review Coordinator, Document Composition/Typography . . Mr. Duke B. Reiber[1]
Technical Writer/Editor, Media Consultant
Satellite Beach, FL

---

[1] Mr. Reiber was managing editor and deputy chief of the Viking Public Affairs Support Office at JPL during the first three years of Viking mission operations. Previously, he served as Viking lander program communication services coordinator (in Denver) and as information administrator for the Viking launch program (in Florida) for Martin Marietta Aerospace, with responsibility for the development and management of engineering, scientific, marketing, news, and expository media. He is a recipient of the NASA Public Service Medal.

NASA MARS CONFERENCE

FOREWORD

The proceedings of the *NASA Mars Conference*, held July 21–23, 1986, were prepared for publishing in a manner that has had a degree of impact on the nature of the document's content. Prepared papers were not solicited or used during the editorial process, and the individual presentations represented as papers in these proceedings were instead developed from transcriptions of the relatively informal orations given at the Conference. Indeed, the character of some of that informality has--in spite of significant editorial modification--carried over into the more formally structured and worded elements of these proceedings, such that they contribute a unique "readability" to the document. However, because the technical content of such talks is subject to error, by accident during the presentation or due to misunderstanding during the transcription and editorial processing, the authors have been allowed the opportunity to improve on the content of the edited presentation transcriptions within the framework of the original context and style. Only two of the twenty-nine Conference participants failed to respond to the review opportunity, and only one presentation was subsequently omitted. And, because an extended period of time has passed since the NASA Mars Conference was held, during which some of the near-term future programs and studies discussed have changed in important ways, e.g., rescheduling of the *Mars Observer* mission, such updates *will* be reflected in the content of those presentations. In this way the proceedings will represent not only a comprehensive overview of America's past, present and future exploration of Mars as discussed at the Mars Conference, but will in fact reflect a more current and accurate representation of the subject material presented at that time.

Finally, much of the informational footnoting offered throughout the document is a product of the editor's experience, and the editor alone is responsible for footnotes concluded with editor identification. The "generalist" information provided in this manner reflects a desire to beneficially broaden or clarify the reader's understanding.

I would like to express my appreciation for the cooperation I was given during this difficult editorial process, and for the opportunity it afforded to again experience Mars -- as was my exciting privilege for ten years as an American Viking.

NMC Editor: D B Reiber

NASA Organizations or Facilities Referenced by Abbreviation: ARC/Ames Research Center, Moffett Field (Mountain View), CA; GSFC/Goddard Space Flight Center, Greenbelt, MD; JPL/Jet Propulsion Laboratory (NASA–Caltech), Pasadena, CA; JSC/Johnson Space Center, Houston, TX; KSC/Kennedy Space Center (Merritt Island), FL; LaRC/Langley Research Center, Hampton, VA; MSFC/Marshall Space Flight Center (Huntsville), AL.

NASA MARS CONFERENCE
Contents: Welcoming and Keynote Addresses

Mars Exploration Based on Leadership and Success

Opening Comments and "Viking Replay"[1] by Conference Chairman, Dr. Geoffrey A. Briggs

> *"Mars exploration will be extremely valuable in developing long-duration space flight experience that will surely be important in the distant future. And, of course, it is represented by a key word in NASA's program planning vocabulary, exploration, which differentiates the planetary program from all other science programs. The planetary program is <u>exploration</u> in its truest and finest sense!"*

~~~~~

> *"I want to reaffirm NASA's commitment to the long-term importance of space science in its space program. Right now, as we face decisions on space systems for the present and for the future, there is no better time to remind ourselves about why we are creating those systems and where we plan to send them. By doing the necessary planning and study now, we are setting the stage so that we can move out fast when we are ready to continue the exploration of the solar system -- particularly Mars."*

---

[1] Dr. Briggs opened the **NASA Mars Conference** by replaying an audio tape of control room events at JPL during the historic moments of the first Viking landing on Mars.

NASA MARS CONFERENCE
Contents: Session 1

Our Current Knowledge and Understanding of Mars

Session Chairman, Dr. Gerald A. Soffen

---

[1] The presentation identifier code is made up of two characters. The first is a number (1, 2 or 3) representing the first, second or third of the Mars Conference day-long sessions. The second is an alphabetic character indicating the presentation's sequence, beginning with "A" each day. This two-character identifier is also incorporated into the figure numbers associated with each presentation.

[2] Dr. Levin's presentation represents extensive work and collaboration with his former Viking biology team (labeled-release) colleague, Dr. Patricia A Straat.

NASA MARS CONFERENCE
Contents: Session 2

The Present and Future Unmanned Exploration of Mars

Session Chairman, Mr. James R. French, Jr.

---

[1] The editor was unsuccessful in his attempts to reestablish contact with Professor Mizutani (Nagoya University, Japan) to conduct a proper and needed review cycle. Due to our concern about language and technical interpretations, we regrettably have--and with our sincere apologies to Prof. Mizutani--elected not to include his presentation in these proceedings of the NASA Mars Conference.

NASA MARS CONFERENCE
Contents: Session 3

Issues and Options for the Manned Exploration of Mars

Morning Session Chair: Dr. Michael B. Duke
Afternoon Session Chair: Dr. Carol Stoker

PRESENTATION   **PAGES**

The Engineering Aspects of Human Exploration

The Human Dimension of Mars Exploration

NASA MARS CONFERENCE
Contents: Appendices

Color Illustrations and Index to NMC Figures

COLOR -- Although the use of color illustrations has been allowed in this document as an exception to AAS publishing convention, it has been constrained to a contiguous Color Display Section (Appendix A). Color illustrations are identified as figures in the conventional manner within the context of each presentation, but each is specifically noted as a *color* illustration and is referenced to its display page number in the **Color Display Section** appendix.

CONVENIENT COLOR DISPLAY -- As a convenience to readers who would like to examine the color illustrations while reading associated text, the illustrations are formatted face up on the outer display space of foldout pages such that they can be displayed in full view of the reader as he/she is reading within the document if desired. Figure numbers, full captions, and page-of-reference are also given.

INDEX TO NASA MARS CONFERENCE FIGURES -- All of the figures presented in these proceedings of the **NASA Mars Conference** (including color) are listed in the index provided as Appendix B. They are listed in the order of their presentation in the context of these proceedings, and each includes a figure number, a caption, and the page number(s) of its location (a text-reference page is also given in the case of color and is further explained on the title page of the **Color Display Section**).

CROSS-REFERENCED ALTERNATE ILLUSTRATIONS -- A few of the presenters were unable to provide hardcopy illustrations for visuals used at the Mars Conference. Some of these references were deleted, but many were editorially translated into textual matter for integration with the content of the presentation in order to pre-serve their informational value. Some of these textual conversions will include cross referenced notations where the editor was able to clearly determine that illustrations used in other presentations were comparable to those being omitted at the point of the notation. In these instances, both the alternate presentation and the page on which the illustration referenced in this manner is presented are included in the notation. We hope this additional editorial feature will benefit readers.

INDEX PAGES

OTHER BOOKS ON MARS

THE CASE FOR MARS I, Volume 57, *Science and Technology Series*, ed. Penelope J. Boston, 2nd printing 1987, 348p

THE CASE FOR MARS II, Volume 62, *Science and Technology Series*, ed. Christopher P. McKay, 2nd printing, 1988, 730p

THE CASE FOR MARS III, To be published in 1988-1989.

Order from Univelt, Inc., Publishers for the American Astronautical Society
P.O. Box 28130, San Diego, California 92128

NASA MARS CONFERENCE

July 21–23, 1986
National Academy of Science
Washington, D.C.

Welcome and Introduction: . . . . Dr. Burton I. Edelson[1]
Associate Administrator

AAS 86-151   Space Science & Applications

~~~~~

Keynote Address: . . . . Dr. James C. Fletcher
Administrator

AAS 86-152   National Aeronautics and
Space Administration

---

[1] Dr. Edelson has since returned to private life and is no longer in the NASA post with which he is identified in this document.

NASA MARS CONFERENCE:
Welcome and Introduction

Dr. Burton I. Edelson
Associate Administrator, Space Science & Applications
National Aeronautics and Space Administration
Washington, D.C.

Welcome to the NASA MARS CONFERENCE. The purpose of this Mars Conference is to celebrate the history of our exploration of Mars. First, we'll endeavor to understand, acknowledge, appreciate and delve into the marvelous level and breadth of scientific knowledge that has been produced by missions of the past -- particularly the Viking landings of which we are celebrating the tenth anniversary during this conference. And then we'll look forward, both near and far, to consider and discuss plans for both unmanned and manned missions to Mars now coming into view on the future's horizon.

IN APPRECIATION

I'd like first to express appreciation to some very important people and recognize their contributions to this conference. Dr. Geoffrey Briggs, the present director of our Solar System Exploration Division at NASA Headquarters and a former Viking science team member at JPL, is the general chairman of these proceedings. Also involved are many other former Viking hands. While I couldn't possibly name all of them, I must recognize James S. Martin, Jr., and Dr. Gerald A. Soffen, the former project manager and project scientist, respectively, of the outstandingly successful *Viking Project* -- which I believe is still the premiere scientific exploration program in NASA's history to date.

Special Welcome

I would also like to recognize [in attendance at the conference] Madeline Mutch, widow of Dr. Thomas A. "Tim" Mutch[1]. Tim Mutch was a very forceful and popular participant--as well

---

[1] Dr. Mutch served as team leader for the Viking Lander Imaging Team. He was later appointed Associate Administrator, Office of Space Science, a post he served in until his death in 1980. A very popular Viking scientist during mission operations, Dr. Mutch has since been honored by having Viking Lander 1 (VL-1) dedicated to his memory as the: Thomas A. Mutch Memorial Station. As such, VL-1 is "Dedicated to the memory of Tim Mutch, whose imagination, verve and resolve contributed greatly to the exploration of the solar system." This memorial is dated January 7, 1981, and is signed by then-NASA Administrator, Dr. Robert Frosch. Dr. Frosch charged future explorers of Mars with the responsibility of placing the memorial plaque on VL-1. Dr. Mutch formerly was professor of geological sciences at Brown University, and among the books he authored or co-authored is The Geology of MARS (1976), written with R.E. Arvidson, J.W. Head, K.L. Jones, and R.S. Saunders (published by Princeton University Press). He was awarded the Exceptional Scientific Achievement Medal by NASA for his Viking work. {Ed.}

as scientific leader--throughout the **Viking** program, who then served as Associate Administrator, Office of Space Science, for NASA until his death. I'm sure he is warmly remembered at this conference and that he is equally known--out of professional respect--to those in planetary science who never had the pleasure of working with him.

We also welcome our foreign colleagues and look forward to participating with them in future projects. Among them are several who, in fact, are participants in our conference: Roger Bonnet of the European Space Agency, Hitoshi Mizutani of Nagoya University in Japan, and Heinrich Wanke of Max-Planck-Institute of Fuer Chemie, Federal Republic of Germany. We are particularly pleased to have overseas visitors with us to consider future possibilities with regard to science and exploration on the **Red Planet.**

We have an outstanding turn-out for the conference, estimated to be about 500 people. Of this number, Jim Martin and Geoff Briggs estimated that about 200 of the old **Viking** team are present, and a number of them are on the presentation agenda. As previously noted, the conference coincides with the tenth anniversary of the first **Viking** landing (July 20, 1976). And now, ten years later, we're still pouring over and analyzing that program's incredible accumulation of scientific data and still finding useful results. While it appears that we have not yet discovered positive evidence of life, an issue that is once again addressed during this conference, it still remains--for some--an open question. Perhaps the answers will be found as we look to the future and carry out the three-phase program reflected in the organization of our conference agenda.

Phase 1: The Near Term

First of all, we have an approved Mars mission. It will be the first of our **Planetary Observers**, *Mars Observer*, scheduled for launch in 1992. It is an on-going program, it is approved, it is funded, and it is on the books. **Mars Observer** is an important mission. Its objective is to provide an essential global database for Mars, based on geoscience and climatological observations (and mapping) over time for at least one martian year. RCA is prime contractor for the development of the **Mars Observer** spacecraft.

Phase 2: Sample Return

The second phase, which we've been studying intently and which has gotten a lot of attention, is **Mars sample return** (MSR). We believe sample return is the kind of mission that will be possible in the reasonably near future, and we would like to attempt it late in this decade. The Solar System Exploration Committee (SSEC) has made a positive recommendation for such a mission, calling it the next major priority for the "red planet." The SSEC report has been published and is available to interested parties. A sample return mission would conduct very essential science and acquire unique data on the compositional history and evolution of Mars. Mars sample return represents an exciting technical challenge. A very sophisticated robot spacecraft would be required, and it would have the ability to carry out--with help from Earth--very complex operations on the surface of this distant world. With it, we will gain access to new territory each day of the mission, thanks to the mobility of the rover vehicle and whatever rover technology we can develop prior to the mission.

Phase 3: Human Exploration

During the third phase, about which there has also been a lot of exciting discussion, we look forward a decade or two into the next century to a manned expedition to Mars. Mars is the only inner planet in our solar system where the potential for human missions and habitation is presently believed reasonable to pursue. The final report of the **National Commission on**

4

Space[2], the outstanding product of Tom Paine's chairmanship, suggests that outposts and bases on Mars could lead to a permanent settlement and serve as a bridge between the two worlds.

THE IMPORTANCE OF MARS

The SSEC has stated that Mars spans the evolutionary gap between primitive and evolved worlds; between Mercury and the moon on one hand and our own highly evolved planet, Earth, on the other. It is significant that different scientific panels with different chairmen, looking at different areas of science, seem to focus on Mars for a wide variety of scientific reasons. Atmospheric scientists, geologists, life scientists, and those from many other disciplines, all find Mars to be a very exciting and interesting planet.

As Technology Driver -- Of course, the exploration of Mars represents more than science, it provides a driver for the development of advanced technology. To the extent that all aspects of the space program and planetary science provide a focus for technological development, there will be outstanding opportunities for the development of automated and robotic technologies. Mars exploration will be extremely valuable in developing long-duration space flight experience that will surely be important in the distant future. And, of course, it is represented by a key word in NASA's program planning vocabulary, *exploration*, which differentiates the planetary program from all other science programs. The planetary program is exploration in its truest and finest sense.

International Relations -- The future exploration and scientific study of Mars is filled with many possibilities for international cooperation, and we hope there will be international participation in all three of the phases I've outlined. There are, for example, opportunities for the coordination of the **Mars Observer** program (and its results) with other planetary efforts through cooperation with other countries and other space agencies toward an unmanned sample return mission. And, when we get around to a manned mission to Mars, international cooperation is virtually essential because of its cost and what it represents in physical technology.

But international cooperation on the scale of a manned mission to Mars could also be important to the future of mankind. We have an outstanding example of that quality in the coordinated quadripartite activities of the IACG (Inter-Agency Consultative Group) and common-ally activities. Cooperation during all of the Halley cometary activities was outstanding. There were missions from four countries plus the International Halley Watch (IHW), and all were very well coordinated in the spirit of the common focus. Though we have seen a lot of international cooperation over the years, I would say that the Halley activity was the best example yet; we're certainly going to try to extend that kind of cooperation into the future. We simply can't afford to wait another 76 years to have a cooperative international program as good and as deep as was this one.

We regret that our Soviet colleagues are not present at this conference, but it followed COSPAR a bit too closely. Several of the leading Soviet space scientists attended COSPAR expecting that they would have the opportunity to come here afterward, but they were not able to do so due to the extent of travel involved. They were in fact well prepared for our conference, and one of the presentations they planned to give, describing the dynamic 1988 Phobos Mission, is being contributed for them by James Head in these proceedings [NMC-2A].

---

[2] The final report of the **National Commission on Space** chaired by Dr. Paine has been published for public interest and consideration under the title: **Pioneering the Space Frontier** (A Bantam Book ... May, 1986). {Ed.}

INTRODUCTION: DR. JAMES C. FLETCHER, NASA ADMINISTRATOR

Now, it is my pleasure to introduce the current NASA Administrator, Dr. James C. Fletcher. Dr. Fletcher provides a special perspective relative to the exploration of Mars, because he also was the NASA Administrator in 1976 when the **Viking** landings took place. I'm delighted that he could join us and assist in our review of the past, to see what we have already accomplished with our scientific exploration of Mars, and then to provide a look ahead to what we hope to accomplish in the future. ■

REFERENCES

This space is otherwise available for notes.

~~~~~~~~

PLANETARY SCIENCE AND MARS:
A Dynamic Future Built on Success

Dr. James C. Fletcher
Administrator
National Aeronautics and Space Administration
Washington, D.C.

Few people--least of all myself--would have predicted early in 1986 that I would address a NASA MARS CONFERENCE that summer as NASA Administrator, and would again be discussing the Viking landers and missions as was my great pleasure ten years earlier in 1976. But, like Mark Twain, I've found that "Predictions are very difficult to make, especially when they deal with the future." The tragedy of the 51-L Challenger mission has produced many unanticipated changes in its wake, while placing before us a task of immense importance.

OPENING COMMENTS

It is good to see so many old friends from the **Viking** program with whom I was fortunate enough to spend my time during those exciting days of the first **Viking** landing on Mars. But it is even better to see so many young and unfamiliar faces. It's a sign that Mars continues to fascinate each new generation of scientists and explorers -- as it always has. The **red planet** certainly holds enough mysteries to keep all of us--and many generations to come--busy for a very long time.

Rebuilding for a New Start

This conference comes at a critical time in the nation's space program. A time of intense scrutiny, of thorough reassessment, and of careful rebuilding. NASA is doing all that must be done to make the program safe, reliable, and a renewed source of pride and inspiration for America. Slowly but surely, we are getting the space program back on track.

As we move forward in that recovery process, I can assure you that our commitment to a vigorous program of space science and exploration will continue to play a central role in our planning and in our actions. The interruptions and delays to many of our programs, as a result of the hiatus in shuttle flights, have created extraordinary pressures on the continuity of our research program and on the stability of the research community. To moderate these impacts and to accelerate our return to space, we will be taking a number of specific actions.

First, we are working hard to prepare a new STS (Space Transportation System) flight manifest that will aid our planning for the next few years. Satisfying the combination of NASA and non-NASA requirements will be extremely difficult. Nevertheless, within the limitations that confront us, space science missions will receive a high priority as that new space shuttle manifest is prepared.

Second, the cancellation of the STS Centaur program has posed a very special problem for the Galileo, Ulysses, and Magellan missions, which were designed around its capabilities. We are evaluating other launch options, including: 1) the use of alternative upper stages with the shuttle orbiter, and 2) the use of expendable launch vehicles (ELV's). Just as there will be great competition for space on early shuttle flights, there will also be heavy demand for ELV's. However, we will make every effort to get these three important missions on their way as soon as we can.

Third, as I reported to the President just prior to the conference, NASA strongly supports the need for a mixed fleet of launch vehicles to satisfy the launch requirements and appropriate actions needed to revitalize America's ELV capability. Thus, we are actively considering the utility of a mixed shuttle and ELV fleet to launch NASA science missions.

Finally, I want to reaffirm NASA's commitment to the long-term importance of space science in its space program. For many years, space science has received about twenty percent (20%) of NASA's budget. I am aiming to maintain at least that level of support in the coming years. Moreover, as we proceed to develop the United States Block·1 Space Station, space science will continue to be a major driver in its design. The relationship between science and the space station is quite symbiotic: without a healthy science program and budget at NASA, the rationale for the station's existence in compromised. At the same time, it is my strong belief that the space station will open up new and powerful opportunities for advancements in space science.

SPOTLIGHT ON CHANGE

I believe this conference is especially important and timely relative to our planning to return to space. Right now, as we face decisions on space systems for the present and for the future, there is no better time to remind ourselves about why we are creating those systems and where we plan to send them. By doing the necessary planning and study now, we are setting the stage so that we can move out fast when we are ready to continue the exploration of the solar system -- particularly Mars.

The Mars Database

The decade that has passed since the Viking landers landed on Mars has given us time to assess what we have learned about that planet from our explorations. Although most of the data have come from American spacecraft, the effort to explore Mars has been an international one that dates back to the early 1960's. All in all, there have been 17 attempts to explore Mars with spacecraft, eight by the United States and nine by the Soviet Union. Not all of them have been successful, but we have amassed an extraordinary amount of information.

Now that NASA has plans to return to Mars in 1992[1] with Mars Observer, it is very important to summarize our knowledge base for the new generation of scientists and explorers represented at this conference. So, in a very real sense, this conference comes at a watershed time. Having learned a lot about Mars in the past, we can use that experience to prepare to learn more in the future by framing the right questions now. This conference will help us do that, and it will also give us some idea of how to find the answers.

---

[1] This date reflects a change in the planned launch date for Mars Observer. At the time of the NASA Mars Conference, the program launch date was scheduled for August, 1990 (as originally stated by Dr. Fletcher). In April, 1987, NASA formally directed the program to replan its schedule for a 1992 launch date. {Ed.}

The Evolution of our Knowledge

Since the dawn of civilization, Mars has tantalized and fascinated the human species. For centuries it was a mystery planet, appearing as a blurred red globe in telescopes and teasing astronomers with changing bright and dark patches on its surface. We debated about whether its so-called "canals" were real and whether it might possibly harbor life. Science fiction writers had a field day with Mars, making it both the destination for space-faring humans and the origin planet for all manner of hostile and friendly extraterrestrial beings.

Now that Mars has become the target of scientific exploration, we know that--like Earth--it has an atmosphere -- albeit, a very thin one. And, like Earth, it is a world with weather, climate, clouds, and icecaps. Through the **Viking** landers, we have seen its pink sky--colored by fine, wind-blown dust--and have probed and sampled its reddish rocks and soil. Spacecraft have flown by or orbited the planet as well. They revealed a vast canyon complex lying along and just below the equator, smooth plains, huge shield volcanoes in an otherwise etched and channeled northern hemisphere, and an ancient, heavily cratered southern hemisphere that has preserved traces of the primordial bombardment that prevailed throughout the solar system's early history.

Even though the last **Viking** spacecraft--Viking Lander 1--finally fell silent early in November of 1982, the exploration of Mars continues in our laboratories and right here at this conference. Indeed, from here **Viking** seems both very close and very far away at the same time. Looking back, the first **Viking** landing was truly an incredible event. It was the first soft landing of an American spacecraft on Mars, and perhaps the most significant landing on another world since Apollo 11. In fact, you may recall that the **Viking** landing came seven years to the day after Apollo 11 astronauts Neil Armstrong and Buzz Aldrin made the first manned lunar landing, as Michael Collins orbited the moon above them in the Apollo command module.

VIKING MISSION RESULTS

As you know, the **Viking** program returned a wealth of data. Indeed, many attending this conference are familiar with that data because they helped to obtain and interpret it. But, because it was such a fantastic achievement to get it at all, from a distance ranging from about 65 million kilometers (a median opposition range) to more than 350 million kilometers (prior to and following superior conjunction), I think the high points of that mission activity are worthy of review.

You will recall that the two **Viking** launches (August and September, 1975) deployed four spacecraft at Mars -- two orbiters and two landers[2]. It is interesting to note that the spacecraft elements were originally designed with the objective of achieving an operational life of only

---

[2] Launch and Mars orbit insertion (MOI) milestones: Viking 1 was launched from Cape Canaveral AFS on August 20, 1975 and Viking 2 on September 9, both by Titan IIIE/Centaur launch vehicles from the same launch pad. Viking 1's MOI occurred on June 19, 1976 while Viking 2 achieved MOI August 7. These two interplanetary spacecraft were each made up of an orbiter and a lander, each of which was identical to its counterpart. Perhaps due to the amount of spacecraft/flight team interaction over a long period of time, the loss of each of the four spacecraft in turn felt to many like the loss of a friend. This invoked a unique common sensitivity among those involved that was later described by Kermit Watkins (JPL), project manager during the **Viking** Continuation Mission, as "the Viking spirit" -- still very much in evidence upon the occasion of the 10th anniversary of the first **Viking** landing and the **NASA Mars Conference** reported in these proceedings. {Ed.}

90 days, but remained in operation for from two to six years. The orbiters served as communication relays for the landers, photographed the surface, and mapped the planet's thermal and water-vapor characteristics over time[3]. During their mission lives, the orbiters obtained global imagery of the planet and watched its polar caps and dust storms come and go.

Shortly after the two flight spacecraft went into their orbits around the planet, their landers separated and descended to the martian surface[4]. Each landed safely on relatively flat plains in the northern hemisphere following an extremely difficult and painstaking site search and certification process, and both sites are below Mars datum (zero elevation) within large basins. Their common mission was to photograph the terrain, measure and monitor the atmosphere and climate, determine the nature and inorganic composition of the soil, and conduct chemical and biological tests on the soil in a search for evidence of rudimentary life forms.

Lander 1 landed on the plains of **Chryse Planitia** (22.27°N, 47.97°W), the "golden plain," July 20, 1976, and functioned well into 1982. We settled lander 2 onto the plains of **Utopia Planitia** (47.67°N, 225.74°W) September 3, 4014 miles from lander 1 and approximately 25 degrees farther north in latitude.

The four spacecraft told a story of a planet with a quiescent present, but with evidence of a very different and active past. The martian soil is apparently composed of weathered basalt lava, and includes materials such as sulfates and clay minerals that may have been deposited by water. Volcanoes two-and-a-half times higher than any on Earth, and great eroded canyons that are much deeper and wider than any on Earth, tell us of times when Mars was very active, producing widespread liquid lava flows and extensive fracturing over much of the planet's surface.

Numerous large channels, complete with flow markings and tributaries, range over the martian surface. Most were almost certainly cut by water hundreds of millions to billions of years ago, but some appear to be younger. There still is water on Mars, some in the polar caps in the form of water ice together with carbon dioxide ice (dry ice). We suspect--based on topographic clues found in **Mariner 9** and **Viking** orbiter images--that more water exists within the martian regolith, probably in the form of permafrost, but we can't be certain about where such reservoirs might be located.

Martian Weather and Atmosphere

Typical Mars' summertime diurnal temperatures range from a low of -92°F at dawn to a high of -13°F in the afternoon[5]. The climate is also very dry; there is no rain and only a very

---

[3] Viking Orbiter 1 (VO-1) continued to operate until August 7, 1980, when its ACS (attitude control system) fuel reached exhaustion following a long but productive battle with a leak; VO-2 had been shut down two years earlier (July 24, 1978) following a more serious but similar ACS fuel-depletion problem. {Ed.}

[4] Viking Lander 1 (VL-1) was the first to touch down on Mars, achieving its historic landing July 20, 1976, and it also was the last Viking spacecraft to be lost. Its final transmission, which included image data as evidence of a continuing dust storm at the site, was received November 5, 1982. VL-2, which landed September 3, 1976, was lost more than two years earlier; its last transmission was received April 11, 1980. {Ed.}

[5] The first temperature data recorded by VL-1 following its summer landing reflected a low/high range of -122°F/-22°F. Similar temperatures were measured by VL-2 during the summer, but its northern latitude site exposed it to much colder winter temperatures that were near the margin necessary to condense CO_2 on Mars (about -190°F). {Ed.}

rare and exotic kind of snow (which produced winter frost at the VL-2 Utopia site). Cirrus clouds in the martian atmosphere are composed of ice crystals as on Earth, but their composition might be water ice or dry ice (frozen CO_2) crystals, depending on their altitude and height-temperature. Other kinds of clouds repeatedly form in association with the high mountain slopes, and there is similarly early morning fog and frost in many low areas.

Mars' present mean atmospheric pressure at the surface is less than 1/100th that of Earth's, but its composition suggests that it may have been denser in the past. Today, Mars' atmosphere is composed of 95% carbon dioxide (CO_2) with minor amounts of nitrogen, argon, and oxygen (and traces of other gases). Water vapor may amount to two-to-three percent (2% to 3%) of the atmosphere during the summer (particularly in the northern hemisphere at high latitudes), but drops to a near-zero fraction during the winter.

Viking's complex biology instruments tested the martian soil for life. Although the biology results were tantalizing, they were deemed inconclusive on the basis that organic compounds could not be detected and because an active, nonbiological chemical process was believed by many to be responsible for the positive biology responses.

MOVING AHEAD

Most important now, although not fully appreciated a decade ago, **Viking** assembled a database essential for planning future missions. Those missions will be coming down the pike in just a few short years, and one of them in already in the queue.

Mars Observer -- In 1992 we plan to launch the **Mars Observer** spacecraft to make a two-year global survey of the planet, with emphasis on geoscience and climatology. The payload has been selected, and it is an exciting one. **Mars Observer** science instruments will not only analyze the planet's atmosphere and map its global surface composition, but will also search for a magnetic field, provide improved topographic and gravity maps, and obtain images of some of its surface at the highest resolution yet.

Sample Return -- Looking beyond **Mars Observer**, the next logical major step would be a mission to collect and return samples of Mars' rocks and soil to Earth. Having analyzed those samples in our laboratories, we would then better understand the composition, age, and evolution of Mars. Perhaps we would even be able to learn more about whether life exists or ever has existed on the planet's surface.

The NASA Advisory Council's Solar System Exploration Committee (SSEC) has released a report which describes such a mission in detail. In fact, the committee considers a Mars sample return mission to be one of the two highest priorities for solar system exploration between now and the end of the century. Interestingly, the other is also a sample return mission, but from the nucleus of a comet.

Support Technologies

These missions represent interesting technology development challenges for the nation. For example, they require the development of surface and subsurface sample collection systems, low-thrust propulsion, aerocapture and aeromaneuvering technology, and lightweight affordable power systems. A Mars sample return mission would also require an autonomous rover with improved position and orientation sensing and a teleoperator/manipulator system for drilling into the surface and gathering samples. Ascent and Earth-transfer storage systems must also be developed to maintain sample material in a pristine condition.

CONCLUSIONS

I think the SSEC is telling us something with its definition of these priorities -- something about why we explore space in the first place. Do you remember what Mike Collins told a joint session of the Congress on September 16, 1969? He said then: "We have taken to the moon the wealth of this nation, the vision of its political leaders, the intelligence of its scientists, the dedication of its engineers, the careful craftsmanship of its workers, and the enthusiastic support of its people. We have brought back rocks, and I think it's a fair trade." And so it remains, to this day. As we continue to unlock the mysteries of those rocks, we are beginning to understand the origin of the moon, the Earth, and the solar system itself.

It is interesting to note that Mars, the ancient's God of War, has become a symbol of productive, international cooperation in the peaceful uses of outer space. In 1988, the Soviets plan to launch a mission called, simply enough, **The Phobos Mission** [see NMC-2A]. While this mission clearly focuses on a detailed investigation of Mars' small, inner moon, Phobos, it also involves an orbital study of Mars itself. The Soviet Phobos Mission program therefore strongly complements the **Mars Observer** program. While there is now no plan to do so, the two missions present a possibility for mutually beneficial data and other exchanges as both countries plan future missions to Mars. This could, as one possibility, lead to a joint sample return mission.

Beyond the possibility of exploring Mars with unpiloted spacecraft lies still another prospect for international cooperation -- that of human exploration. Such a mission could be the focus of an enormous international cooperative program involving many countries, including the Soviet Union and the United States. Indeed, given the scope of such an effort and the resources it would require, and its potential benefits it represents for all mankind, it probably would need the joint efforts of several nations.

As we all know, this is not a new dream. Edgar Rice Burroughs' fictional hero, John Carter, was transported to Mars before 1910 just by wishing it. We can't all be that lucky. Modern writers--including Clarke, Asimov, Heinlein and Bradbury--have used a variety of launch vehicles to get their fictional characters there. But human exploration of Mars has now been moved from science fiction to serious consideration by a Presidential commission; **The Report of the National Commission on Space** lists the human settlement of Mars as a possible long-term goal -- a goal not only desirable but achievable.

I believe it was Robert Goddard who said, "Progress is not a leap in the dark, but a succession of logical steps." Clearly, the exploration of Mars by people will come only after many such steps. This conference is an important step in that direction, then, and I invite you to stretch your imaginations into the future. Visualize another conference like this -- one not <u>about</u> Mars, but taking place <u>on</u> Mars. Perhaps some of *you* will be there, discussing a bold, exciting new mission -- a mission to Planet Earth.

Good luck, and thank you very much. ∎

REFERENCES

This space is otherwise available for notes.

~~~~~~~~

NASA MARS CONFERENCE
Session 1, July 21, 1986

Our Current Knowledge and Understanding of Mars

Session Chairman, Dr. Gerald A. Soffen

PRESENTATION   PRESENTER

---

[1] Dr. McElroy elected not to review his presentation. The transcription presented herein therefore differs in contextual style and technical assurance. It has been included because of the subject's importance to the understanding of Mars and because other NASA Mars Conference participants (presentations) deferred to its content and to Dr. McElroy's expertise on key aspects of atmospheric science relevant to their topics.

[2] Dr. Straat is an executive with the National Institutes of Health.

NMC-1A · DR. JOHN S. LEWIS is a professor in the Department of Planetary Sciences at the University of Arizona in Tucson. He earned degrees at Princeton and Dartmouth and got his Ph.D at the University of California at San Diego under Harold C. Urey. Dr. Lewis has served on the Space Science Board for the National Academy of Science and as a Guggenheim lecturer for the Smithsonian National Air and Space Museum, and he has received the American Geophysical Union's Macelwayne Award. He has a broad view of Mars but has focused on its atmosphere and lithosphere. As an author, Dr. Lewis and his wife teamed up in writing: *Space Resources*, Columbia University Press, 1987. He also has been an active science consultant to major aerospace companies.

~~~~~

NMC-1B · DR. LAURENCE A. SODERBLOM is with the U.S. Geological Survey, Astrogeology Branch (Geologic Division), in Flagstaff, Arizona. He was educated at New Mexico Institute of Mining and Technology and earned his Ph.D at Caltech under Bruce Murray; he was a Fairchild Scholar at Caltech in 1983 and 1984. His primary interests are the global geologic histories of the planets, including the time scales for planetary evolution. He was on the **Viking Orbiter Imaging Team** and is now working with science teams involved in **Voyager** image interpretation, the **Magellan** (Venus orbital radar mapping) Project, the **Galileo** project's **Near-Infrared Spectrometer Team**, and the **Mars Observer** program. He has been an associate editor for the Journal of Geophysical Research.

~~~~~

NMC-1C · DR. MICHAEL H. CARR is with the U.S. Geological Survey (USGS), Astrogeology Branch, in Menlo Park, California. He earned his B.S. at the University of London in England, and his Ph.D at Yale University. He joined the USGS in 1962 and was involved in the **Lunar Orbiter** program (mapping the moon, 1966–67). From 1970 to 1980 he was team leader for the **Viking Orbiter Imaging Team**, and was responsible for the development of its cameras and interpreting orbiter pictures. He was chief of the USGS Branch of Astrogeological Studies from 1974 to 1978. He supports post-mission **Viking** work and has authored books and numerous papers concerned with geologic interpretation based of **Viking** imagery. Of special interest on the subject of Mars is Dr. Carr's book: *The Surface of Mars*, published in 1981 by Yale University Press. NASA awarded him its medal for Exceptional Scientific Achievement for his work on **Viking**. He also is involved with the **Galileo** and **Voyager** programs and chaired the science working group for **Mars Observer**. Dr. Carr is a member of the Geological Society of America and an associate editor for the Journal of Geophysical Research.

~~~~~

NMC-1D · DR. VICTOR R. BAKER is a professor of geosciences at the University of Arizona. He was educated at Rensselaer Polytechnic Institute (Troy, N.Y.) and earned his Ph.D at the University of Colorado in Boulder. Dr. Baker has worked in geological sciences for universities and government agencies worldwide, including the Water Resources Division of the U.S. Geological Survey (USGS) and the Department of Geological Sciences at the University of Texas in Austin. He is a fellow in the Geological Society of America and has been a visiting fellow at Australian National University. His central interests are geomorphology and geomorphological processes with paleohydrology, particularly fluvial histories for Washington, Idaho, Oregon, Utah, Colorado, and Texas, as well as Australia and South America, and he has written extensively about the catastrophic flooding at the end of the last ice age which created the "Channeled Scabland" region of the northwest U.S. Dr. Baker was a guest investigator on the Viking Orbiter Imaging Team in the late 1970's, and applied his expertise to the study of Mars' channels and valleys.

~~~~~

NMC-1E · DR. JOSEPH VEVERKA is professor of astronomy at Cornell University. He attended Queens University in Kingston, Ontario, Canada and earned his Ph.D in astronomy at Harvard University. His interests have long focused on planetary satellites and their evolution. He was a member of the Viking Orbiter Imaging Team, authoring or coauthoring numerous papers about the martian satellites, Phobos and Deimos, and he is a member of the Voyager team involved in studying the satellites of Jupiter, Saturn, Uranus, and--ultimately--Neptune (August, 1989). He is also a member of the Galileo and Mars Observer science teams. His professional memberships include the American Astronomical Society, the American Geophysical Union, the American Meteorological Society, and the Royal Astronomical Society of Canada.

~~~~~

NMC-1F · DR. MICHAEL B. MCELROY is chairman of the Department of Earth and Planetary Sciences at Harvard University, and his work focuses primarily on comparative planetology and atmospheric science. He was educated at Queens University in Belfast, Ireland. Dr. McElroy was a member of the Viking Entry Science Team and participated as well in atmospheric analysis as a product of data provided by the lander GCMS (Gas Chromatograph/Mass Spectrometer). He has worked with the Voyager and Pioneer programs and is active in the Earth System Sciences part of NASA's planetary program.

~~~~~

NMC-1G · DR. CONWAY B. LEOVY is a professor in the Department of Atmospheric Sciences and Geophysics at the University of Washington in Seattle. He was educated at the University of Southern California and earned his Ph.D at Massachusetts Institute of Technology (MIT). He was on the Viking Meteorology Team and was awarded NASA's Outstanding Science Achievement Medal for his work. He is a member of the American Geophysical Union, has been an editor of the Journal of Atmospheric Sciences, and is a fellow in the American Meteorological Society. Dr. Leovy is an active participant on NASA's Advisory Committee.

~~~~~

NMC-1H · DR. FRASER P. FANALE is professor of planetary science at the University of Hawaii, Hawaii Institute of Geophysics, in Honolulu. He was educated at Upsala College and earned masters and doctoral degrees at Columbia University. He has also completed postdoctorals at Brookhaven and Caltech. Dr. Fanale is an interdisciplinary scientist; his interests extend from fossil dating to the evolution of planetary atmospheres, and from the evolution of Jupiter's satellites to the history of planetary volatiles in general. He supervised the Planetary Group at the Jet Propulsion Laboratory in the early 1970's, and he's been involved in near-infrared mapping. He is currently an investigator on the **Galileo**, **Mars Observer**, and **CRAF** (Comet Rendezvous Asteroid Flyby) missions. He's been an associate editor of the Journal of Geophysical Research.

~~~~~

NMC-1I · **DR. NORMAN H. HOROWITZ** is Professor Emeritus in the Biology Division at Caltech (California Institute of Technology). He was educated at the University of Pittsburgh and then earned his Ph.D at Caltech. Dr. Horowitz has been chief of the bioscience section at the Jet Propulsion Laboratory and was an experimenter on both the **Mariner 6/7** (twin Mars flyby spacecraft) and **Viking** programs. He has worked at several prestigious facilities, including the Hopkins Marine Station and the genetics laboratory at the University of Paris. Dr. Horowitz is a member of the National Academy of Sciences, the American Academy of Arts and Sciences, the American Society of Biological Chemists, and the Genetics Society of America. He is eminently known for his classical work in <u>Neurospora</u> genetics. Dr. Horowitz has served on the Space Science Board, and he has written a book based on the controversial **Viking** results produced by the biological experiments on Mars: *To Utopia and Back: The Search for Life in the Solar System*, (1986, W.H. Freeman).

~~~~~

NMC-1J · **DR. GILBERT V. LEVIN** has had an extensive career in public health, medical microbiology, industrial research, and public service. He used his experience in scientific research and business administration to form the Life Systems Division of Hazeleton Laboratories, Inc., and then Biospherics, Inc. of Beltsville, Md. (of which he is now President). He earned his Ph.D at Johns Hopkins University and has done consulting work with the Department of Interior and NASA. Levin's association with the **Viking** program dates back to 1959-60 when he developed a sample acquisition and analysis instrument called "Gulliver" (for NASA) that was to fly on what was then known as the Voyager (Mars) spacecraft. He had argued strongly in favor of a Mars soft-lander during that period. That Voyager program was canceled and restructured as the **Viking Project**, and Gulliver ultimately served as a precursor for the **Labeled Release** (LR) experiment in the **Viking** biology instrument package. Dr. Levin is a member of the American Society of Civil Engineers, the New York Academy of Sciences, and the American Institute of Biological Science. He has introduced many technologies in microbiology and his patents span microbiological identification, waste water treatment, and non-caloric sweeteners.

DR. PATRICIA A. STRAAT is a senior executive with the National Institutes of Health. She was educated at Oberlin College and earned her Ph.D at Johns Hopkins University. Dr. Straat has worked in electron transport, inorganic nitrogen fixation, unicellular organisms, aquatic ecology, and the chemical and biological aspects of water pollution. She worked with Dr. Levin at Biospherics where the **Viking Labeled Release** (LR) biology experiment and investigation strategy were being developed, and then participated on the **Viking Biology Team** in the group headed by Dr. Levin (LR experiment) at the Jet Propulsion Laboratory. Dr. Straat previously worked on the **Mariner 9** IRIS (Infrared Interferometer Spectrometer) experiment team. She is a member of the American Chemical Society and the American Institute of Biological Science.

■■■

THE HISTORY OF MARS
[NMC-1A]

Dr. John S. Lewis
Professor, Planetary Sciences
University of Arizona

The solar system formed out of a common nebula, but there are significant differences among the planets. These may be due to conditions associated with either their origin or the divergent evolution of each. Spacecraft have helped us understand these matters while providing highly detailed information about the planets themselves. But even before their time we began to understand something of how the solar system was formed by comparing the mass and densities of planetary bodies. We now know that there is a progressive trend in the composition of the solar system outward from the Sun, as seen in the increasing abundance of volatiles beyond Earth. In the case of Mars, the core is not as massive and dense as Earth's while the mantle has proportionately greater mass and density, demonstrating that elemental, molecular, and mineralogical compositions differ with formation temperatures. Being farther from the Sun, Mars should have an abundance of volatiles at least as high as Earth's. Instead, the amount of nitrogen is small and the rare gases are depleted. This, with the unusual relative abundances of argon and nitrogen isotopes, leads to a scenario in which much of the primordial martian atmosphere was blown away during the final stage of accretion in-fall. Only the last gases released survived, including the rare gases and a significant reservoir of frozen water.

INTRODUCTION: PERSPECTIVE

My perspective on Mars is a little different from that of most of the people making presentations for this Mars Conference. I was not a member of any of the Viking science teams, although I have long been interested in and excited by the saga of the exploration of Mars. Each of the NASA Mars Conference Session 1 presentations deals to some extent with historical matters. My task, however, is to focus on the history of the planet as it relates to the rest of the solar system, whereas the others deal more specifically with the history of Mars itself. Therefore, my title *The History of Mars* perhaps assumes or suggests too much. More accurately, I will in fact be discussing the *early* history of Mars, the origin of Mars, and--as much as the opportunity permits--the relationship between Mars and the other planets.

SOLAR SYSTEM FORMATION AND PLANETARY DIFFERENTIATION

The first concept with which we need to come to grips is that although the whole solar system formed more or less simultaneously out of the same original material (i.e., the primitive solar nebula out of which the Sun and all the planets, satellites, asteroids, and comets formed), there are very significant differences between all of the planets. Each has its own idiosyncratic properties; indeed, Mars itself is distinctive in a number of ways.

Understanding the differences between Mars and the other planets can be thought of as an exercise in determining whether those differences are due to differences in the conditions of origin (of Mars relative to the other planets) or due to the divergent evolutionary paths the various planets have followed since their formation. Indeed, one often finds debates in the literature about whether differences in initial conditions or differences in evolutionary behavior are more important. By the end of this presentation, I expect that I will have thoroughly confused the issue more so by pointing out how intimately connected these two factors are.

I will be obliged to make reference to a number of scientific results from American and Soviet Mars missions, and I regret that I will not be able to get into detail about how these results were achieved. However, considerably more about these missions is represented in the other presentations contributed to the proceedings of this Mars Conference.

Prespacecraft Observation

To convey the tremendous contribution that spacecraft missions have made to our understanding of Mars, I think it is useful to look back in time. Figure 1A-1 is a Robert Leighton observatory photo [c1965[1]] that shows Mars as we knew it just at the dawn of the space age, the era of spacecraft exploration of the solar system. This photo is what, as of the mid 1960's, passed as a superb observatory photograph of Mars, and it represented the best level of visual

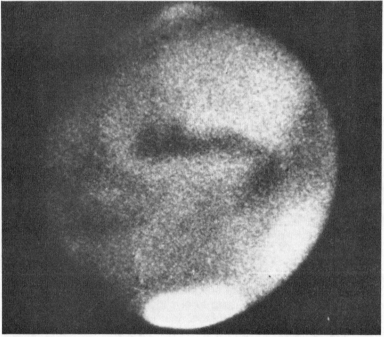

Figure 1A-1: Observatory Photo of Mars by Robert Leighton (c1965)

---

[1] Dr. Robert Leighton, California Institute of Technology, Professor of Physics and eminent astronomer; principal investigator for Mariners 4, 6 and 7 Mars imaging systems and investigations. {Ed.}

detail provided by Earth-based observation at that time. Drawings of Mars made by astronomers who looked through telescopes and then drew what they saw frequently reflected enormously more detail[2]; unfortunately, however, those details often had nothing whatsoever to do with Mars as it has since been photographed. Thus, this picture represents "verifiable truth" about Mars. One can see dark and light areas, the south polar cap, and--just barely visible--one can also see the edge of the frost deposit near the north pole.

By that time in history we knew the orbit of Mars and the radius of Mars to very good precision. Also, through the tracking of the natural satellites, Phobos and Deimos, we knew the mass of Mars to similarly good precision. Regarding the nature of Phobos and Deimos, however, we knew virtually nothing; even their sizes were very poorly known. Since we knew both the mass and radius of Mars, we could calculate its density and then compare the densities of Mars and the other terrestrial planets -- Mercury, Venus, and Earth. What we found was that the density of Mars was substantially less than the densities of the other terrestrial planets. Indeed, when one corrects the densities of these planets to allow for the self-compression of each by its own gravity and the weight of its own material, one still finds that the composition of Mars is intrinsically significantly less dense than that of Earth, Venus, and Mercury.

In fact, we seem to see a progression in the densities of the terrestrial planets; beginning with the most dense planet--Mercury--closest to the Sun, progressing to Venus and Earth (nearly the same uncompressed densities) with less a little farther out, and then to Mars with yet lower density. Since that time we have also been able to place some crude limits on the densities of Phobos, Deimos, and several of the largest asteroids, and all appear--interestingly enough--to have lower densities than Mars.

Compositional Trend

There seems, then, to be a systematic trend in the composition of the solar system. Furthermore, in our studies of Jupiter's satellites, we see that trend continued to yet lower densities. Ganymede, the largest of Jupiter's four Galilean satellites[3] and nearly the size of Mars, has a density lower than any object inside its orbit in the solar system; a density so low that substantial amounts of ice must have been fully condensed at that distance from the Sun for incorporation into Ganymede.

In this progression, Mars appears to link the inner solar system of dense, rocky planets to the asteroid belt, which is as yet a poorly explored zone of preplanetary rubble (material that never successfully accreted into a planet due to the disturbing influence of Jupiter's gravity. Somewhere in that zone, probably a little beyond Mars, is the transition boundary that separates the terrestrial planets from the kinds of solid materials found in the outer solar system, a mixture of rocky minerals and volatile ices (H_2O, CO_2, CH_4, et·cetera).

---

[2] Historically including those of Huygen (1659), Schiaparelli (1870's/80's), and Lowell (1890's). {Ed.}

[3] The four **Galilean** satellites were named for Galileo, who was the first to observe them and suggest what they were. Progressing outward from Jupiter, they are: **Io**, volcanically active; **Europa**, encased in surface ice that may be undergoing slow but constant replacement; **Ganymede**, exhibiting a unique mix of both ancient and younger features (with evidence of at least some previous tectonic activity); and **Callisto**, the most distant of the four (only slightly smaller than Ganymede) with the most cratered and unmodified surface. {Ed.}

THE CHANGING VIEW OF MARS

Long before the spacecraft era, we were familiar with the red color of Mars. In more recent times, the red color has usually been interpreted by scientists as being due to ferric oxides. Although many other imaginative and sometimes plausible explanations were published, they turned out not to have been correct.

Indeed, the very name "Mars" was derived in classical time from its sanguinary appearance[4]. This oxidized appearance of Mars, combined with the obvious absence of oceans, led to much speculation regarding the early history of the planet. It was commonly felt by many scientists (and certainly by many science fiction writers) that Mars at one time had been Earth-like, but--due to low gravity--had lost its hydrogen into space. This theory proposed that the planet's oceans had been decomposed by solar ultraviolet light into hydrogen and oxygen; as the hydrogen dutifully escaped, the oxygen reacted with surface rocks rich in iron to make rust.

This scenario for Mars was essentially that used by many science fiction writers, notably H.G. Wells, Edgar Rice Burroughs, and Ray Bradbury, who pictured Mars as a planet past its prime -- definitely not what it had once been[5]. They portrayed a planet that may have been suitable not only for the origin of life but also as an abode for advanced life forms in the not too distant past. In that imaginative scenario, Mars rapidly degenerated to become either a sterile planet or one populated by the remnants of once-great civilizations.

The Dawn of Factual Reality

Early spectroscopic studies of the atmosphere of Mars had revealed that there was a tenuous carbon dioxide atmosphere. However, the Earth-based astronomical estimates of the carbon dioxide abundance on Mars were not accurate until the era of spacecraft exploration had already begun, although they did NOT get their ultimately accurate answer from the spacecraft data.

At that time (early 1960's) water vapor was understood to be a minor constituent of the atmosphere as well as being a seasonally varying constituent. Carbon monoxide (CO) was detected in the mid-1960's as another minor constituent. This information was the basis of our knowledge as spacecraft and Earth-based, high-resolution spectroscopic techniques became available during the 1960's.

1960's Consensus vs Current Knowledge

The unfortunate truth is that the atmospheric pressure at Mars surface is only, on average, 6/1,000ths that of the atmospheric pressure at Earth's. And, because of its low gravity, low temperatures, and tremendous topographic relief (with extremely deep valleys and high volcanic mountain peaks), Mars experiences a huge range of surface pressures (percentage variations from mean surface pressure) from one region to another and with the changing seasons.

---

[4] The earliest suspected references to Mars are those found in Babylonian history (dating at least to 1500 B.C.), apparently describing a reddish wandering star known then as Nirgal. The Greeks later knew Mars as Ares, giving rise to the word areography, once used as a term for the study of Mars. The names for Mars' two satellites are found in Homer's Iliad; Ares was the god of war or disaster and Phobos and Deimos (fear and terror, respectively) were the horses that drew his chariot. {Ed.}

[5] H.G. Wells wrote the classic: War of the Worlds in which Earth is invaded by martians; Edgar Rice Burroughs (creator of Tarzan) wrote an extensive series of books about an earthling--John Carter--who became a warlord of Barsoom (Mars); and Ray Bradbury is well known for his enchanting, poetic, and ever-popular Martian Chronicles. {Ed.}

A review of the history of that foundation of knowledge reveals a number of even more penetrating generalizations about Mars, and we find that the consensus of the scientific community at that time (early 1960's) was essentially correct. However, the understanding was a bit sketchy. For example, we now find that the internal density distribution of Mars is rather different from what one would expect if Mars were simply a carbon-copy of Earth. If one imagines making Mars out of Earth material, i.e., making the mantle of Mars out of Earth's mantle material and the core of Mars out of Earth's core material, one cannot then fit the observed density distribution for Mars deduced from tracking its natural satellites and artificial satellites. In fact, one finds that the core of Mars must be proportionately *less* massive and dense than Earth's, and that the mantle of Mars must be proportionately *more* massive and dense than Earth's.

This is not unexpected from the perspective of students of meteorites, who know that many classes of primitive meteorites have remained unaltered since the early history of the solar system. Compositional trends can be seen in these primitive meteorites. For example, we find that they were formed at differing temperatures and that we can establish a sequence of elemental, molecular and mineralogical compositions for them -- from those formed at higher temperatures down to those formed at lower temperatures.

This has almost always been interpreted as a sequence that tells us the distance from the Sun at which these objects were formed; those with the lowest formation temperatures originated farthest from the Sun. The reason this is relevant to our understanding of the planets is that the compositional trends seen in the primitive meteorites also predict a decrease in the density of rocky material as one goes out from the Sun, which of course is important to the composition of the planets as well.

Formation/Accretion Relationship

Formation -- Figure 1A-2 gives one an indication of the temperature and pressure conditions in the solar nebula out of which the planets were formed [A.G.W. Cameron]. The temperature and pressure gradients are rather steep, corresponding closely to an adiabatic (convective) radial structure, and drop off rapidly in the direction normal to the symmetry (equatorial) plane. At Mars' distance from the Sun, the pressures are on the order of 10^{-4} atmospheres and the temperatures are a few hundred Kelvin. Close to the Sun, the condensate will consist of a number of silicates of magnesium, calcium, aluminum, sodium, potassium and so on, as well as metallic iron. Farther from the Sun, the metallic iron will be cool enough to react with the preplanetary gases in the solar nebula to make iron sulphide. Farther yet, temperatures will be low enough for water vapor in the nebula to corrode the remaining metallic iron and make iron oxides. As one continues to progress outward from the Sun, then, the amount of metallic iron decreases with distance as the amounts of iron oxides and sulphides (with lower densities) increase.

Figure 1A-3 illustrates the chemistry of the material one would find in such a nebula, and how it would behave. Each of the minerals is present only below its formation curve. $CaTiO_3$ is representative of a family of refractory silicates rich in aluminum, calcium and titanium oxides that condense at similar temperatures. Condensation of $MgSiO_3$ slightly precedes the condensation of Na and K aluminosilicates (feldspars), which are omitted in the figure data to avoid excessive cluttering. Tremolite and talc are hydroxyl silicates which release water upon heating. The symbols along the nebula adiabat stand for (top to bottom) Mercury, Venus, Earth, Mars, Ceres, Jupiter, Saturn, Uranus and Neptune. Ices other than water ice also are present in the outer solar system.

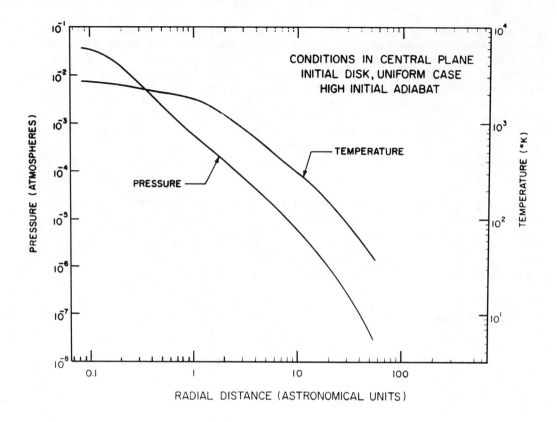

Figure 1A-2: Model of Pressure/Temperature Conditions in Solar Nebula

At Mars' distance from the Sun, it can be seen that temperatures are low enough to fully condense high-temperature minerals. Iron sulfide is formed, and iron itself should be almost fully oxidized. It is also possible to have water-bearing silicates, such as tremolite[6], present at Mars' distance from the Sun at the time of formation -- possibly even talc. As a consequence, the overall density of the material decreases; also, the mass of core-forming material decreases while the mass of mantle-forming material (oxides) increases.

The curve presented in Figure 1A-4 is a representation of the densities of the terrestrial planets, starting with the density of Mercury (near top); it is supposed to level off and then become horizontal just below Mercury. The solid lines are the densities predicted by the chemistry given previously (Fig. 1A-3). They show a trend relative to distance from the Sun (decreasing condensation temperature) that quite faithfully mirrors the observed density of the terrestrial planets. We therefore have reason to believe that the theory is in good shape. Note that the densities generally decrease with increasing distance from the Sun, except that sulfur retention leads to a small density increase near Earth's orbit, and refractory oxides condensed

---

[6] Tremolite is a calcium-magnesium amphibole (a hydrated, layered mineral).

at temperatures too high for metallic iron-nickel alloy to be present are less dense than the metal-bearing condensate present at slightly lower temperatures. The highest predicted density falls somewhat short of the observed density of Mercury, and the density of the Moon can be achieved at two different heliocentric distances -- neither of which does a good job of explaining the refractory-rich and FeO-rich composition of the Moon as we see it.

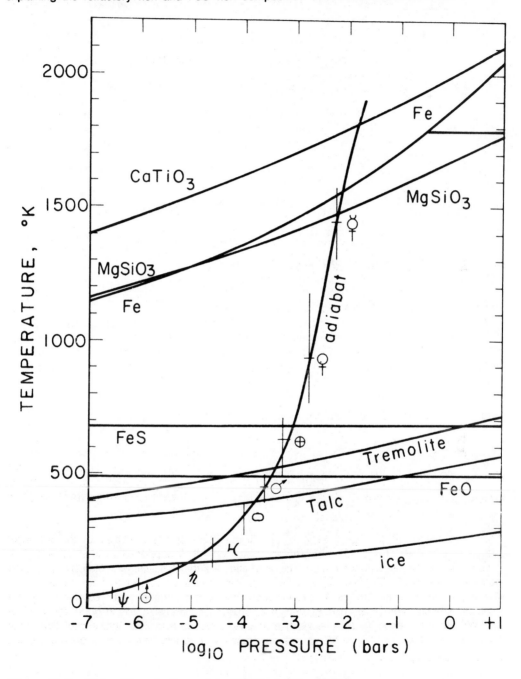

Figure 1A-3: Major Minerals Present in Model of Solar Nebula

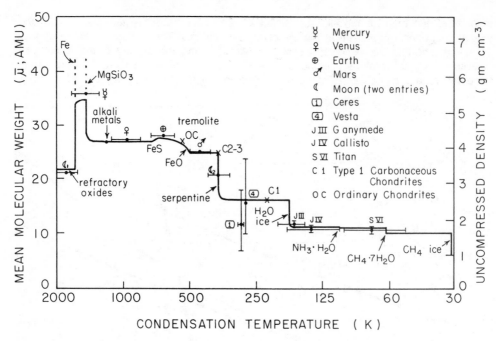

Figure 1A–4: Densities of Solid Bodies in the Solar System

Accretion -- Figure 1A–5 illustrates the nature of the accretion process. In it, the density of accreted preplanetary solids are shown, as a function of distance from the Sun and of the width of a Gaussian statistical sampling function that simulates the accumulation of material originating at different heliocentric distances by growing planets. Note that wide sampling functions wipe out the density variations exhibited by local condensates, as seen in the previous figure (Fig. 1A–4).

As each planet accretes, it attracts material that formed at differing distances from the Sun. It may sample locally or moderately, or even quite broadly; exactly how broadly it samples depends upon the mechanics by which the planet grew from smaller bodies. If the sampling is very local, one sees very sharp planetary density differences wherever there are differences in the composition of the preplanetary material. However, if the sampling is broad, it wipes out the density contrasts caused by the chemistry, and the large observed density differences between the terrestrial planets would not be possible. The truth seems to be that sampling has been moderate, i.e., each planet was assembled from material formed over a range of distances from the Sun and there is at least some compositional diversity in each. For this reason, then, the composition of planets should not be compared precisely and exactly to single classes of meteorites.

Thus, the compositional trends we see in primitive meteorites suggest that the mantle of Mars is proportionately more massive and denser than Earth's because it contains more oxidized iron, and that the core of Mars is less dense and less massive than Earth's because it contains sulphides only. Mars' iron sulphide is not dominated by metallic iron as it is on Earth.

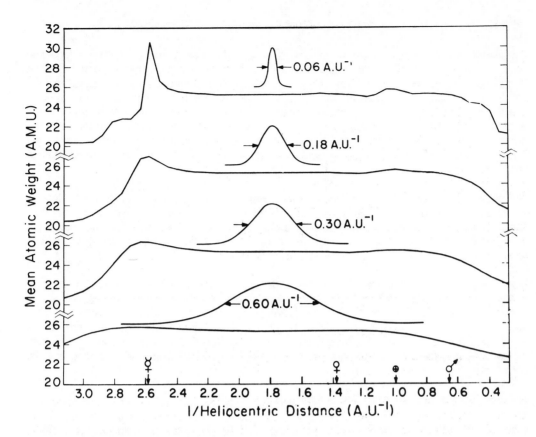

Figure 1A-5:   Density of Accreted Preplanetary Solids

Consequences of Compositional Trends in the Solar System

These compositional trends actually have a number of other consequences. For example, because the minerals that formed Mars were different from those that formed Earth, the melting behavior of Mars material is different. Similarly, because the composition of Mars' core is different from Earth's, the electrical properties of the planets' cores are also different; therefore, the generation of their magnetic fields--by motions in their cores--should not be the same. Indeed, we have learned from spacecraft studies that Mars has only, at best, a minuscule native dipole magnetic field. In fact, its field has NOT been firmly detected.

Thus, many aspects of the planet's history may have been influenced by its conditions of origin. The different mineralogy present at the time of Mars' formation, which is a significant factor in determining the intensity of the present-day magnetic field, has evolutionary significance. Why? Because Earth's relatively massive magnetic field causes the solar wind to stand off and not impact our atmosphere. Conversely, the absence of a planetary magnetic field on Venus and Mars allows the solar wind to impinge upon the tops of their atmospheres, changing both the rate of implantation of solar gases into them and the rate at which their lightest gases are swept away. These are important evolutionary considerations.

VIKING AREOGRAPHY AND SURFACE CHARACTERIZATION

The Viking landers sampled the martian atmosphere during their descents to the surface and while on the surface, using instruments mounted on both the entry capsule elements and the landers themselves. In addition to atmospheric pressure and temperature sensors, accelerometers, and a radar altimeter, each Viking aeroshell was equipped with an upper-atmosphere mass spectrometer and a retarding potential analyzer. For final descent and surface operations, the landers themselves were equipped with pressure and temperature sensing instruments, while elements of each lander's Gas Chromatograph-Mass Spectrometer (GCMS) provided very sensitive and precise atmospheric compositional data at the surface. All of these instruments worked very successfully[7].

The predominance of carbon dioxide, of course, was already known. The Viking lander instruments also provided very precise measurements of the abundances of nitrogen and the rare gases as well. And, because of the peculiar nature of nitrogen and the rare gases[8], they serve as extremely useful tracers of the evolution of the atmosphere. Indeed, they have long been used as tracers of the evolution of Earth's atmosphere.

Since we now have measurements of the abundances and isotopic compositions of these rare gases on Venus, Earth and Mars, as well as in many different classes of meteorites, one would think that we should have solved the whole problem by now. However, I have written a book about planets and their atmospheres [1984ª] that tackles this problem head on and *still* comes to no particular conclusions. We have found that the problem is much more complicated than we had hoped, but perhaps not so complicated as to prevent us from understanding at least something of what we have learned.

Nature of Specific Atmospheric Components

Some of the remarkable results from Mars were, first of all, that the nitrogen component of the atmosphere is only about 2-1/2 percent (2.5%), and this amount is enormously smaller than on Earth[9]. But our study of meteorites leads us to associate lower density rocky bodies with *higher* abundances of volatile elements. If we look at the sequence of meteorites that span the densities and compositions of the terrestrial planets, we find that anything with densities as low as those of Mars or the carbonaceous asteroids should be quite rich in volatile elements like nitrogen. In fact, any reasonable prediction is that Mars, being intermediate between Earth and the asteroid belt (which is chock full of volatile-rich asteroidal material), should have at least as high an abundance of volatile materials--water, carbon, nitrogen, rare gases and so forth--as Earth.

Instead, the amount of nitrogen now found in the atmosphere of Mars is very small, and represents less than one percent (1%) of the amount of nitrogen that Mars would have outgassed if it had a composition identical to that of Earth. And, compounding the puzzle, we have found that the rare gases also are severely depleted. Their abundances are lower by about a factor of 200 relative to the primordial rare gases found on Earth or in primitive meteorites.

---

[7] The Viking Entry Science Team was headed by Dr. Alfred O. C. Nier, University of Minnesota, and surface-based atmospheric science was performed by members of both the Entry Science Team and the Molecular Analysis Team (organic chemistry) headed by Dr. Klaus Biemann, Massachusetts Institute of Technology. {Ed.}

[8] Nitrogen is almost chemically inert, and the rare gases are--for all practical purposes--similarly inert.

[9] Earth's atmosphere is approximately 78% nitrogen.

How one can assemble Mars out of Earth-like or meteorite-like material without it having 100 or 200 times as much of the primordial rare gases is an interesting problem. Of course, one possibility is that Mars, for some mysterious reason, formed without volatile elements; another possible explanation is that the difference is a product of evolution rather than origin, and that for some reason Mars evolved in a way that permitted it to lose its original endowment of the rare gases.

Atmospheric Gases and Their Products -- At the time that Anders and Owen [1977[b]] wrote a review article on the composition of Mars, following the Viking Primary Mission[10], it was believed that known mechanisms for the escape of planetary atmospheres could not explain these observations. These mechanisms, it was thought, would have to be highly selective, highly mass-dependent, and would--instead of reducing *all* of the rare gases in abundance by a common factor of about 200--deplete the lightest rare gases much more severely and produce a significant slope in the abundance pattern. This kind of result is not seen.

In fact, Anders and Owen briefly considered the possibility of a mechanism for removing atmosphere from Mars that could be totally non selective; i.e., remove molecules irrespective of their properties. They were influenced by prevailing opinion to conclude that no such mechanism was feasible. Since no such mechanism was known, they could not consider it further. Now, however, one such mechanism *is* known, and I will discuss it in more detail later in this presentation.

The Argon Puzzle -- I would like to turn to a couple of points involving the isotopes of argon, based on some other results of Viking lander data analyses, that require additional explanation. One of these is that the isotopic compositions of these rare gases on Mars are somewhat different from Earth's. Initially, the data that were perhaps most striking defined the relative abundance of radiogenic argon compared to primordial argon[11]. Viking found ^{40}Ar to be proportionately 10 times as abundant on Mars as it is on Earth. To put this finding in perspective, one should note that the primordial isotopes of argon present at planetary creation (^{36}Ar and ^{38}Ar) are lower by a factor of about 200 relative to Earth, while ^{40}Ar is lower only by a factor of 20 relative to Earth.

There are two ways to consider this argon puzzle. The first offers two alternative explanations for why Mars has *less* ^{40}Ar in its atmosphere than Earth: 1) the planet must be very incompletely outgassed, or 2) it formed with less potassium. The second way to regard the data is to say that, *relative to the other rare gases*, ^{40}Ar is more abundant on Mars and therefore Mars must contain a higher proportion of potassium than the Earth. A puzzling dilemma.

Implications of Surface Phenomena

Evidence of Volatiles -- I would like to contribute a bit more preparation for the presentations that will follow mine by quickly covering a few points I can't pursue in detail but which represent other areas in which spacecraft data completely reformed our ideas of Mars. For example, Figure 1A-6 is a Viking orbiter image in which we have an example of terrain on

---

[10] Because Viking mission operations were repeatedly extended to take advantage of spacecraft availability over a period of more than five years, initial science data published at the time represented only the first and most ambitious phase of mission operations -- the primary mission. That phase of mission operations lasted from July for VL-1 (September for VL-2) into the November conjunction, 1976. {Ed.}

[11] Primordial argon (^{40}Ar) is produced after the formation of a planet by the decay of radioactive potassium (^{40}K) inside the planet.

Mars that is without a precise equivalent on Earth. Martian polar regions exhibit unique features like the swirls and blotches and layers seen in this **Viking** orbiter photo. One can see many layers exposed just above right-center in the picture, composed primarily of condensed volatiles -- notably water and CO_2 ices mixed with darker deposition dust.

Another **Viking** image of the summertime north polar cap is presented in Figure 1A-7. The dark lanes are at lower elevations while the white lanes are higher elevations. The residual polar cap clearly visible in these pictures reminds us that there are large reservoirs of volatile elements on Mars -- particularly water ice. We can see that volatiles exist on the surface of the planet, and there is significant evidence that they also are hidden in its crust--mostly in the porous regolith or "dirt" of Mars--in the form of permafrost or as adsorbed carbon dioxide, water and other gases.

In order to be sure we know how much of each of these volatiles is present on Mars, we need to know how much is hidden in the crust. That can be a very substantial amount. In fact, some estimates for the amount of water that could easily be hidden in the crust of Mars predict an amount equal to from 100 to over 1000 meters of liquid water over the entire surface of the planet. In the meteorite analogy, Mars should have enough water to produce at least several hundred meters of depth over the entire surface. This important issue is discussed in the Mars Conference presentation by Fraser Fanale [NMC-1H].

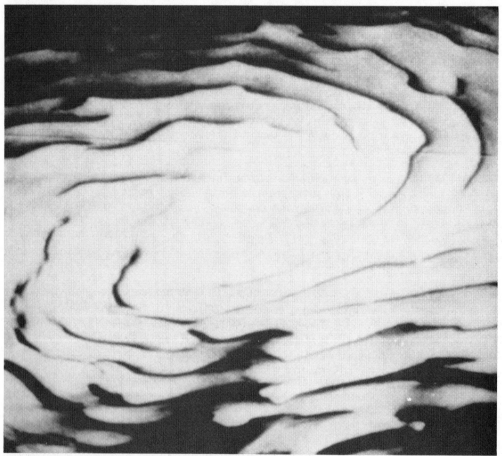

Figure 1A-6: Viking Orbiter View of Polar Cap Coriolis Swirls

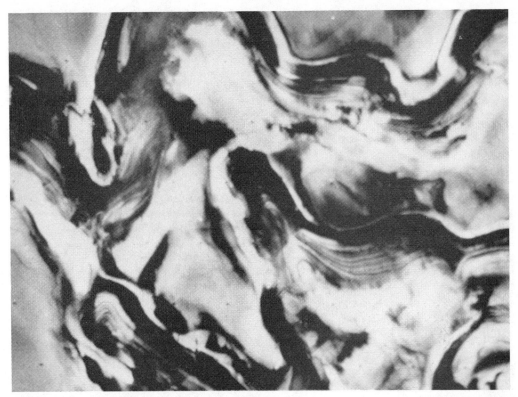

Figure 1A-7: Layered Terrain in Martian Polar Cap

Volcanism -- The photo in Figure 1A-8 illustrates another very striking feature of Mars, the huge, relatively young Tharsis volcanoes. This particular volcano is **Olympus Mons**, as seen with its upper part emerging above most of the atmosphere and morning clouds. While that may sound like hyperbole, this volcano is in fact so high that it sticks out above most of the mass of the atmosphere[12]. The craters at its summit are calderas--volcanic collapse craters caused by inflation and deflation of giant magma chambers beneath the surface--some 70 km across and 3 km in depth at their deepest point. Along the Tharsis ridge itself there are several more very large, relatively young appearing shield volcanoes, while elsewhere there are a number of other volcanoes -- some densely scarred by impact craters and apparently inactive since early in the planet's history.

The sequencing of these volcanoes in time can be estimated from the count of impact craters on their sides. This provides an estimate of how volcanic activity on Mars has varied with time. In turn, this provides crucial evidence relative to the melting behavior and temperature history of the interior of Mars. It is a very interdisciplinary and exciting area in which, unfortunately, we are still rather limited in relevant knowledge. Mike Carr's Mars Conference presentation [NMC-1C] explores martian volcanism in much greater detail.

---

[12] **Olympus Mons**, previously known as **Nix Olympica**, stands approximately 27 km high and is roughly 600 km across at its base. {Ed.}

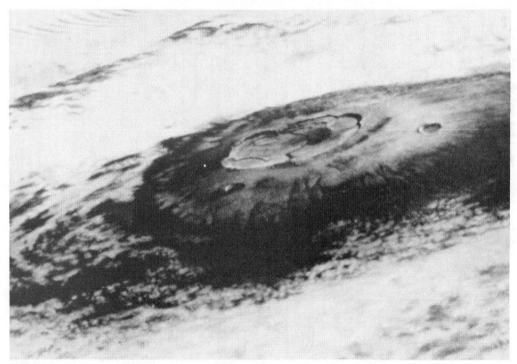

Figure 1A-8: Cloud Enshrouded Olympus Mons Volcano in Tharsis

Erosion and Weathering -- Erosion and weathering have in fact significantly altered the crater-ed, grossly moon-like ancient terrain one sees elsewhere on Mars. And, while it is easy to draw an analogy between the ancient areas of Mars and the Moon, it should not be pushed too far. We should recall at a historical occasion like this[13] that the first three sets of photographs of Mars, provided by Mariners 4, 6 and 7, all *happened* to be of older regions that were heavi-ly cratered, misleading us into thinking of Mars as a geologically dead planet. Fortunately, that turned out *not* to be the case, and we now know that Mars has had a very interesting and dynamic geological history. Indeed, the questions now most likely to arise address those very planetary dynamics, such as: "Until how recently has large scale tectonic activity and volcanic activity on Mars continued?" More can be found on those questions in other Mars Conference presentations.

Channels -- Another interesting feature of Mars is represented by the branched, dendritic channels seen in Figure 1A-9. They are found widely over the surface, although most notably at mid latitudes, and are often seen as sinuous stream beds that cut back and forth inside a valley flood plain. The largest channels are seen as massive river-like features, complete with apparent islands. Explanations for the martian channels suggest that they may be either vol-canic in origin or due to the action of water, the latter flowing either as a liquid or glacially. The large river-like features that seem to be due to the flow of liquid water do not look like well developed terrestrial rivers fed by precipitation, but instead resemble--as Vic Baker ex-plains in his Mars Conference presentation [NMC-1D]--the results of catastrophic and intense flooding episodes. The presence of liquid water is not possible at the present atmospheric pres-

---

[13] The NASA Mars Conference began with the 10th anniversary of the historic first Viking landing on Mars (July 20, 1976). {Ed.}

sure, so these features imply a warmer, denser atmosphere in the past. This evidence, then, leads us to investigate the climatological history of Mars, as well as the course of planetary thermal evolution, in order to understand such events.

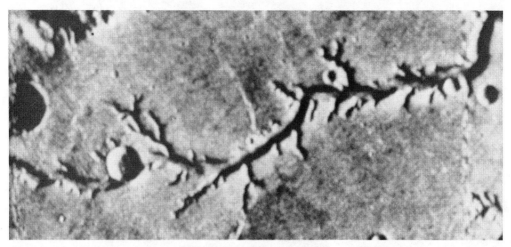

Figure 1A-9: One of Numerous Dendritic Martian Channels (Nirgal Vallis)

Fluid-Like Ejecta Aprons -- Figure 1A-10 shows a heavily cratered region of Mars. But, if one looks closely at the shapes of these craters, their morphology appears quite different and strange compared to lunar or Mercury impact craters. To bear that out, Figure 1A-11 shows yet another of these strange craters. If scaled down, this one looks as though it could have been created when a child threw a firecracker into a sandbox full of mud, producing a big splash. At latitudes greater than about 30 degrees in both hemispheres, one commonly finds impact craters that look as though they were made in mud. They represent one of the most powerful lines of argument for the continuing presence of large quantities of volatiles contained in the regolith over at least have the surface area of Mars, and martian impact craters attest to the important role of these volatiles in their excavation, ejection and weathering.

IMPORTANCE OF EARLY FORMATION AND ACCRETION PROCESS

One thing that has become quite clear is that heavy impact cratering has occurred on Mars, and--if one adopts reasonable models for crater accumulation--the process must have been still going on after the formation of some the atmosphere following its initial release.

Explanation for Current Atmosphere

Our scenario for the early history of Mars is that Mars formed out of material condensed in the early solar nebula and sampled material formed at different distances from the Sun, and that the products of this accumulation were heated by the accretion process itself -- by the energy of in-fall and impact as all of these objects collided with each other. The planet then melted and differentiated, separating into layers according to density (crust, mantle and core) while efficiently and extensively releasing volatiles to form its primordial atmosphere. Then, even though the accretion process tailed off following the formation of its early atmosphere, Mars continued to be subjected to in-fall bombardment for perhaps 100 million years. Thereafter, the escape of gases would mostly follow the process described by Mike McElroy in his Mars Conference presentation [NMC-1F].

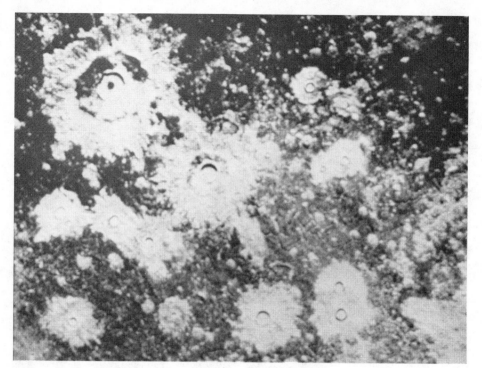

Figure 1A-10: *Cratered Region Exhibiting "Splash" Craters*

Figure 1A-11: *Martian Crater Exhibiting Lobate Ejecta Apron*

Models

Hampton Watkins and I have done some models of this process [°], and Jim Walker [1986] also has recently done some similar work. We have found that the accretion of the last 0.01 percent of the mass of Mars, which occurred after the initial release of atmosphere, generated sufficient energy to blast almost all of that early atmosphere off' of Mars, i.e., accelerate it to escape velocity. The impact explosions created powerful shock waves that traveled faster than the escape velocity for Mars, carrying away the planet's primordial atmosphere.

This means that only the last gases released from the interior of Mars were able to survive to the present time. These included, of course, the radiogenic rare gases (like ^{40}Ar) typically produced *after* the planet's formation. Therefore, the depletion of the abundances of all the volatiles, with the *relative* enrichment of ^{40}Ar, now *do* have a sensible explanation relevant to the formation of Mars, as I suggested earlier.

CONCLUSIONS: VALUE OF FUTURE MISSIONS

Phobos and Deimos

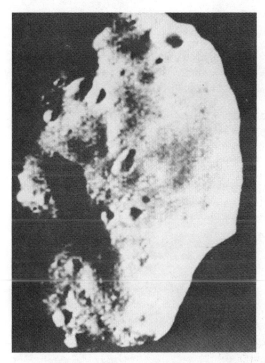

Figure 1A-12 is a Mariner 9 image of Phobos (from a range of 5720 km) that gives one an idea of what the satellite looks like but does not show the complex grooves and crater chains imaged in great detail by the Viking orbiters. We believe Phobos and Deimos are representatives of a group of primitive bodies that were present in the vicinity of Mars at the time of the planet's formation[14].

Figure 1A-12: **Mariner 9 View of Phobos**

---

[14] Phobos, the nearer and largest of Mars' two small satellites, is already beyond its gravitational Roche Limit; i.e., its internal gravity relative to that of Mars is no longer strong enough to hold it together. Therefore, Phobos, with a mean diameter of only 22 km and already weakened by a major impact, may become a martian debris ring at some point in the future. Deimos is slightly smaller at 14 km in diameter and has a somewhat surprising and mobile regolith blanket that subdues surface detail. Both objects appear to be compositionally like Type-1 carbonaceous chondrites, on the basis of their spectral and albedo qualities and their densities. Both of the satellites were imaged at very close range by the Viking orbiters to achieve remarkable resolutions of their surface detail [see Veverka, NMC-1E]. {Ed.}

Whether Phobos and Deimos are in fact made of the same material that formed Mars is quite debatable. Both are heavily mantled by dark, dusty regoliths, and their densities and color strongly suggest that they are made of carbonaceous chondrite material similar to that of the majority of main-belt asteroids. If so, they may contain up to 20% chemically-bound water in hydroxyl silicates and up to 6% organic matter (mostly a complex tarry substance technically known as "gunk"). In any case, they are energetically very accessible to anyone who visits the Mars system, and they have enormous scientific interest as samples of primitive, preplanetary material. In addition, they offer a potentially productive fringe benefit; though not as accessible as the near-earth asteroids, their accessibility and proximity to Mars nevertheless make them interesting potential targets for resource exploitation relevant to future manned missions. They may serve as a ready resource for the extraction of water needed to make hydrogen and oxygen propellants, as well as life-support fluids, and they also may be a source of needed metals. Soon after beginning the exploration of these objects, we could begin using the resources they contain to help sustain operations at Mars.

Joseph Veverka's presentation [NMC-1E] presents some very dramatic high resolution photographs and summarizes our current knowledge of both Phobos and Deimos, and James Head provides a detailed overview of the 1988 Soviet **Phobos Mission** program [NMC-2A]. The **Phobos Mission** is certainly one that I look forward to with tremendous anticipation.

The Expanding Importance of Mars

As interesting as this history of Mars has been, the future may be even more exciting. Both the United States and the Soviet Union are returning to Mars, each after an exploration gap of approximately 15 years. In addition, several European nations and Japan also are emerging with the technologies and commitments necessary to participate in interplanetary exploration.

We have seen two small parts of Mars in great detail -- the local areas seen in panoramic perspective through the eyes of the **Viking** landers. As suggested by Figure 1A-13, Mars seems more reminiscent of the driest, coldest terrestrial deserts -- such as the Atacama in Chile or the dry Antarctic Valleys. But the landers' perspectives represent such a tiny fraction of the surface of Mars, and were biased so strongly in favor of selecting safe and sane landing sites (avoiding so much of the most interesting but more hostile terrain of Mars), that we cannot regard pictures like these as more than a taste of what lies in store for us in the future. ∎

Figure 1A-13: The Harsh, Frigid Reality of Utopia with Winter Frost

REFERENCES

References identified here by alphabetically sequenced lower-case characters are keyed to *specific* references in the presentation text where the same characters are used as identification tags. In the text, however, the identifier will be found within brackets, either by itself or with a name and/or credit date (i.e., when formally published) -- e.g., [a] or [1978a].

This space is otherwise available for notes.

~~~~~~~~

a.   Lewis, J. S. 1984. *The Origin and Evolution of Planets and their Atmospheres.* Academic Press, New York.

b.   Anders, E., and Owen, T. 1977. Mars and Earth: Origin and abundance of volatiles. *Science.* 198:453-65.

c.   Watkins, G. H., and Lewis, J. S. 1985 (August 9-10). *Origin of the Martian Atmosphere* (Paper presented at workshop). Honolulu, HI.

THE GEOLOGY OF MARS
[NMC-1B]

Dr. Laurence A. Soderblom
USGS, Flagstaff, AZ

The history of the exploration of Mars is punctuated by many dramatic changes in man's perception of that planet. However, it is remarkably consistent in its progress toward increasingly detailed, precise knowledge of the planet's environment and geophysical nature. Our image of Mars has shifted from one of science fiction to one of scientific fact, and the most exciting progress has been made possible largely through the contributions of both American and Soviet spacecraft. These explorations were capped by the historic work of four Viking spacecraft, spanning more than five years of operation in orbit and at the surface from mid 1976 through 1982. We now know that Mars is a very complex planet; in some ways ancient like Earth's moon but incredibly dynamic in others. Martian topography offers infinite variety and challenge for the geologist, with terrains that range from the very ancient to the surprisingly young, and is pressing the state of our understanding of Mars forward at every turn. But even as each question is answered, it poses still others for future consideration, establishing a new foundation of inquiry to which our work must now be dedicated.

INTRODUCTION

In his *NASA Mars Conference* presentation, John Lewis discussed the formation of the solar system and its planets in general, defining the solar system's basic chemistry as it relates to Mars [NMC-1A]. The task now at hand is to provide details about the history of Mars and how that planet has evolved through geologic time, which I--as the leadoff--and others will attempt to do during the remainder of the first half of this session [four presentations, NMC-1B through NMC-1E].

Presenting a comprehensive description of the geology of Mars in 25 minutes would be a fairly tall order, so I won't try. Fortunately, there are a number of presentations associated with this first session of the Mars Conference that deal with other aspects of martian geology: volcanos, channels, polar phenomena, and the nature of modification processes in evidence at the landing sites themselves.

Therefore, to provide a framework for the presentations that will follow, I will try to characterize the planet on a global scale. I will discuss the history of the geologic exploration of Mars, reviewing in particular what we learned from the precursor missions (Mariners 4, 6/7, and 9) upon which the planning of the **Viking** exploration ultimately rested. And, finally, I will provide a glimpse of the future by discussing the role that **Viking** data have played and are continuing to play; an extremely important role relative to future explorations of Mars.

FORMULA FOR PLANETARY GEOLOGY

In my formula for dealing with the geology of a planet, three essential areas of information must be established. First, we must develop a chronological understanding of the planet; that is, the order--through time--of geologic events and evolution. For example, we need to know: (a) the age of the oldest terrains and the youngest terrains, and (b) the rate at which they have evolved. I will emphasize the importance of chronology throughout this presentation. The second is comprised of what we know about specific surface modification processes that were active as the planet evolved; for example: impact cratering, volcanism, and wind erosion. I will introduce this issue, but it will then be more fully discussed in subsequent presentations. And, thirdly, we must develop a detailed understanding of the compositional nature of the planet's surface materials: the mineralogical, chemical, and volatile species that evolved from the "primordial soup" to modern-day Mars.

Mars Chronology

The **Viking** landers of course provided highly detailed, direct information from the surface of Mars as their contribution to the formula. However, the **Viking** orbiters also contributed massive amounts of important information on a global scale, which I will present in some detail and without which my formula would not be complete. The orbiters also are important as precursors for the development of future science programs and instruments for Mars-orbiting spacecraft.

From a historic perspective, our concept of Mars has evolved from a time when it was thought the planet might be inhabited and had channels, canals and waterways. Early spacecraft data returned by **Mariners 4, 6 and 7**[1] revealed heavily cratered terrains throughout the southern hemisphere and south polar region. This coincided with what we were learning from the Apollo lunar samples at that time, that the ancient cratered terrains on the moon dated back to the very early history of the solar system. We therefore surmised that the immense crater populations on Mars must also date back to the earliest billion years of planetary time, and the "binary state" of our perception of the planet changed from that of the hoped-for, Earth-like Mars to that of a very ancient Mars.

Mariner 9[2] was next, however, and taught us to be very careful in assuming we know all there is to know about a planet when we have seen less than a few percent of it. For example, we discovered that the planet has a basic dichotomy. That is, its northern hemisphere, lying roughly within a great circle that was not imaged during earlier Mariner missions, differs significantly from its southern hemisphere. This dichotomy is illustrated in the color geologic map of Mars compiled by Scott and Carr [1978[a]], which is presented in Figure 1B-1 (page A-2 in the color display section). [NOTE: All of the color figures referenced in these proceedings are compiled in a **color display section** (Appendix A) on face-up foldout pages at the back of the document.]

---

[1] **Mariner 4** flew past Mars in July, 1965; Mariners 6 and 7, twin spacecraft, flew past Mars in July and August, 1969. {Ed.}

[2] **Mariner 9** began orbiting Mars in November, 1971, and continued to do so throughout most of 1972. It was the first spacecraft to be successfully placed in orbit about another major planet. {Ed.}

The map shows that most of the southern hemisphere is made up of ancient cratered terrains (in color: patchy mixture of blue-grey and olive-green) much like the highlands of the moon. These terrains probably date back to the first billion years of planetary formation when the post-accretional "sweep-up" of debris was most likely occurring throughout the solar system in almost synchronous fashion. In the northern hemisphere, however, the Scott-Carr map shows that older cratered terrains have been replaced by a whole variety of younger geologic units that exhibit the effects of different geologic processes. For example, volcanism appears to be the premier process through which the planet's northern crust has evolved.

Ray Arvidson[3] [b] and his colleagues have prepared color global perspectives of the planet's geology that give one a better sense of how the global dichotomy of old and young terrains are organized into two hemispheres, as seen in Figure 1B-2a/b (presented on page A-2 in the color display section). If one can imagine slicing Mars on a plane that's canted about 30 degrees from the equator, the planet is then divided into the global dichotomy I have described. We can then see the ancient cratered terrains in the southern half. The complex polar layers and residual icecaps of the north and south hemispheres appear quite similar, but they rest on very different geologic terrains; young topography under the north and ancient topography under the south, each in common with its own hemisphere.

Geologic Modification

It was not until the **Viking** exploration, however, that we really became aware of the complexity of martian geology, the diversity of martian phenomena, and the fundamental role that volatiles have played in the evolution of Mars' surface. So, **Mariners 4, 6 and 7** can be credited with having showed us ancient, lunar-like highlands; perhaps somewhat abraded, sanded and filled, but nonetheless ancient terrain. **Mariner 9** then showed us tremendous plains associated with younger volcanic systems and processes. But it was the product of the **Viking** program that revealed the interaction of a far more complex set of processes--involving wind, water, and ice--as playing a fundamental role in the evolution of Mars' geology.

The product of Viking's long-term, detailed observation of the polar regions provides a good example. Even on a diurnal to a seasonal basis, we can see changes in the distribution of frost which appear to be controlled by the martian wind systems and climate. The wind, in fact, has affected geology on a global scale. We find barchanoid dunes throughout the polar regions and at other locations that are very much like barchan dunes on Earth.

In addition to wind processes, there is ample visual evidence of features that appear to have been produced by a denser fluid, most likely water channeled by enormous river systems. The debate still goes on as to whether the water was running as a liquid or instead may have been flowing slowly (versus the extremely slow rate of ice, as in the case of a terrestrial glacier), but these abundant sets of geologic evidence at least indicate that the inventory of volatiles on the planet must have been--and probably still is--enormous. However, the evidence is also a little puzzling, in that it contradicts some of the isotopic limitations calculated for the amounts of volatiles suggested.

---

[3] Dr. R.E. Arvidson (Washington University, St Louis) was a member of the **Viking** lander imaging team. With special interests in aeolian processes, he has authored or coauthored numerous papers focusing on martian aeolian dynamics and their implications. With Thomas A. (Tim) Mutch as principal author, Dr. Arvidson also helped write and prepare **The Geology of Mars**, Princeton University Press, 1976, with James Head (Brown U.), Ken Jones and R. Stephen Saunders (both at JPL). {Ed.}

The ancient cratered terrains at higher latitudes also revealed evidence of volatiles in features below a kilometer or so in resolution, but it is again unclear whether the evidence supports a frozen or liquid form. Water even affected the formation of ancient craters, as I will explain shortly.

When using impact cratering to date surfaces, some of the martian volcanoes above the equator stand in marked contrast with the ancient terrains of the southern hemisphere. At close inspection (in high resolution) of the youngest of the large martian volcanoes, Olympus Mons[4], we can see that the abundance of impact craters is extremely low. So we have found, on one hand, southern highlands that must date back to the very early period of solar system history, but, on the other, northern topography on the same planet that must date back only to "yesteryear" or perhaps "yester-hundred-million-years."

Chronology Models -- On the basis of these kinds of findings, how does one put together a chronology without direct information on the rate of impact cratering or isotopic samples? A number of us have been working with this problem, e.g., Hartmann [1983], Neukum and Hiller [1981[c]], and Shoemaker [1977[d]], and some plausible models have been developed. I think it would be productive to discuss a few of these models, at least in a fundamental sense, to provide an impression of their stability.

I will start with a brief tutorial on how the time scale for a planet's geological history is related to the abundance of impact craters on its surface and to the impact flux history it has endured[5]. We begin with a model that, as shown in Figure 1B-3, is assumed to resurface itself at a rate of only about 10 percent per billion years. The simple model used predicts what percentage of the exposed surface will be in any particular age bracket, based on the assumption that the geologic process will continuously and uniformly bury or destroy older terrains. This produces a percentage distribution of exposed surface ages decreasing exponentially with age. If it is a *very* rapid resurfacing process (as it is on Earth), the exponential distribution tail will be short -- as in the case of the *active planet* model presented in Figure 1B-4. But if the resurfacing process progresses at a very low rate, it in effect provides "windows" into early geologic time; i.e., as a result of the long exponential distribution tail, virtually all of the planet's geologic units will be clearly exposed at one time or another all the back to early time.

We can next model the areal distribution of impact craters that would be expected for a particular crater production curve. For the cases illustrated in Figure 1B-5, we have used an impact flux modeled after the lunar curve. It can be seen that cratering fell off very abruptly during the first 100 to 200 million years, or about 4 billion years ago as the accretional "sweepup" was occurring.

Inactive Planet -- For the *inactive planet* concept, the model (in this case, using the moon) produces a dichotomy in the areal crater populations (upper left, Fig 1B-5). The small square pedestal represents the cratered highlands, then we have an enormous void on the curve in which geologic units located between the highlands do not have crater populations, and finally

---

[4] Formerly known as Nix Olympica in the nomenclature of maps preceding Mariner 9, Olympus Mons is approximately 600 km across at its base and has a summit 27 km above the surrounding plains. The complex summit caldera is nearly 75 km across and has a depth of 3 km in the deepest of its collapse craters. The scarp at the volcano's base includes cliffs up to six km high in some places. {Ed.}

[5] Planetary "flux history" is a history of impact cratering that relates the period of cratering events, impact frequency, and the nature of resulting modification through time for a given planetary body. {Ed.}

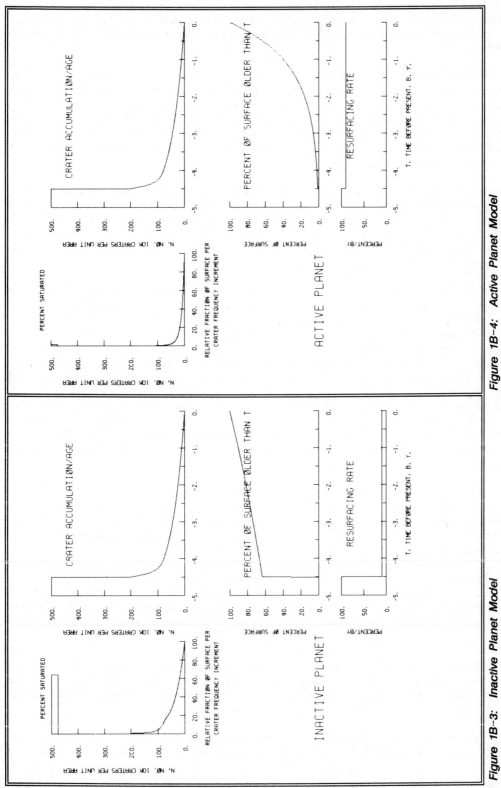

Figure 1B-3: Inactive Planet Model

Figure 1B-4: Active Planet Model

47

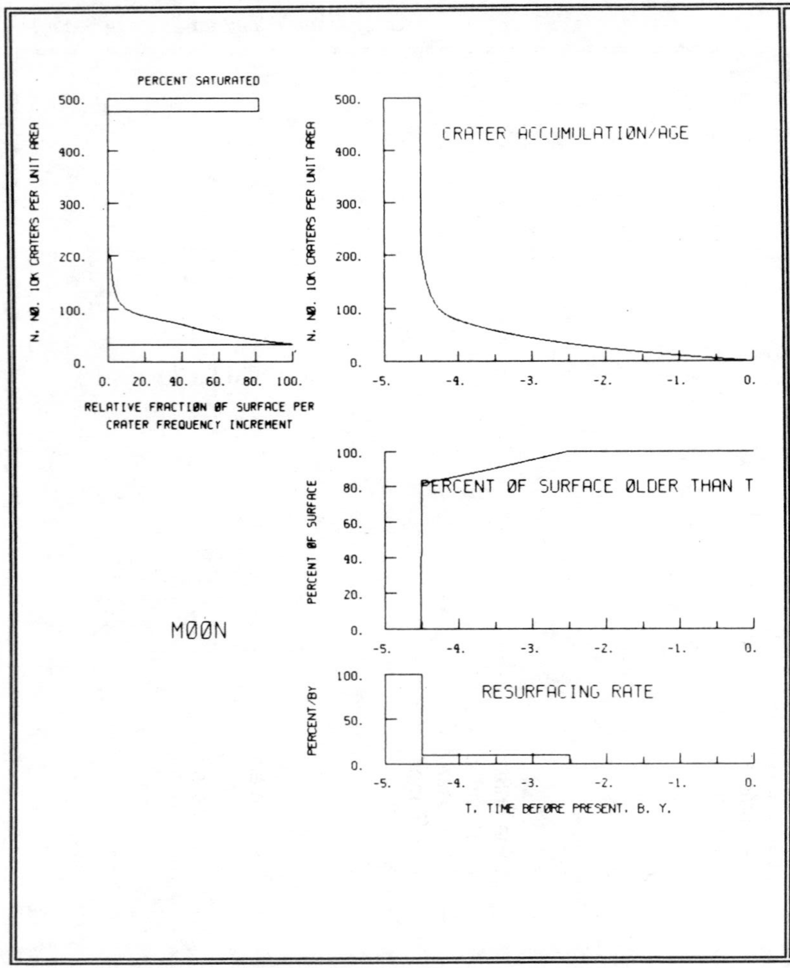

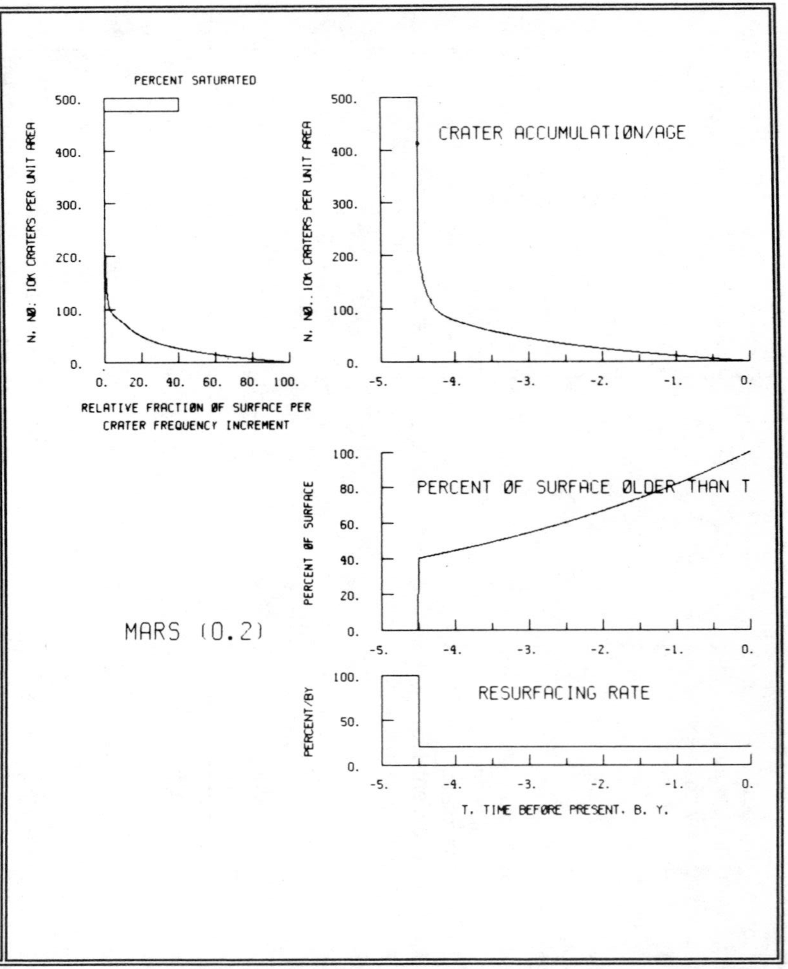

Figure 1B-6: Mars Model

Figure 1B-5: Moon Model

there is a smaller distribution at the bottom of the curve that represents the cratered plains. We find these kinds of crater dichotomies on the moon, on Mars, on Mercury, and in the outer solar system (Jovian and Saturnian satellites); they could provide the basis for a very interesting conference by themselves.

Active Planet -- On the other hand, Figure 1B-6 (opposite page) is a model of a planet (using Mars in this case) that is resurfacing much more rapidly -- 90% per billion years. We cannot in fact see very far back into the planet's evolutionary history because geologic and other processes are rapidly erasing evidence of earlier geology (ref., middle curve in right column, Fig. 1B-6), and this model is much closer to being like Earth than previous models. When we try to apply the crater accumulation curve to it, we find that there aren't any geologic units comparable to ancient highlands because they have simply been destroyed by multiple cycles of the resurfacing processes.

Through fitting such models to currently observed planetary surface data, we can estimate the geologic resurfacing activity through time. Applying this process to the moon, as we did in Figure 1B-5, produces a good example of how it works. The moon experienced a very short resurfacing period that evidently lasted only up until about 2.5 billion years ago. The bottom curve in this illustration was developed largely from the results of analyses of Apollo lunar samples, and it can therefore be used to construct and test the model.

Fitting the process to Mars, which we did in Figure 1B-6, creates another example of the model's utilization. Assuming a uniform level of geologic activity through time, we can reproduce roughly what we see today on Mars by using a resurfacing rate something on the order of ten to twenty percent per billion years. Relative to the case of Mars, we need to make a couple of observations based on the results of using models. If we simply fit the lunar shape to Mars, and assume that the oldest of the plains in the northern hemisphere must date back to the epoch that just followed the rapid fall-off of accretional bombardment, the time frame is set. This process then models a geologic history of Mars that is spread out rather uniformly, with the Tharsis volcanics dating up to the present and the Elysium volcanics or other older processes and volcanics dating back to the highlands.

Models of Crater Flux -- Models developed by Neukum and Hiller [1981, *ibid.*] demonstrate the sensitivity of the inferred martian history to the assumed impact flux history. In the first model (Model 1), as demonstrated in Figure 1B-7a, the crater production rate for Mars is assumed to have been the same as for the moon. One can see that this assumption results in a model in which most of Mars' volcanic activity essentially occurred during the first billion years. The Tharsis region, however, would have been volcanically active throughout most of the planet's history. This would require very different geophysical conditions and thermal histories for different regions of the planet.

The second Neukum and Hiller model (Model 2), similarly represented in Figure 1B-7b, assumes that the impact fluxes for Mars have been about a factor-of-two higher than on the moon. This model appears much more reasonable, in that the volcanic evolution is more uniformly spread through time (gradually decaying in activity), and the Tharsis region no longer appears so anomalous. I favor the second model for these reasons. In any case, either model shows that volcanic processes have acted to some degree throughout martian history and very likely are continuing.

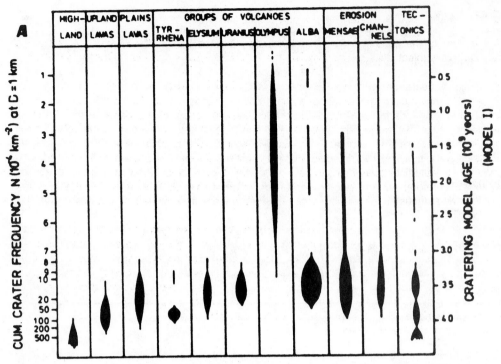

Figure 1B-7a: Neukum–Hiller Model -- Mars Cratering Rate Same As for Moon

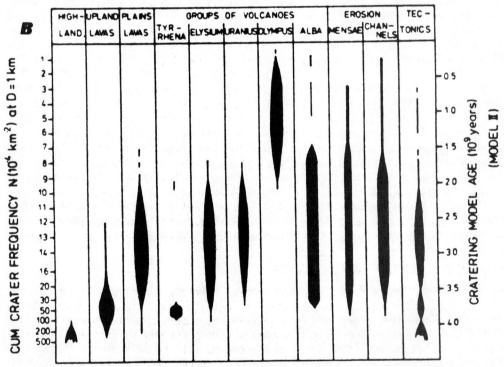

Figure 1B-7b: Neukum–Hiller Model -- Impact Fluxes Factor-of-Two Higher Than on Moon

Surface Materials

I want to turn to the third element of the formula I defined earlier: the importance of surface materials. This subject also relates well to the mission in store for *Mars Observer* (1992).

Mars' global appearance is roughly organized latitudinally. That is, overlying all of these geologic land forms, including the large craters and volcanoes, is a latitudinal distribution of materials that modify land form appearance and attenuate or amplify surface albedo. To illustrate: (1) the equatorial belt is somewhat darker than mean albedo and very changeable through time; (2) the northern and southern mid-latitude regions are brighter, due most probably to deposits of very fine, bright material; and (3) there is a dark collar around the north polar region. These latitudinal characteristics are evidently controlled by wind systems and aeolian processes. Therefore, surface geology is probably not visible as land form (except in a few cases involving dunes) and thereby controls what we are able to see at low resolution.

Figure 1B-8 (page A-2 in the color display section) presents the only global, multispectral map of Mars, which is a mosaic of roughly fifty Viking Orbiter 2 (VO-2) Mars-approach images. In this version, subtle color variations are enhanced. At that time (August, 1976), the south polar region (in its winter season) extended up into the Hellas Planitia basin. The region was therefore rich with volatiles, probably in the form of surface frost and atmospheric ice crystals. For this reason, the area was essentially "contaminated" (masked) for remote sensing of the geologic surface. But one can still see, in the vicinity of the equator, a belt in which very dark materials are exposed.

When we look at the correlation of the visual boundaries, we find that many (located between areas characterized by the dark-bluish and dark-reddish colors) coincide with lithologic boundaries (boundaries between geologic units). Understanding these surface materials, relative to the dynamics of the latitudinal processes that affect them, is important to developing our knowledge of Mars' global mineralogy. So this is a key area in which Mars Observer should, by affording a detailed examination of the planet's foundational geologic materials, gain some important information about its surface chemistry and mineralogy.

The acquisition of Viking high resolution orbital imaging data contributed a dramatic improvement in the quality of visible detail. Figure 1B-9 (page A-2 in the color display section) shows an example of the central equatorial region at a higher resolution and in "exaggerated" color. There is some indication that these darker materials may be in areas where the silicates are somewhat more reduced and richer in ferrous- rather than ferric-silicates. We find that surface areas defined by dominant materials can be roughly classified into three or four categories. This sort of "triad" of materials must be kept in mind as we examine the Viking landing sites. It can allow us to extrapolate the highly detailed knowledge of those sites to a global perspective. The resolution afforded in this illustration, by the way, is about the same resolution we think we will be able to achieve with **Mars Observer**, when mapping the planet using a spectrometer range from 0.35 to 4.30 microns with 320 spectral channels.

NEW WAYS TO USE VIKING DATA

Processing of the Viking data set is far from finished. Ultimately, we will have processed some 60,000 images, and to effectively develop this much data into a coherent picture is a mammoth undertaking. Figure 1B-10 (page A-2 in the color display section) presents a newer map of the western half of the equatorial region, illustrating some ongoing work by Scott and Tanaka [e] relative to the geology of Mars. When comparing this map to the earlier Scott-Carr map (Fig. 1B-1, also presented on page A-2 in the color desplay section), one will find that the definition of detailed geologic units is much more refined.

In addition, we are *re*-processing **Viking** data with ten-year newer technology, and some rather striking, previously undetected surface definition of materials is being discovered in the orbiter database. The images presented in Figures 1B-11 and 1B-12 (both on page A-3 in the color display section) are examples of this ongoing work. The first of these images (Fig. 1B-11) is a reprocessed color image of the south polar ice cap and its surrounding terrain, while the second (Fig. 1B-12) is a sinusoidal projection map of the same polar cap[6]. One of the primary objectives of this work is to organize **Viking** image data into a digital, sinusoidal database to support ongoing Mars exploration.

We can now deal with things like coherent and random microphonic noise patterns in a flash compared to ten years ago when a mainframe computer would have to grind away at a comparable problem for a very long time. Revisiting this data set (using our new processing technology) is in many respects like working with brand new mission data, and Figures 1B-13, 1B-14 and 1B-15 (all on page A-3 in the color display section) are several more examples of what can now be extracted from the **Viking** database.

The striking thing about this combination of improved methodologies is the ease of interpretation now possible relative to the definition of surface materials. Voyager experience, combined with our new ability to merge color information from lower resolution data with high resolution information about surface textures, provides an extremely important tool for the interpretation of what geologic processes are involved -- and what is on top of what.

The end-product of being able to define these kinds of geologic materials will lead to a more precise and better controlled mapping program. That program, in turn, will lay out the geologic framework of Mars and detail the planet's geological history, not only in preparation for **Mars Observer**, but, hopefully, for future manned exploration as well. ∎

---

[6] The sinusoidal projection map of the south polar cap is not a color image, but it has been presented in the color section to provide a side-by-side comparison of its representation with that of the more conventional color image of the same polar cap. {Ed.}

REFERENCES

References identified here by alphabetically sequenced lower-case characters are keyed to *specific* references in the presentation text where the same characters are used as identification tags. In the text, however, the identifier will be found within brackets, either by itself or with a name and/or credit date (i.e., when formally published) -- e.g., [a] or [1978a].

This space is otherwise available for notes.

~~~~~~~~

a. Scott, D.H., and Carr, M.H. 1978. *Geologic map of Mars*. U.S. Geol. Survey, Misc. Inv. Map I-1083.

b. Comments based on personal communication with Ray Arvidson, Dept. of Earth and Planetary Sciences, Washington University, St. Louis, MO.

Also of interest:

Arvidson, R.E. 1979. A post-Viking view of martian geologic evolution. NASA Tech. Memo, 80339, pp. 80-81.

Arvidson, R.E., Binder, A.B., and Jones, K.L. 1978. The surface of Mars. *Sci. Am.* 238:76-89.

c. Neukum, G. and Hiller, K. 1981. Martian ages. *J. Geophys. Res.*

d. Shoemaker, E.M. 1977. *Impact and Explosion Cratering*.

e. Scott, D.H., and Tanaka, K.L. *Geology of Mars*. 1:15M.

THE VOLCANISM OF MARS
[NMC-1C]

Dr. Michael H. Carr
USGS, Menlo Park, CA

Volcanism has played a significant role on Mars throughout the planet's history, some of it with obvious results and much of it with subtle but far-reaching implications. We may have samples of martian volcanic rocks on Earth in the form of meteorites believed to have been ejected from the martian surface by impacts. Martian volcanism appears to have declined and become progressively more restricted in place over time. Its manifestation on Mars has been quite different from that of terrestrial experience, and has resulted in a very distinct global dichotomy. Much of the northern half of Mars is only sparsely cratered and includes most of the major volcanic features, while the southern half is essentially ancient, heavily impact-cratered terrain. Many of the smooth plains found at higher elevations in the north half are clearly volcanic, although many similar but low-lying plains apparently are not. There has also been considerable pyroclastic activity, but the extent is hard to assess. Many of the volcanic features appear to reflect the nature of a reaction between lavas and the materials they encountered while emerging at the surface. Ice may be common near the martian surface at higher latitudes, and features have been identified that can be attributed to the interaction of lava and ice. The character of such features supports the conclusion that Mars is a volatile-rich planet and strongly suggests that water was the key volatile involved in their formation.

INTRODUCTION

My NASA Mars Conference presentation is about volcanism on Mars, which was briefly introduced in the preceding presentation [Soderblom, NMC-1B]. Several themes will be expressed throughout my presentation. One is that martian volcanism appears to have declined with time and has become progressively more restricted in place over time. Another will lend support for the conclusion that Mars is a volatile-rich planet. I will illustrate how this is reflected in the volcanism, since many of the features one can see on the planet can be interpreted in terms of reaction between relatively volatile-poor, mantle-derived (deep) lavas and the very water-rich, volatile-rich materials the lavas encounter at the surface. A third theme is that we may already have samples of martian volcanic rocks on Earth in the form of a unique group of meteorites believed to have been ejected from the martian surface by large impacts. I will show that the properties of these meteorites are consistent with what we have learned about the characteristics of martian volcanics from our mission science data.

Hemispheric Dichotomy

The dichotomy of Mars has been discussed in detail by Larry Soderblom at this conference. To quickly review, one can view Mars as being divided into two halves. The southern half is essentially covered with heavily cratered terrain that looks somewhat like the lunar highlands. In contrast, much of the opposing northern half is only sparsely cratered and includes most of the major volcanic features. Also, while many of the smooth-appearing plains found in the north half (particularly those at higher elevations) are volcanic, many of its smooth, low-lying plains apparently are not.

Northern Hemisphere -- One can clearly see individual volcanoes in the northern hemisphere, notably in the Elysium and Tharsis regions. Much of my discussion will involve Tharsis, which is the most active and persistently volcanic region on the planet, but I will return to both of these important regions in detail later in my presentation.

Southern Hemisphere -- Although most of the volcanics are in the northern hemisphere, I want to emphasize that there are volcanics in the southern hemisphere as well. Smooth plains occur locally within the heavily cratered uplands. While one occasionally finds evidence of flows on the plains, one more commonly sees ridges that appear to be identical to ridges found on the lunar maria. We know that the lunar ridges developed in volcanic terrain, so it is not unreasonable to interpret the martian ridged plains as being volcanic, particularly in view of the occasional flows.

These "intercrater" plains are much more common in the martian highlands than they are in the lunar highlands. It appears that Mars was more volcanically active than the moon during the time of the early heavy bombardment, which was experienced throughout the inner solar system and terminated around 3.8 billion years ago. However, volcanic activity appears to have declined rapidly in the southern hemisphere shortly after the termination of the heavy bombardment, because the cratered uplands in the south are relatively unmodified by younger volcanics.

MARTIAN VOLCANICS

The contrast between the old cratered terrain and the younger volcanic plains is clearly demonstrated in Figure 1C-1, in which one can clearly see younger volcanic flows encroaching on an older cratered surface and into a 130-km-diameter crater (lower right). This further demonstrates an important characteristic of many martian lava flows: they are extremely large compared with most of the terrestrial flows with which we are familiar on Earth. Their large size almost certainly implies large eruption rates.

Lava, Pressure and the Lithosphere -- Some theoretical work has been done, principally by L. Wilson (University of Lancaster, England) and co-workers at Brown University, on various factors that control eruption rates. They have examined, for example, how eruption rates are affected by pressure on the magma column, the dimensions of the conduit through which the magma travels to the surface, and the properties of the magma itself.

Their work indicates that the magma must be forced to the surface under very high pressure in order to achieve high eruption rates. In general, evidence of high pressure on Mars implies a deep magma source. This, in turn, implies that the martian lithosphere (the rigid rind below which melting takes place) was much thicker than Earth's present lithosphere when many of the older martian flows formed quite early in that planet's history. Figure 1C-2 shows what these flows look like in greater detail. Pressure ridges formed on the surface of such flows as they were still moving (after their exposed surfaces had begun to solidify), causing the partly congealed crust to buckle and form the corrugated texture seen in the picture.

Figure 1C-1:   Lava Flows Embaying on Plains and into Crater (lower right)

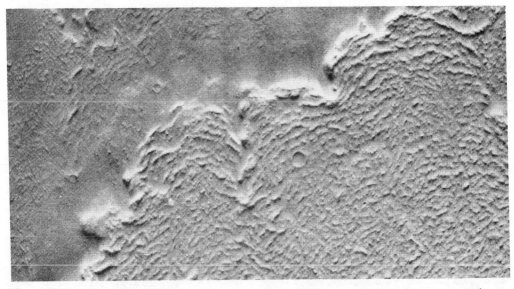

Figure 1C-2:   High Resolution Photo of Lava Flow in Tharsis (frame is 17 km across)

57

Evidence for the Nature of Martian Lavas -- Not all martian plains have such obvious volcanic flows. As seen in Figure 1C-3, many appear smooth except for ridges and are very much like those we see on the lunar maria. Such martian and lunar ridges are believed to have formed by deformation of the surface as a result of compressional stresses, and therefore are not in themselves evidence of volcanic activity. But the identical response of both the lunar and martian plains to such stresses suggests that the features are similar, i.e., that they are in fact volcanic flows. The superimposed impact craters, which are typical of most on Mars, have ejecta arrayed in discrete lobes -- each outlined by a low rampart.

The ridged plains probably formed by eruptions along very long fissures and at very high effusion rates. It is believed that the effusion rates for the volcanic plains of Mars (and the moon) are about 10,000 to 100,000 cubic meters per second. For comparison, the highest rates from fissures in Hawaii are only about 100 cubic meters per second. Clearly, the inferred martian and lunar effusion rates are much higher than those associated with typical flows on Earth.

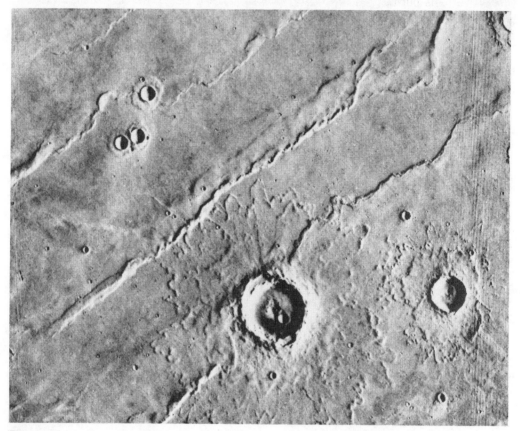

Figure 1C-3: Smooth, Ridged Plains East of Tharsis (240 km across)

There is of course another factor that can contribute to the diminished visibility of flows on the plains -- high lava fluidity. If we get some indication of the fluidity, it can shed some light on the composition of the lavas. High fluidity generally results with what are called **mafic** lavas, which are lavas composed predominantly of iron and magnesium minerals. Fluidity is also enhanced by a high iron-to-magnesium (Fe-Mg) ratio. Lavas with a high Fe-Mg ratio are consistent with the spectral reflectivity of Mars, which suggests the presence of large amounts of iron.

Direct Evidence from Mars -- The high Fe-Mg ratio is also consistent with the composition of SNC meteorites, which are believed to have come from Mars[1]. The reason for suspecting that SNC meteorites had their origin on Mars is several-fold. Initially, it was because they are basaltic and have a young age relative to the age of the solar system (many of them proved to be about 1.3 billion years old). A source therefore had to be found somewhere in the solar system that was volcanically active as recently as 1.3 billion years ago. There aren't many candidates and Mars seemed the most likely[2]. That theory has since been supported by knowledge that the composition of gases trapped in vesicles and glass within the SNC meteorites differ distinctly from Earth's atmosphere and are almost identical in composition to the martian atmosphere (as determined by **Viking** instruments).

The SNC meteorites resemble terrestrial basalts, but have a higher Fe-Mg ratio as just described. One can calculate what the viscosity (fluidity) of basalts was at the time of their formation, and it turns out that the SNC basalts were very fluid compared with terrestrial basalts. In fact, the SNC basalts were 10 to 100 times more fluid that typical terrestrial basalts, which is consistent with the high fluidity implied by the lava flows we see in pictures of the volcanic plains on Mars.

Martian volcanic plains are not always such that they can be neatly fitted into one of the two predominant categories: (1) those with clearly defined flows or (2) those with ridges. We also see plains, like the one in Figure 1C-4 found in Insidis Planitia, that exhibit beaded ridges (lines of domes) with slot-like summit vents. The domes were probably formed by lava fountaining along linear fissures (known as **curtain of fire** on the island of Hawaii), during which spatter ramparts and cones were built adjacent to the fountains as their flows formed the surrounding plains.

Nonvolcanic Plains -- There also are plains that are not volcanic. Figure 1C-5 shows an area at the termination of some very large channels that originate in Elysium (several hundred kilometers southeast of the area pictured). The suspicion is that this kind of plain is comprised of ice-rich sediments left after the formation of the channels. Many of the low lying plains at high northern latitudes lack any evidence of volcanism, but have numerous features thought to result from the action of ground ice. Many probably consist of ice-laden sediments, like the plains in Figure 1C-5. With the knowledge that some of Mars' plains represent the accumulation of sediments and ice, some interesting problems are raised relative to the interaction of volcanic products with the ice-rich materials; I will return to this issue later.

I mentioned earlier that volcanism appears to have declined in intensity with time, and the table presented in Figure 1C-6 shows our best estimate of the ages of different plains. As can be seen, most of these plains are quite old. It is only around Tharsis and Elysium that surfaces younger than about 3.5 billion years old are found. This is consistent with Larry Soderblom's understanding as given in his presentation.

---

[1] SNC meteorites include three unusual types of meteorites: the **Shergottites**, the **Nakhlites**, and the **Chassignites**. These have igneous rock textures and are basaltic in composition. They also have relatively young ages: 1.3 billion years compared to 4.6 billion years for almost all other meteorites. {Ed.}

[2] Although the heavy bombardment rate discussed by Dr. John Lewis [NMC-1A] declined rapidly some 3.8 billion years ago, impacts undoubtedly continued at a much reduced rate into more recent geologic time. It is theorized that one or more of the most powerful of these more recent impacts, occurring within the past 1.3 billion years (perhaps as recently as 10 million years) may be responsible for ejecting the SNC meteorites [Wood and Ashwal, 1981, **Proc. 12B Lunar Sci. Conf.**, pp 1359-1373]. {Ed.}

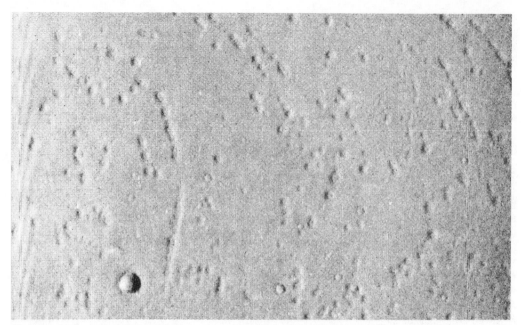

Figure 1C-4: Volcanic Plains in Isidis Planitia with Beaded Ridges (30 km across)

Figure 1C-5: Nonvolcanic Plains (possibly of sedimentary origin)

Geologic Province	Crater Density Relative to Avg Lunar Maria	Estimated Crater Retention Age in billions of years (b.y.)		
		Minimum Likely	Best Estimate	Maximum Likely
Central Tharsis volcanic plains	0.1	0.06	0.3	1.0
Olympus Mons	0.15	0.1	0.4	1.1
Extended Tharsis volcanic plains	0.49	0.5	1.6	3.3
Elysium volcanics	0.68	0.7	2.6	3.5
Isidis Planitia	0.76	0.8	2.8	3.6
Solis Planum volcanic	0.9	0.9	3.0	3.7
Chryse Planitia volcanic plain	1.1	1.2	3.2	3.8
Lunae Planum	1.2	1.3	3.2	3.8
Noachis ridged plains	1.3	1.7	3.3	3.8
Tyrrhenum Patera volcano	1.4	1.8	3.4	3.8
Tempe Fossae faulted plains	1.6	2.3	3.4	3.8
Volcanic plains on Hellas south rim	1.7	2.6	3.5	3.8
Alba shield volcano	1.8	2.6	3.5	3.8
Hellas floor	1.8	2.6	3.5	3.8
Syrtis Major volcanic plains	2.0	2.0	3.6	3.9
Heavily cratered plains				
– small D (<4 km)	1.4	1.8	3.4	3.8
– large D (>64 km)	13.0	3.8	4.0	4.2

Source: Hartmann et al. 1981.

Figure 1C-6: Table of Crater Ages for Different Martian Surface Features (b.y.)

Tharsis and Its Volcanoes -- Of course, plains are not the only product of volcanism on Mars. There are numerous individual volcanoes. Before I discuss the volcanoes themselves, however, it may help to first provide a basic, generalized description of Tharsis (the major volcanic region on Mars). Tharsis is a large rise (bulge) in the martian surface. The contours defined in Figure 1C-7 are at 1-kilometer (km) intervals relative to Mars datum[3]. The elevations at the center of Tharsis are about 10 km higher than those around the periphery.

There is still debate about the cause of this bulge. It may be due to uplift as a result of pressure from below or it may simply be due to sustained accumulation of volcanic deposits, most

---

[3] Mars datum is the reference elevation (zero) used in lieu of a mean-sea-level (MSL) point of reference. The datum baseline represents the martian elevation at which the atmospheric pressure of 6.1 mb has been determined relative to the planet's center. A pressure of 6.1 mb was chosen because it represents the triple point of water, i.e., the lowest pressure at which water can exist in its three possible states -- solid, liquid, or vapor. At partial pressures greater than 6.1 mb, water can exist as a liquid if temperature conditions allow it but will vaporize at lower pressures. {Ed.}

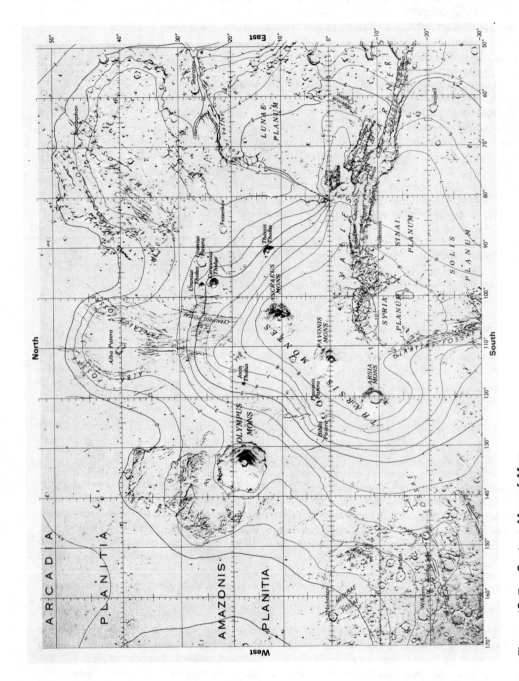

Figure 1C–7: Contour Map of Mars

62

having been deposited before the decline in impact rates about 3.8 billion years ago. The presence of the bulge on the martian surface causes stresses in the crust, and one can determine what those stresses are and then look for the deformation caused by them. Two types of deformation are observed, ridges and fractures. The ridges are generally arrayed circumferentially around the bulge, whereas the fractures have a radial pattern. These orientations are generally consistent with the argument that the ridges and faults formed as a result of the presence of the bulge rather than as a result of uplift.

Olympus Mons -- The most obvious and most spectacular manifestations of the planet's volcanism are its large shield volcanoes, of which the largest are found in the Tharsis region. Olympus Mons, centered at 18°N, 133°W, is the largest and most familiar of the martian volcanoes. The primary central edifice is nearly 600 km across and its summit towers nearly 25 km above the surrounding plains[4]. One of its most obvious characteristics is, of course, its size; Olympus Mons is far more massive than any of Earth's mountains, whether volcanic or nonvolcanic. Its size is a result of the stability and thickness of the martian lithosphere.

Volcanoes grow to great size on Mars for two reasons which have been pretty well understood since the Mariner 9 mission. First, as I just indicated, the martian lithosphere (crust) is very stable, i.e., it exhibits no evidence of plate tectonics. On Earth, volcanoes tend to be limited in size because of the slow lateral movement of large sections of the lithosphere (plates) with respect to each other. The islands of Hawaii, for example, are on the Pacific plate. Because of the motion of the plate, the Hawaiian volcanoes are slowly being conveyed over their stationary magma source below[5]. Ultimately, however, the volcanoes are carried so far beyond the magma source that they become extinct.

As old volcanoes die, new ones form as the magma seeks new conduits to the surface. A new volcanic vent named Loihi has recently been discovered on the ocean floor off the big island of Hawaii. Loihi is just now beginning to grow while the older volcanoes on the big island of Hawaii--Kohala and Mauna Kea--are already extinct, and Loihi will most probably show significant growth as activity associated with Mauna Loa and Kilauea diminishes. In contrast, a volcano on Mars can, in the absence of plate motion, continue to grow so long as magma is available and can reach the surface.

The second reason for the large size of the martian volcanoes is the thick lithosphere. Magma is forced to the top of a volcano as a result of the contrast in pressure created by the dif-

---

[4] Olympus Mons is not on the crest of the Tharsis ridge, and its summit elevation (approximately 26.4 km above the Mars datum) is no higher and may in fact be very slightly lower than those of the three giant shields aligned on the crest. Still, Olympus Mons is by far the most prominent of Mars' large shields, by virtue of its basal diameter and its height above the surrounding plain (which has an elevation of no more than 2 to 3 km above the Mars datum). In comparison, the Tharsis Montes (Ascraeus Mons, Pavonis Mons and Arsia Mons) rise from a basal elevation that is already more than 9 km above the datum. The Olympus Mons caldera is approximately 80 km across, and the volcano's flanks slope away from the summit at an average of 4°. The volume of the mountain's edifice is from 50 to 100 times that of Mauna Loa, Earth's largest volcano. [Elevation information extracted from **Atlas of Mars** and **The Surface of Mars**.] {Ed.}

[5] A magma source is generally believed to be a stable kind of "hot spot" at the base of the lithosphere (50-75 km below the surface), perhaps due to a convective plume originating deep within the mantle that causes a molten reservoir or melting spot to develop in the vicinity of the lithosphere-asthenosphere boundary (one of several models). The magma source should not be confused with a magma chamber, which is part of the conduit system connecting the source to its surface vent and generally forms relatively close to the surface (2 to 5 km). {Ed.}

ference in density between the lava itself and the rocks it must pass through to get to the surface[6]. The greater the vertical distance this pressure contrast is maintained, the greater the pressure forcing the magma to the surface; hence the greater the height the volcano can grow. The Hawaiian lavas are believed to be produced about 60 km below the surface, while (in contrast) the depth of origin for the martian lavas is probably between 150 and 200 km. So these two factors (a stable crust and a thick lithosphere), which are consistent with what we now know about Mars, explain the large size of Olympus Mons and the other large martian volcanoes.

Among the many interesting features of Olympus Mons is its 85-km-diameter central caldera, illustrated in Figure 1C-8, which resembles those of Hawaiian-type volcanoes. As previously noted, magma is pumped to the Hawaiian volcanoes from depths of about 60 km to a shallow magma chamber 2-to-5 km below the summit. When a volcano is about to erupt, this magma chamber actually inflates. When visiting the Volcano National Park in Hawaii, one can see a record of the recent inflation and deflation of Kilauea at the window of the observatory on the volcano's rim. When an eruption occurs on the flanks of a volcano, deflation generally occurs. This deflation may result in some collapse at the summit, and that is how the caldera forms. Its size is probably a reflection of the size of the underlying magma chamber, and we therefore conclude from the large size of the martian calderas that the underlying magma chambers are also large.

Another interesting feature on Olympus Mons is a set of unusual terraces around the caldera on the upper slopes of the volcano (visible in Fig. 1C-8). Such terraces are not found on the Hawaiian volcanoes. We know that the weight of Olympus Mons has caused the lithosphere beneath it to bend, causing the entire volcano to be slightly compressed. Elliot Morris (USGS) has suggested that these terraces form by over-thrusting on the volcanoes flanks as a result of the compression. The sagging of the lithosphere under the weight of the volcano is evident from a broad "moat" that completely surrounds the central edifice. Most of the moat is now filled with relatively young lavas. A similar bending of the lithosphere around other volcanoes has caused circular faults, and the position of these faults has been used by Comer and Solomon to estimate the thickness of the lithosphere. The estimates range from 25 km close to some volcanoes to more than 150 km far from the regions of volcanic activity.

There is another peculiar feature around Olympus Mons that we do not understand at all. It has generally been referred to as the **aureole**[7]. Distinctly visible in Figure 1C-9, the aureole consists of huge lobes of characteristically ridged terrain arrayed asymmetrically around the volcano (particularly to the north and northwest), and it extends out as far as 1000 km from the volcano's basal escarpment. The basal escarpment itself is very sharply defined and includes cliffs up to six kilometers high.

---

6 The similarity of the summit elevations for Olympus Mons and the Tharsis Montes is believed to be related, in that the general height of 27 km may be the maximum possible relative to the magma source and the pressure driving it to the surface; "When the pressure at the base of the column equals the lithostatic pressure, the lava can be pumped no higher." [Eaton and Murata, 1960; The Surface of Mars, Carr, 1981]. The similarity in summit elevations for Mauna Loa, Mauna Kea and Haleakala (Ha-lee-ah-ka-laa) may be a terrestrial example of such a growth limit. {Ed.}

7 By general definition, the word **aureole** is used in reference to an encircling radiance or halo. Looking down upon Olympus Mons from orbit, the peculiar terrain known as aureole does in fact appear as an irregular, nondistinct halo surrounding the mountain, although its most prominent lobate components are found to the northwest and north of the volcano -- the outward and down-slope direction for Tharsis. {Ed.}

Figure 1C-9: Olympus Mons With Surrounding "aureole" Topography

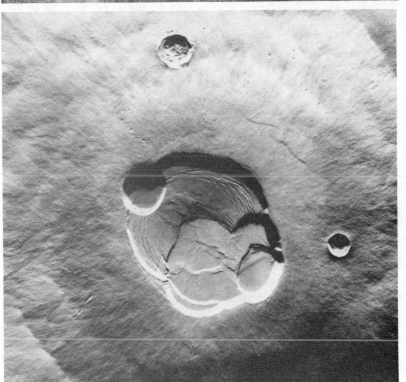

Figure 1C-8: Summit Caldera and Slope Terraces of Olympus Mons

Numerous attempts have been made to explain these lobes. The most plausible hypotheses for the aureole suggest that the disintegration and outward flow of the peripheral parts of the volcano formed the lobes and left the high escarpment that we now see around the remains of the primary edifice. Outward flow to form the lobes may have been facilitated by abundant ground ice or groundwater. A positive gravity anomaly appears to coincide with the aureole.

Arsia Mons –– Another of the giant shield volcanoes of the Tharsis Montes, Arsia Mons[8], is pictured in Figure 1C-10. My USGS colleagues in Flagstaff reconstructed Viking orbiter stereo image data to produce this remarkable dimensional projection of the volcano. Eruptions from fissures on the flanks have built a spur that buries part of the volcano and extends away to the lower right in the image area. Arsia Mons differs somewhat from Olympus Mons, in that late-stage eruptions were concentrated out on the northeast and southwest flanks of the volcano as opposed to the more radially symmetric eruptions on Olympus Mons. Flows from the south-southwest flank have covered the lower slopes of the volcano, extending the elevated spur onto the adjacent plain to the south.

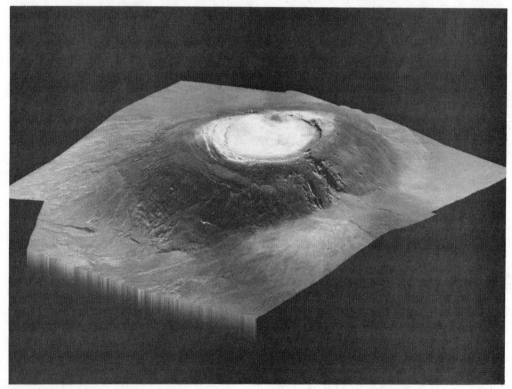

Figure 1C-10: Dimensional Projection of Arsia Mons Viewed from WSW

---

[8] The sparsely cratered caldera of Arsia Mons (9° S, 121° W) is the largest volcanic caldera on Mars. It is 120 km across. Embayments on the northeast and southwest flanks of the volcano, which have produced extensive lava flows, are connected by a row of low mounds across the caldera floor which may mark a rift zone. The most developed of these embayments essentially points down slope toward the south-southwest from the summit. Though lower and less extensive on their respective edifices, embayments also exist at similar orientations on Pavonis Mons and Ascraeus Mons, the other two large volcanoes positioned with Arsia Mons along the Tharsis ridge (Arsia Mons is the only one of the three that lies totally in the southern hemisphere). {Ed.}

On Alba Patera[9], an older and very different appearing volcano to the northeast of Olympus Mons, we see a wide variety of flow forms. It would not be appropriate to go into a detailed discussion of these for this presentation, except to say that some of them have been molded into rather strange and fanciful features. One of Alba's popular features, visible in Figure 1C-11, is Kermit the Frog. Some might say that Alba's profile of Kermit is surely an indication of, if not intelligent life on Mars, at least a friendly reminder of life forms elsewhere in the solar system.

Figure 1C-11: On Alien Alba Patera --
A Friendly, Familiar Profile

Pyroclastic Volcanism -- Everything I have discussed so far has dealt with the result of eruptions in which volatiles in the magma have not torn the magma apart, which is not the case in pyroclastic eruptions like those associated with Mount St. Helens. Typically, as magma rises in the crust, some of the gases that were dissolved in it under high pressure at great depths are exolved (boiled off), and these gases help force the magma out of the vent. If given off in large enough quantities, the gases may cause the magma to be blown apart -- a pyroclastic eruption. In pyroclastic eruptions, the disruption is so severe that mostly ash is erupted rather than discrete flows. We have evidence of pyroclastic eruptions on Mars, but the kinds of features formed by such eruptions are much more difficult to detect. In fact, ash-falls and flows are even difficult to detect on Earth in remote-sensing data.

Associated with their work discussed earlier, Wilson and his coworkers have done some theoretical work to predict what would happen under various circumstances when volatile-laden magma comes to the surface. They demonstrate that, because the martian atmosphere is much thinner than Earth's, a more rapid expansion of the gases takes place in the vent. However, because the atmosphere is so thin, its "carrying power" is diminished. As a result, blocks only centimeters in size can be ejected from a vent on Mars as compared to meter-sized ejecta on Earth.

When the ash is ejected, it can do two things. If the ash and gas mixture escapes the vent at a fairly slow rate, it will entrain and heat large amounts of the atmosphere. The heated atmosphere, in turn, provides buoyancy that results in a rising column of ash above the volcano. In the later stages of its eruption, Mount St. Helens provided a good example of such a column. This kind of column rises into the upper atmosphere, and theoretical calculations indicate that one would rise higher on Mars than on Earth and so result in wider dispersion of the ash. If, on the other hand, the eruption rate is very high, there isn't sufficient time to incorporate the atmospheric gases into the volcanic mixture; the column collapses -- resulting in ground-hugging ash flows.

---

[9] Alba Patera, with its caldera center at 40° N, 110° W, is in some respects the largest volcano on Mars. Although its summit is probably less than 6 km above the Mars datum, this low volcano is approximately 1700 km in diameter, or nearly three times the area of Olympus Mons and 180 times that of Mauna Loa. The volcano has a variety of flow types. But, while younger flows have partly buried an older summit caldera, most have more impact craters than are found on other Tharsis volcano flows -- indicating greater age. {Ed.}

Features that have been interpreted to be a result of pyroclastic activity include obscuration around the summit of Elysium Mons, possible cinder cones to the east of Hellas Planitia, and some highly eroded deposits around Tyrrhena Patera[10]. The degree of dissection of this volcano is surprising because erosion rates are much lower on Mars than on Earth (by factors of 10,000 to 100,000). The simplest explanation of the large amount of erosion around Tyrrhena Patera is that it is surrounded by easily erodible ash deposits rather than lava flows.

In the table presented in Figure 1C-12 I have listed age estimates for the martian volcanoes. All the volcanoes are very old except those in the general region of Tharsis. In contrast, the surfaces of some of the Tharsis volcanoes appear so young that there is reason to believe that some of the large shields may still be active. Impact craters provide an age only for the exposed surface; the large shields may be very old only in the sense that they have been accumulating for billions of years. Other evidence of present volcanic activity, such as plume clouds, is equivocal.

Volcanic Province	No. of Craters > 1 km/10^6 km^2	Crater Age in billions of years (b.y.)		
		Minimum Likely	Best Estimate	Maximum Likely
Olympus Mons	27	<0.1	<0.1	0.3
Arsia Mons	78	<0.1	0.1	0.5
Arsia Mons summit	150	<0.1	0.2	0.9
Arsia Mons flanks	390	0.2	0.6	1.6
Pavonis Mons	350	0.1	0.5	1.4
Biblis Patera	1400	0.7	2.0	3.4
Alba Patera	1850	1.0	2.7	3.6
Jovis Tholus	2100	1.3	3.0	3.7
Uranius Patera	2480	1.7	3.6	3.9
Apollinaris Patera	990	0.4	1.4	3.0
Tharis Tholus	1480	0.7	2.1	3.4
Albor Tholus	1500	0.7	2.2	3.4
Hecates Tholus	1800	1.0	2.6	3.6
Elysium Mons	2350	1.6	3.4	3.8
Uranius Tholus	2480	1.7	3.6	3.9
Ceraunius Tholus	2600	1.8	3.8	3.9
Ulysses Patera	3200	2.6	3.9	4.0
Hadriaca Patera	2100	1.2	3.0	3.7
Tyrrhena Patera	2400	1.7	3.5	3.9

NOTE: Counts are from Plescia and Saunders [1979] and Crumpler and Aubele [1978]. Ages are based on the model of Hartmann et al. [1981].

Figure 1C-12: Table of Crater Ages for Martian Volcanoes

---

[10] Tyrrhena Patera is a low, ancient and very degraded volcano located northeast of Hellas Planitia in Hesperia Planum. An irregular depression at its summit appears to be continuous with at least one and perhaps several channel-like valley(s) descending the volcano's slopes. Hellas Planitia is a very large basin in the southern hemisphere (30° S-to-60° S, 265° W-to-272° W) and is the lowest region on Mars. Like the smaller Argyre basin to the west, the Hellas depression is believed to be a large ancient impact crater. {Ed.}

INTERACTION OF VOLCANISM AND VOLATILES

The last topic of this presentation is the evidence of volcano-ice interaction. Over the last few years, our ideas about how much water has outgassed from Mars' interior have changed substantially. Shortly after the **Viking** mission, it was estimated from the composition of the atmosphere that if all the water produced were spread evenly over the planet, it would form a layer only ten meters deep. The equivalent figure for Earth is three kilometers, so Mars was thought to have outgassed relatively little water. However, these estimates have recently been revised upward. Based on geologic evidence, we now believe that Mars may have outgassed as much as 0.5 km of water and that most of it exists as ice and groundwater within several kilometers of the surface.

Ice is believed to be widespread very near the surface at high latitudes, so we should see evidence of interaction between it and the volcanics -- as indeed we do. The Elysium region has some of the best evidence. Several large channels start northwest of Elysium Mons in Elysium Planitia[11], and extend several hundred kilometers to the northwest. These channels have numerous features that indicate that they are water-worn. Release of the water appears to have been triggered by the Elysium volcanism.

To the northwest of Elysium is another area with numerous features that suggest the presence of ground water ice. Between this area and the region with clearly defined lava flows around Elysium is some highly textured terrain with a peculiar etched appearance, as seen in Figure 1C-13. The peculiarities of this region are believed again to be caused by interaction of the volcanics with ground ice. Flows from Elysium are visible in the lower third of the image area, but the rest of the visible surface has the appearance of having been created by the partial removal of some easily erodible material. Plausible candidates are ice, ash, and Móberg[12].

Figure 1C-14 shows another possible result of ground ice. A coherent volcanic caprock, consisting of flows from Elysium Mons to the south, appears to have broken into blocks and slid down slope to the north as though the volcanics overlie a very fluid subsurface medium. The simplest explanation is that the subsurface material was ice-rich and that the volcanics caused the ice to melt. Fluidity caused by the melting of the ground ice would then undermine the more coherent surface rock. Similar kinds of textures and evidence are found on a much older volcano, Apollinaris Patera (9°S, 186°W), on the southern boundary of Elysium Planitia.

---

[11] Elysium Planitia (10°S-30°N, 180°W-260°W), an extensive plains region mostly in the northern hemisphere, is on another bulge on the martian surface (not as large as that of the Tharsis region). Elysium is perhaps more important for its relevancy to the study of martian channel features and volatiles. Elysium Mons (25°N, 213°W), the region's largest volcano, has a summit elevation of only 11 km or so above the Mars datum, and its primary edifice rises 6 km above the immediate basal elevation [**Atlas of Mars**, Batson, Bridges, Inge; 1979]. Its appearance is quite different from that of the Tharsis shields; the 14-km caldera is very circular and uncomplicated by multiple collapse craters, while the 180-km primary edifice is generally circular and asymmetric but for a broad 200-km ridge extending northward from the volcano center. The average slope on the volcano's flanks is 3.5° [**The Surface of Mars**, Carr; 1981]. {Ed.}

[12] One interpretation of the ridges seen in such pictures suggests that they may be dikes formed when lava intruded into a friable deposit and reacted with ice. On Earth (particularly in Antarctica), the lava-ice reaction forms a characteristic yellow-brown rock found in ridges known as Móberg [Kjartansson, 1960] (an Islandic term for palagonitized tuffs and breccias) which are visually similar to those on Mars. Individual Móberg units are often ridge-shaped with serratic outlines, having been weathered to that configuration, and sometimes exist as straight-lined, nearly continuous rows of such ridges. Móberg landforms on Earth were created by subglacial volcanic fissure eruptions. {Ed.}

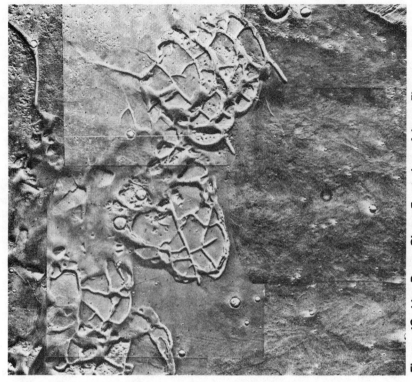

Figure 1C-14: Down-Slope Fracturing of Lava Flows (possibly due to melting of ground ice under flow)

Figure 1C-13: Etched Terrain Indicative of Lava-Ice Interaction (northwest of Elysium)

70

Finally, Figure 1C-15 (next page) shows a vast area in southern Amazonis that is overlain by a thick, friable, easily erodible deposit (smooth feature in upper half of photo) that straddles the boundary between the cratered uplands and the plains. It extends for a thousand kilometers or so along the boundary between the uplands and the plains. This feature has been the object of some interest because its origin is not understood, but there are several interpretations for it. One is that it is a thick pyroclastic ash deposit; another is that it is composed largely of palagonite[13]. Figure 1C-16 (next page) shows what terrain in the same region looks like when the friable deposits have been largely removed. Numerous Móberg-appearing ridges can be seen. These are believed to be places where lava intruded into the deposit. The ridges, being comprised of partly altered but still coherent basalt, are left standing after the intervening softer materials have been stripped away. The scene resembles parts of Iceland where volcano-ice interaction has occurred.

SUMMARY

Mars has had a long and varied volcanic history. Most of the volcanism occurred very early in the planet's history, but it has continued at a lower rate to the present day. The most obvious indications of volcanism are the large shield volcanoes and the volcanic plains, and both probably formed largely of iron-rich basalts (of which we may have samples in the SNC meteorites). There has also been considerable pyroclastic activity, but the extent is hard to assess because of the difficulty of unambiguously identifying its products. Ice may be common near the martian surface, at least at high latitudes, and features have been identified that can be attributed to the interaction of lava and ice. ■

---

[13] Palagonite is a soft, largely amorphous rock formed by interaction between ice and basalt. {Ed.}

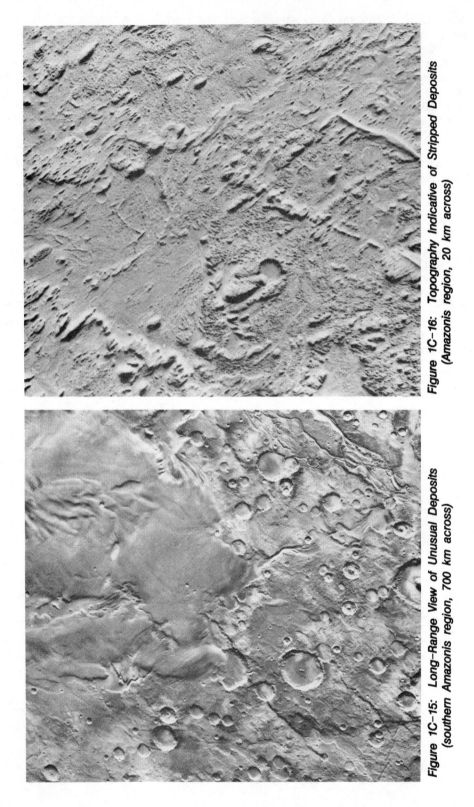

Figure 1C-16: Topography Indicative of Stripped Deposits
(Amazonis region, 20 km across)

Figure 1C-15: Long-Range View of Unusual Deposits
(southern Amazonis region, 700 km across)

REFERENCES

This space is otherwise available for notes.

~~~~~~~~

THE CHANNELS OF MARS[1]
[NMC-1D]

Dr. Victor R. Baker
Department of Geophysical Sciences
University of Arizona

When geomorphology is represented on the surface of a planet, the geomorphic processes are those that shape the planetary surface at the dynamic interface between the atmosphere, the hydrosphere (if one exists), and the lithosphere. These processes are clearly and dramatically represented on the surface of Mars. Etched into the martian surface are large numbers of channels and sinuous valleys of many types, some suggestive of fluid flooding on a massive scale while others are small with tributaries arrayed in Earth-like dendritic patterns. Some of the martian channels are smooth valleys that are more indicative of subsurface flow and sapping while many clearly suggest surface flow -- complete with streamlined islands. The largest of the martian channels appear to be the product of cataclysmic floods comparable to or larger than those that formed the Channeled Scabland of the northwestern United States. While all of the martian channels are very old on what is today a dry and frozen planet, the evidence for liquid surface water in Mars' distant past continues to influence interests in the possibility that life may once have had a chance to gain a foothold on Mars. With much greater certainty it has clearly been important in the formation of topography now present on the planet, and this evidence must therefore be considered a major factor in the planning of future programs and as water becomes increasingly important as a vital in·situ resource for supporting missions to Mars.

INTRODUCTION

Scientists tend to differ a bit from the general public in that we tend to love anomalies. For example, we make preliminary observations on phenomena that lead to the development of theories, but the happiest moments for many of us occur when those theories are dashed by new discoveries. And so it was with our understanding of Mars.

Changes Wrought by Technology

Early Perceptions -- Prior to the beginning of planetary exploration. with spacecraft in the 1960's, we felt we had a good perception of the basic physical parameters affecting processes

---

[1] The author's research on the martian channels was supported by NASA grants NSG-7326, NSG-7557, and NAGW-285.

that operated on the surfaces of other planets. The first flyby missions to Mars (Mariners 4 and 6/7) revealed a heavily cratered surface that seemed (due to the low resolution afforded) to be similar to that of Mercury and the moon. We had in Mars what then seemed to be a cold planet with a very tenuous atmosphere and low surface pressure, and the preliminary indications therefore were that the geomorphic processes on Mars might be rather uninteresting in comparison to other terrestrial planets.

Mariner 9 and Viking Surprises -- Fortunately for geomorphology, however, that concept was rudely dashed by the products of the **Mariner 9** orbital mission, which revealed features on the surface of Mars that could only be economically explained by the presence of flowing fluids. Further, these features provided evidence of fluid properties that had to be so similar to those of water that to invoke any other fluid would be a violation of all scientific knowledge about how such systems operate.

Figure 1D-1 is a **Mariner 9** picture of Nirgal Vallis[2], one of the exotic names provided by the nomenclature committee for the names of valleys when mapping Mars[3]. "Nirgal" is the name historians believe was used by the Babylonians for the red planet. The Nirgal channel is more than 600 kilometers long. Fortunately, about five years or so after the **Mariner 9** picture was taken, the **Viking** orbiters--with vastly superior imaging capability--brought this "anomaly" into much sharper focus, and Figure 1D-2 is a **Viking** image of the same channel feature.

The **Viking** picture is of the downstream part of Nirgal Vallis, and it reveals the channel to be very complex. If we look at the upstream part of Nirgal we find tributaries that appear to represent contributed fluid flows, which fits with the knowledge that Nirgal itself is tributary to another large, much more degraded valley system[4]. In total, the system is organized in a hierarchical branching pattern that we commonly see in many natural fluvial systems on Earth.

Geomorphology and Mars

We tend to examine many pictures when we consider the geology of planetary surfaces, and I want to emphasize that geomorphology is much more than the study of such images although they are certainly an essential element. We are also interested in using the analogy of our own planet, where we can use our hand lenses and our rock hammers to study features first-hand on the ground. It is with this kind of perspective that Mars is perhaps the most exciting "other" place in the solar system, because of its tremendous similarity to Earth and because of the opportunity it represents for intellectual feedback with respect to what we know and are finding out about Earth.

---

[2] Nirgal Vallis is located between $27°$ S–$30°$ S and $37°$ –$47°$ W below the eastern end of Valles Marineris. It appears to have an easterly direction, and, because of the open nature of the network and the lack of a large catchment area, it also is believed to more likely be the result of groundwater sapping than surface runoff [Carr, 1981, The Surface of Mars; Baker, 1982, The Channels of Mars]. {Ed.}

[3] The word "vallis" (as in Nirgal Vallis) is a singular form used in the names of distinctly continuous channel-like valley features. The term "valles" (as in Samara Valles) is the plural form and has been used to name more complicated valley or channel features/systems (often dendritic). "Valles" has been used as a head word in a name only once -- **Valles Marineris**. This massive canyon complex is made up of numerous valleys and canyons linked to each other as a result of their formation process. {Ed.}

[4] Nirgal Vallis appears to flow into Uzboi Vallis, which is located between $27°$ –$37°$ S and $35°$ –$37°$ W at the eastern end of Nirgal. Uzboi has a northeasterly flow and appears to terminate in a large crater -- Holden. {Ed.}

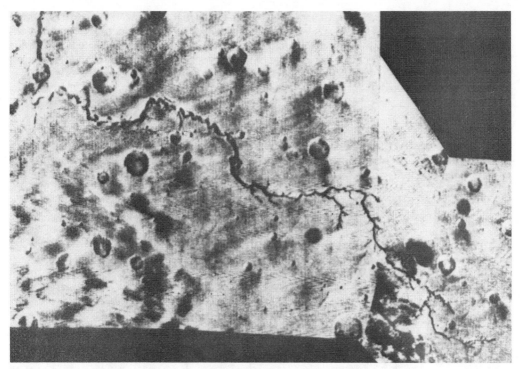

Figure 1D-1: Mariner 9 View of Nirgal Vallis (Joins Uzboi Vallis at Right)

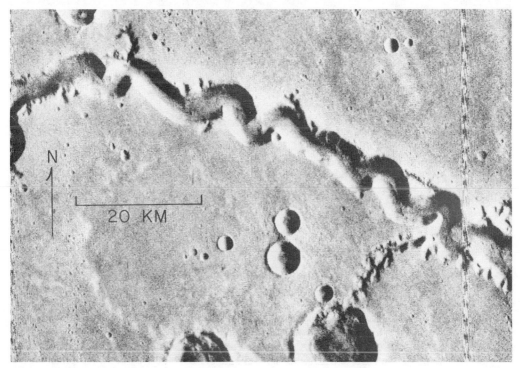

Figure 1D-2: Viking High Resolution Image of a Section of Nirgal Vallis (see Fig. 1D-1)

Remote sensing and imaging provides us with excellent visual material to illustrate presentations, but it fails to demonstrate the "nuts and bolts" associated with complex work -- like mapping. Mapping is often the work of a geologist alone in his laboratory, first integrating and then applying the experience of many years to the processing of topographic data. Morphometry, the quantitative measurement of features, is also applied (along with other kinds of statistical analyses), and we try to integrate this data by formulating abstract mathematical models, laboratory models, or simulations. We also try to consider the temporal changes in phenomena.

Figure 1D-3 presents a table that illustrates (through comparison) why Mars is such an exciting place. In this illustration we can consider, side by side, the geomorphic processes we are familiar with on Earth relative to those of other planetary bodies in the solar system. The table shows that we have processes on Mars--either active or relic, albeit questionable in some cases--that correspond to the entire range of exciting and interesting dynamic processes we are familiar with on the surface of our own planet. Many of these processes involve interaction with volatile materials, particularly water, and the channels are clearly the most interesting manifestation of this dynamic interaction.

Process	Mars	Moon	Galilean Satellites	Venus
Tectonic	R	R	R	R
Volcanic	R	R	A,R	R
Weathering	A,R	R	–	A,R
Karst	Thermokarst	–	–	–
Hill Slope	A,R	R	–	R
Fluvial	R	–	–	?
Glacial (Ice)	R	–	R	–
Aeolian	A,R	–	–	A,?
Coastal	R,?	–	–	–

A = Active process; R = Relict process (landforms)

Figure 1D-3: Table of Geomorphological Process on Planetary Surfaces

THE MARTIAN CHANNELS

The study of channels and valleys is a very complicated business. It may be surprising for some to learn that when one removes from consideration the area of Earth's surface covered by water, the land surface area remaining is nearly the same as the total surface area of Mars. And, just as Earth's land surface area is dissected by a tremendous variety of channels and valleys, the same is true of Mars. The variety of valleys on Mars is so great, in fact, that they cannot be adequately described in the limited space facilitated by this single presentation.

One of the more interesting valley types tends to occur at fairly high latitudes where temperatures are colder, and they are referred to as the fretted valleys. These features are marked by large quantities of debris that has presumably flowed from the valley walls onto valley floors. These masses of debris surround isolated massifs. Their morphologies, which include lobate fronts on debris accumulations, indicate that the material actually flowed (although most probably involving processes that contributed high resistance or viscosity). Using a terrestrial analog, these processes almost certainly involved lubrication by a plastically deforming material such as ice.

The Great Flood Channels of Mars

Another class of valleys and channels collectively represents what are perhaps the most exciting features on Mars. These testify to the release of vast quantities of fluid material from subsurface sources. We can trace and locate these sources because the massive outflow of material released at the headward origins of the channels resulted in great collapse features. These features are referred to as areas of chaotic terrain. Some of the chaotic terrain appears to be intimately associated with volcanism, as Mike Carr indicates in his presentation [NMC-1C]. Other areas of chaotic terrain occur in the highlands of the heavily cratered terrain that dominates so much of the martian equatorial belt and southern hemisphere.

Emanating from these zones of collapse, where perhaps a kilometer or two of material has been removed, are great channel depressions shaped by large-scale fluid flows. Many distinct morphological features are present to provide information on the character of the flows -- even many kilometers downstream. The dimension of some of these channel systems is enormous, reaching 100 or 200 kilometers across and 2000 or 3000 kilometers in length. Downstream, ridges induced ponding; where the fluids managed to breach, break through, or flow around the ridges, they became particularly erosive as the hydraulic gradients increased and as the rate of energy released became more extensive.

The Nature of Martian Channel Erosion

Martian fluid flows etched distinct high-water marks, and evidence of their erosive force is visible on the ejecta blanket around the crater Dromore in Figure 1D-4. The crater was eroded by large-scale flows moving from left to right, and fluid appears to have ponded upstream at the ridge west (left) of the crater before breaching at low points to erode the crater's ejecta. Ejecta blankets are themselves loose flows of material thrown out from impact craters, and they are believed to be more susceptible to erosion. What appears to be an indication of pronounced erosion can be seen up to a discernable maximum elevation on many streamline units.

We also have indications that flows possessed a great deal of turbulence as they approached obstacles, such as the isolated hills visible in Figures 1D-5 (Kasei Vallis) and 1D-6 (Ares Vallis). Flow was from the left (Ares, Fig. 1D-6), and large scale vortices must have been induced to scour depressions visible on the channel floor. The flow possessed sufficient turbulence to produce a horseshoe-shaped scour mark upstream of the streamlined "island" at left-center. In contrast to the powerful turbulence needed to achieve severe scouring, there are areas in which the fluid flowed smoothly around topographic features that may have been relatively less resistant to erosion. Some of the island-like features in Ares Vallis are so tranquil in appearance that one can imagine a school of trout lurking there in wait for the tasty fly that might occasionally float down the channel. Stereo image data show that the uplands rise about 200 to 400 meters above the channel floor and that the "horseshoe" scour hole is approximately 50 meters deep. These streamline forms are a prominent part of large-channel topography on Mars, and the whole assemblage of such landforms is indicative of a fluid with properties very similar to those of water (with perhaps some differences caused by quantities of entrained ice and sediment that may have contributed to the cataclysmic nature of the process).

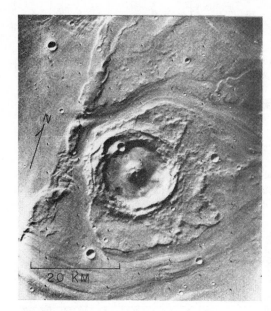

Figure 1D-4: Crater Dromore (20° N, 49° W) with Evidence of Fluid Erosion (area 40x60 km)

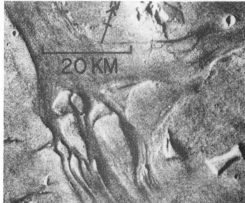

Figure 1D-5: Streamlined Hills in Eastern Kasei Vallis

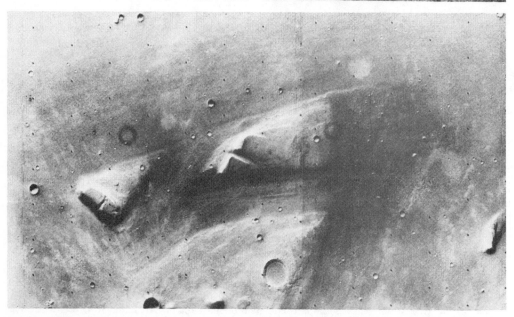

Figure 1D-6: Streamlined "Islands" in Ares Vallis

Analytical Process

A morphometric analysis on the streamline-island forms, which utilizes the dimensions of their shapes, considers an individual unit in terms of its length versus its area. We find that the martian streamline forms follow trends very similar to streamline forms produced by large scale fluid flows on Earth, except that the martian forms are considerably larger.

The next part of the process is to perform quantitative calculations on the character of the fluid flows, and we can begin by looking at the geometry of the channel systems as illustrated in Figure 1D-7. This simple diagram depicts the energy of a flow moving from Section·1 (left) to Section·2 (right).

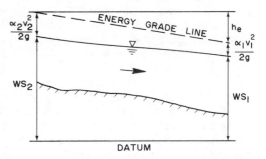

Figure 1D-7: Flow Energy Diagram

In this diagram the flow has a varying water surface (WS) and velocity head ($v_2/2g$). It loses head (h_e) in moving from Section·2 to Section·1. Conservation of Energy principles can be used to calculate velocity (v) and discharge for such flow geometries. In quantifying the geometry for such calculations, we develop cross sections and measure the precise geometry that would control the nature of the fluid.

We then feed this kind of information into an energy-balanced approach that routes the fluid flow through the geometry of the channels. This approach essentially considers the slope of the channel and its geometry, using multiple cross sections (not merely two as shown in Fig. 1D-7), and balances the energy considerations. The bottom line of this kind of calculation is that it yields some interesting figures relevant to the nature of the flow-controlling parameters on Mars, and we find that the discharge volumes are truly immense; the fluxes calculated for the cataclysmic fluid flows are typically tens of millions of cubic meters per second.

We are somewhat limited in this process, of course, because the Viking orbiter images and our knowledge of Mars' global geology are severely constrained by the lack of precise topographic relief data for much of the planetary surface. We could do a much better job with these calculations if we had very precise global topography on the scale of perhaps a few kilometers or so in horizontal resolution and a few meters in vertical resolution. An experiment planned for the Mars Observer mission, using radar altimetry, will give much better topographic data.

An Earth Analogue for the Martian Channels

We have one Earth analogue that yielded flows comparable to those calculated for the martian channels just described. The Earth topography used for this analogy is the region known as the Channeled Scabland, found in the northwestern United States[5]. In this region, layered

---

5 The Channeled Scabland is found primarily in a 300-square-kilometer area of eastern Washington, but it is the end product of a process that also involved southern British Columbia and Alberta in Canada and the states of Montana, Idaho, Utah and Oregon. The area east of the Cascade mountains is sometimes referred to as the Inland Empire. It includes the Columbia Plateau, one of the largest lava beds on Earth, and is marked by strong evidence of both glaciation and massive floods which contributed to the creation of Washington's great depressions called "coulees." Of these, Grand Coulee is the largest and best known. {Ed.}

basalts were once scoured by massive late-Pleistocene flood flows. The photo in Figure 1D-8 is an oblique aerial view of the Dry Falls area where the basalt surface was eroded by these glacial floods toward the end of the last ice age. It illustrates an area that is about 10 to 20 km across and a surface that has been scoured by flood flows of 100 meters and more in depth. What gave rise to the cataclysmic floods is a phenomenon that affected the whole northwestern United States between 12,000 and 17,000 years ago as a result of ice damming.

Two very large lakes were created by ice damming during the Pleistocene period. One of them, known as Lake Missoula, was located in what is now western Montana. This large lake obtained a depth of about 600 meters (2,000 feet) while dammed by glacial ice during the last full glacial period, and contained a maximum volume of approximately 2.5×10^{12} cubic meters. The second lake was Lake Bonneville in Utah, and it too was involved in the flooding that created the scabland during that period. In Figure 1D-9, the grey-screened area defines the entire region affected by the floods while the arrows indicate their flow direction as they came down the Columbia River system. The black areas represent the former extent of glaciers in the region. These floods ultimately carried coarse debris all the way out onto the abyssal plain of the Pacific[6].

Figure 1D-8: Dry Falls Area of Channeled Scabland, Washington

---

[6] The great Missoula floods helped in the excavation of Washington's Grand Coulee, ultimately producing a depression along the Columbia River that is more than 80 kilometers long with depths to 300 meters, ideal topography for the reservoir created by Grand Coulee Dam. Other coulees in the region have been similarly exploited. {Ed.}

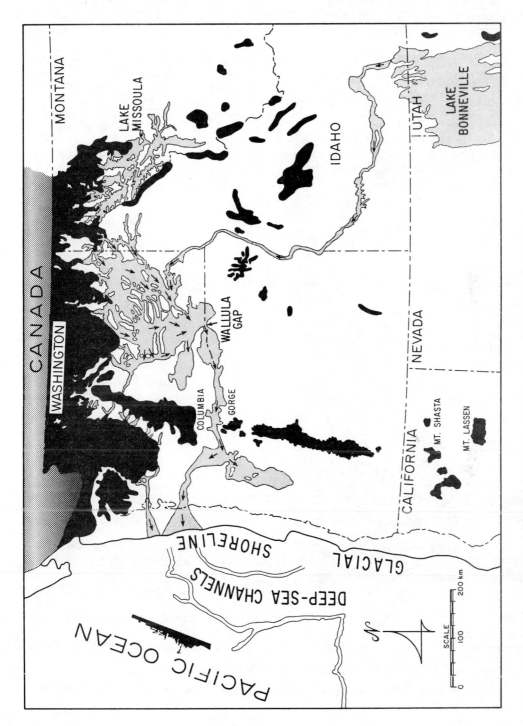

Figure 1D–9: Map of Channeled Scabland Illustrating Flow of Lake Missoula Flood

83

The massive character of this flooding involved the release of 2500 cubic kilometers of water[7]. That is very comparable to releases estimated for some of the channels on Mars, but there are martian flows that had to have been much larger. Some of the features of these flood channels on Earth are also quite similar to comparable features on Mars; e.g., terrain with giant grooves, streamlined hills, scour holes, and cataracts. Indeed, the Channeled Scabland is an assemblage that strongly mimics what we see in the large martian outflow channels.

But there are important differences, as well, and theoretical analysis is needed to understand their influence. One of the differences is that a considerable quantity of ice was entrained in the flows on Mars -- ice that was broken up by the turbulence. The photo in Figure 1D-10[8] is a beautiful example of an ice jam flowing in a Canadian river, scouring the bank and entraining quite a lot of sediment. Note the erosion of the stream bank (right), caused as 1-meter-thick ice blocks being pushed along by the flow are tumbling as they drag and shear past the bank. We think there was a lot of ice and sediment similarly entrained in the martian flows.

Figure 1D-10: Ice and Sediment Entrained in Canadian River, with Ice Erosion of Bank

---

7 Although these flooding events are believed to have occurred a number of times, it has been suggested that one such event alone could have unleashed the entire volume of Lake Missoula in a matter of days, producing discharges as great as 10^7 cubic meters per second and a wall of water up to 120 meters in depth. In comparison, the Amazon's mean annual discharge is 10^5 cubic meters per second. [Baker, 1973, 1981, 1982; Baker and Nummedal, 1978; Baker and Bunker, 1985]. {Ed.}

8 Photo by Dr. Gerald Smith, University of Calgary. The picture is of the Red Deer river in Alberta, and was taken during spring, 1976.

The drawing in Figure 1D-11 presents an idealized cross section of a hypothetical, large-scale martian flood experiencing longitudinal vorticity, icing, and cavitation. It can be seen that the fluid was probably flowing fast enough to reduce flow pressure and form bubbles within the flow. This process of forming bubbles of vapors in a fluid (when its flow reduces the fluid pressure to the vapor pressure) is known as cavitation. Moreover, if the atmospheric pressure was essentially as low at that time as it is today, bubbles were also forming near the surface (due to the effects of low atmospheric pressure) in addition to those forming within the flow due to reduced flow (fluid) pressure. These entrained bubbles were carried by powerful turbulence down to the channel bed, where increased pressure and their impact induced implosion. This process created tremendous erosive forces and was also characterized by the concentration of erosion in the grooved scour lines.

Cavitation Favored Where $\sigma = \min.$

$$\sigma = \frac{Pa - Pv}{1/2\, \rho\, v^2}$$

::: Cavitating Liquid

V Zone of High Velocity Along Vortex Filament

E Zone of Erosion

Ice Particles ?

Figure 1D-11: Cross Section of Martian Flood

As a result of these powerful erosional processes, there was probably a great deal of entrained sediment in some of the flows in addition to the ice and bubbles already described. This complex and unique flow composition results in a tremendous amount of very complicated, very interesting physics. In terms of the scale of geomorphic processes, the kinds of floods in evidence on Mars (of which we have a good example on Earth, the Missoula Flood) produced rates of energy expenditure per unit area larger than almost any other kind of phenomenon. Only when the energy of impact processes (cratering) is considered, which affects relatively small local areas, is the kind of energy expenditure calculated for these cataclysmic martian flows exceeded.

Dendritic Channel Networks

As exciting as the great flood channels of Mars may be, I think the most important scientific questions address much smaller features, the small-scale dendritic valley networks that dissect the heavily cratered terrain of the planet. Figure 1D-12 (a and b) presents a map of channels and valleys within the equatorial belt of Mars [Carr and Clow, 1981]. Its perspective is roughly from 65°N to 65°S, and some of the big outflow channels are highlighted in the stipple pattern. However, the detailed lines are small valley networks that are very interesting because of their omnipresence throughout the heavily cratered terrain. The detail of these small valleys is not visible on the map, but individual **Viking** orbiter frames resolve an immense quantity of these contributory networks dissecting the cratered terrain.

Figure 1D-13 is such a frame (area ≈ 240 by 250 km), and is a picture of Parana Valles (25°S, 26°W). Note the prominent development of tributaries. These networks are associated with a phase of martian history prior to about 3.8 billions years ago. During that period, conditions were favorable--either climatologically, or lithologically, or both--for the development of dendritic networks that are more similar to those on Earth. There are, however, some important differences.

65N

30N

Tharsis

30S 180

65S

Argyre

Figure 1D–12a: Channels and Valleys of Mars (equatorial, northern hemisphere)

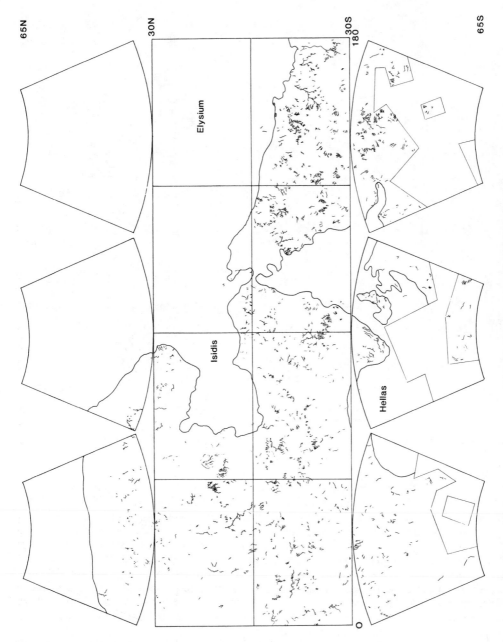

Figure 1D–12b: Channels and Valleys of Mars (equatorial, southern hemisphere)

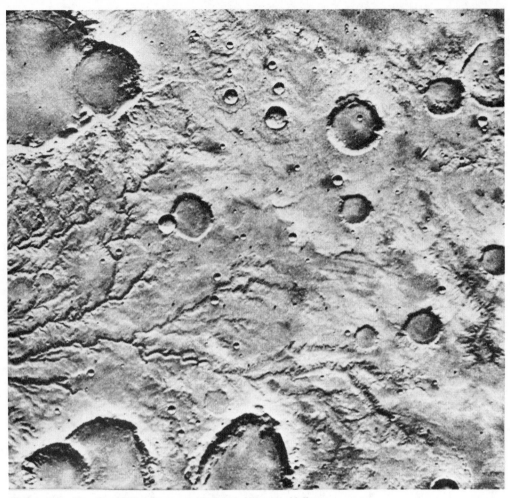

Figure 1D-13: Dendritic Channel Network, Parana Valles

The martian networks evolved over a considerable period of time. Some regions have a very high density (number) of quite degraded channels per unit area. These have been highly modified by erosion of their walls which also has filled them in, and they are believed to have evolved when the atmosphere may have been more conducive to surface flow. In contrast, there are areas exhibiting small channel networks that have a more pristine appearance, wherein the wall boundaries are very sharp. These pristine units are younger and are found in much lower densities, and they have a morphology that is more consistent with that believed to be created by subsurface fluid flow rather than surface fluid flow. These pristine valleys, which tend to be very well developed on the intercratered plains, are believed to have formed about 3.7 or 3.8 billion years ago at the end of the heavy cratering period described in these Mars Conference proceedings by John Lewis [NMC-1A].

If we consider a very simple hydrological balance (the nature of a drainage basin on Earth, for example), the more pristine networks appear to be the result of subsurface water flow. This is the nature of the sapping process in which the emanation discharge undermines overlying material, causing progressive collapse and a headward retreat of the channel. In contrast, the older degraded networks may very well have involved surface runoff as a component. Unfortunately, these higher-density, more degraded networks are much more difficult to interpret.

Potential Early Precipitation Difficult to Determine

The evolution of martian dendritic channel systems can involve other possible processes that must be considered. For example, precipitation may have occurred in the form of rain or snow, and major variations in infiltration capacity and evaporation may have been participatory in the creation of such features. We need to determine the martian water balance with respect to time, such that we can see what it might have been, for example, during the more remote past when Mars apparently had a very active, dynamic hydrologic environment. We could then compare that period to the present. Indeed, present conditions may have persisted for billions of years in areas where other erosion rates have been much lower.

CONCLUSION: WATER CRUCIAL FACTOR IN EVOLUTION, FUTURE

The issue of martian water and its mechanisms has clearly important implications for the entire history of the planet, including the potential for life if it ever occurred. The nature of how this hydrologic system functioned when it was operating more effectively on Mars, during the planet's first billion years or so, has a significant bearing on the evolution of the atmosphere, for example. This, in turn, has very profound implications for the possible development of life, which may in fact have had a considerable period of time in which to evolve before the atmospheric environment deteriorated -- taking with it much of the water resource suggested by the geomorphic evidence I have discussed. I think these implications must and will be effectively explored by some of the future missions discussed at this meeting, thereby providing some exciting prospects for the future exploration of Mars. ■

REFERENCES

This space is otherwise available for notes.

~~~~~~~~

Baker, V.R. 1973. Paleohydrology and sedimentology of Lake Missoula flooding in eastern Washington. *Geol. Soc. Am.*, Spec. Paper 144, p 79.

Baker, V.R. 1979. Erosional processes in channelized water flows on Mars. *J. Geophys. Res.* 84:795–993.

Baker, V.R. 1981. *Catastrophic flooding: The Origin of the Channeled Scabland*. Hutchinson Ross Publishing Co., Stroudsburg, Pennsylvania, p. 360.

Baker, V.R. 1982. *The channels of Mars*. Univ. of Texas Press, Austin, Texas, p. 198.

Baker, V.R., and Bunker, R.C. 1985. Cataclysmic late Pleistocene flooding from glacial Lake Missoula. A review: *Quaternary Science Reviews*, 4:1–41.

Baker, V.R., and Nummedal, D. 1978. *The Channeled Scabland*. NASA, Washington, D.C., p. 186.

Carr, M.H. 1981. *The Surface of Mars*. Yale Univ. Press, New Haven, Connecticut.

Carr, M.H., and Clow, G.D. 1981. Martian channels and valleys: their characteristics, distribution and age. *Icarus* 48:91–117.

THE MOONS OF MARS
[NMC-1E]

Dr. Joseph Veverka
Laboratory for Planetary Studies
Cornell University

The martian satellites, Phobos and Deimos, have been known for only a hundred years because of their small size and low reflective properties. Spacecraft, particularly the Viking orbiters, have contributed dramatic new information about them. But, while many answers to fundamental questions about the martian moons--and perhaps other similar bodies in the solar system--have been provided, new puzzles have emerged to further challenge us. Locked in "synchronous rotation" relative to Mars, the satellites are very small and have very low surface gravities. As a result, escape velocities also are very low and material is readily ejected from their surfaces. But in spite of low escape velocities, both satellites exhibit significant regoliths, particularly Deimos where material appears to be mobile -- moving downhill to fill craters. The most striking features to be found on the tiny martian moons are the grooves of Phobos. These are believed to be the product of extensive fracturing due to a major collision, and they appear to originate at the largest crater on the satellite, Stickney. Phobos is being drawn closer to Mars and is doomed to ultimate destruction, either by breaking up due to increasing Mars tides or by impacting the planet in what most certainly would be a cataclysmic event. Because these satellites have enormous historic significance relative to our understanding of the solar system, and may contain vital resources (like water) for exploitation in support of future missions, the importance of the martian moons should be given a high priority in planning the exploration of the Mars system. The 1988 Soviet Phobos Mission represents the next major step in that process.

INTRODUCTION

Because of their small size, Phobos and Deimos were only first discovered in 1877 by American astronomer, Asaph Hall (see footnote 4, page 116, this presentation), and little could be learned about their physical characteristics via Earth-based observation. But just a hundred years later, Mariner 9 and Viking missions revealed what these two tiny satellites are like in great detail. My objective is to review what was learned about the two moons of Mars from these spacecraft explorations, particularly the two Viking orbiters which conducted observations from 1976 into 1978. I will also summarize the results of some theoretical studies concerning the orbital evolution of the satellites, and will briefly discuss current ideas about their origins. I will conclude with a number of unsolved problems which underscore the importance of future efforts to explore the moons of Mars.

Absent from my talk will be any mention of biology, a fact which may appear an aberration in any discussion of Viking science. But here the case is clear: all experts who have studied Viking data on the martian satellites agree that there is not now, nor has there ever been, life on Phobos and Deimos. Perhaps the only reason that such a categorical conclusion has been reached is that we had no biologists working with us.

GENERAL INFORMATION

For almost a century after their discovery in 1877, we knew little about Phobos and Deimos other than their orbits and apparent magnitudes. Both satellites are very difficult to observe from Earth, not only because they are small and--therefore--faint, but especially because they orbit close to Mars and are almost lost in the planet's glare. Phobos, the inner satellite, circles Mars once every 7^{h}39^m at a distance of 9,380 km (2.76 R♂) from the center of the planet. The outer satellite, Deimos, orbits at a distance of 23,460 km (6.90 R♂) with a period of 30^{h}18^m.

The important aspect of this is that the two orbits straddle the synchronous-orbit-distance (20,490 km or 6.03 R♂) for Mars, the distance at which a satellite has an orbital period exactly equal to the spin period of the planet. Consequently, the evolution of the two orbits differs dramatically [Burns, 1977]. The influence of the gravity tides is very gradually decelerating Deimos and pushing it slowly away from Mars, while Phobos, being inside the synchronous orbit distance, is being accelerated and pulled ever closer to the planet. The rate of orbital change, measured in terms of secular acceleration, is rapid enough to be detected. At the present rate, it has been estimated that Phobos has another 30 million years before it spirals in close enough to be broken up by increasing tidal forces or until it impacts the planet [Burns, 1978; Dobrovolskis, 1982].

Spacecraft Observations and Results

All of our data concerning the physical appearance of the two satellites come from spacecraft observations (mostly Viking) because the satellites are too tiny to be resolved by telescopes on Earth. The first spacecraft glimpse of Phobos was obtained in 1969 by Mariner 7. Although the resolution was very poor by later Mariner 9 and Viking standards, there was enough information in that picture to enable Bradford Smith to estimate the satellite's diameter (~20 km), its albedo (0.065), and conclude that Phobos had an irregular shape [Smith, 1970].

In 1979, Mariner 9 obtained images of both satellites at resolutions approaching a few hundred meters [Pollack et al. 1973]. The Viking orbiters did even better between 1976 and 1978, obtaining frames with resolution as high as a few meters in the case of Deimos [Duxbury and Veverka, 1977; Thomas and Veverka, 1980]. Such images not only make it possible to determine the sizes and shapes of the satellites, but to study the geomorphology of their surfaces in detail. Fundamental parameters for the two satellites (derived from Viking and Mariner 9 observations) are listed in the table presented as Figure 1E-1. Figures 1E-2 and 1E-3 are high resolution Viking images of Phobos and Deimos to illustrate the degree of detail available in the Viking database. Both satellites can be characterized as irregularly shaped, very dark gray objects that are heavily cratered.

General Characteristics

In shape, both satellites can be approximated crudely by triaxial ellipsoids. The average radii is about 10.5 km for Phobos and 6.5 km for Deimos. Notice that the table in Figure 1E-1 contains measurements of mass and determinations of average density for each satellite. During

ORBIT				
	Semi-Major Axis (R♂)	Eccentricity	Inclination	Orbital Period	Orbital Period
Phobos	2.76	0.015	0.01°	7^{h}39^m	7^{h}39^m
Deimos	6.92	<0.001	0.92°	30^{h}18^m	30^{h}18^m

PHYSICAL CHARACTERISTICS				
	Diameter (km)	Density (gm/cm^3)	V$_{esc}$ (m/sec)	g (cm/sec^2)	P$_c$ (bars)
Phobos	20 x 21 x 18	~2	~15	~0.50	0.6
Deimos	25 x 12 x 10	~2	~10	~0.25	0.3

Figure 1E-1: A Table of Basic Parameters for Phobos and Deimos

Figure 1E-2: Viking High Resolution Image of Phobos

Figure 1E-3: Viking High Resolution Three-Frame Mosaic of Deimos

the **Viking** Extended Mission[1], the orbiters (VO-1 and VO-2) made very close flybys of the satellites in a successful effort to determine the masses from the gravitational perturbation on the spacecraft orbit. Since the **Viking** data allow one to determine shapes and sizes, hence volumes, one can find the mean density (mass/volume) -- an important clue to overall composition. In round numbers, the densities of the two satellites are the same within the error bars: approximately 2 gm/cm^3. From the mean density and the dimensions one can readily calculate parameters, such as surface gravity and escape velocity. Surface gravities are tiny (about 10^{-3}g), and escape velocities are slight -- about 10 m/sec for Deimos and 15 m/sec for Phobos (averaged for the variable range of each).

The satellites are only approximately triaxial ellipsoids, which is an unnerving fact to a few theorists who have developed beautiful theories assuming that they are precisely triaxial ellipsoids. Figure 1E-4 gives a truer idea of what the shapes are like as silhouettes. Some irregularities in the silhouettes can be traced to craters. For example, the outlines of the two largest craters on Phobos are evident (Stickney, 10 km across and located near longitude 60°W, and Hall, 5 km across and located close to the satellite's south pole). Other irregularities are due to nearly obliterated ancient craters and perhaps even to "spallation scars" where severe impacts long ago blasted pieces off the satellites.

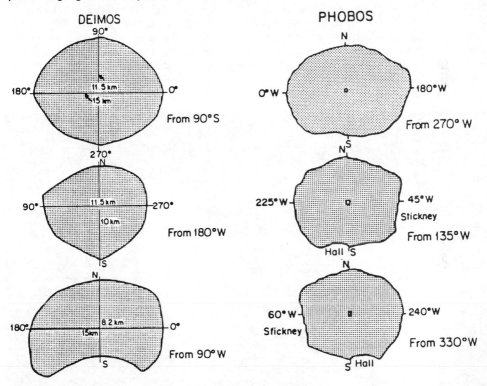

Figure 1E-4: Silhouettes of Deimos (left) and Phobos (right)

---

[1] The **Viking** Extended Mission was the second phase of mission operations. It began following solar conjunction at the end of 1976 as Mars came out from behind the Sun, and it continued for 15 months (ending in April, 1978). The close flybys of Phobos were conducted with VO-1 early in 1977; the spacecraft flew within 88 km of the satellite during its closest encounter in February. VO-2 was used for the high resolution imaging of Deimos later that year (May) when the spacecraft flew within 28 km of the satellite's surface. {Ed.}

The longest axis for both satellites is approximately 1.3 times longer than the shortest. Due to tides, each of the moons is now aligned with its long axis pointed toward Mars, locked in what is termed "synchronous rotation," a situation in which a satellite spins once around its axis every time it orbits the planet (e.g., our own moon is locked in synchronous rotation relative to Earth). Thus, the spin period of Phobos is 7^h39^m, while that of Deimos is 30^h18^m.

Two views of Deimos, taken 19 days apart by **Mariner 9**, are presented in Figure 1E-5. These views were targeted at very nearly the same relative positions for Deimos and the spacecraft, such that if Deimos is in synchronous rotation we should see the same aspect -- as indeed we do. By tracking individual surface features in a number of views, the rate of spin and the orientation of the spin axis can be determined with precision [Duxbury, 1977]. This detailed analysis confirms the synchronous spin state for both satellites, i.e., longest axis pointed toward Mars, the shortest axis perpendicular to the orbital plane, and the spin vector aligned with the shortest axis. If Phobos or Deimos were spun in any other fashion, due perhaps to a large impact, Mars tides would rapidly restore the status quo.

Calculations indicate that the time scale for locking Deimos into a synchronous spin state is some 10^8 to 10^9 years, but the time scale for Phobos is much shorter since the satellite is closer to the planet -- only about 10^5 to 10^6 years [Burns, 1972]. Recently, J. Wisdom of MIT has demonstrated that if one of the satellites was disturbed from a synchronous state, it would go through a period of chaotic motion before settling back into a synchronous lock. Given the heavily battered aspects of the satellites, one should assume that both objects experienced such chaotic states at intervals, especially during the earliest portion of their history when impact fluxes were higher than in recent times.

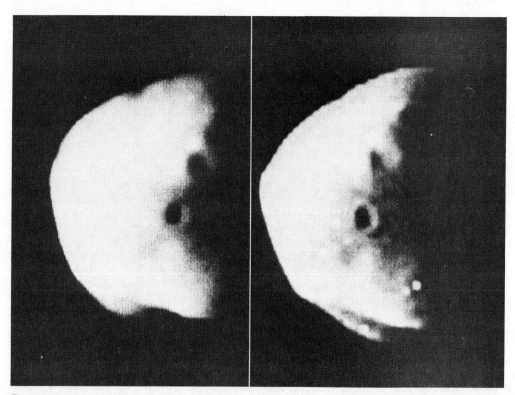

Figure 1E-5: Mariner 9 Images of Deimos Taken 19 Days Apart to Determine Spin Period

KEY CHARACTERISTICS: CRATERS AND REGOLITHS

In the late 1960's there was debate about whether or not recognizable craters rather than "pock-marks" would form on bodies as small and gravitationally as weak as Phobos and Deimos. There was even more controversy regarding the retention of ejecta to form regoliths on bodies with such low escape velocities. Mariner 9 had resolved these issues immediately. The very first pictures showed distinct craters with raised rims, and analysis of Mariner 9 photometry, polarimetry, and thermal measurements showed that--rather than bare, hard rock--the surfaces were covered with a regolith similar in texture to that on the lunar surface.

Satellite Regoliths Offer Puzzles, Suggest Histories

The Viking orbiters provided abundant confirmation of these discoveries. For example, measurements by the orbiters' Infrared Thermal Mapper instruments [Lunine et al. 1982] extended earlier Mariner 9 observations that showed that when Phobos emerges from a Mars eclipse, its surface heats up very rapidly. This indicates a low thermal inertia similar to that of the lunar regolith [Gatley et al. 1974].

But such remote sensing measurements tell us only that the outermost surface has a regolith-like texture; they do not tell us the thickness or depth of the regolith. For that information we must turn to some more ambiguous morphological clues seen in high resolution images. As I will illustrate, there is, for example, visible evidence for significant depths of loose mobile material--perhaps at least 10 meters in places--on Deimos [Thomas and Veverka, 1980]. Similarly, we will see that the morphology of certain features on Phobos (grooves) seems to imply at least comparable amounts [Thomas et al. 1979]. In several instances, detailed views of the interiors of craters on Phobos show layering at depths of 100-220 meters, perhaps additional evidence for a deep regolith [Veverka, 1978].

The regoliths no doubt originated through a long history of comminution of the original surface (possibly solid rock) by impacts. Judging from what has been learned from our own moon, such battering should produce fragments of all sizes, ranging from blocks tens of meters or more across to finely textured powders. Large chunks of presumed ejecta have indeed been identified on the surfaces of both satellites [Lee et al. 1986]; one large block on Deimos is about 150 meters across and is clearly visible just inside the large depression rim (lower right) in Figure 1E-6, a Viking orbiter image.

The puzzle of how small bodies like these may retain regoliths began to be answered through a series of laboratory experiments, including the work of Stöffler et al. [1975]. This work has shown that for a given impact energy, ejection velocities are much lower when the target has the texture of a loose aggregate of individual particles rather than that of a solid rock. This observation has led to the realization that this cushioning effect can play an essential role in building regoliths on small bodies. The process is described by, among others, Hartmann [1978], who coined the phrase "*regolith begets regolith.*" In addition, it has also been suggested that Phobos and Deimos may have an easier time accumulating impact debris than do asteroids of comparable size. The argument for this is that debris blasted off Phobos and Deimos cannot usually escape from Mars orbit, and therefore has a significant chance of reimpacting the satellite from which it came [Soter, 1971]. I will return to the puzzle of how Phobos and Deimos retain regoliths later in this presentation.

Composition: What are they Made of?

The colors and albedos of the satellites were measured using image data acquired by the Viking orbiters' cameras. In these respects the satellites are similar: they are dark (albedo ~0.05) and

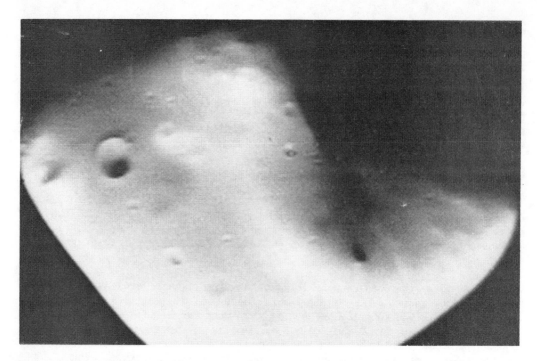

Figure 1E-6: Large Positive Relief Feature (150 meters) on Deimos (lower right)

gray (between 0.4 and 0.6 microns). Combined with additional spectral information obtained from Viking Lander 1 (VL-1) and from the ultraviolet (UV) spectrometer on **Mariner 9**, this information suggests that the surfaces of Phobos and Deimos are made of materials similar to those in carbonaceous chondrite meteorites [Veverka and Burns, 1980]. The low average density (2 gm/cm^3) determined from the **Viking** mass measurements are consistent with this possibility and further restrict the choice of materials, making them similar to either C1 or C2 chondrites (both of which contain water and complex organic material). Thus, the low mean densities are important clues to the composition and, perhaps, even to the origin of the satellites which I will discuss in more detail later. The possibility that Phobos and Deimos are made of C1- or C2-like materials can be verified by difficult (but practicable) infrared measurements from Earth by searching for the presence of a water-of-hydration band at about 3μ.

There have been occasional suggestions [e.g., Hartmann, 1980] that the moons of Mars have low densities because they contain significant amounts of water ice (as opposed to water-of-hydration like that expected in volatile-rich carbonaceous chondrites). Lurking behind such proposals is the vague hope that the moons of Mars may yet turn out to be captured extinct comets. However, it is difficult to build a scenario in which one can preserve ice in a small body at Mars' distance from the Sun for billions of years. It has also been proposed that Phobos and Deimos are examples of the so called "rubble-pile" bodies invented by some theorists. This idea is unlikely to most investigators, given that in many of the images one sees structures (ridges, for example) which suggest bodies of considerable mechanical strength.

It is important to conclude this part of my discussion by stressing that *today* we have only plausible inferences about the composition and internal structure of the moons of Mars. What we need are precise measurements that, in some cases, will require landing instruments on the surfaces of the satellites -- which the Soviets intend to do in the very near future.

SURFACE MORPHOLOGY

It was evident from the **Mariner 9** images obtained in 1971–72 that craters are the dominant landforms on both satellites. With the improved resolution of the **Viking** images, it became possible to study the characteristics of these features in more detail. As suggested by the well defined detail visible in Figure 1E-7, for example, we were able to demonstrate that craters on Phobos have approximately the same depth-to-diameter ratios as small craters on the moon; typically, the depth is about 20% of the diameter [Thomas, 1978].

The Striking Grooves of Phobos

However, the outstanding discovery made by **Viking** was the grooves of Phobos, as seen in Figure 1E-8 (the 10-km-diameter crater Stickney is at the far left). These features had been vaguely glimpsed in some of the better **Mariner 9** images [Pollack et al. 1973], but not well enough to attract general attention or give rise to the realization that they were important clues to the history of Phobos. Early **Viking** images showed that much of the surface of Phobos is striated by long linear depressions. Typically, these grooves are 100–200 meters wide and only 10–20 meters deep; some extend continuously for up to 30 km around the satellite [Thomas et al. 1979].

Figure 1E-7: Phobos Craters

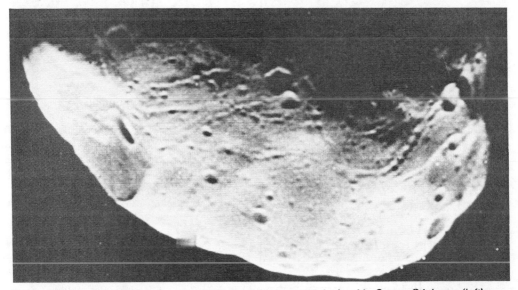

Figure 1E-8: Early Viking Image of Phobos (from 7405 km) with Crater Stickney (left)

101

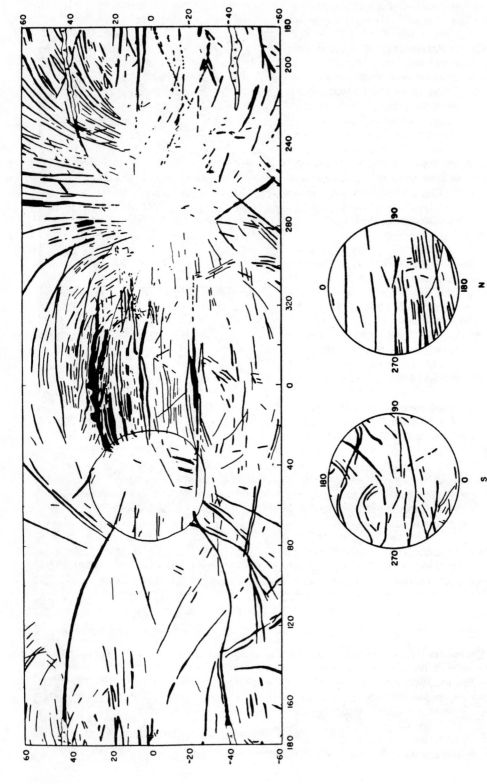

Figure 1E-9: Map of Phobos Grooves Relative to Crater Stickney

As the mission progressed, it became clear that the grooves were associated with Stickney, the largest crater on Phobos[2]. At least three pieces of evidence support this connection:

1. Groove-Crater Relationship -- First, as soon as the first complete map of surface features was put together by Peter Thomas [Thomas et al. 1978], it was seen that many grooves appear to originate at or near the rim of Stickney and converge toward a point that is antipodal to the large crater. Figure 1E-9 (opposite page) is a map reflecting the full scope of the Phobos grooves and their relationship to Stickney; the crater's area (outlined) is centered at about 50°W while the region around its antipodal point is at roughly 270°W in the same latitudinal band.

2. Groove Prominence -- Second, as illustrated in Figure 1E-10 (*a*, top and *b*, bottom), high resolution images later showed that the grooves are best developed (widest and deepest) near Stickney (Fig. 1E-10a) and lose their prominence with increasing distance from the crater (as shown in Fig. 1E-10b; Stickney is just out of the picture to the upper-left). As they approach the antipodal region, the grooves become shallower and narrower, becoming discontinuous and finally disappearing altogether.

3. Relative Ages -- The third piece of evidence of a connection between the grooves and Stickney has to do with the age of the grooves. In many instances it is possible to discern impact craters superimposed on the grooves, as can be seen in Figure 1E-11, and this facilitates the estimate of a crude relative age. From the surface density of these small craters, it can be inferred that the grooves are not recent features and are roughly the same age as Stickney. On the other hand, we know that they aren't older than Stickney because some of the grooves actually cut through the rim of the big crater rather than being obliterated by it.

Possible Implications of Groove Explanation

Given their apparent association with Stickney, it has been suggested that the grooves are manifestations of a nearly catastrophic impact that barely lacked enough energy to disrupt the satellite [Thomas et al. 1979]. If this were true, the grooves would then be surface expressions of fractures within the body of Phobos. However, the appearance of the grooves must have been modified by regolith processes. The very subdued, transverse profiles of the grooves today are consistent with the modification of the original fracture profiles by the slumping of regolith into the depressions. Near Stickney, for example, some grooves have a pitted appearance and somewhat resemble crater pits aligned in depressions. Regolith sifting into deep fractures could produce such features.

Alternatively or in association, internal heating may have led to the venting of volatiles along fractures, forming "sand boils." The transverse profiles of the grooves could give a hint about possible venting of material. In the highest resolution images of grooves obtained by Viking, one of which is presented in Figure 1E-12[3], as well as in a few other instances, grooves

---

[2] As the author has indicated, Phobos is believed to be already inside the Roche Limit, i.e., the satellite's internal gravity alone is not capable of holding the satellite together [Viking Orbiter Views of Mars; NASA SP-441; C. Spitzer, ed., 1980; Dobrovolskis and Burns, 1980]. Therefore, early speculation about the grooves on Phobos focused on the fact that tidal forces might slowly be pulling the satellite apart, thereby producing fractures. This speculation was strongly dispelled by later findings explained by the author. {Ed.}

[3] The Viking photo shown in Figure 1E-11 was imaged at a range of only 140 km (less than 90 miles). The largest crater visible in the image area is about 1000 meters across. {Ed.}

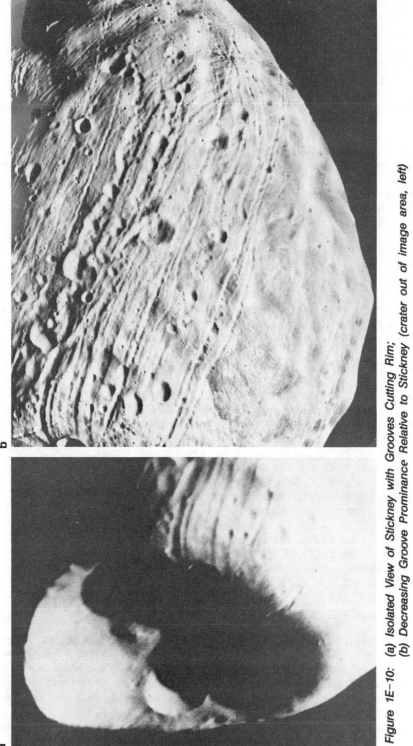

Figure 1E–10: (a) Isolated View of Stickney with Grooves Cutting Rim; (b) Decreasing Groove Prominance Relative to Stickney (crater out of image area, left)

appear to have raised rims -- possible evidence of ejection of material from the grooves. Unfortunately, detailed photoclinometry has so far failed to establish that any of the grooves really have raised rims [e.g., Thomas et al. 1986]. Higher resolution images could certainly resolve this issue.

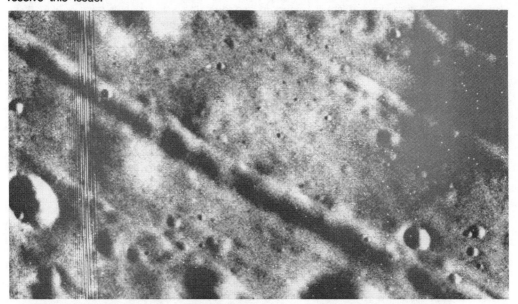

Figure 1E-11: High Resolution Image of Grooves with Superimposed Craters

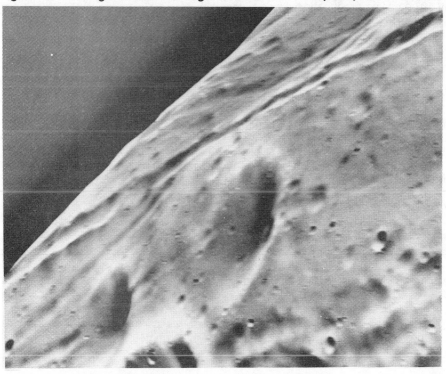

Figure 1E-12: High Resolution Photo of Phobos Grooves

Scenario for a Phobos Collision

What may have happened to Phobos has been elucidated by a series of laboratory experiments carried out by various investigators during the past decade. These experiments were spurred in part by the realization that small bodies in and near the asteroid belt have a high probability of suffering a catastrophic collision during the age of the solar system. One estimate, by Housen et al. [1979], indicates that the time scale for fragmentation-by-impact (in the asteroid belt) for a body 100 km in radius may have been as short as 1.5 billion years; for a body 10 km across, the estimate is 0.5 billion years!

Collision Modeling -- While no one claims that the catastrophic breakup of a 10-km or 100-km body can be simulated adequately in the laboratory, a series of carefully designed experiments have been conducted to define the crucial parameters that determine the outcome of a high velocity collision between a target and an impacting body. A fundamental difference between experiment and reality, however, is that laboratory targets (usually only centimeters in size) are made of mechanically homogeneous, previously unfractured materials. A real asteroid or satellite is likely to be very different in these respects.

Beginning with the pioneering work of Gault and Wedekind [1969], and continuing with the extensive investigations of Fujiwara and his colleagues in Japan [e.g., Fujiwara, 1986], these investigations show that the essential parameter determining the outcome of a collision is the energy delivered by the impact per unit mass of target (ergs/gm). For high velocity impacts (above 1 km/sec), Figure 1E-13 illustrates that a progressive sequence of target damage is observed as the impact energy increases. At the top is a table that characterizes hypervelocity impact damage [Fujiwara et al. 1986] while the sketch (bottom) suggests the nature of the damage for the four stages. Based on the summary compilation of Fujiwara et al. [1977], impacts below about 10^6 ergs/gm produce individual craters with no distal damage. Between 10^6 and 10^7 ergs/gm, fracturing (possible Phobos groove formation) and some spallation is observed, with severe spallation setting in at about 10^7 ergs/gm. Finally, "complete destruction" occurs for impacts above 10^8 ergs/gm.

Results of Collision Modeling -- There are four distinct stages of target damage set out in this particular scheme. The grooves on Phobos probably correspond to a stage 2 event in which fracturing has begun but no large scale spallation has taken place. Note that Thomas et al. [1979] specifically searched for but did not find evidence that spallation occurred near the Stickney antipode. [See also Thomas and Veverka, 1979.] We can conclude, therefore, that an impact energy of 10^6 to 10^7 ergs/gm was involved in the Stickney/groove event. Using a typical impact velocity at Mars of ~10 km/sec, this impact energy corresponds to an impacting body that is roughly 200 meters across.

One of the general results of these investigations is that catastrophic impacts at high velocity lead to a wholesale spallation of surface layers, leaving behind a significantly smaller, irregularly shaped core. In view of the fact that early impact rates in the solar system (during the first half-billion years or so) were significantly higher than they have been since [e.g., Hartmann, 1972], it is conceivable that Phobos and Deimos represent "cores" of larger satellites that suffered catastrophic impacts early in the solar system's history. The surface features that we see today (including Stickney and the grooves) may have been formed mostly during the tail end of the heavy bombardment period some 3.5 to 4 billion years ago. Unfortunately, lacking an absolute cratering time scale for the Mars system, we cannot convert counts of the number of craters of a given size (per unit surface area) into a reliable age.

IMPACT PHENOMENA (v > 1 km/sec)	E/M_t (ergs/g)
Simple Cratering (no spallation)	$\leq 10^6$
Cratering Plus Some Spallation	$10^6 - 10^7$
Spallation/Formation of Core Remnant	$10^7 - 10^8$
Complete Destruction	$\geq 10^9$

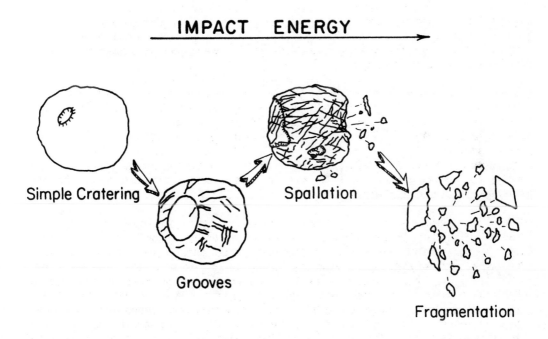

IMPACT ENERGY →

Simple Cratering

Grooves

Spallation

Fragmentation

Figure 1E-13: Characterization of Hypervelocity Impact Damage

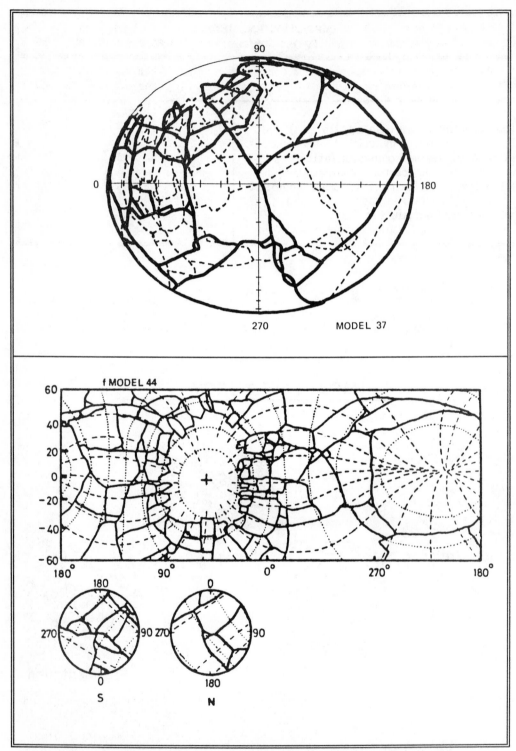

Figure 1E-14: Simulation if Phobos Impact (see Fig. 1E-9 Phobos groove map)

Simulation of Phobos Impact -- In an effort to better simulate the groove-forming process on Phobos, Fujiwara et al. [1983] have expanded their studies to include impacts on ellipsoidal targets. The resulting pattern of fractures produced in Phobos-shaped clay targets bears a general resemblance to what is seen on Phobos. Figure 1E-14 (opposite page) presents drawings that illustrate fracture patterns produced by hypervelocity impacts into two Phobos-shaped models. In the upper diagram, thick lines represent northern hemisphere fractures and dashed lines represent those in the southern hemisphere.

The lower diagram has been prepared such that it can be compared directly to Figure 1E-9. The (+) mark indicates the point of impact and the heavy lines trace out the fractures. As on the satellite, major sets of fractures run radially away from--and concentrically to--the impact site. While the usual cautions about extrapolating experimental results on tiny clay targets to real satellites must be borne in mind, the investigations of Fujiwara et al. suggest that we are on the right track to understanding the grooves on Phobos. An intriguing implication is that one might now expect to find groove-like features on some asteroids as well as on some other small moons [Thomas and Veverka, 1979].

The Puzzle of Deimos

A natural question to ask, based on the case of Phobos, is: "Are there any grooves on Deimos?" The answer is "no!" The most likely reason for their absence is that there are no craters preserved on the surface large enough to have involved sufficient impact energies. Based on the work of Fujiwara et al. [1977], one can estimate that groove formation requires an impact that delivers between 10^6 and 10^7 ergs per gram of target body. The largest crater visible on Deimos is less than 2.3 km across; its formation involved some 6×10^5 ergs/gm [Thomas and Veverka, 1979], too little to have produced grooves.

Even without grooves, Deimos is a bizarre satellite. Its shape is distinctly nonellipsoidal and the surface has a surprisingly subdued look about it. As can be seen in Figure 1E-15 (an unusual contrast-stretched, high Sun angle picture), patches of brighter material are quite conspicuous. There is evidence suggesting that these bright patches may represent loose surface material that is slowly filling depressions (such as the large crater at the top). Similarly, one can also see the ghost-like outlines of smaller depressions that may be filled or partially filled by regolith.

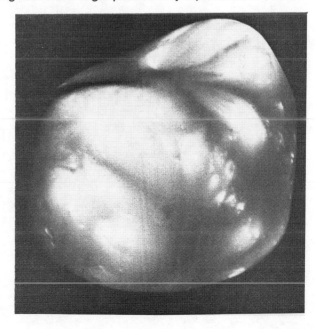

Figure 1E-15: Bright Patches on Surface of Deimos

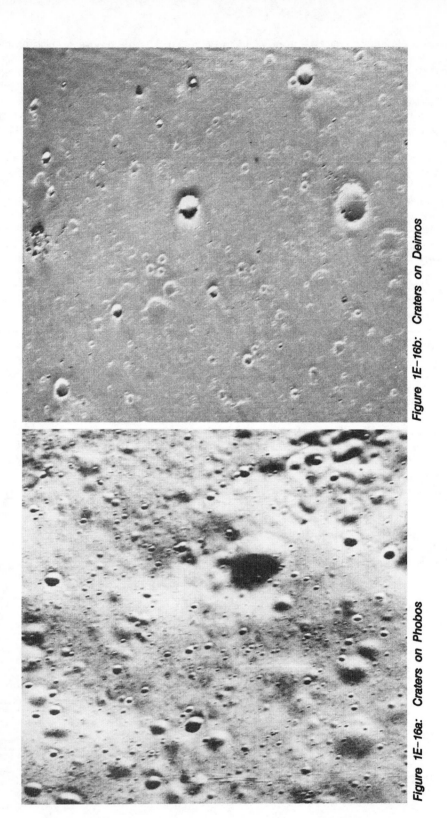

Figure 1E-16b: Craters on Deimos

Figure 1E-16a: Craters on Phobos

Unexpected Regolith Features -- At almost any resolution, Deimos looks strikingly different from Phobos, not only due to the absence of grooves but principally because of an overwhelming impression that surface morphology has in many places been blanketed by finely textured material. Figure 1E-16 (opposite page) presents a pair of very high resolution **Viking** images of Phobos (*a*, top) and Deimos (*b*, bottom). Compared to the relatively crisp appearance of craters on Phobos, many of those on Deimos appear to have been filled by loose material. Moreover, not only does this blanketing material fill many craters, but there is evidence that it is mobile and migrating downhill.

In the simplest terms, one would expect that Deimos, being smaller than Phobos, would not retain ejecta as well. This expectation is confirmed by actual calculations, such as those of Keven Housen [see Veverka et al. 1986], which yield model regolith depths about five to ten times larger for Phobos than for Deimos. These calculations, as demonstrated in Figure 1E-17, agree with observations in predicting that both satellites should be covered by meters of regolith. The curves reflect regolith depth distributions on both satellites as predicted by Monte Carlo simulations of regolith production. The problem is that they indicate that there should be a lot less on Deimos (with 1-to-10 meters) than on Phobos (with 10-to-100 meters), a conclusion strongly at variance with what our intuition tells us when we compare pictures of the two satellites. Are we being deceived by appearances? ...or is the problem of modeling regolith retention on these bodies much more complex than we realize!?

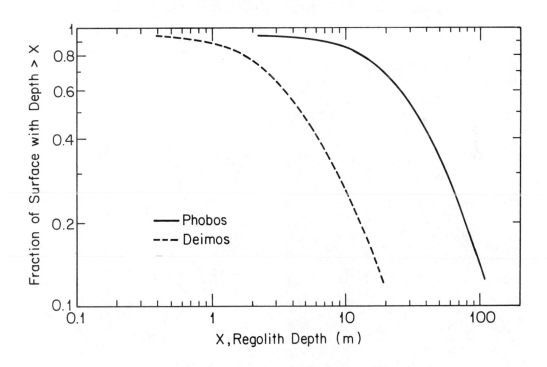

Figure 1E-17: Predictions for Regolith Depths on Phobos and Deimos

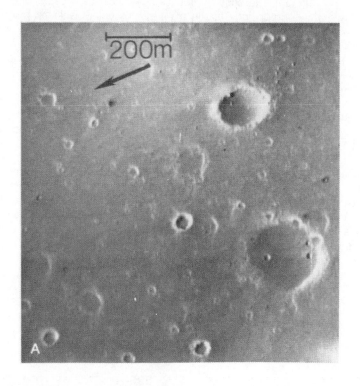

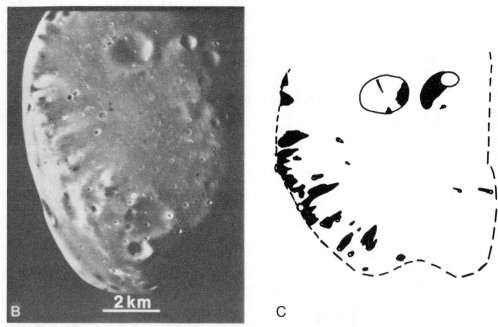

Figure 1E-18: Details of Downslope Movement of Surface Material on Deimos

Evidence of Surface Material Migration -- The conspicuous patches of slightly brighter material can be understood as having a somewhat finer texture than the surrounding regolith. One finds that these patches are often tapered and elongated, and that inevitably they trend in the local downhill direction [Thomas and Veverka, 1980]. Figure 1E-18 (opposite page) illustrates details of this downhill migration. In the top photo (A), faint bright streamers can be seen in addition to the sediment-filled craters. The arrow indicates the slope direction. The area imaged is on the flank of a subtle ridge located on the terminator in the bottom photo (B) and outlined in drawing C. In photo B, the generally smooth surface of Deimos is surrounded by ridges and bright streamers; drawing C is an interpretation diagram of its topography in which the dashed line is the ridge crest and the black areas are bright streamers. This result is quite unexpected, given that surface gravity on Deimos is minuscule -- only 10^{-3} g. Clearly, something we do not understand is happening.

What Happens to Ejecta?

The problem of determining the fate of ejecta on Phobos and Deimos is highly complex. Since both satellites are small, escape velocities are very low. The escape velocities that I gave earlier, about 10 m/sec for Deimos and 15 m/sec for Phobos, are only rough approximate values based on simplified assumptions that the satellites are spherical, spin slowly, and are far away from any planet. But in fact, Phobos and Deimos are irregular, spin quickly, and are found deep in the gravitational well of Mars. Thus, actual escape velocities depend on the specific location being considered on each satellite, and of course on the direction in which material is ejected.

This difficult problem has been analyzed in detail by Dobrovolskis and Burns [1980] and by Davis et al. [1981]. Davis found that on the face of Phobos oriented toward Mars, material propelled at only 3.5 m/sec can escape from the equator, whereas a velocity of 15 m/sec is required for ejecta to escape near the poles. Figure 1E-19 is taken from the work of Dobro-volskis and Burns [1980], and illustrates the trajectories of test particles launched at different velocities from two particular locations on Phobos: one for an impact near Stickney (top left) and the second for an impact on the trailing hemisphere (bottom). The debris is ejected at 45° to the local vertical at initial velocities shown on the trajectories.

Given the complexity of the situation, it is perhaps small wonder that our models of regolith evolution on the satellites of Mars are in rudimentary shape at best. Compounding the difficulty is an important fact already alluded to, and noted almost twenty years ago by Soter [1971], that ejecta escaping the satellites will generally stay in orbit around the planet and have a high probability of reimpacting the surface.

ORIGIN THEORIES

Our understanding of what Phobos and Deimos are like increased substantially as a result of Viking. But, can we at this point say anything useful about their origins? The possible identi-fication of a composition similar to carbonaceous meteorites is an important clue. If it is true that Phobos and Deimos are made of material similar to composition of C-asteroids, and if it is true that such material was formed only in the outer half of the asteroid belt, then the moons of Mars are most likely captured objects. For example, Donald Hunten [1978] has out-lined a scheme whereby the satellites could be captured very early in Mars' history when the planet was still surrounded by a distended atmosphere. However, detailed attempts to trace back the orbital evolution of the satellites lead to possible difficulties with a capture scenario.

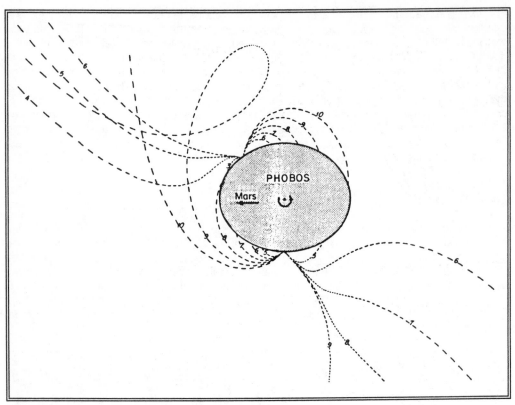

Figure 1E-19: Predicted Ejecta Trajectories for Two Impact Points on Phobos

Orbit Evolution Models Encounter Problems

Today, the satellites orbit in almost circular paths (very low eccentricity) which almost coincide with the planet's equatorial plane (very low inclination). As indicated earlier, the effect of Mars tides on Phobos is pronounced, i.e., the satellite is being gradually accelerated and pulled into the planet. Another effect would be to circularize an initially more eccentric orbit. One possible assumption is that the current eccentricity (about 0.015) is a tidally damped remnant of an initially higher value. Working the orbital evolution of Phobos back in time, as has been done (for instance: by Cazenave et al. [1980] and by Szeto [1983]), one finds that the orbital eccentricity of Phobos increases dramatically. The curves in Figure 1E-20 represent the work by Cazenave et al., and they trace the past evolution of the eccentricity of Phobos as a function of its semi-major axis. Model 1 includes tides in Mars only; Model 2 represents tides in Phobos only; Model 3 represents simultaneous tides in Mars and Phobos. The time evolution is 10^9 years.

A very eccentric past orbit for Phobos appears consistent with a capture origin, but there is a difficulty -- perhaps a severe one. The difficulty is that, when compared to the orbital evolution of Phobos under the influence of tides, that of Deimos is very slight. The result is that, as the orbit of Phobos evolves back in time, a stage is reached when the orbit of Phobos crosses that of Mars. In such a case, a collision is inevitable. Consequently, capture models for Phobos and Deimos are currently in disfavor, but alternate scenarios (e.g., original accretion in Mars orbit) have not been worked out, to my knowledge.

114

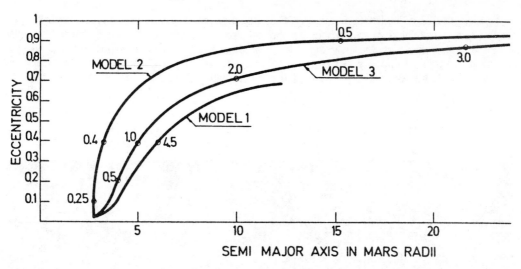

Figure 1E-20: *Evolution of Eccentricity for Phobos as Influenced by Tides*

In a detailed analysis of the orbital evolution of Phobos, Yoder [1982] demonstrated clearly that for the eccentricity calamity to be a problem, significant tidal dissipation must occur within Phobos (i.e., it cannot be a very rigid body in its response to martian tides). If Phobos has substantial internal strength, a large change in eccentricity going back in time is not expected. Yoder also points out that the current eccentricity may not be a "tidally damped remnant" but rather the result of relatively recent "resonances" with respect to the planet's gravity field as Phobos has been drawn closer to Mars.

The Fickle Fate of Prevailing Theory

Some eight years ago, when Joe Burns wrote a review article on the origin of Phobos and Deimos [Burns, 1978] (at a time when capture origins were much in vogue), he wisely added the following sentence to the end of his paper:

> *"The pendulum appears to have swung towards the CAPTURE ideas,*
> *. . . but pendulums usually swing back."*

That was in 1978. A decade later his prediction has come true and the origin of the moons of Mars remains an issue of contention and debate, a situation that will not surprise those of our colleagues who have devoted their careers to unraveling the origin of our own moon!

CONCLUSIONS

In closing, I would like to make a few comments about future studies of Phobos and Deimos. There is considerable discussion in the proceedings of this conference of NASA's **Mars Observer** mission. **Mars Observer** will tell us many new things about the planet, but it will not be able to study the satellites. As suggested by Figure 1E-21, the next important phase of Phobos and Deimos discoveries is expected from the 1988 *Phobos Mission* (PHOBOS·88), a Soviet project in which France and other nations are participating. It will involve multiple, independent space-craft elements and is scheduled for arrival at Mars in 1989. A major objective will be an ambitious investigation of the satellites, which will involve hovering close to and landing on the surface of Phobos (and possibly Deimos if mission success permits).

Based on available information, the PHOBOS·88 spacecraft will carry analytical instruments capable of determining once and for all the precise composition of the satellite[4]. Will this information tell us unambiguously the origins of the satellites? Probably not. But we should not hold it against Phobos and Deimos that their origins remain an enigma; that fact simply places them in the same category as all the other planetary satellites in the solar system. ■

Figure 1E-21: Target of Next Soviet Planetary Program (Phobos Mission, 1988–89)

---

[4] The Soviet multidisciplinary PHOBOS·88 mission involves the launch of two identical Mars orbiters, as in the case of Viking, each equipped with two different kinds of landers -- one a unique hopping rover. The mission and its spacecraft elements are described by James Head in these proceedings [NMC-2A]. Deimos could become the target for the second spacecraft elements if the first is highly successful at Phobos. The two orbiters will be equipped to conduct comprehensive interplanetary and Mars investigations, as well. The Soviets plan to include a plaque recognizing American astronomer Asaph Hall, who discovered first Deimos and then Phobos on August 11 and 17, respectively, in 1877 (26-inch refracting telescope). It is not yet known whether the plaque will be mounted on the PHOBOS·88 orbiter or one of the landers, but it will contain a photo-transfer (on aluminum) of the page dated August 17, 1877, from Hall's logbook and will include two citations: "U.S.S.R. Phobos Mission 1988" and "Discovery of Phobos – Asaph Hall – U.S. Naval Observatory – August 17, 1877." {Ed.}

REFERENCES

This space is otherwise available for notes.

~~~~~~~~~

Burns, J. 1972. Dynamical characteristics of Phobos and Deimos. *Rev. Geophys. Space Phys.* 10:463-482.

Burns, J. 1977. Orbital evolution. In *Planetary Satellites.* (J. Burns, ed.), pp. 113-156. U. Ariz. Press.

Burns. J. 1978. The dynamical evolution and origin of the Martian moons. *Vistas in Astronomy.* 22:193-210.

Cazenave, A., Dobrovolskis, A., and Lago, B. 1982. Orbital history of the Martian satellites with inferences on their origin. *Icarus.* 44:730-744.

Davis, D.R., Housen, K.R., and Greenberg, R. 1981. The unusual dynamical environment of Phobos and Deimos. *Icarus.* 47:220-233.

Dobrovolskis, A.R. 1982. Internal stresses in Phobos and other triaxial bodies. *Icarus.* 52:136-148.

Dobrovolskis, A.R., and Burns, J. 1980. Life near the Roche limit: Behavior of ejecta from satellites close to planets. *Icarus.* 42:422-441.

Duxbury, T.C. 1977. Phobos and Deimos: Geodesy. In *Planetary Satellites.* (J. Burns, ed.), pp. 346-362. U. Ariz. Press.

Fujiwara, A. 1986. Results obtained by laboratory simulations of catastrophic impact. In *Catastrophic Disruption of Asteroids and Satellites.* (D.R. Davis, P. Farinella, P.C. Paolicchi, and V. Zappala, eds.), 57:47-64. *Mem. Soc. Ast. Italinia.*

Fujiwara, A., Kamimoto, G., and Tsukamoto, A. 1977. Destruction of basaltic bodies by high velocity impact. *Icarus.* 31:277-288.

Fujiwara, A., and Asada, N. 1983. Impact fracture patterns on Phobos ellipsoids. *Icarus.* 56:590-602.

Gatley, I., Kieffer, H., Miner, E., and Neugebauer, G. 1974. Infrared observation of Phobos from Mariner 9. *Astrophys. J.* 190:497-503.

Gault, D.E., and Wedekind, J.A. 1969. The destruction of tektites by micrometeoroid impact. *J. Geophys. Res.* 74:6780-?

Hartmann, W.K. 1972. Paleocratering of the Moon. *Astrophys. and sp. sci.* 17:48-64.

Hartmann, W.K. 1978. Planet formation: Mechanism of early growth. *Icarus.* 33:50-61.

Hartmann, W.K. 1980. Surface evolution of two-component stone/ice bodies in the Jupiter region. *Icarus.* 44:441-453.

Housen, K.R., Wilkening, L.L., Chapman, C.R., and Greenberg, R.J. 1979. Regolith development and evolution of asteroids and the Moon. In *Asteroids*. (T. Gehrels, ed.), pp. 601–627. U. Ariz. Press.

Hunten, D.M. 1979. Capture of Phobos and Deimos by protoatmospheric drag. *Icarus*. 37:113–123.

Lee, S.W., Thomas, P., and Veverka, J. 1986. Phobos, Deimos, and the Moon: Size and distribution of crater ejecta blocks. *Icarus*. 68:77–86.

Lunine, J.I., Neugebauer, G., and Jakosky, B.M. 1982. Infrared observations of Phobos and Deimos from Viking. *J. Geophys. Res.* 87:10297–10305.

Pollack J.B. 1977. Phobos and Deimos. In *Planetary Satellites*. (J. Burns , ed.), pp 319–345. U. Ariz. Press.

Pollack, J.B., Veverka, J., Noland, M., Sagan, C., Duxbury, T.C., Acton, C.H., Born, G.H., Hartmann, W.K., and Smith, B.A. 1973. Mariner 9 television observations of Phobos and Deimos. 2. *J. Geophys. Res.* 78:4313–4326.

Smith, B.A. 1970. Phobos: Preliminary results from Mariner 7. *Science*. 168:828–830.

Soter, S.L. 1971. The dust belts of Mars. *Cornell Ctr. Rad. Sp. Res. Rep.* p. 462.

Stöffler, D., Gault, D.E., Wedekind, J., and Polkowski, G. 1975. Experimental hypervelocity impact into quartz sand: Distribution and shock metamorphism of ejecta. *J. Geophys. Res.* 80:4062.

Szeto, A.M. 1983. Orbital evolution and origin of the Martian satellites. *Icarus*. 55:133–168.

Thomas, P. 1978. The Morphology of Phobos and Deimos. Ph.D. Thesis. Cornell University, Ithaca, N.Y.

Thomas, P., and Veverka, J. 1979. Grooves on asteroids: A prediction. *Icarus*. 40:394–405.

Thomas, P., and Veverka, J. 1980. Downslope movement of material on Deimos. *Icarus*. 42:234–250.

Thomas P., Veverka, J., and Dermott, S. 1986. Small satellites. In *Satellites*. (J. Burns, ed.), pp. 802–835. U. Ariz. Press.

Thomas, P., Veverka, J., and Duxbury, T.C. 1978. Origin of the grooves on Phobos. *Nature*. *273:202–204.*

Thomas P., Veverka, J., Bloom, A., and Duxbury, T. 1979. Grooves on Phobos: Their distribution, morphology and possible origin. J. Geophys. Res. 84:8457–8477.

Veverka, J. 1978. The surfaces of Phobos and Deimos. *Vistas in Astronomy*. 22:163–192.

Veverka, J., and Burns, J. 1980. The Moons of Mars. *Ann. Rev. Earth Planet. Sc.* 8:527–558.

Veverka, J., and Thomas, P. 1979. Phobos and Deimos: A preview of what asteroids are like? In *Asteroids*. (T. Gehrels, ed.), pp. 628–651. U. Ariz. Press.

Veverka, J., Thomas, P., Johnson, T.V., Matson, D., and Housen, K. 1986. The physical characteristics of satellite surfaces. In *Satellites*. (J. Burns and M.S. Matthews, eds.), pp. 342–402. U. Ariz. Press.

Yoder, C.F. 1982. Tidal rigidity of Phobos. *Icarus*. 49:327–346.

THE ATMOSPHERE OF MARS
[NMC-1F]

Dr. Michael B. McElroy
Harvard University

Due to the unavailability of an author-reviewed manuscript of this presentation (see FOREWORD), the following is essentially an unedited transcription of Dr. McElroy's NASA MARS CONFERENCE discussion [NMC-1F]. In addition to the apparent differences in format and language, NMC-1F also lacks supporting visuals identified and discussed in its context as viewgraphs. Because Dr. McElroy's expertise on the subject of Mars' atmosphere is widely acknowledged, his presentation is of significant importance within the framework of these Mars Conference proceedings and has been specifically referenced in other Conference presentations. As a precaution, readers should understand that the translation of an oral presentation into publication form, via phonetic transcription from audio tapes (without the aid of an author's review), must be considered with special care. Errors may have been made during the presentation, or, transcribers may have had difficulty understanding technical terms and language or interpreting syntax and structure (e.g., punctuation, which can have a significant impact on context and implied meaning, is very difficult to identify in speech). Readers interested in scientific specifics discussed or defined by Dr. McElroy in this presentation are therefore urged to confirm their understanding of pertinent information given before applying it to their own work. The bibliography at the end of the presentation includes some of Dr. McElroy's papers (or those he contributed to) concerned with the martian atmosphere. {Ed.}

OPENING COMMENTS AND INTRODUCTION

Thanks, Jerry [Dr. Gerald Soffen, session chairman]. This is going to be an interesting experiment for me because this is the first time I think I've ever talked to viewgraphs without having control of the viewgraphs. If I get a little disorganized and don't quite know what comes next, you'll have to bear with me.

What I want to do is to cover a number of topics which excited us during and after, and--in some cases in fact--before the Viking mission. I'll talk a little bit about what we know: what we learned about the chemistry of the atmosphere and why the atmosphere is the way it is. I'll talk a little bit about the processes that control the stability of the atmospheric composition, with particular emphasis on the time scales that are involved in driving potential change in the atmosphere. And then I'll say some words about the isotopic composition of the atmosphere, and about some of the things that one can learn and some important constraints that one can place on the evolution of the atmosphere based on the isotopic composition.

I guess I should say right away, [since] Jerry mentioned Al Nier, that a great deal of what I'll talk about here I really learned from Al Nier. [Dr. Nier] was of course pivotally involved in the acquisition of the atmospheric data that I'll be discussing for the most part[1].

(Projection Cue -- signifies speaker-request for viewgraph change)

OVERVIEW AND REVIEW

I'll begin this by summarizing what we know about the composition of the atmosphere of Mars with a comparison to the composition of the Earth's. In the first part of this talk I'll be talking about the first three gases in the left hand column -- the first three gases on [under the column heading for] Mars. What is it that determines the oxidation state of the atmosphere? What is it that determines the relative abundances of CO and O_2? As we go along, I'll say a little bit more about nitrogen. And finally, if time permits, I'll say something about the noble gases, and I'll say quite a bit about water.

Atmospheric Chemistry

(Projection Cue)

The next [current] viewgraph attempts to summarize some of the chemistry that is involved in controlling the concentration of CO_2. In particular, it points to a little bit of a problem that we were aware of for a number of years--indeed, before the **Viking** mission--that, if you have carbon dioxide and you photodissociate carbon dioxide, you form CO and O. There is a little bit of a problem putting the molecule back together again. The problem is that carbon monoxide is a singlet molecule, oxygen is a triplet, and CO_2 is a singlet. So the direct path [for] reforming CO_2 is quite inefficient. Under planetary conditions, you have to do it by some subtle technique. The subtle technique, as best we understand it, involves the interaction with radicals from the hydrogen species drawn ultimately from photolysis of water. I've written down just one radical chain; a rather simple one which you can see is chemically equivalent to a reaction of CO with atomic oxygen, reforming CO_2. But it goes through these three steps in which the hydrogen species are recycled, so we say this is catalytic. It's analogous to some of the processes that are very important in the chemistry of the Earth's stratosphere.

Where [do] the hydrogen species come from? These radicals come ultimately from reaction, the second reaction from the top [on viewgraph]. The dissociation of water vapor can occur either directly by sunlight or it can occur by reactions involving metastable oxygen atoms. That water is reconstituted by a radical-radical reaction down here. There is another radical-radical reaction which has the effect of forming molecular hydrogen, and we believe that to be very important in controlling the hydrogen escape from Mars. In order to get hydrogen atoms up to the upper levels of the atmosphere in a very cold environment, you have to get it up as molecular hydrogen. You can't get it up in the form of water. This reaction is very important in regulating escape and so [thus] the relative apportionment between these two channels here -- if there is a critical role in governing the escape and evolutionary history of the atmosphere.

---

[1] Dr. Alfred O.C. Nier, University of Minnesota, was team leader for the **Viking** Entry Science Team and is one of the nation's true pioneers in scientific spectrometry. Dr. Nier is thought of as the senior **Viking** alumnus, and was awarded NASA's Exceptional Scientific Achievement Medal for his **Viking** work. Dr. McElroy was a member of Dr. Nier's team and also supported atmospheric analysis work based on data provided by the **Viking** landers' GCMS [Gas Chromatograph/Mass Spectrometer] instruments. {Ed.}

Processes, Escape Rates, and Reactions

(Projection Cue)

[This viewgraph is a] summary of some numbers here. The rate at which you decompose CO_2 by photolysis (by sunlight) is pretty well known. It depends on the total number of photons coming in from the Sun below some energetic threshold. Every photon effectively dissociates CO_2, whether directly or indirectly. There's the atmospheric abundance of CO. I can do a very simple calculation and ask: how long does it take to make the observed concentration of carbon monoxide in the atmosphere? And the answer is: only a few years. So the photolysis of CO_2 has the potential to change the concentration of carbon monoxide exceedingly rapidly. That's another way of saying that, given that there's a relatively small amount of carbon monoxide in the atmosphere, one has to be able to put the molecule back together again quite efficiently.

Molecular hydrogen has not been measured [on Mars], so the number that I'm quoting here is based on--essentially--model studies. We expect the martian atmosphere to contain perhaps 10 parts-per-million of molecular hydrogen. The hydrogen escape rate, as measured back on the **Mariner** missions [by] Charlie Farth and his colleagues. The oxygen escape rate, which I'll come back to, is something that we actually expected or predicted in advance of **Viking**, and the **Viking** data are consistent with the escape rate of oxygen of this magnitude. It occurs by a very interesting process which involves the recombination of a molecular ion; it's not a thermal process, it's a non-thermal process. As such, we can calculate the oxygen escape rate with considerable precision, I believe, and can do so all the way back in the history of Mars. So this is a fairly certain number. Notice the interesting comparison between these numbers. We believe (under current conditions) that escape of hydrogen from Mars occurs stoichiometrically [i.e.,] with the stoichiometry of water -- 2 hydrogen and 1 oxygen. The question that occurs is why and what controls that?

(Projection Cue)

[This viewgraph] points out some of the simple things that occur in some of the outer regions of the martian atmosphere. In the ionosphere, the photons can tear the electrons off the atoms and molecules which are present, and ion-molecule processes reasonably well understood make molecular oxygen ions. Molecular oxygen ions recombine to form O atoms, and so also the CO_2^+ to combine to form O atoms. The point is that we can calculate the energy of these O atoms; that's determined by atomic physics. The O atoms formed on Mars have sufficient energy to escape the gravitational field of the planet so long as there are not collisions that sort of impede that escape process. It's a rather sensitive process because it occurs on Mars and it does not occur at all on the Earth (nor does this process occur directly on Venus). It is a Mars-specific process in which the atoms simply have enough energy to develop speeds approaching or exceeding 5 kilometers per second. So, they can escape from Mars but not from the other two planets with considerable atmospheres in the inner solar system.

Time Scales

(Projection Cue)

This is a summary of some of the time scales I've been alluding to. The time [needed] to produce all the CO in the martian atmosphere by photolysis is about four years. The martian atmosphere contains a little bit more oxygen than carbon monoxide, so it's oxidizing -- mildly oxidizing. You can form that molecular oxygen by photolysis [of] CO_2 in about 13 years (very, very short times). Molecular hydrogen, on the other hand, depends on the escape process; the time scale to build up molecular hydrogen is about 1000 years. If I consider the excess oxygen that's present in the atmosphere and attribute that to differential escape of hydrogen, I can calculate how long it takes to build up that excess -- and it's about 300,000 years. That's interestingly poised for the topic that will be discussed is some of a later papers in this session, the possibility of climatic cycles driven by the orbital changes of the planet due to obliquity, eccentricity and so on.

(Projection Cue)

I'm going to switch time scales a little bit here, beginning by talking about the fast processes first. What we want to do is to ask: how does this hydrogen-oxygen escape balance really work? Is it an accident that you have two hydrogen atoms escaping for every one O? What controls what, and what maintains this curious balance? From an atomic physics point of view, we believe that the constant of the problem is the escape rate of atomic oxygen. That depends on the total number of photons coming into the atmosphere below about 900 angstroms -- photons with enough energy to produce ions which then recombine, producing energetic oxygen atoms. So the constant in the problem is the escape of atomic oxygen at a rate of about 6 times 10 [6x10]--sometimes 6 to the 7th [6^7]--atoms per square centimeter per second. How, then, does the hydrogen know about this situation? How does the hydrogen keep up with the oxygen escape rate?

(Projection Cue)

The next chart attempts to back up and have a look at the chemistry and the controls that are conserving quantities in the atmosphere, so that, for example, the total rate of which you'd associate CO_2 is the first (determined by the number of photons coming in from the Sun below--in this case--about 1800 angstroms). I can think of alpha as a conserved quantity. You must put CO_2 back together again at the rate of which you form it. The first step in putting it back together again is [the] reaction of CO with OH. If I also fix the water abundance in the planet, and insist on putting water back together again at more or less the rate at which I'd associate water vapor (this is another "conserve" quantity called a beta), that's the rate at which you reform water. If I now say that the hydrogen escape rate is fixed by the oxygen escape (otherwise the oxidation escape will evolve rapidly to some new value), I can fix the product of H times HO_2.

As result of the three fixed quantities that we can now look at, [which represent] some very simple relationships, and--in particular--we can conclude: the hydrogen abundance is inversely proportional to oxygen, the OH is inversely proportional to CO, and so on back down to here. This product here is proportional to the molecular hydrogen source and therefore to the escape rate of hydrogen. [Thus,] we can say that the escape rate of hydrogen depends on the ratio of CO to O_2 in the atmosphere, and the gross oxidation reduction state of the atmosphere controls the production of molecular hydrogen and therefore the escape of H.

(Projection Cue)

Here's how it goes. Suppose we have a transient in which a little bit more hydrogen escapes than is tolerated by this equilibrium, what's going to happen? That excess escape of hydrogen will generate some excess molecular oxygen [and the] excess molecular oxygen will adjust this ratio downward. When that ratio goes down, as we saw in the previous chart, the molecular hydrogen production rate goes down; when the hydrogen production rate goes down, the hydrogen escape rate goes down. You can see that we have a sort of stabilizing situation which works in both directions. It's reasonable to expect--on fast time scales--that the hydrogen escape will adjust to the oxygen escape, and one will indeed preserve the stoichiometry of water in the escaping gases from Mars.

Potential Instabilities in Atmospheric Composition

(Projection Cue)

I want to say a few words here about instabilities or potential instabilities in the composition of the atmosphere. This is a complicated mess, but what we're trying to do is look at: keep CO_2 fixed at its current value [of] 1. Then what we're going to do is look at what happens as you change the abundance of water relative to CO_2. As the abundance of water relative to CO_2 decreases, we move up in this direction. This is a calculation which is done assuming a constant oxidation state, recognizing that the oxidation state can adjust only in time scales of a few hundred thousand years. [Therefore,] we're looking at fast responses to rapid variations in water, if you will.

As the water abundance goes down relative to CO_2, what we're effectively doing is choking off this recombination chemistry for CO_2. The response is to build up carbon monoxide and molecular oxygen. At some point you could get into an interesting dilemma [in which], if you don't have enough water present, carbon monoxide and molecular oxygen will build up very rapidly. And to the extent that carbon monoxide is controlled by thermal equilibrium, for example with the regolith, there is no such balance for CO and O_2. In principle, at least, it's possible to acknowledge an atmosphere with vastly different chemical properties in which the major constituents might be carbon monoxide and molecular oxygen. Finally, you can reach a new equilibrium in which molecular oxygen effectively shields CO_2 and turns off the dissociation of CO_2 -- that occurring under very obviously different conditions.

(Projection Cue)

How does the oxidation state (now looking at the longer time scales) respond on longer time scales to changes in the water-to-CO_2 ratio? This [viewgraph data] is a result of some calculations looking at that. The oxidation state goes down as the oxidation state [given] here, measured as the excess oxygen over CO (assuming the stoichiometry of CO_2 as the reference point) goes down with low water-to-CO_2 [ratio] -- up with high. It's a fairly sensitive curve.

Escape Processes Involving Isotopes

(Projection Cue)

I want to switch attention back to escape processes and say a few words about isotopes, and [about] things that we can learn by looking at the isotopes of oxygen (initially) and then also nitrogen. I'll say a few words perhaps about carbon, as well. [A velocity of] 4.9 kilometers-per-second is required to escape from Mars, and [the result of] translating that into energies for the atoms [is given here in a summary of that process]. The process which produces energetic oxygen atoms has more than enough energy to escape both oxygen-16 and oxygen-18, so let's see what the implications of that are.

(Projection Cue)

This is a schematic of what you'd expect generally in a planetary atmosphere. What we're looking at is a region of the lower atmosphere which is well stirred (well mixed). Let's think about oxygen-16 and oxygen-18. The relative abundance of oxygen-16 to -18, or oxygen-18 to -16, will stay effectively constant up to some level in the atmosphere above which molecular diffusive processes are able to separate out the light gases from the heavy; the heavy gas will tend to fall off with altitude relative to the light. So we build up, near this region which we call the exculpation [?] region (above which collisions are not terribly effective), an enriched situation of oxygen-16 relative to oxygen-18, or nitrogen-14 relative to -15, [or] carbon-12 relative to -13. It is easily predicted what that trend is going to be, at least for the present martian atmosphere; we can do this with some precision.

Suppose the escape process that occurs is able to energize oxygen-16 and oxygen-18 equally, and that [is what] we believe to be the case. What's going to happen, then, is that--over time--there will be differential escape of oxygen-16 relative to oxygen-18. The differential escape will be predictable. What would you expect [in this case]? You would expect to see some response in the atmosphere, in the sense that the isotopic oxygen composition of the atmosphere would become heavier with time. The magnitudes of the escape rates involved are quite large relative to the atmospheric reservoir, and so this differential escape for oxygen is a significant process if the oxygen reservoir involved only includes the oxygen present as carbon dioxide in the atmosphere.

(Projection Cue)

What do we actually know about this? The evidence is that the oxygen isotopic composition on Mars is more or less normal, in the sense that it is more or less similar to the Earth. That means that the reservoir which is exchanging with the atmosphere has to be quite large. We can readily, then, predict a minimum size for that reservoir in order to keep the oxygen isotopic composition fresh, and the size of that reservoir corresponds to no less than several tens of meters. Let's assume that the oxygen comes from water in the regolith; how much water do you have to flush chemically through the atmosphere, exchanging with CO_2? You have to flush at least ten meters worth -- ten, twenty or thirty meters worth of ice. That's an absolute lower bond to the amount of oxygen that has to process chemically and isotopically equilibrate with CO_2. But the point is that you are driven immediately to the conclusion that the abundance of volatile oxygen on Mars is very, very much larger than the abundance of oxygen that is present in the atmosphere as CO_2. The minimum concentration of oxygen per square centimeter must be about 5 times 10 to the 25th [5×10^{25}] molecules per square centimeter. We're talking about several bars worth of oxygen equivalent carrying capacity.

Nitrogen: A Link With the Past

Let me turn a little bit to the nitrogen story. It's an interesting one because there are a number of processes that can produce energetic nitrogen atoms. One is **dissociative recombination**, which can go by two channels -- [although] there's some uncertainty what the actual branching ratio really is. There are a number of processes involving electron dissociative events, too, and so on, which we can calculate with greater or lesser accuracy. This is interesting because in this case one of the channels develops enough energy to escape nitrogen-14 [^{14}N] but does not develop enough energy to escape nitrogen-15 [^{15}N] -- which requires 1.9 volts. Unless we play some games and look to some vibrational excitation in the molecular nitrogen ion, there are some interesting atomic physics problems that are involved in the nitrogen story that are still not totally resolved.

(Projection Cue)

[Based on] the observations from Al Nier (upper atmosphere), and also [those] from Toby Owen (and I think Klaus Biemann's experiment) on the surface[2], oxygen-16 and -18 [appear to be] pretty similar to the Earth. But nitrogen-15, -14 very, very strongly enriched from the heavy isotope on Mars; carbon-12, -13 similar. Let me say a few words about this. What this means is that any model for the evolution of Mars is now very strongly constrained. If we assume that Mars began with the isotopic composition that the Earth did, as far as nitrogen is concerned, [and] if you make an assumption about the initial nitrogen reservoir size and you assume you know the atomic physics that is regulating the differential escape of ^{14}N, it has to come out right. [That is,] you have to get the right answer -- Al Nier doesn't allow too much tolerance in it. That's a fairly heavy constraint that has to be satisfied by any particular model.

You can do a calculation in which you ask: what's the minimum nitrogen reservoir size required to equilibrate and give this result? And that minimum size roughly contains more nitrogen than the current CO_2 content of the atmosphere. The minimum size would, if it were all in the atmosphere, account for about 7 millibars or so of nitrogen partial pressure, but that is strictly a minimum and quite dependent on your model for evolution. As such, it provides a very good test of ideas and models about evolution.

If, for example, there's a nonexchangeable nitrogen reservoir, suppose--and we believe this to be the case--you can form some nitrates at the surface of Mars photochemically in the atmosphere. If you form nitrates, that pushes the requirement on the initial reservoir up considerably. Without stretching the imagination too far, you can drive the requirements for nitrogen up to the point where one is talking about perhaps hundreds of millibars of total nitrogen that has exchanged through the atmosphere and been isotopically processed to leave this rather large enrichment behind.

I'll say a few words about another interesting constraint that you can impose on Mars based on this single measurement. It would be nice to have more of these data, but, based on this single measurement, suppose you have the idea that Mars is currently degassing and the nitrogen is coming out of the interior for the first time. Let's also suppose it's reasonable that fresh "juvenile" nitrogen would have an isotopic composition that was sort of like the Earth's. That observation doesn't allow you to play very many games; it forces you to conclude that there is an upper limit to the current degassing rate of nitrogen which cannot be more than a few percent of the planetary mean degassing, otherwise you would not be able to account for such a heavy nitrogen content of the atmosphere. There are many such constraints that can be imposed based on this single measurement, exploiting the fact that the escape process can be quantified. It can be quantified even better with a little bit more atomic physics.

---

2 Dr. Klaus Biemann, Massachusetts Institute of Technology, was team leader of the Viking organic chemistry investigation (Molecular Science Team). The team's instrument, one on each lander, was the GCMS (Gas Chromatograph/Mass Spectrometer), and it could be used for atmospheric analysis when not processing soil samples in search of organic compounds. Dr. Tobias (Toby) Owen, State University of New York at Stony Brook, spearheaded the GCMS atmospheric analysis work. He was instrumental in developing techniques to facilitate the selective enrichment of trace elements in the GCMS to afford more precise identification. In this way, the ratios of rare gases, e.g., isotopes of the noble gases, could be determined to help provide a better basis for understanding the evolution of Mars and its atmosphere. Dr. Biemann and Dr. Owen were both awarded NASA's Exceptional Scientific Achievement Medal for their Viking work. {Ed.}

CONCLUSIONS

(Projection Cue)

I want to conclude with a few remarks, then, about what the possibilities are. There's still a wide region of space available for speculation about the original composition of the martian atmosphere and the evolutionary track, but the constraints imposed--particularly by these isotopes--are quite important and quite serious. The martian concentrations here are expressed in terms of grams per gram of total planet and, correspondingly, the terrestrial columns over here. This is current atmosphere. What you can see is that it's possible, without stretching credulity too far, to drive this number up perhaps by a couple of orders of magnitude or more, so it's possible to bring the nitrogen inventory of Mars up rather closer to the inventory of nitrogen on the Earth.

In the case of CO_2 or carbon, of course, the big unknown is how much carbon is present in the regolith or still bound in the planet. You can speculate quite a bit about that, and surely this number is too low -- the total planetary content is surely higher. Perhaps Fraser Fanale [NMC-1H] will say a little bit about that later. [The] same remark [can be made] about water. Obviously, what you see in the atmosphere is just the thermally-allowed trivial tip of the iceberg, literally, and once again it's possible to imagine a Mars which has an inventory of water that's not unlike the concentration on the Earth.

Noble Gasses

When you get down to the noble gases, life gets a little bit more complicated, as was eluded to in John Lewis's talk earlier [NMC-1A]. In the case of argon-36, you can see we're talking about a small concentration relative to the Earth. [The] same [is] true of the other solar-component noble gases; argon-40 [is also] lower but not quite as low; the ratio is, well, John [Lewis] made this point earlier.

The question is: how do you make sense out of this? There are a number of possibilities, and John [Lewis] mentioned one which would have to be of course tested against the isotopic constraints very carefully to make sure that it satisfied them -- and my guess is that it would be a pretty tough job. I'm struck by the fact that Venus [Projection Cue] shows a noble gas pattern that goes in the other direction. If we look at argon-36 on Venus, [we] see that Venus is very rich in argon-36 relative to the Earth [while] Mars appears to be poor. [The] same remark [applies] for the heavier noble gases. So, without wondering or worrying too much about why, I can speculate that the abundance of solar-component noble gases contained by planetary objects appears to be decreasing -- a decreasing functional distance from the Sun. Venus has more (a lot more) relative to the other gases, CO_2 and nitrogen. Mars appears to have less.

A number of us--George Witheral, for example--[have] suggested that the Venus data might reflect implantation of solar-wind noble gases, perhaps in a preplanetary phase. You get excess injection of noble gases. The planetary material is rich then in the solar-component noble gases when it forms, and then it is able to release more. That would ultimately depend on a model for the formation of the planets in which the surface-area-to-volume (of the particles) would play a very critical role. We speculated (it was a little more than speculation and is still a little more than speculation) that in the case of Mars the surface-area-to-volume consideration might go in the opposite sense. Perhaps you [might] even have particles large enough (surface area small enough) so that there could be some melting with the short-lived radioisotopes (which are believed to be present in the solar system) so that you could actually get rid of noble gases in this preplanetary phase there. In any event, these are open issues that are still as yet unresolved.

128

Summary

Okay, so let me wrap it up and summarize very briefly what I've tried to cover here. I believe that in the case of the atmosphere of Mars we have a first-order understanding of the photo-chemistry of Mars. It's a very interesting situation in which escape processes appear to play a very important role in determining the bulk composition of the atmosphere. That, from the viewpoint of a terrestrial "aeronomer" [aerologist] is a very curious situation. You are forced to contemplate the possibility that the processes occurring at the very top of the atmosphere, affecting a very, very small fraction of the total atmosphere mass on relatively short time scales, have the ability to change the composition of the entire bulk atmosphere. That's interesting in itself.

It's also interesting to notice the possibility for instabilities, in the sense that the abundance of CO_2 is surely, in some sense (although in detail we don't understand), controlled by thermal considerations. It's not obvious that it's controlled directly by the polar caps, though perhaps by the regolith. There is no obvious such constraint limiting the abundance of molecular oxygen or the abundance of carbon monoxide. Given that the atmosphere processes carbon monoxide and molecular oxygen very rapidly, we can at least contemplate the possibility at times in the past where the relative abundance of CO to CO_2 could have been vastly larger than it is today.

Finally, there is also the interesting conclusions and opportunities for research that present themselves, given careful measurement of the isotopes. I've said a little bit about oxygen, a little bit about nitrogen; I could have gone on and talked about some of the others, including carbon and also hydrogen deuterium for which we have unfortunately no experimental data currently that allows you to draw some conclusions.

The Cost of Frozen Technology

I was learning about mass spectrometry in the days of planning for **Viking**. Al Nier was kind enough to give me frequent tutorials on this topic. I was very impressed by one statement Al made [about the fact] that the instruments we were flying in **Viking** were state-of-the-art -- circa 1968[3], wasn't it Al!? Given a few more bucks, Al assured me that he could improve the precision by a couple orders of magnitude. Now it's 1986, Al, and you're not that old yet; so it's time to go back and do it properly. Anyway, that's my story. ■

---

3 Technology "frozen" into spacecraft instruments is dictated by the amount of time they must be available during the final development, test, and installation procedures for the spacecraft itself (when modifications and/or their perturbations might severely impact schedule performance). For this reason, it is rarely possible to utilize the very latest instrument technologies on spacecraft systems as critical and advanced in concept as were the **Viking** landers, which demanded considerable lead-time relative to instrument parameters and assurance. Although the **Viking** spacecraft were launched in 1975, schedule delays due to funding considerations, as well as development problems associated with some of the more complex instruments, had driven the launch out to that point from an earlier launch date (1973). As a result, the technology of many "mature" **Viking** instruments was essentially frozen at state-of-the-art capabilities achieved for and according to the earlier schedule. It is not surprising, then, that some of them reflected state-of-the-art technology for the late 1960's. Nevertheless, the **Viking** instruments penalized in this manner proved extremely well prepared for their tasks and performed accordingly. Newer, immature instruments may not have fared so well. {Ed.}

REFERENCES

This space is otherwise available for notes.

~~~~~~~~

Note: In lieu of author-provided references,
the following is a brief list of references containing
pertinent information on the atmosphere of Mars as derived from
Viking or Mariner data and other relevant studies.

Biemann, K., Owen, T., Rushneck, D.R., Lafleur, A.L., and Howarth, D.W. 1976a. The atmosphere of Mars near the surface: Isotope ratios and upper limits on the noble gases. *Science.* 194:76-78.

Carr, M.H. 1981. *The Surface of Mars.* Yale Planetary Exploration Series. Yale University Press.

Cunningham, N.W., and Schurmeier, N.M. 1969. *Mariner-Mars 1969: A Preliminary Report.* Early results of the Mariner 6/7 missions. NASA SP-225.

Fanale, F.P., Banerdt, W.B., and Saunders, R.S. 1981. The Mars atmosphere-cap-regolith system and climate change. *3rd International Colloquium on Mars.* pp. 74-76.

Hartmann, W.K., and Raper, O. 1974. *The New Mars: The Discoveries of Mariner 9.* NASA SP-337, GPO Stock No. 3300-00577.

Hunten, D.M. 1971. Composition and structure of planetary atmospheres. *Space Sci. Rev.* 12:539-99.

Leighton, R.B., and Murray, B.C. 1966. Behavior of carbon dioxide and other volatiles on Mars. *Science.* 153:136-144.

Louis, J.S. 1984. *The Origin and Evolution of Planets and their Atmospheres.* Academic Press.

McElroy, M.B., and Yung, Y.L. 1976. Oxygen isotopes in the martian atmosphere: Implications for the evolution of volatiles. *Planet. Space Sci.* 24:1107-13.

McElroy, M.B., Kong, T.Y., and Yung, Y.L. 1977. Photochemistry and evolution of Mars' atmosphere: A Viking perspective. *J. Geophys. Res.* 82:4379-88.

Mutch, T.A., Arvidson, R.E., Head, J.W., Jones, K.L., and Saunders, R.S. 1976. *The Geology of Mars.* Princeton University Press.

Nier, A.O., and McElroy, M.B. 1976. Structure of the neutral upper atmosphere of Mars: Results from Viking 1 and Viking 2. *Science.* 194:1298-300.

Nier, A.O., and McElroy, M.B. 1977. Composition and structure of Mars' upper atmosphere: Results from the neutral mass spectrometers on Viking 1 and 2. *J. Geophys. Res.* 82:4341-49.

Nier, A.O., Hanson, W.B., McElroy, M.B., Spencer, N.W., Duckett, R.J., Knight, T.C., and Cook, W.S. 1976. Composition and structure of the martian atmosphere: Preliminary results from Viking 1. *Science*. 193:786-88.

Nier, A.O., Hanson, W.B., McElroy, M.B., Seiff, A., and Spencer, N.W. 1972. Entry science experiments for Viking. *Icarus*. 30:200-11.

Owen, T. 1966. The composition and surface pressure of the martian atmosphere: Results from the 1965 opposition. *Astrophys. J.* 146:257-70.

Owen, T., and Biemann, K. 1976. Composition of the atmosphere at the surface of Mars: Detection of argon-36 and preliminary results. *Science*. 193:801-03.

Owen, T., Biemann, K., Rushneck, D.R., Biller, J.E., Howarth, D.W., and Lafleur, A.L. 1977. The composition of the atmosphere at the surface of Mars. *J. Geophys. Res.* 82:4635-39.

THE METEOROLOGY OF MARS
[NMC-1G]

Dr. Conway Leovy
University of Washington

Although tenuous, the martian atmosphere can support dynamic meteorological processes. It has played an enormous role in shaping the planet's surface over time, and is the one other atmosphere in the solar system most likely to be meteorologically similar to Earth's. The southern summer pattern of winds due to dust storms reveals that most have a localized origin at low latitudes in the northern hemisphere. These move across the equator and then apparently bend eastward in a pattern analogous to a monsoon-like circulation in Earth's atmosphere. Global dust storms tend to occur when the planet is receiving maximum solar radiation near perihelion. Atmospheric pressure was found to vary significantly with the seasons. The annual minimum is recorded during the period of maximum CO_2 condensation at the South Pole (southern winter), and a weaker minimum is recorded during northern winter. Annual cycling repeats with little or no variation, but there are substantial variations in individual cycles over a period of years. Traveling cyclones and fronts were observed from orbit and were also detected by lander instruments, and lander imaging provided evidence of increased atmospheric opacity during such events. While the martian atmosphere is in many respects a laboratory for studying the meteorological processes associated with planetary atmospheric dynamics in general, there is particular interest in the cycling of water vapor between the martian atmosphere and surface. Future mission science should contribute important new knowledge to our understanding of this process.

INTRODUCTION: THE ATMOSPHERE AND ITS METEOROLOGY

The *NASA Mars Conference* presentation sequence is organized such that mine is at exactly the right place on the agenda: between Michael McElroy's [NMC-1F], which discusses that part of the atmosphere that has escaped and the resulting implications for the current atmosphere, and Fraser Fanale's [NMC-1H], which is concerned with that part of the atmosphere that has been retained in reservoirs on and in the surface. I am going to discuss only that part of the atmosphere that is actually in the atmosphere at the present time. This amounts to only two percent (2%) of Earth's atmosphere in terms of mass per unit area.

Relevancy of Martian Atmosphere

One might wonder why a meteorologist would be interested in an atmosphere that is so thin. There are a number of reasons. One important reason is that it is enough atmosphere, as on Earth, to play a role in shaping the surface. As explained and illustrated during other conference presentations, the martian atmosphere has in fact played an enormous role in shaping the planet's surface over time. However, just how that role has operated still remains, in many respects, to be deciphered.

The martian atmosphere is also of interest to meteorologists because it is the atmosphere in the solar system most likely to be similar in its dynamical processes to the atmosphere of Earth. Therefore, if we want to extend our Earth-bound views of weather systems or of atmospheric dynamics in general, Mars is the place to study next because it stretches our imaginations. But it does not stretch them so far that we find ourselves on extremely thin ice. Its processes are not nearly as difficult to understand as those driving the atmospheres of Venus or very large planets like Jupiter and Saturn.

Integration of Historic and Modern Work

One of the first meteorologists to take a serious look at the atmosphere of Mars was Seymour Hess[1]. In 1950, Dr. Hess spent some time at the Lowell Observatory working with temperature data acquired by Coblentz and Lampland in the 1920's. These were very difficult observations to make at that time and represented temperature distributions on the surface of Mars. Dr. Hess used these data to infer distributions of atmospheric pressure, and he produced the crude surface weather map presented in Figure 1G-1. He combined several bits of ephemeral data, primarily the temperature data and, secondarily, wind data. The latter were obtained by assuming that observed drifts of yellow clouds[2] and other martian clouds corresponded with wind drift.

Figure 1G-2 presents a somewhat later example of compilations of apparent wind drift measurements at the surface of Mars, as documented by Gifford in the mid 1950's. As in the case of the data used by Hess, Gifford's data also represented Earth-based observatory measurements of the drift of yellow clouds. While it was not noted at the time, Gifford's work already revealed a very systematic pattern of winds. The southern summer pattern of wind drifts due to dust storms reveals that most of the storms originate in a localized band of longitude at low latitudes in the northern hemisphere during the southern summer. Most of these dust clouds move across the equator and then apparently bend eastward, following the wind drift. This could be very analogous to a monsoon-like circulation in the Earth's atmosphere. On Earth, during northern hemisphere winter, a monsoon circulation flows southward across the equator from the Asian sector, then bends eastward toward Australia. If we had recognized it at the time, we might have understood that a similar process was occurring in the martian atmosphere. Only recently have we begun to see the older circulation data as evidence of typical monsoon behavior for large martian dust storms.

---

[1] Until his death, Dr. Seymour L. Hess was a highly respected and noted authority on the meteorology of Mars. His work, which matured while he was in the Department of Meteorology at Florida State University, earned him the position of principal investigator and team leader for the Viking Meteorology Science Team. {Ed.}

[2] Prior to the detailed knowledge of Mars afforded by Mariner 9 and Viking data, martian clouds were often categorized according to color. Yellow clouds were believed to be dust clouds close to the surface, and therefore most indicative of near-surface meteorological conditions like wind. Mariner 9 and Viking observations dramatically demonstrated that this assumption was correct.

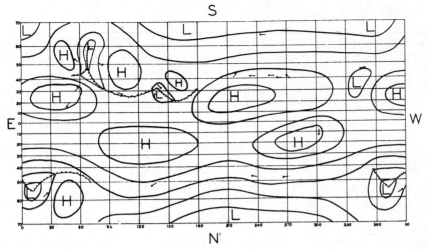

A schematic streamline map for Mars in northern hemisphere winter. This is the telescopic view with south at the top. To obain the usual meteorological view, merely invert the page. The arrows represent observed cloud drift directions.

Figure 1G-1: Early Map Based on Temperature and Wind Data by Seymour Hess (J. Met., 1950)

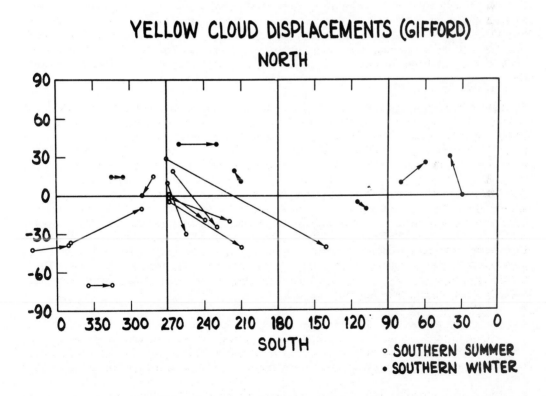

Figure 1G-2: Early Map of Wind Drift Measurements

135

General Comparison: Mars and Earth

As a reminder of those qualities in which a meteorologist must be interested when comparing Earth and Mars, Figure 1G-3 is a listing of comparative data. One of them is the size of the planet. Mars is of course much smaller than Earth, being only a little more than half of Earth's diameter. The length of a martian year is almost twice that of ours, a fact that is not as crucial for meteorology as are other aspects of the martian orbit. The seasonal variations are important because Mars' axial tilt is large -- almost identical to that of Earth's. The martian surface gravity is three-eighths that of Earth's. At Mars' greater distance from the Sun, sunlight averages only about forty-four percent (44%) of the incoming solar radiation on Earth. The length of a martian day ($24^{hr}38^{min}$) is very similar to that of an Earth day. For this reason, we might expect the martian diurnal cycle to be somewhat similar to that of a terrestrial desert, as indeed it is. Finally, the surface pressures are very different, with a mean pressure on Mars of only seven millibars compared to roughly one-thousand millibars (one bar or one atmosphere) at Earth's surface.

SIMILARITIES:		
	EARTH	MARS
ROTATION Length of Day (sec)	86400	88640
SEASONS		
Obliquity	23.5°	25.0°
Solar Orbit Period (Earth days)	365	687
OPTICAL THICKNESS		
Visible Transmissivity	60%	90%
IR Transmissivity	30%	80%
FROUDE NUMBER f^2d^2/gH	11	6
ATMOSPHERIC DEPTH Scale Height	7.5	10

DIFFERENCES:		
ATMOSPHERIC MASS (Heat Capacity)	10^4 kgm/m^2	140 kgm/m^2
SURFACE THERMAL PROPERTIES	Maritime	Continental
TOPOGRAPHY	Moderate, Regional	Large, Global
RADIATIVE TIME SCALE $f\tau_r$	300	30

Figure 1G-3: Comparison of Earth and Mars

Another important general characteristic of Mars that is relevant to the meteorologist is that the martian orbit is highly elliptical. The amount of radiation received near perihelion is about forty percent (40%) larger than the amount of radiation received from the Sun at aphelion. This variability in solar heating produces a long, cool northern summer and a short, relatively hot southern summer. This proximity between perihelion and the southern summer solstice is characteristic of the current phase of Mars' long-period orbital cycle, which varies with a period of approximately 10^5 years. The precession of perihelions through that cycle produces important climatic cycling on Mars. The major global dust storms, which I will describe in detail later, tend to occur during that part of the cycle near perihelion when Mars is receiving maximum solar radiation.

Martian Meteorology as Revealed by the Mariners

The global dust storms are Mars' most famous meteorological phenomenon. When **Mariner 9** began orbiting the planet, the meteorologists were extremely happy because they had the spacecraft camera system to themselves. Mars was engulfed in a massive global dust storm and nobody could see the surface. The only visible features were some curious looking (apparent) mountain tops that correlated with features observed earlier by Mariners 6 and 7 (highlighted in the **Mariner 7** photo presented in Figure 1G-4[3]). Points O, P and Q are the Tharsis Montes, but pictures taken during these 1969 flybys afforded inadequate resolution and did not allow the features to be recognized for what they were.

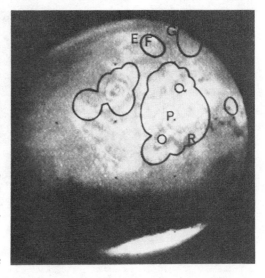

Figure 1G-4: Mariner 7 Image of Mars

Olympus Mons and the tallest mountains of Tharsis protruded through the higher layers of dust and were about the only surface features that could be seen by **Mariner 9** as that spacecraft began its orbital mission. Because there was little else to look at, they and their associated atmospheric features were closely studied. As the atmosphere slowly cleared, there was no longer any question about what the features were. Framed by the dust in the atmosphere, their summit calderas clearly identified them as giant volcanoes. The streak-like features extending southward and from one of the summits seen in Figure 1G-5 were actually atmospheric wind streaks in the lee of a volcano, and were an indication of the very strong prevailing northerly winds at the time.

Other kinds of meteorological evidence were provided by **Mariner 9**, and one of the possibilities suggested was the probable existence of terrestrial-type storm systems on Mars. Seymour Hess had speculated on the possibility of the occurrence of such phenomena, and Yale Mintz

---

[3] Mariners 6 and 7 flew past Mars within a five-day period during the last week of July and the first week of August, 1969. The pictures they acquired of the Tharsis region were generally from more distant approach ranges (low resolution). The most striking of these were frames 7F73 and 7F74, taken by **Mariner 7** on August 4, the first to reveal Olympus Mons (outlined "bull's eye" feature) and the Tharsis Montes, although with insufficient resolution to determine what they were. It wasn't until **Mariner 9**'s mission that brightnesses associated with these features were recognized to be mid-day clouds forming along elevated crests. {Ed.}

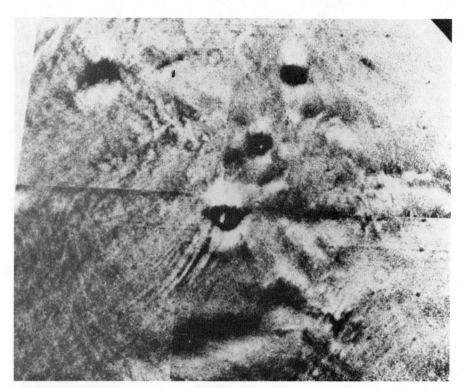

Figure 1G-5: Early Mariner 9 Image During Global Dust Storm

Figure 1G-6: Progress of Cold Front Imaged by Mariner 9 (each mosaic 1500 km across)

[1976[a]] later did a detailed analysis of the possibility. The implication of Mintz's work was that weather systems (such as fronts embedded in moving cyclones) were likely in the middle latitudes during winter.

Figure 1G-6 (opposite page) presents a series of **Mariner 9** pictures (taken Feb. 11-14, 1972, with longitude and latitude lines included for 45°N, 145°W). They illustrate the progress of a feature that looks very much like a terrestrial cold front. The front is the bright band extending from the lower left to the upper right just above the middle of the central mosaic. On the second day of this sequence, there is evidence of an intense dust storm associated with strong northerly winds in the lower-left portion of the frame. On the third day of the sequence, a large crater rim is seen to be producing wave clouds (presumably composed of water ice) that resemble a sonic boom's shock wave. The clouds were being produced by extremely strong low-level winds passing over the crater, and the day-to-day variations are indicative of day-to-day weather changes and frontal systems. These weather extremes were observed by **Mariner 9** following the global dust storm described earlier.

Viking Meteorology Instrumentation

Figure 1G-7 is an illustration of the primary **Viking** lander meteorology instrument. The instrument package was located at the end of an arm-like boom that was equipped with an elbow hinge so that the instrument could be deployed from its stowed position after landing; it is in fact clearly visible in some of the **Viking** lander pictures taken at the surface of Mars. The device consisted of thermocouple units to measure the atmospheric temperature, and wind speed sensors. Each lander also had an atmospheric pressure sensor which, because it had to be shielded from wind, was mounted inside the lander rather than in the boom assembly.

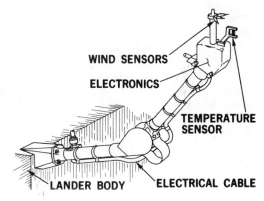

Figure 1G-7: Viking Meteorology Sensors

Measuring wind speed required the use of two orthogonally oriented, heated sensors, together with an unheated reference-temperature sensor. The power required to maintain a constant overheat of the two orthogonal sensors relative to the reference is measured, and from this the cooling due to wind--and thence the wind speed itself--can be determined. One cannot uniquely determine the wind direction with two orthogonal sensors, so a thermocouple array (hot-film anemometer) was used to determine the wind directional characteristics. It worked very much like a wet finger: heat it, expose it to wind, and then measure which side is cooler. Wind speed could also be crudely determined from the magnitude of the cooling. This meteorology package was mounted on each of the landers, and, for the most part, they performed quite well on the martian surface[4].

---

[4] The **Viking** landers were so well instrumented that some of the instruments did double duty, helping out other areas of science when possible. Another thermocouple sensor was mounted on one of each lander's foot-pads. Its primary function was to help complete an atmospheric temperature profile during the landing and then to support physical properties science by measuring temperatures at the surface boundary or in the soil if the footpad was buried (as it was with VL-1); this sensor was also helpful to meteorology. It was also found that the seismometer on VL-2 (VL-1's seismometer failed to uncage and could not be used) could provide some rough but reliable data on winds. {Ed.}

Figure 1G-8: VL-1 Panorama at Chryse Planitia Revealing Deployed Meteorology Boom

Initial Viking Meteorology Results

The picture presented in Figure 1G-8 (above) provokes a strange emotional response for those of us who worked on the meteorology team. This remarkable picture was produced by camera·1 three days after touchdown; the boom had been similarly imaged in the camera·2 panorama automatically acquired and transmitted within minutes of the historic landing by the first **Viking** lander (VL-1). it shows our instrument package deployed on the surface of Mars at the end of its one-meter boom.

First Day -- Figure 1G-9 (*a* through *d*) is a representation of the meteorology data that VL-1 produced for us that first day (VL-1/Sol 1)[5]. VL-1 landed July 20, 1976, at a time comparable to 4:13 in the afternoon at the site. The lander touched down at 22.27°N, 47.97°W in a basin known as Chryse Planitia and began acquiring data almost immediately. The first reduced meteorological data were available to us that same day at 5:30 pm, PDT. Seymour Hess then gave the following martian weather report[6]:

> *Light winds from the east in the late afternoon,*
> *changing to light winds from the southwest after midnight.*
> *Maximum winds were 15 miles per hour.*
> *Temperature ranged from minus 122 degrees Fahrenheit*
> *just after dawn to minus 22 degrees Fahrenheit.*
> *Pressure steady at 7.7 millibars.*

The data produced reflect the temperature (Fig. 1G-9a), wind speed (Fig. 1G-9b), wind direction (Fig. 1G-9c), and pressure (Fig. 1G-9d). These are of course the most conventional meteorological quantities, but *un*conventional instrumentation was required to be able to measure them under conditions occurring on Mars. No one knew with certainty exactly what those conditions would be, of course, and the instruments therefore had to work in an environment that could only be theorized in advance.

---

[5] A "sol" is a martian day; VL-1/Sol 1 is the first day of VL-1 operations (the day of its landing). {Ed.}

[6] Taken from <u>Viking: The Exploration of Mars</u>. NASA EP-208 [JPL 400-219 5/84; U.S. GPO: 1984-784-698]. {Ed.}

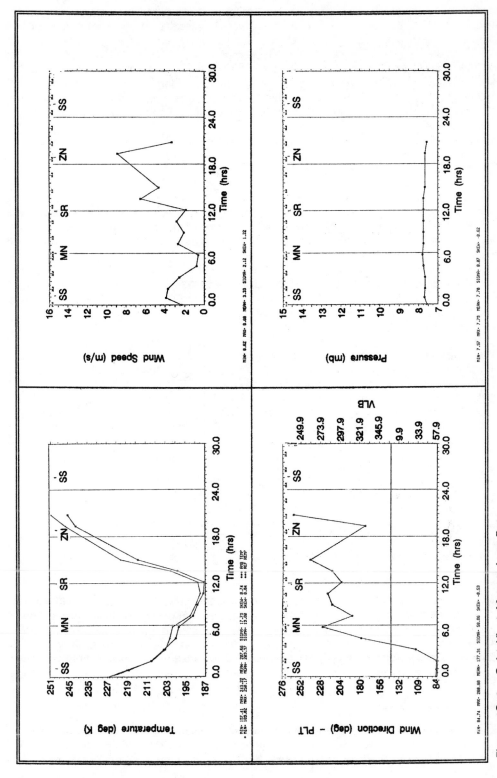

*Figure 1G-9: Sol-1 VL-1 Meteorology Data
(top, left/right) a. Temperature, b. Wind Speed, (bottom, left/right) c. Wind Direction, d. Atmospheric Pressure*

141

Wind Behavior -- The VL-1 landing occurred during northern summer at its latitude (the martian subtropics). Winds were found to be light but increasing in the afternoon. A diurnal rotation of the strongest winds was noted, with the prevailing winds coming from the west to southwest (more or less following the local topographic slope). This wind behavior is similar to the way winds tend to behave in the western great plains of the United States. During the summertime when southerly winds tend to prevail on the western plains, they are influenced by topography -- particularly by a kind of thermal low pressure area produced by heating over the uplands. The near-surface winds will tend to follow the topographic slope, with low pressure (uplands) on the left.

OVERVIEW OF VIKING MISSION RESULTS

We didn't know how long VL-1 would last. Technically, the landers were designed for only a three-month mission, but VL-1 continued to operate for the better part of four Mars years (more than six Earth years) at the Chryse site[7]. The extended mission time made it possible to develop extremely valuable long-term profiles of the martian meteorological processes and systems.

Long-Term Pressure Profiles and Site Comparisons

The data shown in Figure 1G-10 demonstrate the character of surface pressure variations over the martian seasons and clearly illustrate the cycling of the atmosphere between the polar caps. The minimum in the pressure cycle occurs during southern winter when the CO_2 mass condensed onto the south polar cap is at its maximum. Later, as the seasonal CO_2 sublimes out of the south polar cap, the pressure rises until the north polar cap starts to form. A much smaller north cap forms, as indicated by the weak minimum in atmospheric pressure during the northern winter period, and the CO_2 then begins to reform at the south polar cap as winter ends in the northern hemisphere. This cycling sequence repeated without perceptible variation during each annual period observed by Viking, even though there were substantial variations in the characteristics of individual cycles.

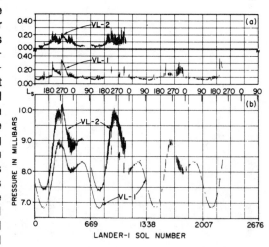

Figure 1G-10: Pressure Variations and Standard Deviations for both VL Sites

There are a number of other interesting features reflected in the long-term pressure curve. The difference in pressures between the two lander sites (Chryse vs Utopia) is simply a question of the elevation difference between them, but one can see that there is a lot of apparent noise (looks like "grass") on the pressure curves. This "spikiness" is *NOT* noise; it is due to

---

[7] The meteorology and imaging systems were the only science instruments left in operation for long-term science, and these systems continued to operate without difficulty until each of the landers was lost (VL-2 on April 11, 1980, and VL-1 on November 5, 1982). {Ed.}

day-to-day variations in pressure. The variations are associated with traveling cyclones of the kind that had been speculated on (as mentioned earlier) by Seymour Hess and by Yale Mintz. They occurred only during the winter season and were detected at both sites, but they were much more prominent at the more northerly VL-2 site (at 47.67°N, 225.74°W) in Utopia Planitia. The intensity of these pressure fluctuations varies from winter to winter.

Atmospheric Tides

The curves presented in Figure 1G–11 illustrate some very useful information. These data represent atmospheric optical depth computed from pressure variations for two different models by Richard Zuruk (solid and dashed lines), as compared with plots (vs time in sols) of optical depth observed at the VL-1 site [Pollack], and illustrate the level of variability within a single day. This measures, among other things, what meteorologists call *atmospheric tides*. Atmospheric tides should NOT be confused with the **gravitational tides** Joe Veverka discussed in his presentation [NMC-1E]. In meteorology, atmospheric tides are those wind and pressure variations that are produced by the daily cycle of heating over the whole atmosphere. What results from the daily loading cycle, among other things, are traveling waves that follow the Sun and have both diurnal and semidiurnal periods.

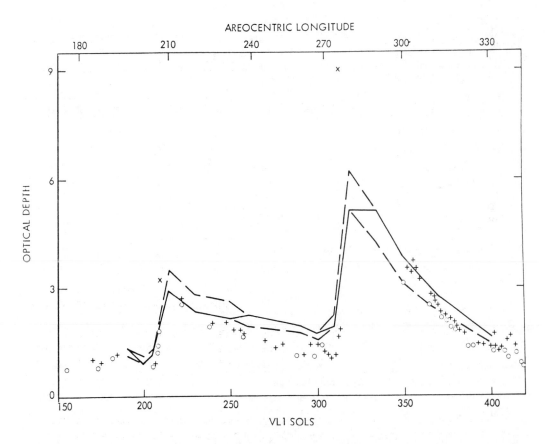

Figure 1G-11: Optical Depth Computed from Pressure for Models in Comparison with VL-1 Data

143

One can see marked variations in the amplitudes of these tides. The variations are reflected in patterns, such as the repeated sharp rise and gradual fall reflected in both the modeled curves and during the first VL-1 year. Such variations are associated mainly with the semidiurnal component of the atmospheric tide. As reflected in Figure 1G-12, one can see in the Viking data (recorded over time by both landers) that there are substantial variations in the atmospheric tides from winter to winter, and that the largest tides (the largest daily variations) occur during the northern winter season -- the period when Mars is closest to the Sun (perihelion). The solid and dashed curves correspond to models with different scattering ratios (forward-to-back) by dust.

MARTIAN DUST STORMS

I want to move now to a discussion of martian dust storms, and in particular to what was learned from various Viking instruments about storm characteristics. I am going to focus on global dust storms. Mars in fact experiences many localized dust storms, and many were observed by the two Viking orbiters. But the most spectacular meteorological events are those large storms during which the atmosphere over essentially the entire planet appears to fill with dust for periods of at least a few weeks or longer.

Atmospheric Dust and Optical Depth

Figure 1G-12, referred to briefly above to demonstrate the amplitude of semidiurnal tides on Mars, represents measurements of atmospheric dust over time (determined from imaging data acquired by both landers). The measurements are presented as atmospheric optical depth, based on data developed by Jim Pollack and his colleagues [1977[b]]. These data demonstrate that there is some dust in the atmosphere at all times; even when the atmosphere was at its clearest, atmospheric dust was strongly in evidence at both lander sites. However, a sharp rise in dust was noted quite suddenly early in the winter season, followed by a gradual fall-off and then another quick rise.

Indeed, the amount of dust in the atmosphere at the VL-1 site increased to such an extent during the first sharp rise that the Sun was obscured and optical depth measurements were for a time impossible. After this, a gradual fall-off occurred and the atmosphere cleared. A similar pattern occurred at the northern VL-2 site, but was not as regular; at this site, dust variations are further evidence for travelling cyclones. The VL-2 data dramatically illustrate a period during which no direct sunlight could penetrate the dust shroud engulfing the site.

Of special interest (in Fig. 1G-12) are the sawtooth features associated with two of the global dust storms, and one of these (for VL-1) is again illustrated in Figure 1G-13. When one compares the feature in Figure 1G-13 with the most similar feature in the VL-2 data represented in Figure 1G-12 (upper part), a great similarity can be seen: the very sharp rise followed by a gradual fall, and then another very sharp rise followed by a gradual fall. The dashed and solid curves correspond to models with different ratios of forward-scattering to back-scattering by dust.

When we examine the semidiurnal pressure oscillation at the VL-1 site during that first year of measurements, we find that the atmosphere becomes dusty very suddenly and that the sudden increase in dustiness over the entire globe absorbs solar radiation, greatly enhancing atmospheric heating and exciting this particular mode of atmospheric response. The heating excites the process so well, in fact, that one can use this pressure oscillation as a measure of the globally varying atmospheric dust (the time-varying, globally-averaged amount of dust).

144

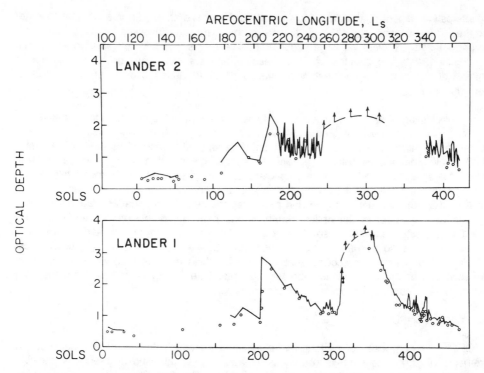

Figure 1G-12: *Relationship Between Semidiurnal Tidal Amplitude and Optical Depth*

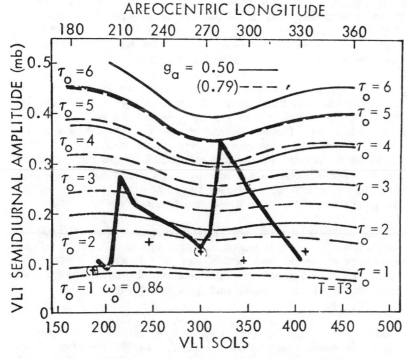

Figure 1G-13: *Detail of "Sawtooth" Variation in Tides*

Another interesting aspect of the data acquired at the VL-1 site shows up in a comparison with data for the following year. The VL-1 data were more sparse the following year, but are sufficient to show that there were no large semidiurnal tidal peaks indicative of global dust storms. This is consistent with the findings of Earth-based observers, who have noted a variability in these global dust storms from year to year. **Viking** provided dramatic evidence in support of those observations.

The theory explaining these oscillations is fairly well understood and has been worked out specifically for the Mars case by Zuruk. He was able to establish a relationship between the solar radiation reaching the surface and the global dust amount measured by the optical depth, τ_d[8]. The semidiurnal tidal amplitude associated with a given value of τ_d varies seasonally because of Mars' elliptical orbit and varying distance from the Sun. The connection between the amplitude of the semidiurnal tidal oscillation and the optical depth of the atmosphere can be seen (Fig. 1G-13). In other words, if one measures the semidiurnal tidal oscillation, one gets a very good measurement of the optical depth. If one plots the global optical depth, inferred from the tidal amplitudes rather than directly from the optical data at the lander sites, one sees variations of the sort I have described with an increase to τ_d = 5 in the second dust storm. Further, if one plots the second annual cycle, in which there was no visual evidence of a global dust storm, the data points representing optical depth, again inferred from the tidal amplitudes, clearly indicate the absence of storm activity.

Year-to-Year Variability

Figure 1G-14 is a composite of all four winter seasons for which the **Viking** landers acquired measurements of atmospheric pressure and its variations. With this kind of data, it is possible to say a little bit more about the interannual variability. The first year--the one with what we can think of as two calibrating storm systems--is represented by the dotted curve. The data points representing the second year show no evidence for a global dust storm, not unlike the third year (indicated by the dashed line) which again gives us no evidence for a dust storm during the northern winter season. But finally, during the last martian year in which measurements were made by a **Viking** lander, there is evidence for an enormous dust storm more intense than anything seen during the first year of **Viking** operations on Mars.

Characteristics -- These year-to-year comparisons provide strong evidence for interannual variability in global dust storms, which is of considerable interest for a couple of reasons. The first is the character of the storms themselves; they are distinguished by a very sharp rise in the optical depth factor when they occur. That is, the onset of these storm systems is extremely sudden, either in the onset of global transport of the dust or in the onset of excitation of the associated semidiurnal tides. Perhaps more importantly, this process suggests something that is of even greater interest to meteorologists. There appears to be a bifurcation (splitting) of the dynamical behavior of the martian atmosphere between two distinctly different atmospheric circulation regimes, one dominated globally by dustiness and one that is relatively clear.

Wind Behavior -- To expand on that a bit, wind hodographs representing diurnal wind behavior at both landing sites are presented in Figure 1G-15 (*a*, VL-1, left) and (*b*, VL-2, right, with two successive years). The way one should look at a wind hodograph is to imagine a wind vector drawn from the origin point to a point that corresponds to wind direction and speed (following the usual directional convention for a compass) for each hour of the day.

---

[8] Optical depth τ_d is defined such that $\exp(-\tau_d)$ is the fraction of the incident sunlight that reaches the surface. Thus τ_d = 5 means that the direct solar radiation is attenuated by a factor of approximately 150 as a result of the dust in the atmosphere.

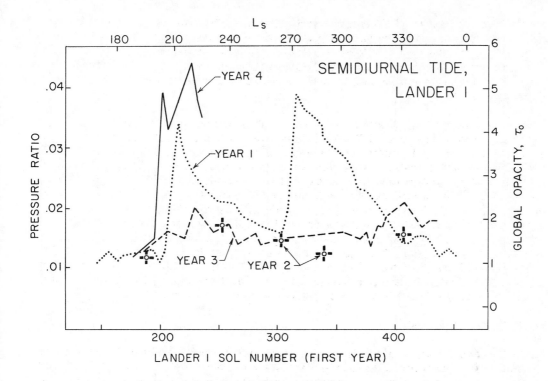

Figure 1G-14: Four-Year Composite of Semidiurnal Amplitude at VL-1 Site

Using VL-1 as an example (Fig. 1G-15a), the hodograph shown in the upper-left panel is representative of the summer season at the time of the VL-1 landing. One can clearly see the prevailing southerly winds, much like those one would see in the western great plains where there would be high elevations and a thermal low to the west of the measurement site. The hodograph in the upper-right panel reflects a seasonal change to autumn and also the period during which the first of the global dust storms occurred. One can see that the wind at that time went through two cycles in the course of a day, tracing out a pattern of two loops. The double-loop signature is evidence of the semidiurnal tide, and the relationship between the wind and the pressure clearly indicates that what we are seeing with the pressure variation is the planetary, westward travelling, semidiurnal atmospheric tides. As that dust storm decayed, the semidiurnal oscillation decreased (as seen in the lower-left hodograph panel). Finally, the largest dust storm announces itself with strong winds and a pronounced double-loop (shown in the lower-right hodograph panel). In this case it is an important point to note that these winds are quite strong, and the tides could be a major contributor to the atmosphere's ability to raise dust. These explanations and interpretations can be used to understand the VL-2 hodographs (Fig. 1G-15b); except that additional panels are used to define wind behavior over two years.

Travelling Phenomena

Another set of features observed in the meteorology data demonstrates the passage of travelling cyclones and frontal systems. As I have indicated, cyclones and fronts were the subject of speculation by a number of observers prior to the **Mariner 9** and **Viking** missions, and the **Viking** lander pressure data provide good evidence for their existence.

By combining pressure, wind and temperature data, wherein one can see day-to-day oscillations in their variables (as shown for VL-2 in Figure 1G-16), the nature of these disturbances can

147

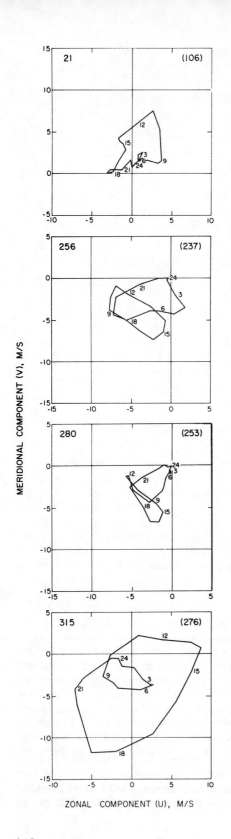

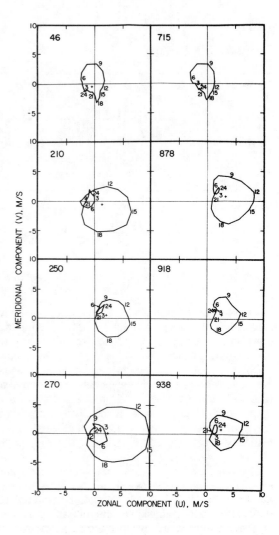

Figure 1G-15:

Wind Hodographs
Reflecting Wind Velocity/Direction
Patterns at Viking Lander Sites

- *Left Column – VL-1*

- *Above – VL-2*

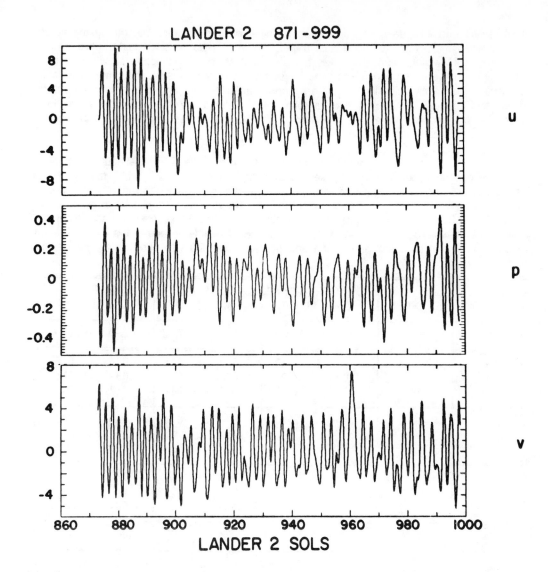

Figure 1G-16: Pressure and Wind Data for VL-2; u = West to East Wind, p = Pressure, v = Temperature Data

be deduced. Note that the maximum negative meridional wind (the maximum wind from the north) precedes just slightly the highest pressure point in each oscillation. That is typically what we see on Earth when systems travel from west to east. It indicates travelling wave cyclones moving from west to east in the strong westerly winds of Mars' northern hemisphere. This is very similar to the behavior of such features in Earth's atmosphere. We can even measure and assess the spatial scales of these martian systems: there are about two to four waves moving around the planet in the westerly stream, and the mid-latitude VL-2 site lies in its path.

Figure 1G-17 is a table of some data that are indicative of the year-to-year variations in these travelling waves, as well as the variation in the tides that I have already discussed. All four years measured by Viking are represented according to season, beginning with: autumn (A) and progressing to winter (W) and spring (S). Winter and spring are combined during that period in which we had little data. Of importance in these data is the fact that during the autumn season, when there was no major dust storm, the VL-2 data revealed fairly active synoptic disturbances. However, when there was a major dust storm, the synoptic activity was totally suppressed. In the spring, when again there was no dust storm present, the synoptic variability rose again. During the following year, the synoptic variability was large throughout the autumn, winter, and spring. The same pattern is reflected in data at the VL-1 site, but it should be noted that the synoptic variability measured at the VL-1 site was smaller than that recorded at the VL-2 site.

The third year (during which there was no dust storm) is quite easily distinguished because of the very large synoptic variability seen that year at the VL-1 site, while the fourth and last year (which produced a massive storm) shows practically no synoptic variability whatsoever. One can see the dust storm characteristics in both the VL-1 and VL-2 tidal amplitudes, and the anticorrelation between the tidal and synoptic amplitudes. These characteristic features of interannual, meteorological variability suggest a bifurcated regime. In one of the regime branches, the semidiurnal tides are very strong and probably contribute to dust-raising themselves.

YEAR	SEASON	SYNOPTIC RMS AMPLITUDE		DIURNAL AMPLITUDE		SEMIDIURNAL AMPLITUDE	
		VL-1	VL-2	VL-1	VL-2	VL-1	VL-2
1	A	.008	.017	.022	.018	.036	.010
	W	.003	.007	.031	.020	.039	.014
	S	.007	.020	.016	.011	.014	.009
2	A	–	.020	.019	.014	.015	.011
	W,S	(~.015)	.021	.020	.013	.016	.006
3	A	–	–	.024	–	.020	–
	W,S	.018	–	.027	–	.021	–
4	A	(~.004)	–	.024	–	.044	–

A = Autumn, W = Winter, S = Spring

Figure 1G-17: Table of Pressure Variations

As they become very strong, the atmospheric circulation (as a whole) intensifies except for the synoptic features; these travelling cyclone waves were suppressed, at least at the latitudes where we measured them.

This leaves the question of the initiation mechanism for the global dust storms unanswered. Why, for example, do they occur in some years and not in others? One is also led to many new questions connected with the occurrence of these systems which must reflect major factors in modifying the martian surface through wind erosion and particle transport. One is led to wonder, for example, about how global dust storms, atmospheric tides, and synoptic waves may have varied in the past with different orbital parameters or different mean surface pressures.

Meridional Circulation

Figures 1G-18 and 1G-19 demonstrate another aspect of the dust storm/bifurcation phenomenon by schematically illustrating the average circulation in the meridional plane. The general flow is from the summer pole when CO_2 is subliming to and condensing on the winter pole. On Earth, there is a familiar rising motion in the tropics and a descending motion in the subtropics, and there is a meridional flow pattern that connects them. The schematic drawings in Figure 1G-18 depict the typical speeds for the horizontal component of this circulation in meters per sec (m/s). There is a strongly-seasonally varying circulation rather than one centered about the martian equator, such that in the summer the air rises near the subsolar point in the southern hemisphere subtropics and crosses the equator to a point where it can descend. Instead of a two-cell circulation (as on Earth), it is more of a one-cell circulation with a strong descending motion or subsidence in the winter hemisphere.

Figure 1G-19 includes four panels representing meridional cross-sections. The panel in the upper-left corner represents zonal mean wind speed (again in m/s), with east-west being positive, and the upper-right panel depicts pressure on a horizontal plane near the surface. The lower-left panel is meridional wind (southward is positive); note the strong northward motion near the surface and strong southward motion aloft. By mass continuity, these flows imply a rising motion in the southern hemisphere (left) and a sinking motion in the northern hemisphere (right). Temperature data in degrees kelvin is given in the lower-right panel, in which a tendency for symmetry can be noted. The high temperatures given for the winter hemisphere (right) are due to compression in the southern hemisphere. The calculations are for an atmosphere in which the circulation is driven strongly by dust heating in the southern hemisphere.

There is evidence for very remarkable expansion and strengthening of this meridional circulation system as dust is raised during a global dust storm. The large scale meridional system evidently intensifies during such storms, extending up to 50 kilometers or so in altitude and to high latitudes beyond VL-2 (i.e., beyond 50° N). This intensification appears to be contributing to the raising of dust during such periods.

This is a very interesting phenomenon, particularly when we can see it occurring in other planetary atmospheres. We think of Mars almost as a laboratory experiment with respect to this process. It is an experiment in which the atmosphere responds to the amount of heating, the amount of heating depends on the amount of dust in the atmosphere, and the amount of dust in the atmosphere varies with the season and from year to year. From this perspective, it seems that Mars turns its own dial up or down to control the alternation of its circulation regimes.

A significant impact of this enhanced meridional circulation is that the same kind of feature has been observed in terrestrial general-circulation model studies, and has been demonstrated in simulations of the **nuclear winter** phenomenon. Given enough atmospheric absorbing material (smoke and dust) to strongly and sufficiently absorb solar radiation, a greatly enhanced meridional circulation analogous to that observed on Mars can develop quite suddenly on Earth. The nuclear winter simulations suggest a bifurcation between weakly absorbing and strongly absorbing circulation regimes.

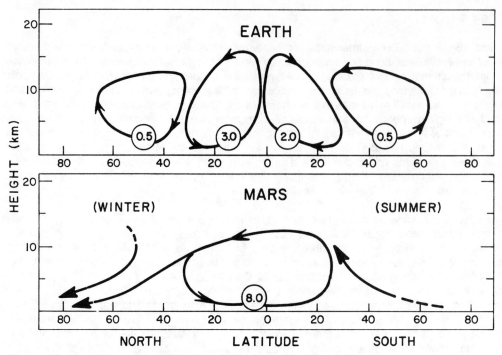

Figure 1G-18: Schematic of Meridional Winds in Meridional Plane

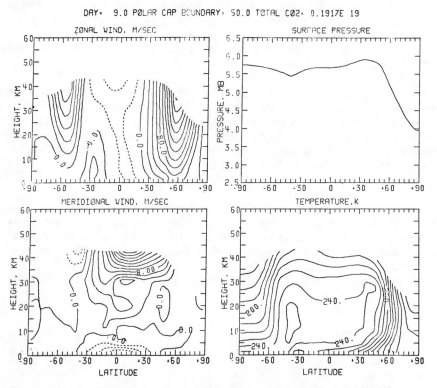

Figure 1G-19: Meridional Plane Cross Section Calculations [Haberle]

CONCLUSION

There is much that we would yet like to learn about the meteorology of Mars. In particular, of course, there is considerable interest in the cycling of water vapor between that atmosphere and the surface. The **Mars Observer** mission represents, perhaps for the first time, an orbiter mission that is ideally suited to making synoptic atmospheric measurements. Its data products will allow us to look at Mars in a manner that will facilitate good transport measurements, and will allow us to begin to assess the sources and sinks for water vapor. ■

REFERENCES

References identified here by alphabetically sequenced lower-case characters are keyed to *specific* references in the presentation text where the same characters are used as identification tags. In the text, however, the identifier will be found within brackets, either by itself or with a name and/or credit date (i.e., when formally published) -- e.g., [ª] or [1978ª].

This space is otherwise available for notes.

~~~~~~~~

UNSPECIFIED REFERENCES

Hess, S.L. 1973. Martian winds and dust clouds. *Planet. Space Sci.* 21:1549-1557.

Hess, S.L., Henry, R.M., Tillman, J.E. 1979. The seasonal variation of atmospheric pressure on Mars as affected by the south polar cap. *J. Geophys. Res.* 84:2923-27.

Hess, S.L., Henry, R.M., Leovy, C.B., Ryan, J.A., and Tillman, J.E. 1977. Meteorological results from the surface of Mars: Viking 1 and 2. *J. Geophys. Res.* 82:4559-74.

Hess, S.L., Ryan, J.A., Tillman, J.E., Henry, R.M., and Leovy, C.B. 1980. The annual cycle of pressure on Mars measured by Viking 1 and 2. *Geophys. Res. Lett.* 7:197-200.

Leovy, C.B. 1977. The atmosphere of Mars. *Sci. Am.* 237:34-43.

Leovy, C.B., and Mintz, Y. 1969. Numerical simulation of the weather and climate of Mars. *J. Atmos. Sci.* 26:1167-90.

Leovy, C.B., and Zuruk, R.W. 1979. Thermal tides and martian dust storms: Direct evidence for coupling. *J. Geophys. Res.* 84:2956-68.

PRESENTATION REFERENCES

a.   Pollack, J.B., Leovy, C.B., Mintz, Y.H. and Van Camp, W. 1976. Winds on Mars during the Viking season: Predictions based on a general circulation model with topography. *Geophys. Res. Lett.* 3:479-82.

b.   Pollack, J.B., Colburn, D., Kahn, R., Hunter, J., Van Camp, W., Carlston, C.E., and Wolf, M.R. 1977. Properties of aerosols in the martian atmosphere, as inferred from Viking lander imaging data. *J. Geophys. Res.* 82:4479-96.

THE WATER AND OTHER VOLATILES OF MARS
NMC-1H

Dr. Fraser P. Fanale
University of Hawaii

Some of the volatiles believed present on Mars, water in particular, are not adequately accounted for by what has been found in the atmosphere or on the surface. Water should be abundant on Mars and, in fact, the summertime residual (permanent) northern ice cap is known--on the basis of temperature--to be composed of water ice even though plated over with an extensive veneer of CO_2 frost part of the year. The southern ice cap differs by being too cold to be water ice once its own plate of winter CO_2 has sublimed, inferring that the residual cap is itself composed of frozen CO_2. However, there is reason to believe that the southern cap is a periodic rather than permanent feature, triggered by some as yet unknown phenomenon and sustained over a period of time by a self-preservation process. Layered terrain of ice and dust at both poles indicates a cyclic, climatic process over time, produced by large oscillations in Mars' obliquity and eccentricity -- much as a similar but less pronounced cycle has produced glacial and interglacial periods on Earth. No other reservoirs of water ice have been detected on the surface of Mars to help explain where the predicted water might be, but Viking data include evidence for the probability that much of the missing water is in the martian regolith -- with the most significant amounts located poleward of latitudes ±50°. Numerous features indicative of permafrost between those latitudes and the ±30° latitudinal bands have also been identified. A variety of models have been developed to study the possible emplacement and distribution histories of water ice or the location of brines. It seems clear, in any case, that the dry martian surface almost certainly harbors significant amounts of water and other volatiles humans may one day need.

INTRODUCTION

Planetary volatiles are, after all, very important. They can change the melting temperature of the interior of a planetary object by a very large factor. They can effortlessly and naturally combine to create a self-replicating molecule, and can then almost completely control the delicate environment on the surface of the solid object from which they have been degassed to create an atmosphere-surface interface environment where the progeny of those replicating molecules can cavort and evolve. Of course, the level of activity of the local star is also critical, but the volatiles are capable of taking a perfectly reasonable amount of solar insolation and turning the environment into an inferno via a gas greenhouse effect or by obscuring the star's light with suspended dust. Finally, volatiles can destroy an entire civilization as punishment for the excessive use of underarm spray[1]. Planetary volatiles are, therefore, of great interest.

VISIBLE MARTIAN VOLATILES

Figure 1H-1 is a picture in which martian volatiles can be seen at the north pole of Mars during the spring. The central portion of this cap is a massive deposit of water ice that remains throughout the year. Plated over the permanent H_2O ice cap in the picture, making the north polar cap appear much more extensive than the water ice cap itself, is a thin veneer of CO_2 frost. The CO_2 cap is receding at this time of year (spring) and will disappear entirely in the summer. These volatiles are particularly important because they can be seen. That is, they aren't the product of a calculation by a theoretician who says that they should be there, hidden from sight in the regolith. These volatiles can actually be seen and identified, and we **know** that they are there.

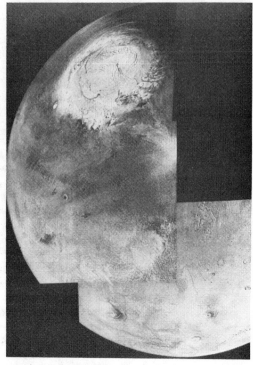

Figure 1H-1: Ice Cap at Mars' North Pole

The Polar Caps

Although some of the things I will be talking about are difficult to see in a picture of this type (Fig. 1H-1), we can identify no fewer than three different polar cap units. The first is a broad, thin plating of CO_2, which is a seasonal cap and appears only during the winter (at each pole) in the form of a surface deposition of atmospheric CO_2. In the north, the second polar cap unit is a centrally-located, permanent residual cap of water ice that I will discuss and illustrate shortly. The south pole's residual cap differs, however, and I will explain those differences as well. The north residual cap is visibly smaller than the seasonal CO_2 cap, but it is in fact far more massive than the superficial seasonal plating of CO_2. The third polar cap unit at the north pole is the most massive of all. It underlies much of the area shown in Figure 1H-1, extending down about 10 degrees, and it is comprised of a series of overlain fossil caps composed of various amounts of water ice and dust; it is known as the "layered terrain."

Figure 1H-2 is a picture of the residual north polar cap as seen during the summertime. It is relatively hot at that time -- almost as hot as North Dakota in January! In fact, it is far too hot (>200°K) for any of the wintertime CO_2 visible on the surface in the previous picture (Fig. 1H-1) to exist. That winter veneer of CO_2 has sublimed back into the atmosphere, leaving behind what basically appears to be a small, warm and dirty water ice cap. The residual water

---

[1] Chemical reactions between volatiles in the atmosphere and industrial fluorocarbons (used, for example, as pressurizers and air conditioning coolants) are known to be having an alarming and destructive impact on upper atmosphere ozone (producing "holes" in the ozone layer above the poles). The ozone is the atmospheric component that protects life on Earth from ultraviolet radiation, which is known to cause skin cancer. {Ed.}

ice cap has a volume of ~3×10^7 km^3, which is a small but not trivial fraction of what we believe to be the Mars surface H_2O inventory [Carr, 1986[a]].

Features representative of the giant cap that underlies the whole region around each of the poles are portrayed in Figure 1H-3. The reason such topography is called *layered terrain* is clear in such pictures. This evidence suggests that such terrain is comprised of a complex series of layers representing varying mixtures and amounts of ice and dust. This is very important for several reasons. The most important is that it is excellent evidence of a cyclic process occurring on Mars; indeed, it is the only evidence for a cyclic process on Mars involving varying climatic conditions (H_2O pressure, CO_2 pressure, dust storms, et·cetera). We are convinced that this kind of cyclic process is almost certainly the result of oscillations in the orientation of Mars' axis. Since a similar process with similar amplitude and period is responsible

Figure 1H-2: Residual North Polar Cap

for climate change on Earth, causing the glacial and interglacial periods, the exposed evidence of the process on Mars is of great importance. Furthermore, if the H_2O ice in the layered terrain of both poles could be spread over the entire planet, its depth alone would be in excess of 10 meters; this amount is much greater than the inventory represented by the north residual cap alone but is still small compared with the inferred H_2O inventory for Mars.

Figure 1H-3: Layered Terrain Associated With North Polar Cap

Differences at the South Pole

The residual (summertime) south polar cap is pictured in Figure 1H-4. This polar cap is apparently composed of CO_2 that remains largely in place throughout the year and is much colder (even during its summer season), thus giving it characteristics that differ significantly from the composition and behavior of the north polar cap. When the winter CO_2 plate goes away, for example, the residual cap still exhibits temperatures that are consistent with frozen CO_2 being in equilibrium with the atmosphere -- i.e., ~150°K. So, this obviously raises a rather challenging interrogative, which I shall call "stinker" number 1: Why is the residual north polar cap made of water ice while the residual south polar cap appears to be made of CO_2?

A quasi-acceptable answer to that question is contained in the hypothesis I am now going to discuss, but it has to be worked through bit by bit. Note that the interior of the polar cap (Fig. 1H-4) is somewhat darker than near its edge where it receives more insolation. The theory holds that in areas where the cap is brighter, the deposition of CO_2 that formed it also originally contained the same proportions of dust that darkens the interior ice. However, exposure to greater solar radiation allowed the dust particles near the outer perimeter of the polar cap to get hot relative to the frozen CO_2 matrix, such they sank into it -- thereby decontaminating the surface and causing it to brighten. As it brightens it reflects more of the solar radiation back into space, preventing sufficient warming to sustain CO_2 sublimation [Paige, 1986[b]].

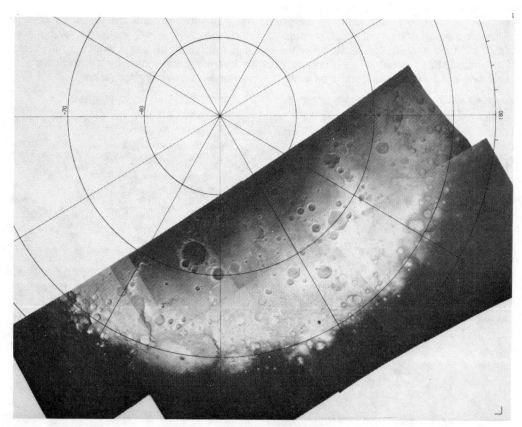

Figure 1H-4: Residual (summertime) South Polar Cap

160

The evidence, therefore, appears to represent a negative feedback effect through a kind of bootstrap process, in which the ice of the south polar cap survives and essentially protects its own existence by developing a high albedo. This, of course, is the opposite of the more familiar phenomenon known as the "Boston snowball," wherein dirty water ice sublimes and leaves a coating of dust on the surface. So, in terms of a philosophically bothersome but physically reasonable explanation, we can say that CO_2 is not residual at the north pole quite simply because it is not consistently cold enough there. Further, we can say that the south cap is significantly colder than the north cap largely because it is brighter, and that it is brighter because it is made of CO_2.

However, this explanation does not address the issue of explaining how the situation came about. How did it come about? Perhaps my intuition is wrong, but it seems apparent in these pictures (without doing any calculations) that the current martian residual caps are not indicative of a stable situation. If the present circumstances represented a stable polar situation, one that persisted throughout much of the planet's history, the water in the residual north cap would have migrated to the south pole. I think that is abundantly clear. This has not happened, how-ever. And, because of the albedo differences, it is reasonable to conclude that the current polar conditions represent a rational situation -- but one that is metastable rather than stable. That is, it is colder at the south pole today simply because it is brighter today, and it is brighter today because the ice is a mixture of CO_2 and dust rather than H_2O and dust, which is due to the fact that it is colder, et·cetera, et·cetera.

We have, therefore, a metastable situation, and there are a couple of reasonable explanations for how it came about. One explanation is facilitated by a stochastic phenomenon, such as some physically explicable thing like a dust storm obscuring the pole. In that case, the CO_2 is depos-ited on the polar cap a few years ago and maintains itself there for a period of time -- and then goes away. However, we know of major climatic cycles that have resulted from changes in obliquity and eccentricity on Mars [Ward et al. 1974c]. At some point, via this scenario, there is perhaps a repeated phenomenon that involves the appearance--at least for a time--of a CO_2 cap at the south pole. That initial CO_2 cap then establishes the conditions favorable to its own temporary preservation. This might be a preferable explanation, philosophically, since it does not involve an isolated individual event. And, no matter what the explanation, the south polar cap cannot be a permanent feature; if it were, it would have "cryopumped" the entire north polar cap down to its own polar region by now.

The Layered Terrain

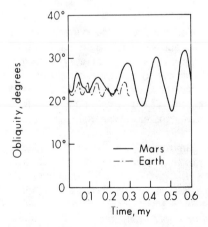

Figure 1H-5: Obliquity vs Time (Earth/Mars)

As I indicated previously, the process respon-sible for the various layers in the layered terrains is particularly interesting in its own right; it is almost certainly associated with large oscillations in the obliquity and eccen-tricity of Mars. In comparison, we know that Earth also has experienced such oscillations, albeit on a much smaller scale, and Figure 1H-5 illustrates the nature of these obliquity oscil-lations for the two planets over time.

Although the oscillations are smaller in Earth's case, the consequences have been quite severe in terms of their affect on Earth's climate because of positive feedback effects. Indeed, the consequences have been significant enough

161

to have been recorded in temperature profiles, deep sea cores, and oxygen isotope temperatures, such that they can be related to the onset of glacial and interglacial ages [Imbrie, 1979[d]]. In the case of Mars, the oscillations are very much larger and more dramatic. In addition, the record of these oscillations is laid out bare in the martian layered terrain for our examination. This offers us an opportunity to better understand the nature of climate change on both planets by studying its record in the "fossilized" layered terrain on Mars. Therefore, this is a legitimate example of what we refer to as comparative climatology.

To simplify the explanation of what happens when Mars' axis oscillates as I have described, I would like to start with what we know to be an UNtrue condition for the model of Mars: that the north polar cap is a great mass of frozen CO_2. If that were so, the polar insolation would increase with each cyclic increase in axis obliquity; as the polar temperature went up, the CO_2 vapor pressure (the atmospheric pressure) would then increase. Then, whenever the obliquity decreased, it would get colder at the pole and the vapor pressure would drop. These factors would then oscillate back and forth in this fashion with each cycle [Ward et al. 1974].

In truth, however, not only is the residual north cap too warm to be CO_2, Mars does not have a massive residual CO_2 cap at the south pole. That cap is believed to contain only a millibar or so of CO_2. It is, in fact, somewhat discontinuous, and its mass is nothing like that of the residual north polar cap. Clearly, then, the exact process I have described does not happen. On the other hand, there is good evidence to suggest that something like it happens, bringing us to yet another challenging interrogative. "Stinker" number 2 is: How can the atmosphere be buffered by a cap that amounts to only twenty percent of the total mass of the atmosphere?

The paradox to which I refer is that the theory given above seems to suggest the existence of a very unnatural system. We frequently deal with systems in which a block of a volatile (in its condensed phase) is contained in association with a certain amount of gas, and in which a change in the temperature of the condensed volatile buffers the gas phase pressure because the gas pressure follows the vapor pressure curve. What makes this system "unnatural" is that the tiny fraction of condensed volatile known to exist in it can't function as a real buffer because it can be made to go away too easily. In other words, the buffer is much less massive that the phase it is buffering. This is not, therefore, the kind of buffer that exists in nature. So, the question in another sense might be: why did God put just enough CO_2 on Mars to build the atmospheric pressure up to a certain value, with just barely enough left over to form a condensed phase that mimics a buffer but can't serve as one?

Effects of Obliquity Oscillation

The answer to this problem is contained in a description of what really happens when the obliquity oscillates, and it involves the planet's regolith. To achieve equilibrium for 5 mb of atmospheric pressure at the temperature of the martian regolith, based on laboratory measurements and other considerations, the deep soils of Mars would have to contain a certain predictable amount of adsorbed CO_2. Of course, it isn't exactly predictable because we do not know what kind of mineralogy exists deeper in the regolith, but the Viking lander data provide a reasonably good generic basis for extrapolating to what that mineralogy might be.

As a function of the regolith's predicted thickness, the total amount of CO_2 that can be hidden in the adsorbed-CO_2 "ocean" within the regolith is very much greater than that contained in the martian atmosphere [Fanale and Cannon, 1971[e]]. Moreover, a great deal of it is exchangeable when thermal waves caused by the obliquity cycle penetrate to a certain depth (several hundred meters), possibly even more so when solar variations force the thermal effect to a much greater depth (several kilometers). The characteristic depth of thermal wave penetration will of course be proportional to the square root of the period.

Because a lot of regolith CO_2 can be mobilized on the high side of the obliquity cycle, it will somewhat mimic the kind of pressure increase one might expect to see when heating an imaginary polar cap of CO_2 [Fanale and Cannon, 1978[f]]. This kind of result is not quite as dramatic as the process I described earlier. Figure 1H-6 shows the affect of obliquity on atmospheric CO_2 pressure due to some kind of feedback mechanism with the dust, as Conway Leovy suggests [NMC-1G]; it presents the theoretically predicted variation in the Mars atmospheric CO_2 pressure resulting from the obliquity variations. The actual variation may be even greater than shown in the figure, and higher pressures may be achievable at high obliquity as a result of positive feedback effects that are still poorly understood. In truth, we don't really know what happens once this process gets underway, because of the uncertainty represented by possible feedback mechanisms, but it is probably something similar to what I have outlined. On Earth, positive feedback effects greatly exaggerate the climatic affects of obliquity variations, and this could be the case for Mars as well.

It is easier to predict the nature of what happens on the low side of the obliquity cycle because a large CO_2 cap would surely cryopump much of the adsorbed CO_2 out of the regolith. That CO_2 would then plate out in a gigantic polar cap. The pressure would be buffered by the CO_2 as this process evolved, and would follow the vapor pressure curve all the way down to a point where the atmospheric pressure would be less than 1 mb. Nitrogen and argon would then constitute the bulk of the atmosphere.

In this way, then, obliquity oscillation is perfectly capable of producing the layered polar terrains. To begin with, there are almost no dust storms on the low side of the cycle because it is virtually impossible to generate dust storms at such pressures. However, one does not have to progress very far up on the high side of the obliquity cycle before pressure is great enough to make nearly continuous dust storms possible. Clearly, then, these oscillations (not only in the planet's obliquity but in its cyclic manifestation of alternating global dust storms and volatile deposition periods) can very capably produce the layered features we see in the martian polar regions [Toon et al. 1979[g]].

In Figure 1H-7, one can again see the predicted oscillations of water (vapor pressure) at the base of the atmosphere (dotted line), and the enormous variation involves orders of magnitude [Fanale et al. 1986]. The figure shows the expected variation at latitude ±50°. The smaller variation (solid line) shows the expected variation at the buried-ice interface in the topsoil where the atmospheric variations are dampened. However, these actual numbers should not be taken seriously. The problem with such models is that one can formulate these factors very precisely, calculating to five decimal places and incorporating all of the predicted insolation variations and a variety of other variables, for example, and still get into qualitative trouble. Because, in doing so, one inevitably must encounter a modular element of the model wherein the feedback effects related to albedo must be predicted. In fact, we can't even determine the sign of the albedo effect, which is needed to help us find out if the cap's nature is that of a Boston snowball or a self-cleansing bootstrap mechanism.

If we can't determine the sign of the albedo change, it is then very difficult to factor the element it is associated with into a quantitative model with any real reliability. In spite of such weaknesses in models of this kind, however, they are generally illustrative of the kind of process one could expect to see, with enormous water and CO_2 pressure variations occurring cyclically. It is the magnitude of the oscillations during the obliquity cycle that are in question, due to the uncertainty associated with polar albedo change and dust feedback effects.

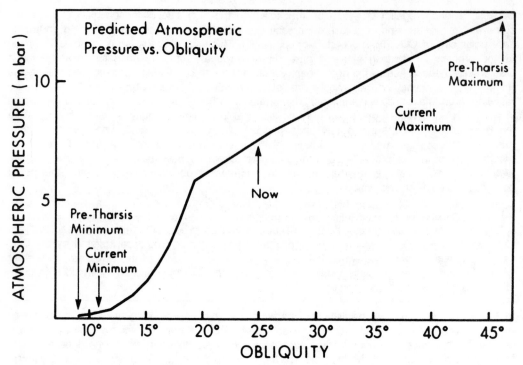

Figure 1H-6: Predicted Atmospheric Pressure vs Obliquity

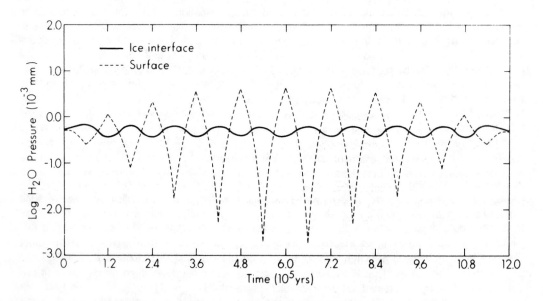

Figure 1H-7: Predicted Oscillations in Water Vapor Pressure (at base of atmosphere)

164

Hidden Volatiles

Perhaps the most interesting part of my subject involves the other 95 percent of the volatile inventory on Mars, believed to be dispersed and buried to an unknown extent in its surface. I will use the next series of figures to suggest what a cross section of the outer few kilometers of the planet might be like internally if we could slice Mars in half from one pole to the other. This is entirely theoretical, of course, because we have no direct knowledge of what the deep regolith of Mars is like and only a superficial knowledge of its surface material. But, while it is not very dependable, this model is about the best we can do.

We start building the model by determining rational, meaningful estimates for the degassed water inventory and soil porosity, and we then try to combine the resulting numbers to produce a reasonable initial condition. The mass of the original ice is then scattered randomly within the planet's regolith so as not to bias the system. This modeling procedure is reflected in Figure 1H-8, wherein a putative initial random distribution of regolith ground ice is assumed at the onset of Mars' history. A computer program is used to predict what would then happen during the course of martian history, based on the model, tracking the redistribution of the ice in response to solar insolation and internal heat flow.

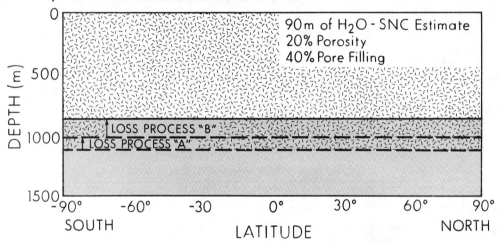

Figure 1H-8: Predicted Redistribution of Ice in Response to Solar Insolation and Internal Heat

The total H_2O inventory is scaled to be consistent with expectations for the total degassed H_2O inventory, based on studies of the SNC meteorites (believed to have originated on Mars) and other cosmochemical considerations. The product is a series of curves, presented in Figure 1H-9, that show a recession of ice from the equatorial zone over time, such that the ice is clearly struggling with finite kinetics to get out of the regolith and migrate to the poles. These contours depict the calculated recession of the top surface of Mars' ground ice as a function of time and latitude. The exact rate of recession of low-latitude ice is stable at latitudes higher and lower than 50 degrees ($\pm 50°$).

While this procedure is sound, it encounters a couple of problems. The first is that the exact position of these curves, as a function of time, depends on what soil parameters are factored into the computer model. That is the catch. How does one determine which set of curves, from among all of those representing different soil permeabilities, is the right set? There is, of course, no way this particular determination can be made, but it is important to note that the model does predict a qualitatively reasonable process.

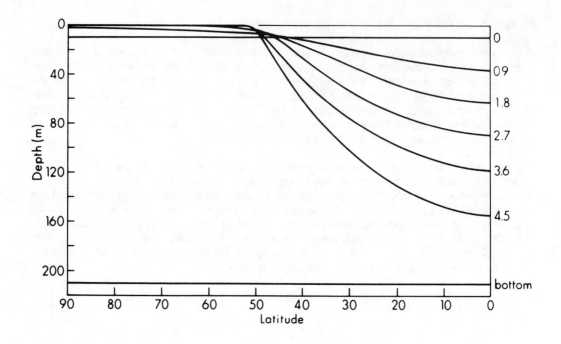

Figure 1H-9: Predicted Recession of Ice From Equatorial Zone Over Time

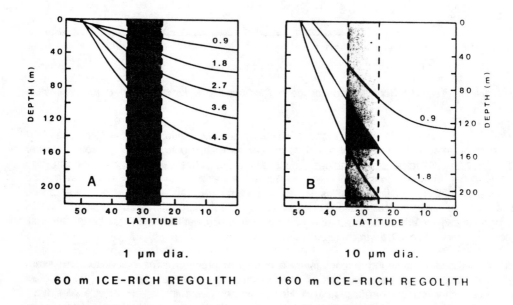

1 μm dia.   **10 μm dia.**

60 m ICE-RICH REGOLITH   **160 m ICE-RICH REGOLITH**

Figure 1H-10: Permafrost Distribution Model [Fanale et al. 1986]

There is, in fact, a way to get at the soil permeability via an empirical approach. However, we must first explain one seemingly puzzling aspect of both the regolith CO_2 buffering model and the H_2O redistribution model shown in Figure 1H-9. This is another of my series of challenging interrogatives, "Stinker" number 3: Why is all the CO_2 adsorbed into regolith when it should be drawn to the poles where it is colder? Indeed, the general behavior of volatiles is such that they are normally drawn to the coldest cold trap. Similarly, why does it appear that much of the water ice we have scattered about the planet is going to stay put for eternity, and not just because it is slow? Figure 1H-9 clearly indicates that the deep subsurface water (poleward of latitude ±50° in each hemisphere) just doesn't want to go to the poles. Why?

The answer to the first part of this problem is that CO_2 molecules want to be in the regolith at a higher temperature because CO_2 has a stronger affinity for the surface crystal field of the rocks than for other CO_2 molecules in a dry-ice cap. The adsorption of CO_2 on other CO_2 molecules is called condensation. But, when adsorption takes place on the surface of rock material, the interaction of the CO_2 is stronger with the rock than with its own kind. For this reason, it occurs on a higher temperature adsorbent substrate in equilibrium with a lower temperature condensate.

The second part of the problem is more complicated, and I obviously can't use the same explanation. A piece of ice is a piece of ice! But, while one can assign a temperature to the poles and say that they are colder than other places, in the real world they get surprisingly warm during the summertime. At that time, a great mass of water molecules spurts forth from the north pole and struggles across the disc of the planet through the atmosphere, although it isn't clear where they end up: the regolith, the south pole, et-cetera [Jakosky, 1981[h]]. Because of the nonlinearity of the water vapor pressure curve, the time-weighted average H_2O pressure, as "felt" by a pore deep in the soil is the same as if the cap were always at a much higher temperature than it really is most of the time.

The average water pressure experienced by a block of ice buried a couple of meters down makes the ice behave as though the poles are at a temperature of 205°K all the time, which isn't the case. This explanation defines 205°K as the average (virtual) temperature of the mythical poles for the Mars model being used. If all of the factors claimed work as defined, the results confirm that most of the ice buried at latitudes poleward from ±50° would in fact remain in place pretty much as previously suggested.

The pore (permeability) issue is not easy to deal with because we do not know the properties of the deep soil. However, a problem one of my colleagues was involved in led to some estimates. The work was focused on some troughs at a certain latitude in Elysium, some of which appeared to be filled with material at a certain depth while others were not. My colleague foolishly took one of my models seriously and tried to figure out which set of these curves would give him the right answers for both cases (filled and not filled), representing different latitudes and depths for the time that the filling took place.

This work is reflected in Figure 1H-10 (opposite page), which represents an attempt to fix the unknown parameters in the model using empirical geomorphological studies. The latitudinal distribution of filled and unfilled troughs of various depths in Elysium are recorded, and an attempt is made to correlate the filling of troughs with the presence of ground ice (which causes fluidization) in some cases, and the lack of filling with its absence [Mouginis-Mark, 1984 [i]]. Fortunately, and perhaps somewhat surprisingly, a reasonably respectable result was produced by the work. If a reasonable, effective pore size of about 1 μm was assumed, the edge of the global ice wedge would be near the boundary between the filled and unfilled troughs.

There also has been some applicable work performed by Squyres and Carr [1986 j] in which a global geomorphological study revealed numerous differing kinds of "softening" features, possibly indicating the presence of more fluid ground related to the existence of ground ice. These features are typically concentrated at latitudes higher and lower than 30 degrees (±30°), which is in general agreement with the model.

The fact that these features are restricted in latitudinal occurrence presumably indicates that sub-surface rigidity is less outside the ±30° latitude band. That would be in harmony with the soil parameters inferred from the Elysium study. Even though there is actually a distribution of pore sizes rather than a specific pore size in the soil, and even though there are all kinds of intercalated sills, buried volcanoes, and other such subterranean features that make the result less than realistic, these morphological studies allow the derivation of an "effective" pore size that can be inserted into the reference model. Thus, the overall model is based on basic physics, but contains a "fudge factor" designed to make it generally consistent with the observations.

Figure 1H-11 illustrates what then happens with the model, and it appears to be almost--but not quite--to where one might expect it to be. The first thing one sees is the recession of the ice (the kinetics), demonstrating how the low latitude ice moves to the poles. It is interesting and important to note that there is about the same amount of ice missing in this exchange as is needed to make the layered terrain. It may also be of interest that the magic number (representing the soil pore size) one needs to put the ice wedge in the right place isn't an unrealistic number, being something like a couple of microns. In summary, 4.5 billion years appears to be about long enough to allow initial low latitude ice to migrate to the poles, but ground ice at higher and lower latitudes than ±50° is stable and in equilibrium with the north polar cap.

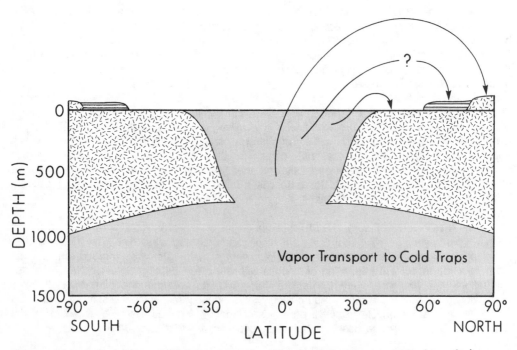

Figure 1H-11: Recession of Ice From Low to High Latitudes (layered terrain, poles)

The next thing one might want to do with this model is determine what the regolith's thermal regime is like and what the location and composition of any existing brines might be, given the composition of the rocks and soil. To work on these issues with respect to those factors already defined, one needs an estimate of thermal conductivity. A gradient must be estimated and an isotherm drawn, such that--within that isotherm--brines could exist. This process establishes a basis for working on the problem, but the nature of Mars makes it difficult. In general, for example, Mars offers plenty of water molecules and there are also sufficient calories available; we know where the water molecules are and we know where there are enough calories to interact with the water to bring about the result we are looking for relative to brine occurrence. The difficulty is that the water and calories have a diabolical tendency not to be in the same place at the same time.

The nature of this model therefore takes the form illustrated in Figure 1H-12. Internal heat flow will operate on the ground ice and determine its state at any given time. It will also affect the rate of transfer of H_2O molecules. In lieu of a better approach, we assume a chrondritic heat flow and a thermal conductivity typical of garden soil, and the position of a resultant contour at $-52°C$ is indicated[2]. Note that an overlap exists between the thermal zone where brines could exist and the zones where water molecules might be abundant. Thus, brine might exist in this kind of buried zone if the water molecules are somehow trapped, such that they do not flow out despite their molten condition. Or, brines might exist in another type of buried region because a batholith was occasionally intruded, such that a substantial amount of water was supplied in one place and hasn't yet had a chance to escape. Intrusives are much more important than extrusives on Earth, and the same may be true on Mars.

Although a viable source of water at any time or latitude, intrusives are not a good heat source for stabilizing brines because the heat goes away much too quickly. However, this concept is useful as a mechanism for emplacing water molecules. It may very well be that some local transient brines could actually be created and stabilized as a product of this process, but only locally and for a short time. Any major brine reservoirs must be consistent with the global model, which is displayed in its complete form in Figure 1H-13. This figure presents the calculated current state and distribution of free water molecules in the martian regolith. The expected distribution of stable ground ice is indicated, as are the zones in which stable or metastable brines might conceivably exist.

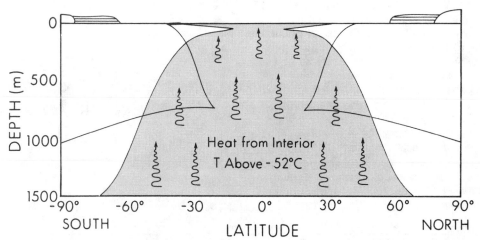

Figure 1H-12: Effect of Internal Heat Flow on Ground Ice (chondritic heat flow, garden soil)

---

2 Brines could be stable on Mars above this temperature, but higher temperatures are likely to be required.

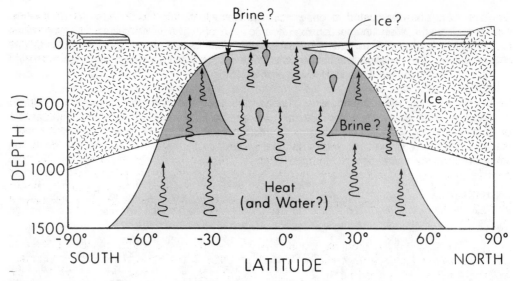

Figure 1H-13: Calculated Current State and Distribution of Free Water Molecules in Regolith

Also still to be resolved is the question of what becomes of the CO_2, nitrogen, and other volatiles that were surely released with the water. Once again the repository may be the martian regolith in which one is likely to find chemically bound volatiles in the soil. Obviously, a volatile like ^{36}Ar would be a "sitting duck" for any atmospheric loss process, but H_2O, CO_2, and other volatiles might possibly be able to hide in the regolith and be at least partially exempt from such loss.

Figure 1H-14 is a picture of a sample that might be the best spectral match for Mars soil found on Earth -- palagonite[3]. The identification unfortunately does not represent something that is clearly and directly indicated, as in the case of simply matching five infrared absorption band centers. The reasoning behind this analogy is more difficult; it is more that nearly all of the other candidates have something wrong with them spectrally. An exception is the fact that a Mg-O-H strech band was also detected, which would be consistent with palagonite [Singer, 1985k]. However, palagonite also makes sense because it is exactly the kind of weathering product one would expect on Mars. This is so because it is a product not of rainfall attack, but rather of the interaction of igneous bodies with permafrost.

Figure 1H-14: Palagonite Sample Particles

---

3 Palagonite is "an altered tachylite, brown to yellow or orange, and found in pillow lavas as interstitial material or amygdules." [taken from A Geologic Guide to the Island of Hawaii, R. Greeley, 1974, and credited to Gary et al. 1972]. It is more generically defined as a weathered basaltic tephra [Terrestrial and synthetic analogs to martian weathering products, 3rd Colloquium on Mars, [Evens et al. 1981]. {Ed.}

On Earth, sandstone is created in one place due to a given set of conditions and properties, limestone is produced in another location due to a different set of conditions and properties, and flocculent precipitates are created in still another location due to another set of conditions and properties. The common factor associated with all of these formations, and which helps to produce them, is abundant liquid H_2O or ion mobility.

On Mars, weathering by vast amounts of surface water is counterindicated. However, the regolith is comprised of rock debris mixed with a significant amount of water ice, and a sill or some other subterranean igneous body could be intruded into a volume of this mixture. The rock could then be essentially allowed to evolve in this isochemical environment, wherein it would be altered to produce a melange of "garbage minerals." This isochemically weathered assemblage, perhaps including phyllosilicates and zeolites, would be so poorly crystallized that one could not get reliable X-ray lines from it. It is impossible to tell a basalt from a palagonite based on an elemental analysis alone, because the chemical composition is conserved during isochemical weathering.

Although such weathering products could have evolved by intrusion into permafrost, there is also independent evidence from intense fluvial dissection of the oldest terrain for a "warmer-wetter" period in Mars' earliest history. It is not appropriate at this time to get into the geomorphological evidence for early abundant water erosion on the surface of Mars, which includes the possibility of rainfall during one or more periods of the planet's early history, because those topics are adequately discussed at this conference [Baker, NMC-1D]. However, the possibility produces my final interrogative, "Stinker" number 4: If there was rainfall during the planet's early history, what could explain it?

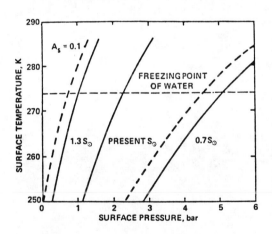

The most popular explanations take into account the possible contribution of a CO_2 greenhouse. Figure 1H-15 was taken from a very recent paper [Pollack and Kasting, 1986[i]], and it shows the surface temperature as a function of atmospheric CO_2 surface pressure for various values of the incident solar flux. The solid curves represent a surface albedo (A_s) of 0.215 and the dashed curves are for A_s = 0.1. The data included also provide various constants to represent the CO_2 greenhouse and a variety of values for the solar constant. There are various arguments that suggest that even though the solar constant may have been lower during that period, particularly favorable orbital and obliquity conditions may in fact have allowed a substantial greenhouse effect to develop periodically.

Figure 1H-15: Surface Temperature for Values of Solar Flux

A martian greenhouse effect sufficient to allow liquid H_2O to exist on the surface would require upwards of a couple of bars of CO_2 pressure, and even that amount would be quite marginal. In rationalizing this amount of Mars atmospheric CO_2, the problem is not so much imagining that one cannot get the required inventory degassed from a Mars that is the parent body of the SNC meteorites; rather, the problem is that a couple of bars of CO_2 is a bit hard to hide in the martian regolith because carbonate has some very deep infrared spectral adsorptions that we have not yet detected.

Even if the regolith is assumed to be quite thick and to contain abundant opaque minerals, and that it is all intimately mixed together by some mechanism, finding a way to hide two bars of CO_2 without seeing the carbonate spectral signature gets to be a very worrisome thing to try to accomplish realistically. In particular, it should instead be possible to detect two bands at $4.0\mu m$ and $3.6\mu m$. Several of my colleagues at the University of Hawaii are working on this problem at the Mauna Kea observatory, so perhaps the bands will be detected[4]. Alternatively, the carbonates may be hidden in a cemented layer deep in the regolith. A third possibility is that the greenhouse was much weaker and only worked together with high early heat flow to raise the water table, but with no stable surface water.

For those who are very skeptical (I can't imagine why) and don't believe us when we say that the perversity of nature has probably hidden these huge amounts of water, CO_2, and other volatiles from our sight on Mars with a diabolical sense of contradiction, one has only to imagine the Viking lander sitting on a surface like that shown in Figure 1H-16, scraping samples from the top few centimeters and leading us to wonder where the ice is. This photo (courtesy of R. Goody) reveals terrestrial permafrost below a thick layer of topsoil. This visible terrestrial evidence underlines the fact that no matter how great the inventory of ground ice might be, the summer seasonal thermal wave can effectively cause a net annual sublimation down to a depth of tens of centimeters or meters (depending on the soil's thermal conductivity), thus depleting ice in the topsoil.

Figure 1H-16: Terrestrial Permafrost Below Layer of Top Soil

---

[4] While editing this presentation manuscript, Dr. Fanale was informed that two of his colleagues had detected a weak feature centered between 3.8 and 3.9 μm, slightly short of the expected position for $MgCO_3$. Laboratory studies are being conducted in an effort to determine if it is in fact due to carbonate. [Blaney, D., and McCord, T.B. 1987. **Mars: An observational search for carbonates.** Submitted to J. Geophys. Res.] {Ed.}

CONCLUSIONS

There is in fact a real danger in what one assumes based on the most obvious data acquired by a spacecraft, as the early **Mariners** have shown. Our conclusion from the analysis of data acquired by a lander sitting on a surface like the one shown in Figure 1H-16 might well be: "There's no water on this planet, I can tell you that!" We have an intuitive response to barren looking landscapes like those we have seen on Mars, and Mars is indeed a very dry looking planet. It is therefore important to remember that the dry martian surface is almost certainly harboring significant amounts of water and other volatiles humans may one day need. As in the case of many terrestrial deserts, hot and cold, where water often wells up to within only inches of the parched soil that overlies it or exists as permanent ice just below a cold, rocky, desiccated surface, the fact that much of the water we think exists on Mars can't be seen does not mean it isn't there. We just have to dig a little deeper. ∎

REFERENCES

References identified here by alphabetically sequenced lower-case characters are keyed to *specific* references in the presentation text where the same characters are used as identification tags. In the text, however, the identifier will be found within brackets, either by itself or with a name and/or credit date (i.e., when formally published) -- e.g., [ª] or [1978ª].

This space is otherwise available for notes.

~~~~~~~~

a.   Carr, M.H.   1986.   Mars: A water rich planet?   *Icarus*.   86:187, 217.

b.   Paige, D.A., Kieffer, H.H., and Stephens, J.B.   1985.   Nonlinear frost albedo feedback on Mars: Observations, models and experiments.   *Trans. Am. Geophys. Union*.   66:945.

c.   Ward, W.R., Murray, B.C. and Malin, M.C.   1974.   Climate variation on Mars 2: Evolution of carbon dioxide and polar caps.   *Jour. Geophys. Res.*   79:3387-3394.

d.   Imbrie, J.   1982.   Astronomical theory of Pleistocene ice ages: A brief historical review.   *Icarus*.   50:408-423.

e.   Fanale, F.P. and Cannon, W.A.   1971.   Adsorption of the martian regolith.   *Nature*.   230:502-504.

f.   Fanale, F.P., and Cannon, W.A.   1978.   The role of the regolith in determining atmospheric pressure and the atmosphere's response to insolation changes. *J. Geophys. Res.* 83:2321-2325.

g.   Toon, O.B., Pollack, J.B., Ward, W., Burns, J.S., and Bilski, K.   1979.   *Icarus*.   44:552-607.

h.   Jakosky, B.M., and Farmer, C.B.   1982.   The seasonal and global behavior of water vapor in the Mars atmosphere: Complete global results of the Viking atmospheric water detector experiment.   *J. Geophys. Res.*   87:2999-3019.

i.   Mouginis-Mark, P.J.   1985.   Volcano/ground ice interactions in Elysium Planitia, Mars.   *Icarus*.   64:265-284.

j.   Squyres, S.W., and Carr, M.H.   1986.   Geomorphic evidence for the distribution of ground ice on Mars.   *Science*.   231:249-252.

k.   Singer, R.B.   1985.   Spectroscopic observation of Mars.   *Adv. Space Res.*   5:59-68.

l.   Pollack, J.B., and Kasting, J.F.   In press, 1987.   Partitioning of carbon dioxide between the atmosphere and lithosphere of early Mars.   *Icarus*.

THE BIOLOGICAL QUESTION OF MARS
[NMC-1I]

Dr. Norman H. Horowitz
Professor Emeritus
California Institute of Technology

The study of Mars was for a long time dominated by one man, Percival Lowell. He created an earth-like Mars -- a dream world that survived into the space age. Ultimately, and chiefly through spacecraft, the real Mars was revealed as a planet that once had running water on its surface but that now is made extremely inhospitable to life by its dryness and radiation flux. The Viking mission found no organic matter in the martian soil at the parts-per-billion level, and the surface material is chemically reactive -- probably due to OH radicals formed in the atmosphere by solar UV. We have learned many lessons from the exploration of Mars, the most important of which is the uniqueness of Earth.

INTRODUCTION: THE LOWELLIAN MARS[1]

For most of the 20th century, the study of Mars--especially its biological aspects--was dominated by the ideas of one man. That man was Percival Lowell. Lowell died in 1916, and with him died the martian canals and the civilization that he imagined had built them. These particular features of the Lowellian fantasy were controversial, even in his day, and, although they were embraced enthusiastically by the public, they had no scientific standing. These Lowellian features essentially disappeared from view following Lowell's death except in the popular literature.

Lowell's interest in Mars did not end with the heroic martians, however. He was interested in all aspects of Mars, and he used the observatory he founded with single-mindedness and passion to study the planet during the last 22 years of his life. Indeed, he left--as his legacy--a physical image of Mars that survived well into the space age. Its major elements were: first, polar caps composed of water ice; next, the semi-annual transfer of water vapor from one pole to the other, concurrent with a wave of darkening that spread across the planet. In Lowell's mind, this wave demonstrated the growth of vegetation. Finally, Lowell's Mars had an atmospheric pressure of 85 millibars.

These Lowellian inventions not only survived him but took on new life as, one after the other, they were apparently confirmed and reconfirmed by another generation of planetary observers.

---

[1] This presentation was based on the author's book: To Utopia and Back: The Search for Life in the Solar System. W.H. Freeman. 1986. Utopia Planitia, which inspired the name for Dr. Horowitz's book, is the martian region in which Viking lander 2 (VL-2) landed on September 3, 1976. {Ed.}

(Lowell's name was rarely mentioned in connection with these confirmations, and then only in a patronizing manner). The story of these supposed validations of what we now know were illusions, reported by well-known scientists using modern instruments, is one of the strangest in the history of modern science. I will not review the painful record here, and will simply say that it is hard to avoid the conclusion that self-deception played an important role in this monumental series of blunders. Without exception, their effect was to make Mars appear more Earth-like and therefore more hospitable to life than it actually is.

Whatever caused it, the misinformation had large consequences. It guaranteed that the generally accepted concept of Mars, as the space age dawned in 1957, would be a fiction. This fictional image influenced the United States planetary program in important ways. First of all, it underlay the decision to give the highest priority to the search for martian life. It also influenced the selection of experiments for the Viking landers and strengthened the case for a planetary quarantine program policy [a]. Finally, it explains the remarkable fact that NASA's historic series of martian expeditions (three flyby spacecraft, three orbiters, and two landers) was completed within a mere 20 years of the launch of Sputnik 1.[2]

THE DE-LOWELLIZATION OF MARS

It was not until 1963 that the fabric of Lowell's Mars began to unravel with the discovery by Kaplan, Münch and Spinrad [b], during one clear night on Mt. Wilson, that the martian surface pressure was considerably less than 85 millibars. In his book, *Mars as the Abode of Life*, published in 1908, Lowell had applied photometric arguments to the surface pressure problem and arrived at an estimate of 64 millimeters of mercury -- or 85 millibars.

In the years following Lowell's death, a dozen or so further attempts were made to apply photometric and polarization methods to this problem. The results were reviewed by G. de Vaucouleurs in his influential book, *Physics of the Planet Mars*, the English edition of which appeared in 1954 [c]. In it, de Vaucouleurs concluded that the best value of the surface pressure was 85 (±4) millibars, in agreement with Lowell's estimate. A panel of the Space Science Board reviewed the evidence again in 1962 and concluded: "It is unlikely that the true surface pressure differs by as much as a factor of 2 from 85 millibars."

In fact, as we now know, the surface pressure differs from 85 millibars by a factor of more than 10. The errors had their origin in a variety of unverifiable assumptions that had to be made before pressure could be derived from photometric and polarimetric data. For example, the amount of particulate scattering in the martian atmosphere had to be assumed. The estimate of Kaplan, Münch and Spinrad, on the other hand, was based on the absorption lines of CO_2, and from these the martian pressure could be calculated by known physical principles without recourse to arbitrary assumptions. The Kaplan, Münch and Spinrad estimate was 25 millibars. A large error was attached since it was based on a single infrared plate. This plate and the resulting estimate of atmospheric pressure marks the beginning of the de-Lowellization of Mars.

By 1969, just six years later, an avalanche of new observations on Earth and by spacecraft all but completed the job. The mean surface pressure was found to be six or seven millibars -- in an atmosphere dominated by CO_2. Even the winter polar hoods were determined (pre-**Viking**) to be frozen carbon dioxide, not water ice. There was no seasonal transfer of water across the surface of the planet. The wave of darkening that seemed to suggest the movement of

---

2 Flyby Missions: **Mariner 4**, July, 1965, **Mariners 6/7**, July/August, 1969; Orbital Missions: **Mariner 9**, 1971-72, **Viking** orbiters 1 and 2, 1976-79; Surface Investigations: **Viking** landers 1 and 2, 1976-82. {Ed.}

water vapor became much less interesting when it was discovered that the spectral evidence thought to indicate organic matter in the darkening regions was actually produced by deuterated water in Earth's own atmosphere rather than plant life on Mars. By 1970, for these reasons, Mars presented so unpromising a biological picture that it would have been hard to rationally justify a biological emphasis for the landers being developed for the **Viking Project.**

THE REAL MARS

A radical reversal took place in 1971-72, however, with the orbiting of **Mariner 9**[3]. **Mariner 9** showed that Mars had been geologically active in the past, with surface features clearly indicative of extensive volcanism and running water. Evidence of liquid water, even if in the remotest past, improved the biological outlook considerably. If conditions were so temperate as to permit water to exist on the martian surface at some time, then life may have originated as well. If so, it was conceivable that it still survived. The probability was not high, but in matters of this sort a·priori judgments carry little weight when an empirical test can be performed. It was decided that the **Viking** landers would, after all, emphasize biology.

Centuries of myth-making on the subject of Mars and its inhabitants came to an end 10 years ago with the **Viking** landings (July 20 and September 3, 1976). Each lander carried six instruments involved in the search for life: two cameras, a gas chromatograph-mass spectrometer (GCMS) for identifying organic matter in the martian surface, and three experiments (packaged in a single instrument) designed to detect the metabolic activity of microorganisms in the soil. The microbiological experiments assumed--reasonably I think--that martian life would be carbon based. That these instruments gave us a plausible and consistent answer to the question of martian life was due not only to the proverbial skill of the **Viking** scientists and engineers but also to a lot of good luck, as I will shortly explain.

Viking Results

As everyone must know by now, neither the cameras nor the GCMS found evidence of martian life. As far as the cameras are concerned, this was no surprise. Microbial life was the only form of life most of us believed even remotely possible for Mars. However, the failure of the GCMS to find organic matter in the martian surface material was a different matter. This instrument was capable of detecting organic compounds at abundances as low as parts-per-billion. To find not even a trace of organic carbon at this level of detectibility was a real surprise.

Expectation for Organics -- Because Mars is close to the asteroid belt, it was thought that even a lifeless Mars would have accumulated enough meteoritic organic matter over the ages to register a signal in this ultra-sensitive instrument. In addition, it had been shown that carbon dioxide and water vapor exposed to ultraviolet wavelengths shorter that 300 nanometers can yield simple organic molecules on surfaces. It seemed possible that, over time, significant formation and polymerization of these molecules might have occurred in the martian surface. I well remember that one of the problems that worried us at JPL during the years preceding the **Viking** landings was whether we would actually be able to tell the difference between

---

[3] **Mariner 9** went into orbit at Mars on November 14, 1971, and operated successfully there for 349 days. It observed a massive global dust storm upon arrival, but ultimately transmitted 7329 images as a component of 54 billion bits of science data. **Mariner 9** exhausted its ACS gas on October 27, 1972, such that it was unable to maintain solar orientation for power replenishment -- thus ending its mission on that date. {Ed.}

biologically produced organic compounds and the meteoritic or other non-biological organic matter we fully expected to find.

Biological Implications -- The GCMS discovery of no organic matter on Mars was the single most important result of the Viking mission. This finding immediately changed the terms of reference of the biology investigation. It seemed to me at the time, and I said so at a press briefing, that with the GCMS result the three microbiological experiments had lost their purpose. With no detectable organic carbon in the martian surface, it was no longer a question of searching for life in the sands of Mars, but rather one of reconciling a lifeless Mars with whatever chemical activity these experiments might observe. As it turned out, the results of two of the experiments reinforced and helped explain the GCMS finding.

The Viking Biology Experiments

The two experiments I am referring to were Lowellian in concept. That is, they were designed for a Mars with liquid water on its surface. Both incubated martian soil samples in an aqueous solution of organic compounds. These experiments could not be run under martian conditions but had to be heated and pressurized to prevent the liquid from freezing or boiling. Before the mission, I had opposed sending two aqueous experiments to Mars. It seemed to me that just one would be sufficient to cover the remote possibility that liquid water somehow survived on the planet. In its wisdom, however, NASA did not take my advice, and the two Lowellian experiments were made a part of the biology package. This turned out to be a truly serendipitous decision, because these instruments agreed in showing that, in the presence of water, the martian surface is chemically active. Their unanimity on this point is important.

Gas Exchange -- One of the instruments, the gas exchange experiment, mixed a sample of martian surface material with a solution of nutrient in a pressurized chamber at a temperature of about 10°C. The production and disappearance of gases in the chamber were then monitored. Before the sample was actually wetted, however, it was sealed in the chamber with the nutrient solution--but separated from it--so that the soil sample was first exposed to water vapor only (at a pressure that had not been seen on Mars for many millions of years).

A surprising effect was observed. The sample released four gases: nitrogen, argon, carbon dioxide, and oxygen. The first three of these gases appeared in small amounts, suggesting displacement of adsorbed gases by water vapor. But the quantity of oxygen evolved was too large to be accounted for in this way. Thus, while oxygen comprises only 0.1 percent of the martian atmosphere, it made up 7.4 percent of the gas evolved in the gas exchange experiment at the Chryse Planitia (VL-1) site. The production of oxygen at the Utopia Planitia site was less but still highly significant. Preheating of the surface sample did not abolish the gas production, confirming its non-biological origin.

Oyama and Berdahl [1977[d]] pointed out at the time that the production of oxygen implied the presence of peroxy compounds in the martian soil. Neither Oyama nor Berdahl, nor anyone else on the Viking biology team, was yet aware that a source of such compounds had already been proposed by D. M. Hunten [[e]]. Hunten had been interested, as had other aerologists at the time, in the stability of CO_2 in the martian atmosphere. CO_2 is photolyzed to carbon monoxide (CO) and O in the martian ultraviolet flux, as Mike McElroy explains in these proceedings [NMC-1F]. And, in the steady state, CO is expected to be the dominant form of atmospheric carbon, obviously a different state of affairs from that which actually exists.

Clearly, a strong oxidant must be reoxidizing CO in the martian atmosphere. Hunten's calculations indicated that the oxidant is the hydroxyl radical, as McElroy showed, formed by the photolysis of water vapor. Further gas-phase reactions lead to hydrogen peroxide (H_2O_2) and then to hydrogen superoxide (HO_2). All three species can diffuse into the very dry martian soil where they would destroy organic matter. Their presence there would, in addition, explain the production of oxygen in the gas exchange experiment.

Labeled Release -- The other aqueous experiment carried to Mars was the labeled release experiment. In this case, the martian sample was wetted with a solution containing radioactively labeled organic compounds. The appearance of radioactive gas, effectively CO_2, was then monitored. There was no provision for exposing the sample to water vapor before wetting it, but the volume of solution initially injected was made too small to wet the entire sample. The unwetted portion of the sample thus came into contact with water vapor only during the first phase of the experiment.

Within moments after the addition of the solution there was a surge of radioactive gas. When it ended, the radioactivity in the head space was approximately that expected if just one of the 17 labeled carbon atoms used as nutrients in the solution had been converted to CO_2. The carbon atom most likely to be released from the various nutrients is that of formic acid, which is easily oxidized. When the evolution of gas had nearly ceased, a second shot of solution was added to the chamber. This time, no gas was released.

This silence of the instrument was eloquent, like that of the *dog in the nighttime* in the classic Sherlock Holmes story: "'But the dog did nothing in the nighttime,' said Dr. Watson. 'That was the curious incident,' remarked Sherlock Holmes." If microorganisms had caused the gas production, the second injection of medium should have yielded more gas. Its failure to do so must have meant that the oxidant in the unwetted portion of the surface sample was destroyed by a reaction with water vapor.

Aqueous Experiment Results Related -- The two aqueous experiments apparently discovered the same oxidant. Clearer agreement between the two experiments could hardly have been imagined. They differed in one respect, however; heating the martian sample destroyed its activity in the labeled release experiment but not in the gas exchange experiment. It is possible, of course, that different substances in the martian surface are responsible for the destruction of organic compounds on the one hand and the release of oxygen on the other. But I prefer the explanation advanced by Oyama.

Oyama notes that the chamber of the gas exchange experiment was kept open during the heating of its sample, with a stream of helium running through it, while the labeled release chamber was kept closed when its sample was heated. The GCMS data had shown that up to one percent of their weight in water is evolved from martian samples during brief heating to several hundred degrees centigrade[4]. Retention of this water vapor in the labeled release experiment but not in the gas exchange experiment may explain the observed difference between these two experiments.

There is reason to believe that the chemical reactivity detected in the martian surface by these experiments is only a fraction--perhaps a small fraction--of what would be found in uncontaminated (pristine) martian samples. In their descent to the surface of Mars, both Viking landers moved across ground (with their retro engines firing) that was later sampled. The ex-

---

[4] The molecular analysis experiment (GCMS) contained a carousel of small ovens in which samples of approximately 100 milligrams were packed. The samples were heated to differing temperatures to vaporize specific organic compounds (resulting vapors are carried into the gas chromatograph column by a CO_2 carrier gas, where organic compounds of similar nature are grouped for delivery to the mass spectrometer to be ionized and recorded). {Ed.}

haust gases included 0.5 percent water vapor and about 30 percent ammonia. It seems inevitable that this exhaust plume would have reacted with peroxide compounds or other oxidants in the surface material. Indeed, we are lucky that enough remained to be detected. A prime goal of any future Mars lander should be to examine the chemistry of pristine surface material.

Pyrolytic Release -- The third microbiological experiment, called pyrolytic release, is the one for which I was responsible. With it, George Hobby, Jerry Hubbard, and I attempted to perform a biological test under actual martian conditions. The plan was to expose a martian surface sample to martian atmosphere enriched with labeled CO and CO_2, and then to measure the incorporation of radioactive carbon atoms into organic matter in the sample. The incubation was to be at martian pressure and temperature in a martian atmosphere, using simulated martian sunlight. After 120 hours, the sample would be analyzed for labeled organic matter. These conditions were in fact met, with but two exceptions.

Owing to heat sources within the spacecraft, the temperature of the experiment ranged from 8°C to 26°C, whereas the actual ground temperature outside was below 0°C for both missions at both landing sites. Since martian equatorial temperatures reach 25°C, however, the chamber temperatures were not altogether unmartian. In addition, the light source of the experiment was not the Sun, but a xenon lamp that simulated the Sun's radiation at Mars and from which wavelengths shorter than 320 nanometers were removed. The reason for the short-wavelength filtering was to avoid the photocatalytic synthesis of organic compounds I referred to earlier. This departure from martian conditions was, I believe, justified because wavelengths shorter than 300 nanometers are so destructive to organic matter that martian organisms would have to avoid such radiation themselves in order to survive.

Of the nine pyrolytic release experiments performed on Mars, seven showed fixation of trace amounts of carbon into what was presumably organic form. The levels of fixation were below the detection limit of the GCMS, but they were significant on the basis of criteria established in pre-flight laboratory tests. However, the heat stability of the reaction indicates a nonbiological source, as does the fact that subsequent laboratory tests on Earth have produced results similar to those of the **Viking** experiments on Mars[5].

Although it is usually assumed that the product associated with such results is organic, this has not been established. The synthesis of organic matter on Mars would seem to contradict the considerable evidence for the destruction of such matter on the planet. It is possible, however, that one effect of each lander's plume gas, which I discussed a moment ago, was to destroy peroxy compounds on some of the soil grains it contacted, leaving clean surfaces suitable for organic synthesis.

CONCLUSIONS

I'd like to conclude with a few general observations. The **Viking** mission completed the de-Lowellization process started in 1963. It is sometimes said that the results of the mission were ambiguous on the question of martian life. I believe that this opinion is mistaken. Indeed, it seems to me that the results were quite clear, and my reason for reviewing them here was to demonstrate this fact. **Viking** brought to an end not only the age of Lowell but also the long history of the so-called Copernican world view -- that unwarranted extension of the Copernican revolution that held that the Earth occupies no special place in the solar system. It was this

---

5 Such laboratory tests have been conducted using a variety of soils containing iron-rich minerals believed to be similar to the martian soils sampled at both **Viking** landing sites. {Ed.}

view that led Huygens[6], for example, in the 17th century and Kant[7] in the 18th, to the strongly expressed belief that all of the planets are inhabited.

As time went on, however, it became clear that the planets weren't all alike. By the time the space age dawned, only Mars was regarded as an Earth-like body with a reasonable likelihood of indigenous life. Venus had been a mystery because of its cloud cover, but it was eliminated from consideration when its very high surface temperature was revealed. The revelation soon to follow, that the perceived description of Mars was also an illusion, was mind boggling. It had seemed to rest on genuine evidence. Looking back, one can say that this bizarre chapter in the history of science had at least one reassuring aspect; it proved that, given time, science discovers and corrects its own errors.

Viking not only found no life on Mars, it showed why there is no life there. The two sites sampled by the Viking landers were 25 degrees apart in latitude and on opposite sides of the planet. Yet, despite this separation, they were very similar in their surface chemistry (in both organic and inorganic terms). This reflects the importance of global forces in shaping the martian environment -- forces such as the extreme dryness, the pervasive short-wavelength ultraviolet radiation, and the planet-wide dust storms.

Viking found that Mars is even drier than had previously been thought. The highest abundance of atmospheric water vapor measured for a given column of martian atmosphere was about a hundred precipitable microns. This amounts to a pressure at the surface of only four or five microbars, and water vapor was found in this abundance only around the edge of the North polar ice cap in mid-summer. The equatorial latitudes, where temperatures rise well above $0\,°C$ and where--for this reason--conditions were thought to be especially favorable before Viking, were found to be desiccated (the driest part of the planet). The dryness alone would suffice to guarantee a lifeless Mars; combined with the planet's radiation flux, Mars becomes almost moon-like in its hostility to life.

For some, Mars will always be inhabited, regardless of the evidence. One does not have to search far to hear the opinion that somewhere on Mars there is a Garden of Eden -- a wet, warm place where martian life is flourishing. This is a daydream. The Garden of Eden would reveal itself in photographs by a permanent water cloud above it and snow on the ground. Nothing like this has been seen, and it is most unlikely that such a place exists or can exist on Mars.

Another theory holds that the Viking instruments actually found life but that the martian microorganisms are living in the surface at population densities below the GCMS detection limit. This view, unlike the previous one, assumes that martian life does not need water. This is a form of the "blue unicorn" theory, which says that a blue unicorn is living in a cave on the moon. The theory can't be disproved because its author provides the unicorn with what-ever attributes are needed to survive on the moon. On Mars, these would include the ability to live without water or any other solvent, and immunity from the processes that destroy all other forms of organic matter on the planet.

---

[6] Christian Huygens, the 17th Century Dutch astronomer, mathematician and physicist, is best known as the founder of the wave theory of light. However, it was as an astronomer that he drew (November, 1659) the earliest existing sketch of Mars. {Ed.}

[7] Immanuel Kant, the great German philosopher of the 18th Century, is considered to be among the most prolific, wide ranging, and important thinkers associated with the threshold of modern thought. {Ed.}

The failure to find life on Mars was a disappointment, but it was also a revelation. It now seems certain that the Earth is the only inhabited planet in the solar system[8]. We have come to the end of the dream. We are alone -- we and the other species that share our planet with us. Let us hope that the Viking findings will make us realize the uniqueness of Earth and thereby increase our determination to preserve it. ■

---

[8] This topic is discussed at length in the author's book. The conclusion expressed is that even should a water resource exist, as some believe may be the case on Jupiter's ice-surfaced moon Europa, such "bodies of water cut off from the Sun, the ultimate biological energy source, are not suitable habitats [for life]." {Ed.}

REFERENCES

References identified here by alphabetically sequenced lower-case characters are keyed to *specific* references in the presentation text where the same characters are used as identification tags. In the text, however, the identifier will be found within brackets, either by itself or with a name and/or credit date (i.e., when formally published) -- e.g., [ᵃ] or [1978ᵃ].

This space is otherwise available for notes.

~~~~~~~~~

a. Horowitz, N.H., Sharp, R.P., and Davies, R.W. 1967. Planetary Contamination I: The problem and the agreements. *Science* 155:1501-1505.

Murray, B.C., Davies, M.E., and Eckman, P.K. 1967. Planetary Contamination II: Soviet and U.S. practices and policies. *Science* 155:1505-1511.

Sagan, C., Levinthal, E.C., and Lederberg, J. 1968. Contamination of Mars. *Science* 159:1191-1196.

b. Kaplan, L.D., Münch, G., and Spinrad, H. 1964. An analysis of the spectrum of Mars. *Astrophys. J.* 139:1-15.

c. de Vaucouleurs, G. 1954. *Physics of the Planet Mars*. London, Faber and Faber, Ltd.

d. Oyama, V.I., and Berdahl, B.J. 1977. The Viking gas exchange experiments from Chryse and Utopia surface samples. *J. Geophys. Res.* 82:4669-4676.

e. Hunten, D.M. 1974. Aeronomy of the lower atmosphere of Mars: Rev. Geophys. *Space Phys.* 12:529-35.

A REAPPRAISAL OF LIFE ON MARS
[NMC-1J]

Dr. Gilbert V. Levin
Biospherics Incorporated

Dr. Patricia A. Straat
National Institutes of Health

A decade has passed since the first labeled-release (LR) Viking biology experiment produced an astonishing positive response on Mars. But that response was deemed unconvincing when no organic compounds was found. As a result, many attempts have been made to explain the LR data without invoking life. The dominant theory expounded hydrogen peroxide as a chemical agent, suggesting that it reacted with one of the nutrient compounds to mimic a biological response. This theory was tested and essentially disproved on Mars. There is in fact no evidence that it exists on Mars, and even if it formed it would be destroyed by the environment long before it could affect an experiment. We have carefully tested all of the nonbiology theories and have found none to be scientifically adequate. We also verified that the GCMS organic detection sensitivity may have missed very low densities of organic matter. It is now our contention that the survival of the LR data, together with other information not previously considered (including Viking lander image and spectral data that suggest the possible existence of martian lichen), justifies the conclusion that it is now more probable than not that the LR experiment did in fact detect life on Mars.

OPENING COMMENTS AND REVIEW

It is great to get back to the subject of life on Mars. I took issue with the title originally proposed for my presentation[1]--A Second Opinion--because I think all the opinion and philosophy one needs on this topic has already been delivered and it is our intent to stick strictly to the facts in presenting our own conclusions.

It has been a decade now since martian soil was wetted with labeled nutrient (radio-carbon, ^{14}C) in the **Viking** labeled-release (LR) experiment, yielding the astonishing result of a rapid and continuing evolution of radioactive gas over a period of eight days. Figure 1J-1 is a table

---

[1] Because the original title alluded to was incorrectly reflected in the conference agenda, it was not used in this proceedings document. {Ed.}

of the nutrients used in the LR experiment [Levin and Straat[2], 1976]. When the positive signal came back from Mars, we immediately applied the control procedure we thought we would never have to use -- we heated a control portion of the same sample to 160°C (320°F). It produced no response. Therefore, the preflight criteria published for the experiment (blessed and approved by NASA, the National Academy of Sciences, the Space Sciences Board, et·cetera) were fully demonstrated.

SUBSTRATE	STRUCTURE AND LABEL POSITION (*)	CONCENTRATION $(\times 10^{-4}M)$	μCi ML^{-1}	SPECIFIC ACTIVITY (Ci/Mole)
^{14}C-glycine	$NH_3 \cdot {}^*CH_2 \cdot {}^*COOH$	2.5	4	16
^{14}C-DL-alanine	${}^*CH_3 \cdot {}^*CH(NH_3) \cdot {}^*COOH$	5.0	12	48
^{14}C-sodium format	H^*COONa	2.5	2	8
^{14}C-DL-sodium lactate	${}^*CH_3 \cdot {}^*CHOH \cdot {}^*COONa$	5.0	12	48
^{14}C-calcium glycolate	$({}^*CH_2OH \cdot {}^*COO)_2Ca$	2.5	4	16
			34 (see note)	

Note: 6.8×10^7 dpm ml^{-1}

Figure 1J-1: Table of Nutrients Used in Viking LR Experiment

Level of Confidence Prior To and During Mission

In spite of the initial positive results, however, our Viking LR data were deemed unconvincing because of the lack of supporting data from other experiments on the landers, notably the search for organic matter. This reflected a change in scientific disposition, in that during preparations for the mission we had been told by NASA that it had selected the three experiments for the Viking biology package[3] such that each would test a different model for potential martian life. It was understood at that stage, then, that only one of the three experiments might return a positive response, were there truly life on Mars, and that such independent data would most probably be strong enough on its own merit to substantiate the detection of life.

---

[2] Dr. Patricia A. Straat was Dr. Levin's principle colleague during the development, testing, and mission operation/analysis work associated with the Viking biology LR experiment. In addition, she worked with Dr. Levin and others on the Mariner 9 IRIS experiment. {Ed.}

[3] The Viking biology package was remarkably miniaturized. It contained--in one small complex box--three independent experiments designed to detect possible life processes in the martian soil (based on different models of Mars). The package itself, weighing only about 33 pounds and measuring 11.5 x 13.5 x 10.75 inches (with a volume of 1,669 in.^{3}), was built by TRW. In addition to the labeled-release experiment, the instrument also contained gas-exchange and pyrolytic-release experiments, a complex array of plumbing and electrical circuitry, and the necessary mechanics to facilitate soil processing and distribution. Each of the experiments was capable of four cycles, and all of the experiments in both landers were successfully operational on Mars, experiencing few instrument problems and no failures. {Ed.}

Failure to Find Organic Matter -- The molecular analysis experiment, utilizing a gas-chromato-graph/mass-spectrometer (GCMS) specifically designed for the **Viking** mission, failed to find organic compounds in the martian soil. This indeed created a problem for our results, because it is difficult to suggest that life has been detected where no organic matter can be found. With no organic chemistry detected to support the positive conclusions suggested by the LR responses, many attempts were immediately made to explain the results without invoking life.

However, none of these explanations has been adequate to the task over the ten-year period since our data were acquired. Indeed, it is our contention that the survival of the LR data in the face of these attempts to discredit them, together with possible visual (photographic) evidence of martian life produced by the lander cameras (and other data heretofore not considered with respect to this question), now justifies the conclusion that it is more probable than not that the LR experiment did in fact detect life on Mars.

Instrument Performance Integrity -- The **Viking** LR experiment represented a fifteen-year test and development program[4]. It was extraordinarily sensitive and could detect as few as several cells per cc of soil. In hundreds of laboratory and field tests, it never once provided a false positive response.

The test program was concluded with a full-up experiment in a proof-test instrument identical to the one flown to Mars; with the exception that, because we had no martian soil, we instead used some California soil provided by NASA. Prior to the experiment, the soil was exposed for three days to a simulated environment modeled on the best-available experience for what was anticipated on Mars.

Labeled-Release Results on Mars

In Figure 1J-2, representing results of the pre-**Viking** tests using the California soil, a response is seen to rise quickly to about 10,000 counts per minute (cpm) during the eight-day period of a single experiment cycle, which represents the standard cycle period pre-programed into the LR instrument. Elevating the temperature of a duplicate (control) sample to 160°C for three hours prior to testing it produced essentially a negative response. The test results illustrated were produced under simulated Mars environmental conditions with a moderately populated terrestrial soil. However, results were equally dramatic following the actual landing on Mars.

Figure 1J-2: Responses From California Soil; Natural (active) and Sterilized (control), Performed in Simulated Mars Environment

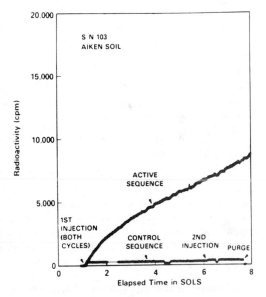

---

4 The labeled-release (LR) experiment had its origin in 1959 when Dr. Levin first proposed and then--in 1960--began development of a self-sufficient sample acquisition and analysis instrument called Gulliver. It was to fly on what was then called the Voyager (Mars) spacecraft. It didn't, of course, because that Voyager program was canceled and then scaled down to become the **Viking** Project. However, Gulliver was not wasted; its essential elements served as the precursor for the **Viking** biology instrument's labeled-release (LR) experiment discussed in this presentation. {Ed.}

VL-1 LR Results -- Four cycles of **Viking** experiments conducted by the first lander (VL-1) on Mars[5] are illustrated in Figure 1J-3 (opposite page): two actives and two controls. The active cycles reflect the same kind of response seen in the terrestrial experiment previously illustrated (Fig. 1J-2), with measurements in the 10,000 to 15,000 cpm range. The on-Mars control cycles produced responses essentially at the background level for the instrument, although background levels on the Mars landers were somewhat higher due to radiation from the two nearby nuclear power generators[6].

VL-2 LR Results and the UV Theory -- At the second landing site[7], the same sort of result evolved as we processed the first VL-2 sample. Initial VL-2 results are illustrated in Figure 1J-4 (opposite page). One of the nonbiological explanations almost immediately theorized by some was that ultra-violet (UV) light striking the surface of Mars somehow "activated" the soil to produce the positive response. At the VL-2 site, however, we were able to get additional data to test that particular theory by manipulating the experiment from Earth. The lander control engineers cleverly extended the soil sampler before dawn one morning (to avoid UV radiation exposure for a sample being retrieved), moved a small rock that had provided UV shielding for the small area of soil beneath it for perhaps a few hundred thousand or even millions of years, and acquired a sample of that protected soil for analysis. Only slightly weaker, that sample also responded in the active area at about 8,000 or 9,000 cpm.

Adapting the Experiment to Mars -- Based on **Viking** experience, we then further modified the pre-mission experiment criteria such that our on-Mars results would be more acceptable to the scientific community. Initially, for example, if we had gotten a zero response after heating a duplicate soil sample (as a control for one that had produced a positive response) to 160°C, the result would have been construed as evidence that the positive response had in fact evolved from living organisms that could be destroyed by the high temperature used during the control procedure.

To improve the basis of that control for Mars, we wanted to adjust the test criterion to a more conclusive temperature. We ultimately succeeded in reducing it to just 50°C (122°F), a relatively severe (warm)--though not necessarily or immediately destructive--temperature on Mars[8]. Our chemistry-oriented colleagues agreed that this temperature would likely be able to damage Mars organisms without inhibiting the chemical reactions they believed were mimicking life and producing our results. If our data were in fact due to a chemical reaction, the response should not have been reduced at that temperature. Instead, as seen in Figure 1J-4, the result was a response reduction of about 65%, which is more in keeping with what one might expect had a

---

[5] Viking Lander 1 (VL-1) landed in Chryse Planitia (at 22.5°N, 48°W) on July 20, 1976. {Ed.}

[6] Each **Viking** lander was equipped with two plutonium-fueled, AEC SNAP-19 Radioisotope Thermoelectric Generators (RTG's), mounted atop the landers and protected by windscreens. The RTG's were capable of producing 35 W each (total output of 70 W per lander), and were used to provide both direct power and recharging power for two internal wet-cell, nickel-cadmium battery packs (two 28vdc, 24-cell batteries in each pack). RTG's facilitate non-solar-dependent power for very long periods, and VL-1 survived into late 1981. {Ed.}

[7] Viking Lander 2 (VL-2) landed in **Utopia Planitia** (at 44°N, 226°W) on September 3, 1976. {Ed.}

[8] A temperature of 50°C (122°F) is very warm for an otherwise cold planet (compared to Earth). The **Viking** landers recorded daytime temperatures in the -25°C to -35°C range, dropping at night to -85°C to -90°C during the martian summer. However, it is believed that temperatures at the very surface may, in some areas (e.g., equatorial regions), be significantly warmer (25°C+). This potential is enhanced when correlated with higher atmospheric pressures and/or the presence of water vapor associated with some topographic features. {Ed.}

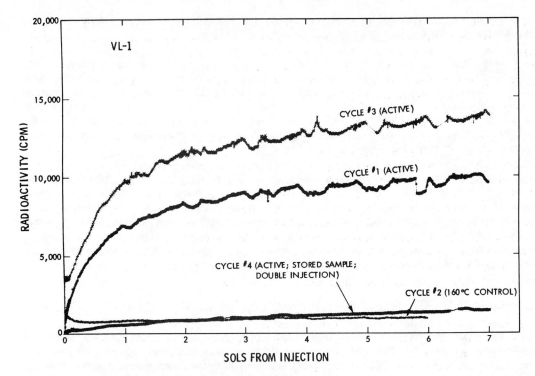

Figure 1J-3: Results From LR Experiment Cycles for VL-1

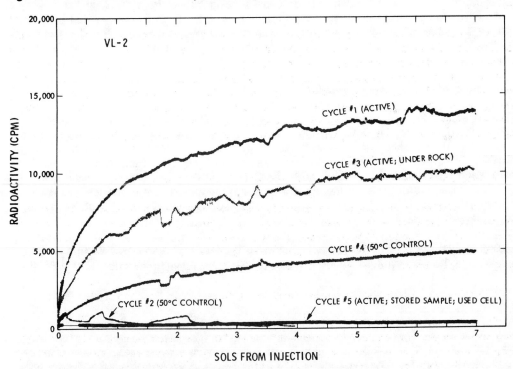

Figure 1J-4: Results From Experiment Cycle for VL-2

quantity of living organisms been attenuated at that modest temperature. Clearly, then, this experiment provided further support for a biology rationale.

Effects of Long-Term Sample Storage -- Another analysis opportunity, however, afforded through what might be called "inadvertence" or serendipity, proved to be even more important. We were able to test a sample that had been held in the sample collection hopper for two to three months. A portion of the sample had been tested at the time of its acquisition and had produced a positive response. Aside from the fact that it was then retained in a dark box (inside the lander a couple of feet above the surface of Mars) and maintained there at a temperature of about 10°C, nothing had happened to that sample.

The negative results of this analysis are reflected as Cycle 5 (Fig. 1J-4). At ambient temperature, the disappearance of an indigenous chemical merely placed in a dark box is hard to explain. Because of the nature of this particular sample and the results of its analysis, one is led to conclude that something mysterious was indeed going on, e.g., the evaporation of hydrogen peroxide (after causing the LR response). In the case of hydrogen peroxide, however, it would have been necessary for it to reform every day in amounts large enough to produce the response. The alternative, of course, is that something had died.

MARS AND EARTH EXPERIMENT SAMPLES IN COMPARISON

Had the strong positive response we got on Mars been produced by Earth samples, as indeed similar results often were, the data would have served as unquestionable proof for the presence of organisms in the soil. At the very least, the scientific community should grant that the **Viking** LR data are evidence for life on Mars. It has been stated by others that "there is no evidence" for life on Mars. Evidence is defined as being distinct from proof, in that evidence is something to be considered when trying to determine whether proof exists. While I have never claimed the latter, I DO submit that our experiment produced scientifically sound evidence for martian life, and that it is for the future to determine whether it is or is not associated with proof of martian life.

Differing Earth and Mars Results

The LR results were not without some disappointments. When a second dose of nutrient was applied to martian samples (following the completion of cycles during which positive responses were detected), we anticipated--as was usually seen in terrestrial samples--a renewal of gas evolution as a result of new growth. However, there was no increase when the second injections were made on Mars after the eighth sol[9], as Cycle 1 at the VL-1 site demonstrates in Figure 1J-5. Indeed, a decrease in the amount of gas in the atmosphere of the test cell is clearly defined. The lack of gas production following the second injection of nutrient led some to conclude that no life was present in the soil.

Terrestrial Analogue -- We more recently reviewed our library of results with terrestrial samples in search of data reflecting a response behavior similar to our Mars results, and found Antarctic

---

[9] A "sol" is one solar day for Mars. While a martian day is comparable in length to an Earth day, it is somewhat longer ($24^h37^m22^s$). This requires that sols rather than Earth days be used as an independent chronological reference when the planet's axial rotation is a pertinent factor in a given consideration. This, of course, is particularly true at the surface of Mars where darkness prevails at night just as on Earth. {Ed.}

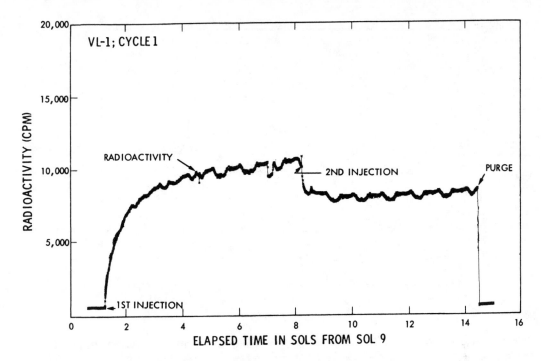

Figure 1J- 5: Results of Cycle 1 (1st and 2nd Injections, VL-1)

test soil No. 664[10]. In that soil (pH = 8.1), which initially produced a positive response with the first injection, we found that some of the gas had been readsorbed following the second injection and that the analysis essentially mimicked the Mars result. The pertinent data from this experiment are reflected in Figure 1J-6. In this soil, then, which had been shown to contain microorganisms [Cameron et al. 1970[a]] in classical microbiological tests, we demonstrated the death of those organisms after eight days (in **Viking** LR test instrument at 23°C). Could this not have happened on Mars?

The Hydrogen Peroxide Issue

Factors Involved in Formate-Peroxide Theory -- After the initial positive results were known, the theory expounding hydrogen peroxide (H_2O_2) was almost immediately proposed as a non-biological explanation. It suggests that hydrogen peroxide continually forms in the martian atmosphere and precipitates to the surface, where--when taken into an experiment like our LR instrument--it can react with one of the labeled nutrient compounds, ^{14}C-sodium formate. Indeed, the theory held that formate was the only labeled compound in the LR nutrient that would react with peroxide. Each of the seven substrates (includes the D and L forms as separate compounds) comprising the LR nutrient (ref., Fig. 1J-1) is capable of generating up to about 15,000 cpm, and since the positive responses on Mars were typically at or below that count, they were viewed as evidence of the formate-peroxide reaction.

---

[10] Antarctic Sample No. 664 was a bonded and certified NASA-supplied soil given to experimenters to test their instruments during the development of the **Viking** biology experiments. Such samples were stored, sealed, and maintained in their pristine condition for testing purposes.

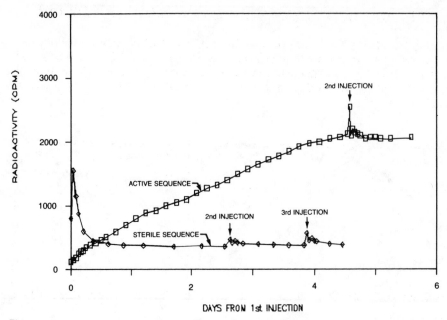

Figure 1J-6: Antarctic Soil No. 664 (pH 8.1) in Viking LR Flight-Like Instrument (23°C)

Testing Peroxide Theory on Mars -- Fortunately, we were able to test that theory on Mars. Using a sample that had yielded a low but positive result, we heated it terminally to 50°C. Doing so drove an additional 15,000 cpm out of the soil, as illustrated in Figure 1J-7, for a total yield of some 20,000 cpm. Clearly, then, the result exceeded the count possible if only the formate-peroxide reaction were taking place and implies that more than one nutrient was involved in the Mars response.

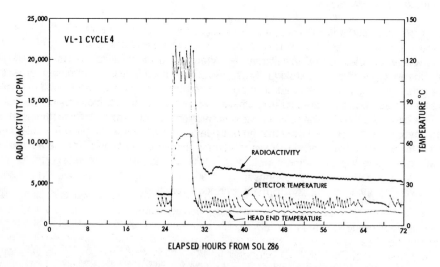

Figure 1J-7: Results From Heated Mars Sample

194

CONSIDERATIONS AND CONCLUSIONS BASED ON MARS LR RESULTS

It is worth exploring the Mars LR data to see how the facts can best be explained using biological and/or nonbiological arguments. For example, radioactive gas did in fact evolve from the medium, and while this can be attributed to life it might also be due to an unusual physical condition or compound in the soil. In this case, then, each scores a definite Yes as illustrated in Figure 1J-8.

VIKING LABELED-RELEASE (LR) DATA	POSSIBLE EXPLANATION	
	LIFE	NONBIOLOGICAL
Radioactive Gas Evolved from LR Medium by Mars Soil	YES	YES
Heating to 160°C Destroyed Mars Soil Activity	YES	Questionable
Heating to 46°C Attenuated Mars Soil Activity 65%	YES	Highly Questionable
Sample Held Months Inside Lander at Mars Temperature Lost 90% Activity	YES	NO
Sample from Under Rock was Active	YES	Questionable
More than One LR Compound Reacted	YES	NO (for one-compound theory)
Same Results at Sites 4000 Miles Apart	YES	YES

Figure 1J-8: *Comparison of Life vs Nonbiological Explanations for Viking LR Data*

Control Opportunities -- Temperature obviously plays a role in the nature of the response, and control temperatures are particularly worthy of note. Heating a duplicate sample to 160°C destroyed the activity that caused the initial positive response. This satisfies the criteria originally established for the experiment's ability to detect life and it puts a Yes in that column. However, some of the chemical arguments cast some doubt on those initial criteria, hence the nonbiology column can be marked as "questionable" rather than with an absolute No. Heating a sample only to 46°C attenuated the soil activity by 65%, which again is completely compatible with the argument for biology and does not reflect the anticipated behavior for a nonbiology explanation. One would have to score this control a Yes for life and "highly questionable" for the nonbiology argument.

Long-Term Sample Quarantine or Protection -- A sample held for a long period of time inside the lander at ambient Mars temperature lost ninety percent (90%) of its previous active response. That kind of reduction does not appear to correlate with any nonbiological theory yet offered, and therefore earns a Yes for the biology argument and a No for the nonbiology argument. The response obtained from soil acquired from beneath (and previously shielded by) a rock again records a Yes for the biology argument, and, at the very least, seriously questions the theory that UV or ionizing radiation was responsible for the positive LR response.

Causal Factors Concerning Type of Response -- I believe the probability that more than one LR nutrient compound had to be involved (to produce the total test count already discussed) overpowers the theory that formate alone, reacting with hydrogen peroxide, could have produced our results. Indeed, it renders the nonbiological argument virtually impossible when based on only one nutrient compound. However, the fact that the same results occurred 4,000 miles apart can be equally attributed to living organisms or to a uniform, global distribution of a suspect soil constituent[11]. Microorganisms are distributed with surprising uniformity on Earth, relative to similar soils and conditions, and the Viking landing sites on Mars were so similar that they might well support equal microbial populations. In the same sense, however, the similarity of the soil chemistries determined at those sites might allow for similar nonbiological reactions.

The Lack of Organic Chemistry in the Soil

While all of the factors I have just reviewed seem to add up to strong support for a biological explanation, a major **negative** result remains to be dealt with: the failure to detect organic carbon in Mars' soil (at both Viking lander sites). This fact, coupled with the belief that UV radiation may have precluded the development or existence of organic compounds, suggests that the martian environment is simply too hostile to life.

We believe we have established a sound answer to that constraint, as reflected in Figure 1J-9. Central to our argument is a body of evidence indicating that the sensitivity of the GCMS was too low to detect a very low density of organic compounds in another viable Antarctic soil sample provided by NASA. Further, experimental evidence already alluded to by Dr. Horowitz [NMC-1I] strongly supports the probability that organic matter not only forms but accumulates on the surface of Mars in the very face of the UV flux to which it is deemed so vulnerable.

Figure 1J-9: Questionable GCMS Sensitivity for Detection of Low-Density Organics

MAJOR BAR TO BIOLOGICAL INTERPRETATION OF VIKING LR RESULTS:
Finding: No organic matter detected in Mars "Soil"
Rationale: Organics, if formed, could not accumulate because of UV
AVAILABLE DATA NOW REMOVE THIS CONSTRAINT:
1. Sensitivity of Viking GCMS instrument too low to detect organics or organisms in viable Antarctic soil;
2. Experimental evidence supports current formation and accumulation of organic matter on Mars

Organic Matter and Life in Test Sample Undetected by GCMS -- While reviewing the Mars data, we found still another interesting Antarctic soil sample, Number 726. This sample, also provided for the experimenters by NASA, was brought back and maintained--as in the case of all such samples--in a sealed, pristine condition to preserve its original quality for testing. It had revealed no organic material during such tests in a test-standards GCMS instrument, but a wet-chemical analysis of the soil showed that it in fact contained 0.03% organic carbon. We had received an aliquot of the same sample in our laboratory, and I found that we indeed had tested it in the LR instrument.

Figure 1J-10 illustrates the results of that work. Clearly, we had detected living organisms in that soil, and the results of the second injection as well as the sterile controls verified that the response was produced by those organisms. These results, then, reflect a compatible truce between the negative findings of the GCMS and the positive LR results.

---

[11] It is believed that weathering and meteorology over time on Mars is responsible for a homogeneous and generally uniform distribution of the planet's surface chemistry. {Ed.}

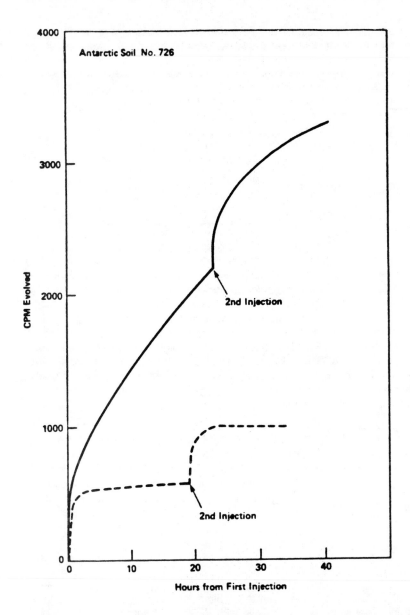

Figure 1J-10: Reactivity of Antarctic Soil No. 726 With LR Nutrient

A Link Between UV and Organic Generation -- Turning to the questions posed by possible UV effects, an answer has been found in the literature where we should have found it long ago. In his Mars Conference presentation, Dr. Horowitz has described a significant problem he and his group [b] encountered while developing the Viking pyrolytic release (PR) biology instrument, which threatened to remove the experiment from the mission. The substance of the problem, as reviewed in Figure 1J-11, was that, with Mars light (including UV) shining on Mars atmosphere under simulated martian conditions, organic matter formed and was deposited on the simulated soil particulates. In addition, the organic material continued to accumulate over time. The authors of the paper from which Figure 1J-11 was taken [Hubbard, Hardy, and Horowitz, 1971]

197

stated, "Our findings suggest that UV presently reaching the martian surface may be producing organic matter. ... *the amount of product formed could be considerable over geological time.*"

Production of CO_2 and organics by UV irradiation of different concentrations of CO				
CO %	% Conversion* of CO to		Relative production rates	
	CO_2	Organics	CO_2	Organics
0.02	48.7	5.1	0.4	0.042
0.37	5.0†	1.15†	0.76	0.18
5.5	0.44‡	0.12‡	1.0	0.27

Quartz chambers contained 30 mg of sterilized soil in a gaseous mixture of the indicated CO concentration, 2.8% water vapor, and the balance CO_2.
* Based on the yields of [^{14}C]CO_2 and [^{14}C]organics after 17 hr of irradiation with a xenon lamp.
† Calculated from Fig. 1.
‡ Calculated from determinations made after irradiation for 135 and 216 hr.

UV-dependent conversion of CO to CO_2 and organics in various gas mixtures					
Diluent gas*	Vycor* sub-stratum (mg)	Xenon UV (hr)	% Conversion of [^{14}C]CO to		
			(A) [^{14}C]CO_2†	(B) [^{14}C]-organics	A/B
97% N_2 { none	none	17	77.6	0.91	85.2
30	30	17	51.0	1.05	48.5
150	150	17	10.6	1.14	9.3
97% [^{14}C]CO_2 { none	none	17	89.0	1.14	78.0
+ 0.03% O_2 { 30	30	17	34.6	2.12	16.3
150	150	69	82.0	2.46	33.5

* Chambers contained 0.06% [^{14}C]CO, 2.8% H_2O vapor, and the indicated additions.
† Corrected for the concentration of $^{14}CO_2$ present in un-irradiated controls.

Figure 1J-11: Photocatalytic Production of Organic Compounds in Mars Simulation (1971)

The problem was resolved in the PR experiment by interposing a filter to screen out UV light below 3,000 angstroms, but that of course does not prevent the phenomenon from occurring on Mars. The formation and accumulation of organic matter under UV light was subsequently confirmed in a paper published in the JACS [Farris and Chen, 1975], in which they demonstrated the production and accumulation of organic material by shining UV light on a mixture of methane and water vapor. They concluded that this phenomenon--the accumulation of organic matter under UV--has "pre-biological significance."

Clearly, then, there is a good case not only for the formation of organic matter in the martian UV environment, but for its accumulation. Indeed, if there is a problem suggested by the body of data produced by the Viking GCMS on Mars, it is the question of why the instrument was unable to detect organic matter that must certainly be there. When considered in perspective with the proven need for greater sensitivity than the instrument could muster during the analysis of Antarctic sample 726, one must at least agree that the GCMS may have been unable to detect similar densities of organic matter on Mars -- remembering, too, that the terrestrial sample contained living organisms.

Hydrogen Peroxide: Pro and Con

Why, then, were no martian organic chemicals detected when, even without life, their formation on Mars seems so probable? Could they have been overlooked as I have just suggested? Most probably not, if the hydrogen peroxide theory prevails with its soritical capacity for converting the LR experiment into a hydrogen peroxide meter. However, we again have found empirical evidence that argues against hydrogen peroxide, this time a product of NASA's Mariner 9 mission (1971-72).

Dr. Straat and I were co-experimenters with R.A. Hanel and others on the Mariner 9 infrared interferometer spectrometer (IRIS) experiment team. While wrestling with the hydrogen peroxide issue associated with Viking results, I realized that there could be some pertinent data hidden in the product of the Mariner 9 mission. I called Bill Maguire [Goddard Space Flight Center], a NASA scientist formerly associated with the Mariner 9 IRIS work, and asked if there may have been a window provided by the IRIS instrument that could detect hydrogen peroxide. After a brief review of the data, he informed me that there was indeed an excellent window for hydrogen peroxide -- and that none had been found.

A Review of Nonbiological Hypotheses

Figure 1J-12 (a and b on the following pages) presents a compilation of all the hypotheses that attempt to explain the LR results in a nonbiological manner, and which have been published or otherwise called to our attention. The first, of course, is the hydrogen peroxide theory (which also represents several subset theories), and it includes a rationale for how peroxide forms and exists on Mars.

Because the Mariner 9 IRIS instrument found no trace of hydrogen peroxide on Mars, one can at least presume that if it is there at all it must be present only in extremely small amounts. And, because of the amount of atmospheric water vapor detected by the Viking orbiters in the vicinity of the landing sites, any small amount of hydrogen peroxide that may have been missed by Mariner 9 would somehow have to survive a confrontation with up to 5,000 times its own volume in atmospheric water[12]; if water vapor immediately reacts with hydrogen peroxide, as Dr. Horowitz has explained, the hydrogen peroxide would not have been present. If water vapor doesn't do the job, one should remember that hydrogen peroxide is light-labile and especially vulnerable to virtually immediate destruction by UV radiation[13], or that contact with iron or other possible metal catalysts in the surface material (aided by atmospheric water vapor) is also capable of destroying it. Assuming either or all of these destruction scenarios to be active on Mars, one is driven to conclude that any hydrogen peroxide that somehow managed to find its way into the martian soil would certainly have been destroyed long before it could be picked up by a Viking surface sampler and exposed to our LR nutrient.

Other nonbiology theories involve: minerals that catalyze the reactions with the LR nutrients, ultra-violet radiation (already discussed), ionizing radiations of various kinds, finely divided and desiccated oxygen-rich minerals, and large surface areas of fine particulate generating heat of hydration upon wetting. We and others have tested all of these proposals, and none of them could provide a fully satisfactory explanation for--or a duplication of--the provocative responses we got from the LR experiment on Mars. If we can't evoke a tenable hypothesis for a chemical or physical reaction that can duplicate the Mars results after ten years, perhaps it is finally time to consider the biological explanation we were looking for in the first place. Perhaps, after all, we did indeed detect life on Mars.

A CASE FOR LICHEN ON MARS

We have a natural model for one possible martian life form close at hand -- lichen. In fact, the lichen possibility was offered as a model for martian life long ago, and such organisms possess all the characteristics necessary for survival in the harsh martian environment. They serve as a good model for Mars life because their known characteristics satisfy virtually all of the criteria imposed by the nature of the results in both Viking LR experiments. On Earth, lichens are the pioneers of vegetation and are frequently credited with the initiation of weathering over large areas.

---

[12] The limit of sensitivity for the Mariner 9 IRIS experiment, relative to its ability to detect hydrogen peroxide, was 1×10^{-2} precipitable microns. By comparison, the water vapor measured above both Viking landing sites by the orbiters' IR water vapor mapping instruments ranged up to about 50 precipitable microns. {Ed.}

[13] The reaction coefficient for the UV destruction of H_2O_2 exceeds that of its formation by 10^6.

HYPOTHESIS	DEFECT
■ H_2O_2 on Mars Reacts With LR Medium: a. H_2O_2 Forms in Atmosphere, Rains onto Surface b. H_2O_2 Generated by Frost Weathering of Freshly Exposed Minerals c. H_2O_2 Reaction With LR Medium Catalyzed by γ-Fe or Other Metals or Minerals	A) Mariner 9 IRIS found no H_2O_2 on Mars. B) UV Flux on Mars would destroy H_2O_2. C) Mars sample held several weeks at upper ambient temperature became inactive. Thus, if chemical, agent would have to be replenished frequently and could not have accounted for LR response from sample under the rock.
■ Mars Minerals Catalyze Degradation of LR Medium	None can account for heat sensitivity of Mars agent in LR experiments.
■ UV Activates Mars Surface, Which Reacts With LR Medium	Under-the-rock sample would not have responded.
■ Ionizing Radiation Activates Mars Minerals To Catalyze Degradation of LR Medium	A) Gamma radiation of early Mars analogue "soil" produced positive response, but exposure of sample to UV eliminated the effect. B) Thermal sensitivity of Mars sample not duplicated. C) Gamma radiation of updated Mars analogue "soil" failed to yield response to LR test.
■ Ionizing Radiation of Finely Divided, Desiccated, Oxygen-Rich Mars Minerals Produces Disjunctions Which Degrade LR Medium on Contact	A) Silica gel exposed to gamma radiation yielded LR type positive response, but UV irradiation eliminated response. B) Thermal sensitivity of Mars sample not duplicated.
■ Large Surface Area of Finely Divided, Desiccated Minerals Generates Sufficient Heat on Wetting to Degrade LR Medium	Silica gel heat desiccated under vacuum, failed to produce LR response.

Figure 1J–12: Defects in Nonbiological Hypotheses for Viking LR Results (1 of 2)

200

HYPOTHESIS	DEFECT
■ Superoxides on Mars React With LR Medium	A) No evidence for superoxides. B) Superoxides putatively formed in dust storms, but LR tests of under-the-rock sample was positive. C) Superoxides do not have correct thermal stabilities.
■ Polymeric Suboxides on Mars React With LR Medium	A) No evidence of polymeric suboxides. B) Would be accompanied by related compounds which Viking GCMS would have detected.
■ Smectite Clay Catalyzes Decomposition of LR Medium	A) No evidence on Mars for ionic forms of smectite. B) Observed responses attributable to pH effects. C) Observed responses not elimitated by 160°C.
■ Palagonite Clay (newer hypothesis than smectite) Catalyzes Decomposition of LR Medium	No reaction with LR medium.
■ Formate in LR Medium in Contact with Mars Catalyst Decomposes Thermodynamically to Yield CO	Extensive experiments failed to find the needed catalyst; experiments in flight-like LR instrument showed end-product is CO_2 and not CO.
■ Lactate in LR Medium in Presence of Fe^{+++} Decarbonylates to Produce CO	Response reported only under very acid conditions and fails to simulate LR Mars data; experiments in flight-like instrument showed end-product is CO_2 and not CO.
■ All of the Above!	No laboratory tests of any of these hypotheses have duplicated the LR Mars test and control results; properties of data do not support candidacy of any suggested chemicals.

Figure 1J-12b: Defects in Nonbiological Hypotheses for Viking LR Results (2 of 2)

Lichen can: 1) grow on bare, unweathered desert and mountain top rocks in extremely hot or cold environments, 2) protect themselves from UV radiation, 3) withstand long-term desiccation, and 4) absorb water from atmospheric vapor. In addition to being able to grow on bare rocks in inhospitable regions, they exhibit "endolithic" capability -- growing and surviving *within* rocks.

It should be noted that many forms of terrestrial biota could inhabit Mars but for the apparent lack of liquid surface water and the difficulty in husbanding water from the atmosphere against the low vapor pressure. However, we should not close our eyes to the latter possibility. In juxtaposition, one might well imagine martian scientists looking down on Earth and saying: "There can't be any plants or trees on that poor planet, there is only 125 parts-per-million of carbon in the atmosphere!" I think the arguments are fairly comparable. We have found that lichen organisms can essentially--quite satisfactorily, in fact--replicate the LR results in all respects. And, while it is not appropriate to get into the details of these studies here, I will present some photographic material that suggests to us (under very close scrutiny) that the **Viking** lander images may in fact have provided evidence of lichen growth on Mars.

Possible Evidence of Lichen on Mars

Figure 1J-13 presents two images of a rock known as Patch Rock at the VL-1 (Chryse Planitia) site. The image was produced at JPL by the image reproduction system developed specifically for the precise production of **Viking** lander images from digital data transmitted back to Earth from Mars. In this picture pair, one can see for the first time a hint of something not monotonously orange-red, as initially reported about Mars. Instead, greenish patches are seen on some of the small rocks. We found that the configurations of the patches changed over a period of time. The picture on the left was taken on sol 28/VL1 and the one on the right was taken sol 615/VL1[14]. A lot had of course happened over the intervening 587 sols which affected the character of the site; there had been some digging by the lander's surface sampler and there had been a dust storm. However, the intensification of the greenish color and a change in the appearance of the rock's patch was not readily explained by these events.

Spectral analyses were made of these images at JPL, and the patches proved to be the greenest and least color-saturated objects in the field of view. Figure 1J-14 illustrates the results of a color-saturation analysis involving the same rock, this time using pictures taken sol 1/VL1 and sol 302/VL1. A change in the response over time was evident, and, of all the features in the field of view (of whatever color), the spots and patches highlighted were the least color saturated. I should point out, incidently, that lichens come in many colors: red, yellow, pink, gold -- even white, black, and colorless. The presence of greenish colored patches were subsequently verified by others [1979c].

Lichen Detection Tests With Viking STL Imaging System

Because **Viking** mission operations were still underway as we were conducting these initial analyses, we took some lichen-bearing rocks to JPL and put them in a simulated Mars terrain

---

[14] Because martian days (sols) are longer than Earth days, they do not correlate directly with Earth days. For example, a martian year is 669 sols compared to 687 Earth days. Viking lander operations are typically considered in terms of the number of sols elapsing from the first day of each landing, such that the sol count for each lander is different. Sol 1 for VL-1 is July 20, 1976, while Sol 1 for VL-2 is September 3. And, because there is a difference of more than 40 sols, it is necessary to know which lander is in reference for a given sol number -- hence the use of "/VL1" or "/VL2" with sol numbers used in this presentation. Sol 615/VL1, for example, represents the 615th martian day of VL-1 operations. {Ed.}

Figure 1J-13: "Patch Rock" at VL-1 Site (left, sol 28; right, sol 615)

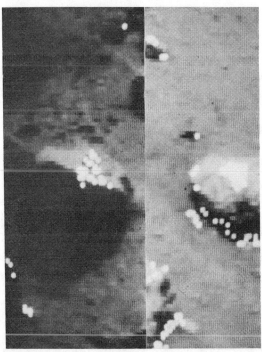

Figure 1J-14: Areas of Lowest Color Saturation (based on computer analysis of data)

model that had not yet been dismantled[15]. We took pictures of them using the science test lander (STL) imaging cameras (identical to the flight cameras), and Figure 1J-15 presents the results of that work (top); the picture at the bottom is a conventional photograph of the modeled terrain for the landing site, which included highly detailed foam duplicates of rocks at the site (painted to look like the real rocks). The greenish lichens on our sample rocks were less clearly visible than in Earth sunlight and resembled the Mars images. When we subjected these pictures to a color-saturation analysis similar to the one I spoke of in reference to the actual Mars images, the Earth lichens were again the least color-saturated objects in the field of view.

Figure 1J-15: Viking Test Image of Terrestrial Rock Lichen (top) STL Test Scene (bottom)

Antarctic Rock Correlates With Mars Results -- Working in the Antarctic, Friedmann and Ocampo [1976] discovered that there were living organisms--algae and lichen--below the surface of many porous sandstone rocks[16]. These organisms are typically found at a depth of about 1 to 1.5 cm and are somehow "eking" out a living inside these Antarctic rocks, with very little sunlight or water able to penetrate to their depth. Indeed, they seem almost to be hiding from it, and it is in fact likely that they have evolved their retreat from rock surfaces to escape the harshness of the Antarctic environment. In Figure 1J-16, a relatively common sample of rock with surface lichen is compared with a sample of Antarctic rock, the latter split to reveal the lichen-bearing layer beneath its surface (dark band along top of exposed face).

---

[15] The Viking science test lander (STL) was used at JPL during mission operations to program and test actual command sequences (for directing the cameras and soil samplers on the Mars landers) before transmitting them to the specified lander. Each landing site, therefore, had to first be modeled very precisely with accurate reproductions of the surface material, very small elevation variations (including trenches ultimately dug by the samplers) and rocks -- particularly those within range of the soil sampler. Each of the STL's two cameras could image like its counterpart on each of the Mars landers, such that complete imaging sequences could be tested for errors prior to transmission. {Ed.}

[16] Dr. E. Imre Friedmann, a biology professor and director of the Polar Desert Research Center at Florida State University (Tallahassee), is credited with the first discovery of organisms living within rocks [1964] while studying rocks from Israel's Negev desert. His Antarctic research was funded jointly by NASA and the National Science Foundation. {Ed.}

Drs. Friedmann and Ocampo were kind enough to bring one of the Antarctic rocks to our laboratory, where we carefully split it to reveal a green band about 1.5 cm below the surface. We then aseptically scraped out about 25 milligrams for testing in the LR experiment. To facilitate a comparison, Figure 1J-17 (bottom, right) illustrates the results of these with plots reflecting both active and sterile Mars data acquired by VL-2 in Utopia Planitia. The similarity is remarkable, and we contend that these data must certainly be considered in the realm of contributing evidence.

Figure 1J-16: Lichen Examples --
(above right) Antarctic
Subsurface Lichen;
(below) Common Rock Lichen

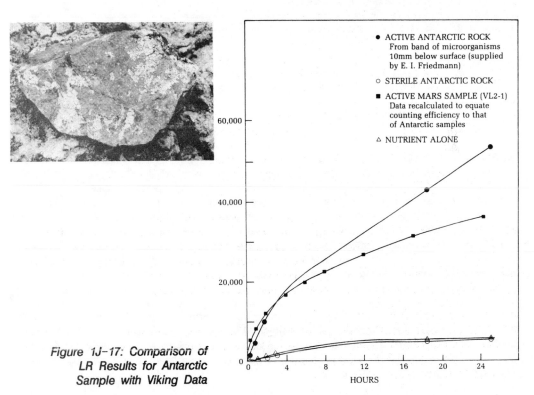

Figure 1J-17: Comparison of
LR Results for Antarctic
Sample with Viking Data

- ● ACTIVE ANTARCTIC ROCK
  From band of microorganisms
  10mm below surface (supplied
  by E. I. Friedmann)
- ○ STERILE ANTARCTIC ROCK
- ■ ACTIVE MARS SAMPLE (VL2-1)
  Data recalculated to equate
  counting efficiency to that
  of Antarctic samples
- △ NUTRIENT ALONE

HOURS

205

Summary Results of Imaging Experiments -- In Figure 1J-18, we have summarized the results of the imaging study in the same manner as the previous summary of the LR data (Fig. 1J-8). The greenish spots or patches could be attributed to life; alternatively, chemicals or minerals could be responsible. The changing configuration of the patches with time, however, is strongly suggestive of life and is not easily explained by mechanical or meteorological events.

IMAGING ANALYSIS	POSSIBLE EXPLANATION	
	LIFE	NONBIOLOGICAL
"Greenish" Spots Observed on Mars Rocks	YES	YES
Shapes of Spots Appear to Change with Time (note 1)	YES	Very Unlikely
Spots are Greenest Objects in Field of View (as are terrestrial rock lichen when test-imaged by Viking system)	YES	YES
Spots on Mars Rocks Exhibit Color Saturation Characteristics Similar to Lichen on Terrestrial Rocks (imaged by Viking system)	YES	Questionable
Spots on Mars Rocks are Elevated about 1 mm above Rock Surface	YES	Very Questionable
Possible "Desert Varnish" Seen on Mars Rocks (note 2)	YES	Very Unlikely

Note 1: Additional data for this determination probably exists in Viking lander imaging database.

Note 2: Based on private communication.

Figure 1J-18: Summary of Image/Spectral Analysis Relative to Lichen Possibility

In summary, then, we feel that the evidence produced during the imaging studies thus far undertaken is more indicative of life than of chemistry or physics. In the near future we hope, and have so proposed, to try to resolve this problem through access to additional image data being maintained in the Viking database.

SUMMATION AND CONCLUSIONS

We have waited ten years for all of the theories, experiments and results produced by the many scientists investigating our experiment to be reviewed before voicing a committed conclusion of our own. After examining these efforts in great detail, and after years of laboratory work trying to duplicate our Mars data by nonbiological means, we find that the preponderance of scientific analysis makes it more probable than not that living organisms were detected in the LR experiment on Mars. This is not presented as an opinion, but as a position dictated by the objective evaluation of all relevant scientific data.

In conclusion, then, we submit that this real possibility for martian biology should be an important--even dominant--consideration in the future exploration of Mars. This clearly is not the case at the present time, according to published NASA plans for continuing the unmanned exploration of Mars which neglect the biology issue. The search for life on Mars, when evidence for its possible existence offers such important potential, should have much greater significance for the planning of such missions. ■

REFERENCES

References identified here by alphabetically sequenced lower-case characters are keyed to *specific* references in the presentation text where the same characters are used as identification tags. In the text, however, the identifier will be found within brackets, either by itself or with a name and/or credit date (i.e., when formally published) -- e.g., [a] or [1978a].

This space is otherwise available for notes.

~~~~~~~~

NOTED PRESENTATION REFERENCES (see following page for additional references)

a.   Cameron, R.E., King, J., and Daird, C.N.   1970.   *Antarctic Ecology*,   Vol. 2, p. 202 (Holdgate, ed.), London.

b.   Hubbard, J.S., Hardy, J.P., and Horowitz, N.H.   1971.   *Proc. Nat. Acad. Sci.*   3:574-578.

c.   Strickland, E.L., III.   1979.   NASA Tech. Memo. 80339.

    Guiness, E., et al.   1979.   NASA Tech. Memo. 80339.

UNSPECIFIED REFERENCES OF INTEREST

Cooper, H.S.F. 1980. *The Search for Life on Mars: Evolution of an Idea*. An Owl Book. Holt, Rinehart and Winston, New York.

Levin, G.V., and Straat, P.A. 1976a. Labeled release--an experiment in radiospirometry. *Origins of Life*. 7:293-311.

Levin, G.V., and Straat, P.A. 1976b. Viking labeled release biology experiment: Interim results. *Science*. 194:1322-29.

Levin, G.V., and Straat, P.A. 1976a. Recent results from the Viking labeled release experiment on Mars. *J. Geophys. Res.* 82:4663-67.

Levin, G.V., and Straat, P.A. 1977b. Life on Mars? The Viking labeled release experiment. *Biosystems*. 9:65-74.

Levin, G.V., and Straat, P.A. 1979a. *Status of interpretation of Viking labeled release experiment*, (abs). NASA CP-2072, p. 52.

Levin, G.V., and Straat, P.A. 1979b. Completion of the Viking labeled release experiment on Mars. *J. Mol. Evol.* 14:167-83.

Ed. Note: Vol. 82, No. 28 (pages 3959-4681) of J. Geophys. Res. (item 4 above) was a complete publishing of the Viking Primary Mission results (with early Extended Mission results), i.e., those results of Viking mission activity during the last half of 1976 and early 1977 that could be properly reported by mid 1977 (publish date was September, 1977). The results of--and early conclusions associated with--the Viking biology experiments are included. The results and conclusions are also modestly reviewed by H.S.F. Cooper (The Search for Life on Mars, item 1 above) as the author describes the activity and interaction of the Viking biology team's key characters, as well as the evolution and product of their work, against the backdrop of Viking mission activity. Similarly, ON MARS: Exploration of the Red Planet - 1958-1978 (E.C. Ezell and L.N. Ezell, 1984, NASA SP-4212 in the NASA History Series), presents a thorough historic account of the Viking Project and includes a conservative, nontechnical overview of its science results.

NASA Mars Conference
Session 2, July 22, 1986

The Future Unmanned Exploration of Mars

Session Chairman, Mr. James R. French, Jr.

---

[1] The processing of Presentation NMC-2E could not be completed for the publishing of these proceedings; (see note on Session 2 contents page for further explanation).

[2] Dr. Hilton was a graduate student in pursuit of his Ph.D at the time of his presentation and Mars Ball demonstration for the 1986 NASA Mars Conference, studying under Dr. Donald M. Hunten at the University of Arizona. He has since completed his degree requirements and assumed new career responsibilities.

NMC-2A · DR. JAMES W. HEAD, III is professor of geological sciences at Brown University, where he earned his Ph.D studying under Thomas A. (Tim) Mutch. Dr. Head has consistently participated in NASA lunar and planetary activities for many years. He was involved in the Apollo program, participating in the landing site selection process and in astronaut training activity, and he served as interim director (1973-74) at the Lunar Science Institute in Houston. He has since participated in science planning work associated with the SIR-B, Magellan and Galileo programs, and he is a member of the President's transition team for space. As a specific qualification for his contribution to the NASA Mars Conference, Dr. Head has been active in cooperative studies and mission data analysis associated with the Soviet planetary exploration program, and he is a guest investigator on the Soviet Venera 15 and 16 missions.

~~~~~

NMC-2B · MR. WILLIAM I. PURDY has been involved in planetary work at the Jet Propulsion Laboratory in Pasadena, California since 1966, and he is currently project manager for NASA's Mars Observer program. He earned his degree in engineering at the University of California at Los Angeles (UCLA) and has pursued graduate work in engineering and engineering management at California State University and California Institute of Technology (Caltech). Mr. Purdy has had technical responsibilities on numerous programs, including: Mariner 4 (Mars, 1964), Mariner 5 (Venus, 1967), Mariners 6 and 7 (Mars, 1969), and Mariner 10 (Venus/Mercury, 1973). In addition to technical responsibilities, he has served as assistant satellite system manager for SEASAT-A, spacecraft system manager for spacecraft development in support of the International Solar Polar Mission, spacecraft system manager for the Venus Orbiting Imaging Radar (VOIR) study, and as project manager for JPL's Talon Gold Project. Mr. Purdy is a member of the American Institute of Astronautics and Aeronautics, and he is a recipient of NASA's Exceptional Service Medal and the United States Air Force Space Division Excellence Award.

~~~~~

NMC-2C · DR. ARDEN L. ALBEE is professor of geology and dean of graduate studies at the California Institute of Technology (Caltech), and received his Ph.D from Harvard University. He served as chief scientist (1978-84) at the Jet Propulsion Laboratory, and is currently project scientist for the 1992 Mars Observer mission. He has studied developmental partitions between coexisting rock-forming minerals in the development of correction procedures for electron microprobe analysis, and he led the development of a scanning electron microscope for the forthcoming comet rendezvous mission. Dr. Albee has been a member of the terrestrial bodies science working group, the Mars '84 science working group, the space science advisory committee, and the Solar System Exploration Committee (SSEC); he chaired the lunar science review panel for five years, and he has chaired a number of working groups and studies on martian and cometary missions. He is currently a member of the solar system exploration management council as well as the task group on planetary and lunar exploration. He serves as an editor for the bulletin of the Geological Society of America and for the annual reviews of Earth and planetary sciences. He is a recipient of NASA's medal for **Exceptional Scientific Achievement.**

~~~~~

NMC-2D · DR. DONALD M. HUNTEN is professor of planetary science in the Lunar and Planetary Laboratory at the University of Arizona, Tucson. A Canadian by birth, he earned his Ph.D at McDill University in Montreal in 1950 and then taught physics at the University Saskatchewan prior to taking a position at the Kitt Peak National Observatory (near Tucson) in 1963. Dr. Hunten has been--or is currently--an interdisciplinary scientist on the **Pioneer Venus** and Galileo programs, and is chairman of the science working team responsible for developing **Mars Aeronomy Observer** recommendations. He is a member of the National Academy of Science.

~~~~~

NMC-2F · MR. JAMES R. FRENCH, JR. is currently with the American Rocket Company in Camarillo, California, but his expertise on the topics he discussed at the **NASA Mars Conference** [NMC-2F and NMC-3E] was developed during nineteen years of work at the Jet Propulsion Laboratory. Mr. French earned his degree in engineering at Massachusetts Institute of Technology (MIT) in 1958. He then worked in engine design (Rocketdyne and TRW) for nearly ten years before joining JPL. His work at JPL ultimately focused on defining and resolving problems associated with sample return missions (spanning about ten years) and in·situ propellant production (ISPP), the latter in support of both unmanned and manned missions. He left JPL early in 1986 to become chief engineer and vice president of engineering at American Rocket, but he continues to be actively associated with work related to his previous tasks at JPL.

~~~~~

NMC-2G · DR. DOUGLAS P. BLANCHARD is chief of the Planetary Materials Branch at NASA-JSC in Houston, Texas. He earned his Ph.D in geochemistry at the University of Wisconsin. Dr. Blanchard has applied his education to lunar science and sample processing, and he is particularly interested in trace-element chemistry. He has been curator of the Apollo lunar samples since 1981, and has responsibility for reviewing proposals for their analysis and for coordinating access to them by researchers and research institutions around the world. His multi-faceted sample management and science expertise is also being contributed to the development of Mars sample return scenarios and mission planning, in association with others at JSC and JPL now studying sample return in preparation for future mission activity.

~~~~~

NMC-2H · DR. DOUGLAS A. HILTON was involved in his Ph.D program at the University of Arizona in Tucson at the time of the **NASA Mars Conference**. He has since completed the program, which included dissertation work (under Dr. Donald M. Hunten) involving a numerical study of the hydrogen torus of Titan (largest satellite of Saturn). At the time of the conference, Dr. Hilton had been student administrator of the university's **Mars Ball Project** from its inception in August, 1983. The **Mars Ball** is a surface rover vehicle (research concept) based on the application of inflatable sectored tires, and it was developed by the department of planetary sciences at the University of Arizona. Dr. Hilton brought a working model of the vehicle to the conference for demonstration purposes [see photos, NMC-3H].

■■■

THE 1988-89 SOVIET PHOBOS MISSION[1]
[NMC-2A]

Dr. James W. Head, III
Professor, Geological Sciences
Brown University

A multidisciplinary Soviet planetary mission, which features an exciting contact investigation of Phobos as well as new orbital studies of Mars, will get underway in mid 1988 with the launch of two identical spacecraft. Each will feature an advanced orbiter, which at one point will hover within 50 meters of Phobos, and two types of small landers. One of the lander designs incorporates a unique concept in mobility, such that it will be able to "hop" about on the surface of Phobos, while the other is designed for a longer life and will essentially bolt itself to the surface. Though focusing on Phobos, the comprehensive mission strategy includes extensive studies of Mars, the Sun, and the interplanetary medium. Mission plans for Mars include studies of the surface, atmosphere, ionosphere, and magnetosphere. The study of Phobos will include investigations of the satellite's surface, its internal structure, and its orbital motion. The internal structure is of interest because it relates to the origin of Phobos' grooves. Other objectives include determining the satellite's mineralogical and chemical composition, and a capability will also be provided for obtaining surface images with a very high spatial resolution. Radiometric characteristics of the surface layers will also be analyzed, and seismology will be conducted to improve on the understanding of possible tidal distortion in Mars' orbit.

INTRODUCTION

Academician Roald Z. Sagdeev, the director of the Institute for Space Research in Moscow (also known as IKI) and head of the scientific council responsible for the 1988-89 multidisciplinary

---

[1] The short name for the 1988 Intercosmos Phobos mission--the **Phobos Mission** or the **Phobos Project**--is somewhat generic and can lead to some confusion within a broader discussion of the martian satellites or the mission itself. In a previous Soviet paper [1985, see references], the program was more formally identified as: the **Phobos Multidisciplinary Space Mission**. To simplify and clearly define specific references to the program, mission, or spacecraft (for consistency in these proceedings only), we have elected to use **PHOBOS·88**. One should remember, however, that none of these program names accurately portrays the full scope of the mission plan or the capabilities of the spacecraft. The two Mars orbiters will, for example, be instrumented for a much broader science investigation than is suggested, including studies of Mars, the Sun, and the interplanetary medium. The Soviets hope to quickly follow up on their 1988-89 PHOBOS·88 mission with a **Vesta** mission, possibly as early as the 1994 opportunity, and long-range plans include a goal for Mars-sample-return before the year 2000. {Ed.}

investigation of Phobos and Mars (the 1988 Intercosmos Phobos mission, PHOBOS·88), sends his regrets for not being able to attend the *NASA Mars Conference*. He has asked me to fill in for him and to present a summary of the program, including its scientific objectives, a mission profile, and details of the scientific instruments that make up the payload. The report will include, as well, a few comments on the post-PHOBOS·88 exploration plans of the Soviet Union. Last Friday, Academician Sagdeev sent a cable in which he stated:

> *. . . . I wish to express to the Mars Conference [that] the Viking mission data demonstrated the interesting results that come from exploring Mars. We hope that the Phobos Project [PHOBOS·88] will begin the post-Viking studies which may become a precursor of an extensive and comprehensive Mars program. We believe in the international character of this program.*
>
> *Regards to all attendees from Roald Sagdeev*

Mission and Program Overview

Two PHOBOS·88 spacecraft, like the one roughly illustrated in the drawing presented in Figure 2A-1, will be launched in mid-1988[2]. They will represent the first Soviet spacecraft to be sent to Mars in about 15 years; **Mars 4, 5, 6 and 7** were launched in 1973. The PHOBOS·88 spacecraft concept is representative of a new generation of planetary spacecraft for the Soviet program; it has been described as being analogous to our planned **Mariner Mark II** spacecraft[3]. It has a constant 3-axis, Sun-star orientation with a stabilization accuracy of ±1 degree in its PHOBOS·88 configuration. It also has a mode in which it can achieve a constant 1-axis orientation to the Sun with a stabilization accuracy of 0.5 angular degrees.

The **Mariner Mark II** concept under study at JPL is essentially designed for missions beyond Mars to the realm of the outer planets, but it reflects some of the same fundamental philosophy represented in the **Planetary Observer** concept and the design of the PHOBOS·88 spacecraft and mission. That is, it is designed as a flexible, reconfigurable vehicle that can be utilized for a variety of deep-space planetary missions. A set of drawings of possible **Mariner Mark II** configurations, as originally illustrated in the NASA Advisory Council's report, are included in Figure 2A-1 to provide an idea of its physical characteristics relative to those of the PHOBOS·88 spacecraft.

---

[2] Dr. Head was unable to provide hardcopy illustrations or graphics for the visuals he used during his presentation. To compensate where appropriate, black-line illustrations extracted from the Soviet science paper, **The Phobos Mission: Scientific Goals** [Sagdeev et al. 1986], have been used in their original form; sequential figure numbers have been assigned to these in the conventional manner within the context of Dr. Head's presentation. All other identification references to specific visuals have been edited out of the presentation, and related discussions have been rewritten in explicatory form. Because Dr. Head originally used a number of **Viking** photographic products, cross references to comparable figures in other Mars Conference presentations have been provided (when possible) as a convenience to readers. These references are most frequently to figures in presentations given by L. Soderblom, M. Carr and J. Veverka [NMC-1B/1C/1E]. {Ed.}

[3] The **Mariner Mark II** spacecraft concept is briefly described and illustrated in **Planetary Exploration through Year 2000**, Part 1 of the **Solar System Exploration Committee** (SSEC) report for the NASA Advisory Council (1983). It is described as a modular, reconfigurable multi-mission spacecraft for exploration beyond Mars [Ref. pp 74-80]. {Ed.}

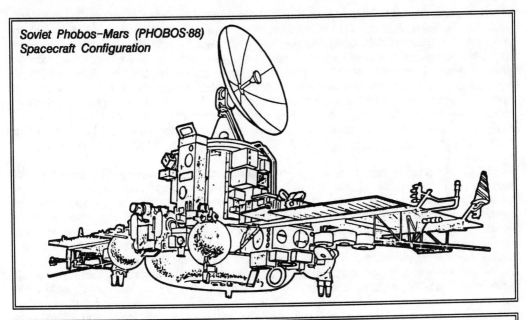

**Soviet Phobos–Mars (PHOBOS·88)
Spacecraft Configuration**

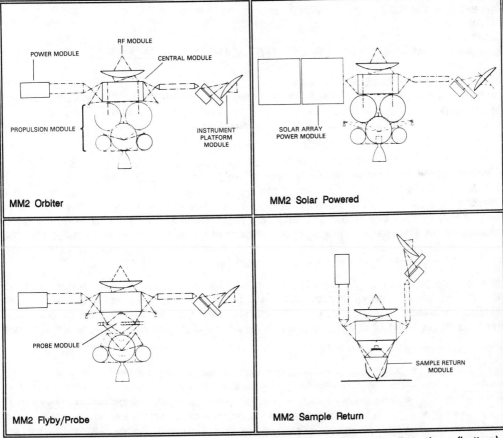

POWER MODULE

RF MODULE

CENTRAL MODULE

PROPULSION MODULE

INSTRUMENT
PLATFORM
MODULE

MM2 Orbiter

SOLAR ARRAY
POWER MODULE

MM2 Solar Powered

PROBE MODULE

MM2 Flyby/Probe

SAMPLE RETURN
MODULE

MM2 Sample Return

Figure 2A–1: Soviet PHOBOS·88 Spacecraft (top); Mariner Mark II Configurations (bottom)

The Intercosmos (Soviet/East European team) PHOBOS·88 program is an international project. A number of agencies, scientists, and specialists from many countries are involved in developing the scientific program and in designing and manufacturing the scientific instrumentation, and many will of course be involved as well in conducting the experiments. Participants include (in alphabetical order) representatives from Austria, Bulgaria, Czechoslovakia, the European Space Agency, the Federal Republic of Germany, Finland, France, the German Democratic Republic, Hungary, Poland, the Soviet Union, Switzerland, and Sweden. Some key project personnel include: Academician Sagdeev, who is head of the international scientific council of the PHOBOS·88 project; V.M. Balebanov, mission director; Alexander (A.V.) Zakharov, scientific coordinator and project manager; and V.M. Kovtuneneko, project scientist for the landers.

The PHOBOS·88 program represents a multi-faceted mission that will include many objectives not associated with the martian satellite for which it is named. Although one gets the impression that it is predominantly a Phobos-oriented mission, there is a significantly broader range of objectives. In addition to the investigation of Phobos, studies of Mars, the Sun, and the interplanetary medium will be conducted during the flights to Mars and while operating in orbit. However, the Phobos investigation amounts to quite a bit of exciting activity by itself, and includes fascinating events during which the orbiters will "hover" just above the satellite and landers will hop about on or attach themselves to its surface. I will discuss these spacecraft elements and activities in more detail during the course of this presentation. And, while I will of course focus primarily on the studies of Phobos and Mars, out of respect for the purpose of this **Mars** conference, I will also briefly describe the scientific objectives and instruments designed to study the Sun and the interplanetary medium.

FUNDAMENTAL IMPORTANCE OF THE MARTIAN SATELLITES

Let me begin by reviewing a few reasons why the martian satellites are important. As Joe Veverka explained during his **NASA Mars Conference** presentation [*The Moons of Mars*, NMC-1E], providing an excellent and detailed overview of Phobos and Deimos, these two satellites are believed to be small primitive bodies that are of extreme interest in terms of understanding the formation and evolution of the solar system.

A sense of the size of the martian satellites can be grasped, for example, by driving around the beltway of Houston, Texas. Should one visit NASA's Johnson Space Center, landing at Houston's Intercontinental airport, a lap around the I-610 loop gives one a rough idea of the size of Phobos at a familiar scale. Phobos is in an orbit approximately 9,378 km from the center of Mars, while Deimos is in a higher orbit at 23,459 km. Phobos is approximately 27x21x19 km in size, and the satellite has an average escape velocity of about 12.3 m/s (which varies over the surface depending on location relative to Mars[4]). Deimos is smaller at about 15x12x11 km and has a similarly variable low average escape velocity of about 7 m/s.

The basic characteristics of the satellites, as outlined by Joe Veverka, include irregular shapes and heavily cratered surfaces, aspects that are obvious in the pictures and drawings he used [see Fig. 1E-3 (NMC-1E), pg 96]. Indeed, both are heavily cratered and appear to have thick regoliths (although that of Deimos appears to be thicker and surprisingly mobile. Phobos has

---

[4] Because of the considerable variability in calculated escape velocities over the surfaces of Mars' two satellites, depending on location relative to Mars (both satellites are locked in synchronous rotation such that the same face is continuously oriented toward the planet), numbers given for "average" escape velocities tend to vary somewhat as discussed by different individuals. So long as these differences are minor and safely within the range of escape velocities calculated for each satellite [e.g., by Veverka, Duxbury, and others], they are not of significant importance. {Ed.}

a density of about 1.9 g/cm³ and Deimos a density of about 1.5 g/cm³. Both satellites are characterized by low albedos (about 5%), and there is evidence that they have surface compositions similar to that of carbonaceous chondrites.

Joe Veverka also reviewed the questions of both the origins of the satellites and the evolution of their orbits. These are major questions associated with Phobos and Deimos, and the possibility that they may be captured asteroids is an important issue. As Joe Veverka pointed out, there are difficulties with the orbital histories implied by the captured-asteroid hypothesis, but alternate models (such as their accretion in the martian environment) also have problems. As a result, the detailed study of these satellites--their orbital histories, their present orbits, their surface characteristics--are important objectives in trying to understand the nature of small primitive bodies in the solar system.

Phobos and Deimos are very exciting from a geological point of view as well. Phobos, for example, is characterized by a set of grooves that Joe Veverka has already described in detail [see Fig. 1E-10 (NMC-1E), pg 104]. One can see that the grooves are parallel, linear depressions. They are generally 100 to 200 meters wide although a few may be nearly 700 meters across, and most are probably 10 to 20 meters deep. Their origin is not well understood. For example, they may be a manifestation of primary layering, they may be due to deformation associated with tidal distortion or the drag associated with capture, or--as strongly suggested--they may be dramatic evidence of impact fracturing (accompanied perhaps by the release of volatiles). There is also a possibility that at least some of them are due to secondary ejecta impact.

Cratering has in fact played an important role in shaping the surfaces of these satellites, which is obvious in spacecraft images [see Fig. 1E-7 (NMC-1E), pg 101]. The crater Stickney, a 10-km diameter crater on Phobos, illustrates in particular the effects of a large impact that may be associated with many of the groove-like striations [see Fig. 1E-9 (NMC-1E), pg 102]. Therefore, in terms of what they may reveal about surface processes or the tidal evolution of the satellite (perhaps even something about volatile release), the origin of the grooves represents a major question.

Objectives for the Investigation of Phobos

The scientific objectives relative to Phobos itself, as outlined in Figure 2A-2, include the study of the surface, the inner structure, and the satellite's orbital motion. Some of the important surface aspects that are going to be studied specifically include determining chemical and mineralogical composition. A capability will also be afforded for obtaining surface images with a very high spatial resolution, such that the evolution of the surface can be studied. One of the PHOBOS·88 goals is to develop a thermal map to facilitate the study of the physical properties of the surface in a tremendous variety of ways, and another is to understand something about the radiometric characteristics of the surface.

There is a desire to study the large-scale internal structure of Phobos to understand what the interior is like, and this of course relates to the origin of the grooves and whether or not they represent deep fractures or some other aspect of the satellite's surface. Radiometric characteristics of the surface layers will be analyzed and seismology will be conducted, the latter in fact representative of a major mission objective. One of the applications for seismological data will be to improve understanding of the possibility for tidal distortion in Mars' orbit, and another will be to provide a measurement of the frequency of meteorite impacts. Also, one will be able to look at the free forced librations and secular decelerations of the satellite's orbital motion.

```
┌─────────────────────────────────────────────────────────────────┐
│  SURFACE                                                         │
├─────────────────────────────────────────────────────────────────┤
│  ■  Chemical Composition                                        │
│  ■  Mineral Composition                                         │
│  ■  Surface Imagery with High Spatial Resolution                │
│  ■  Thermal Map                                                 │
│  ■  Physical Properties                                         │
│  ■  Radiometric Characteristics                                 │
├─────────────────────────────────────────────────────────────────┤
│  INNER STRUCTURE                                                 │
├─────────────────────────────────────────────────────────────────┤
│  ■  Large-Scale Structure                                       │
│  ■  Seismology                                                  │
│  ■  Radiometric Characteristics                                 │
├─────────────────────────────────────────────────────────────────┤
│  ORBITAL MOTION                                                  │
├─────────────────────────────────────────────────────────────────┤
│  ■  Free and Forced Libration                                   │
│  ■  Secular Deceleration                                        │
└─────────────────────────────────────────────────────────────────┘
```

Figure 2A-2: PHOBOS·88 Scientific Objectives for Phobos Investigation

Objectives for the Investigation of Mars

It isn't news that Mars itself has always been a body of extreme interest because of its significance in comparative planetology. This is particularly true in terms of the formation and evolution of its interior, its solid surface, and its atmosphere. Furthermore, interaction of the planet's atmosphere and lithosphere, the role of volatiles, and the interaction of tectonics and volcanism (in its thermal evolution) represent fundamental questions concerning Mars. Indeed, they concern all of the terrestrial planets. The relationship of Mars to its inner-planetary environment (and the nature of its ionosphere and magnetosphere) are of great interest and pose fundamental questions that the PHOBOS·88 mission can and will address. Relative to Mars, the scientific objectives of the PHOBOS·88 program focus on the four areas outlined in Figure 2A-3: surface studies, atmospheric studies, studies of the ionosphere, and studies of the magnetosphere.

The Surface -- Elements of our current perspective of the surface of Mars provide the general basis for many of the questions that are to be addressed by the PHOBOS·88 mission relative to Mars itself. These goals include learning more about the nature of the surface, trying to understand its chemical and mineralogical composition, and obtaining images of the planet at high resolution (both spatially and spectrally) to study the thermal and radiometric properties of the planet.

On the surface, the **Valles Marineris** canyon region of Mars [see Fig. 1C-7, pg 62 (NMC-1C)] exemplifies some of the kinds of things one might want to study on Mars. One can see in that region volcanic plains cut by extensive tectonic graben to produce giant rift valleys like **Valles**

SURFACE
■ Chemical Composition ■ Mineral Composition ■ Orbital Imagery of Planet ■ Thermal Map ■ Radiometric Characteristics
ATMOSPHERE
■ Chemical Composition ■ Density Profile ■ Temperature Profile ■ Dust Density ■ Dynamics, Evolution
IONOSPHERE
■ Density Profile ■ Dynamics
MAGNETOSPHERE
■ Parameters of Magnetic Dipole: Its Moment and Orientation ■ Plasma: Ion and Energy Composition ■ Three-Dimensional Function of Plasma Distribution ■ Characteristics of Plasma Waves (Electric and Magnetic Fields) ■ Structure of Magnetosphere and Its Boundaries

Figure 2A-3: Scientific Objectives for PHOBOS-88 Mars Investigation

Marineris, and these in turn are modified by mass wasting that one can see along the flanks of the valleys. Channel formations and "small" tributaries the size of Earth's Grand Canyon are typical of this one incredible surface feature. The balance of much of the surface is further modified--quite obviously--by impact cratering (particularly evident in the ancient terrain of the southern hemisphere), and one also finds evidence of eolian deposits associated with many of the martian surface features. There also is evidence for chemical reaction of the surface and the atmosphere.

Atmosphere -- The atmosphere is also a topic of extreme interest, as has been described at length by other NASA Mars Conference participants and documented in these proceedings. The objectives for this aspect of the PHOBOS·88 science program include determining the atmosphere's chemical composition as well as its temperature and density profiles. Mission goals also reflect interests to improve understanding of the atmospheric dust density while learning more about the atmosphere's dynamics and evolution. The density profile and dynamics of the ionosphere will also be studied and reflect major objectives of the science mission.

Magnetosphere and Plasma Studies -- The magnetic field making up the martian magnetosphere will be studied in detail, as will many aspects of the interaction of the interplanetary medium and Mars itself. The experiments developed to support this aspect of the science program are also used to study the interplanetary medium during the flight to Mars.

Solar Studies

One might find it useful to understand the objectives associated with those aspects of the PHOBOS·88 science program that focus on questions that do not directly involve Mars and its satellites. For example, the Sun will be a major object of study during the mission. I won't address these objectives in detail, except to again emphasize the multi-objective nature of the PHOBOS·88 science plan. These aspects of the science activity are outlined in Figure 2A-4, and the characteristics of the solar investigation are reflected in Figure 2A-5.

CORONA AND UPPER CHROMOSPHERE
■ Large-Scale Structure ■ Active Regions ■ Bright Spots ■ Coronal Holes
SOLAR FLARES
■ Precursors and Transient Sources ■ Explosion Stage ■ Localization of Acceleration Region
SOLAR OSCILLATIONS
■ Helioseismology ■ Spectrum of Pressure and Gravity Nodes
SOLAR ACTIVITY FORECASTING
■ Direct Observations of Solar Activity Centers Invisible from Earth (opposite side of Sun) ■ Updating of Spectrum Distribution of Solar Constant

Figure 2A-4: PHOBOS·88 Scientific Objectives for Solar Investigation

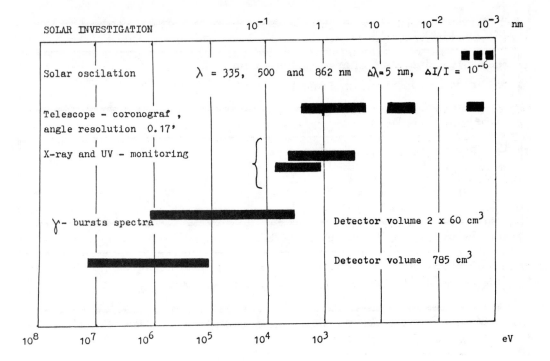

Figure 2A-5: Aspects of Solar Investigation

Specifically, corona and upper chromosphere, large-scale structure, coronal holes, and bright spots will be observed and studied. Examinations will be made to determine if the precursors of solar flares can be detected and to search for transient sources. One would hope that the explosion stage of these solar flares can be seen, such that scientists can attempt to localize and study the acceleration regions.

Relative to solar oscillations, helioseismologic investigations will be undertaken so that pressure and gravity nodes can be analyzed. These studies will of course attempt to improve understanding of solar activity in terms of forecasting its long-range effects. In fact, one of the more interesting aspects of the mission plan is that it will make possible direct observations of solar activity centers on the side of the Sun that is not visible from Earth. Simultaneous measurements can be made from the vicinities of Earth and Mars, and continuous measurements will be taken in order to update the spectral distribution of the solar constant. This is an ambitious set of objectives, and a surprising number of experiments have been incorporated to achieve them.

Interplanetary Medium

As Figure 2A-6 indicates, the interplanetary medium will not be neglected by the PHOBOS·88 science program; its characteristics are further reflected in Figure 2A-7. The solar wind will be analyzed to determine its energy, ion and charge composition, and its velocity distribution function. The measurement of solar cosmic rays, along with spectral analyses and acceleration mechanisms, will also be undertaken. The structures of interplanetary shocks will be investigated, and shocks near Mars will of course be closely studied. In addition, localization and spectral analyses (associated with studies of cosmic and solar gamma bursts) will be undertaken as a component of the study of the interplanetary medium and beyond.

SOLAR WIND
■ Energy, Ion, and Charge Composition ■ Velocity Distribution Function
SOLAR COSMIC RAYS
■ Spectra ■ Anisotropy ■ A-Shock Near Mars
COSMIC AND SOLAR GAMMA BURSTS
■ Localization ■ Spectra

Figure 2A-6: PHOBOS-88 Scientific Objectives for Investigation of Interplanetary Medium

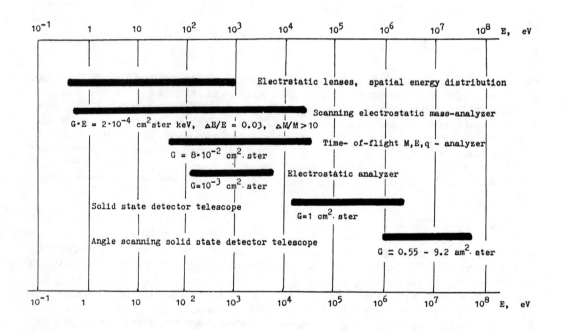

Figure 2A-7: Aspects of Investigation of Interplanetary Medium

Mission Profile

Some of the characteristics of the mission itself are reflected in the basic mission profile illustrated in Figure 2A-8. Data given in the illustration reflect the basic flight parameters, beginning with a mid-1988 (summer) launch. The Earth-to-Mars flight duration will be 200 days and the spacecraft will have to catch Mars from its transfer orbit before rendezvous and MOI (Mars Orbit Insertion) can be accomplished. At that point, the Earth-Mars-Sun angle is about 40 degrees.

The spacecraft orbital plan will be characterized by four basic working orbits during the mission. Figure 2A-9 is an enlargement of a specific portion of the flight and of its orbital geometry. The first orbit, identified in the illustration as Orbit 1, is elliptical and has a pericenter at 4200 km and an epicenter at 79,000 km. Orbit 1 involves a very close approach to Mars, affording a valuable, high resolution remote sensing opportunity during this phase of the mission. The spacecraft remains in the first orbit for about 25 days. The second orbit (Orbit 2) is also elliptical, with its pericenter at 9700 km and it epicenter again at 79,000 km. The spacecraft remains in this orbital configuration for about 30 days.

The third orbit (Orbit 3) is a circular orbit at 9700 km. The spacecraft will initially be in this orbit for 35 days, after which it will go into a fourth orbit that is both circular and synchronous (with a pericenter at 9400 km) to facilitate the rendezvous with Phobos. The fourth

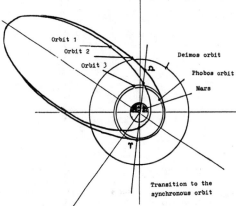

Figure 2A-8: Interplanetary Flight Plan

Figure 2A-9: Orbital Plan and Geometry

orbit is identified as the Phobos orbit in the illustration (Fig. 2A-9). The PHOBOS·88 orbiter spacecraft will stay in the fourth (Phobos) orbit for about 30 days, during which it will rendezvous with Phobos and close to a hovering range of about 50 meters, as illustrated in Figure 2A-10[5]. At this point in the mission activity, a detailed analysis and examination of the satellite's surface will be undertaken and two lander vehicles will be deployed.

Subsequent to the Phobos encounter (over a period of approximately 30 days), the spacecraft will return to Orbit 3 (circular) and will remain there for an additional 140 days. During this period, science investigations will again concentrate on the other program objectives already discussed -- Mars and the solar observations. The total duration of the scheduled PHOBOS·88 program activity is approximately 460 days.

---

5 The hovering period for the PHOBOS·88 orbiter is short-lived, lasting only 15 to 20 minutes. Its position is not fixed relative to the satellite, but instead drifts at a low relative velocity (2-5 m/s) as it deploys its landers and conducts its science program. {Ed.}

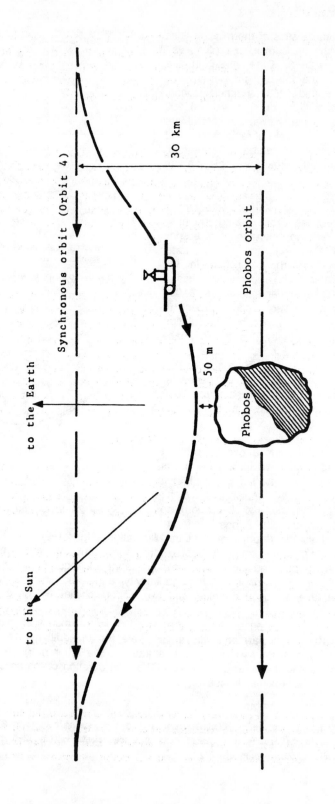

Figure 2A–10: PHOBOS-88 Orbiter, Phobos–Hovering Operation

PRIMARY EXPERIMENTS

At this point I will provide a general description of the most significant experiments and explain their basic objectives. There are at least 25 major experiments on the spacecraft which have been designed to achieve the scientific objectives I've already outlined. While I will discuss the experiments that are focused on Mars and Phobos in greater detail, I will first briefly review the solar-oriented and interplanetary medium experiments[6].

Solar Experiments

Experiment TEREK is a joint experiment between Czechoslovakia and the Soviet Union, and utilizes a telescope coronagraph designed for X-ray solar radiation studies. Experiment IPHIR is a joint project involving a number of countries, and the instrument is designed to make continuous measurements of solar radiation intensity in three narrow spectral channels. A joint Czech-Soviet experiment, RF-15 and SUFR, will monitor X-ray and UV solar radiation as well as solar flares. In addition, two French-Soviet experiments--VGS and LILAS--are designed to conduct studies of cosmic and solar gamma bursts.

Interplanetary Medium Experiments

The experiments dedicated to the study of plasma and the interplanetary medium include (for particle studies) experiment ASPERA, which also supports investigations of Mars' magnetosphere and was designed as a cosmic plasma scanning analyzer. In addition, the experiment SOVIKOMS is a spectrometer for studying the energy, mass, and charge composition of the solar wind. Experiment TAUS was designed to perform solar wind and alpha particle spectrometry, and experiment APV-F will study the spectrometry and analyze the angular distributions of low density electrons and ions. Experiment LET is designed for high energy solar cosmic ray spectrometry while experiment SLED is for low-energy solar cosmic ray spectrometry. Finally, experiment PLASMA will be used to conduct radio sounding of the martian ionosphere.

Experiments for the Study of Mars and Phobos

The Mars/Phobos investigation segment of the mission represents both active and passive techniques for the analysis of the martian surface and atmosphere and the surface of Phobos. In addition, some very interesting contact technology will be employed during the surface investigation of Phobos. These three techniques, then, make up the basic components of the PHOBOS·88 planetary science profile.

Active -- The active techniques include the LIMA-D experiment, which facilitates a remote-laser, mass spectrometric analysis of soil composition, and the DION experiment, which performs the remote mass analysis of secondary ions. In addition, the Soviets have designed a specialized radar package that will be used in a radio sounding experiment. Its purpose is to make some determinations about subsurface structure and the electrophysical characteristics of the regolith. I will discuss these active experiments in a little more detail, following this overview, to provide a better understanding of their design and how they will be implemented.

---

6 Some of the experiments described may not be supported by illustrations, and their descriptions may be unusually brief because they were originally described "on screen" via detailed visual illustrations that were not provided for publishing. In addition, the experiment and instrument names are often unusual (contextually) or unfamiliar terms (and/or acronyms) for which we are unable to provide explanations or definitions. {Ed.}

Passive -- A series of experiments utilize passive techniques. Among them is the FREGAT experiment that is essentially an imaging system designed for the study, survey, and mapping of the surfaces of Phobos and Mars. There is also a series of experiments--a radiometer, spectrophotometer, spectrometers, and scanning radiometers--that will make thermal and spectral measurements of the surfaces of Phobos and Mars. Experiment GS-14, for example, is an experiment that will study the spectrometry of gamma emissions from those surfaces. The IPNM experiment will make measurements of the neutron radiation of Phobos' surface, and experiment AUGUST will study the spectrometry and chemical composition of the martian atmosphere. And, as I have already indicated, there are additional experiments to support the study of fields and waves in the planetary medium (primarily concerned with Mars' magnetosphere) that can also be classified as passive, including experiments APV-F (mentioned earlier) for measuring plasma waves, and ASPERA -- described as a cosmic plasma scanning analyzer. Two magnetometer instruments are provided, which are identified in Soviet documentation as MAGMA and FGMM-1.

Contact: Phobos Landers -- Two surface contact techniques will be used on Phobos, representing different types of landers. One of these is the long-term automated lander -- or LAL. (I really believe, as one may have already concluded at this point, that the Soviets have definitely closed the acronym gap!) The second surface vehicle is a unique hopping lander designed to explore Phobos by--quite literally--leaping about on its surface.

DETAILS: THE INVESTIGATION OF PHOBOS

The PHOBOS·88 orbiter spacecraft will descend from an orbit about 30 km from Phobos, achieving a closest-approach distance of only about 50 meters where it assumes a hovering orientation relative to the satellite. A series of remote experiments will then be undertaken using instruments on the orbiter. More importantly, however, two different landers will be deployed to conduct a variety of experiments on the surface of the satellite.

Orbiter-Based Phobos Experiments

LIMA-D -- One of the most interesting experiments to be conducted is a remote-laser mass spectrometric analysis of soil composition (LIMA-D). This experiment is illustrated and further defined in Figure 2A-11. The objective of the LIMA-D experiment is to determine the elemental and isotopic composition of the regolith of Phobos.

To begin, a laser beam is focused on a small area on the surface of Phobos with the help of a laser altimeter. The material analyzed will represent a one- to two-micron layer derived from a surface area of only one or two square millimeters. The laser beam irradiates the surface area from the orbiter's range of approximately 50 meters, and some of the vaporized (ionized) material is then essentially captured and measured on board the spacecraft as it scatters. The time-of-flight for ions so measured is also recorded. The mass spectrometric analysis of the scattered ions is made in a reflectron equipped with a retarding field. Approximately 10^6 ions will be counted during one cycle of the experiment. The pulse energy of the laser is 0.5 joules and the pulse duration is about 10 nanoseconds. This is an incredibly interesting experiment and it is designed to make a continuous set of measurements across the surface of Phobos.

DION -- Figure 2A-12 illustrates the nature and function of another experiment, DION, which facilitates the remote mass analysis of secondary ions. This experiment's purpose is again the study of Phobos' surface composition. The analysis will represent a layer about ten angstroms thick, and it will determine what elements are present that have been implanted by the solar wind. The methodology is as follows: there is an injection of an ion beam from the spacecraft

228

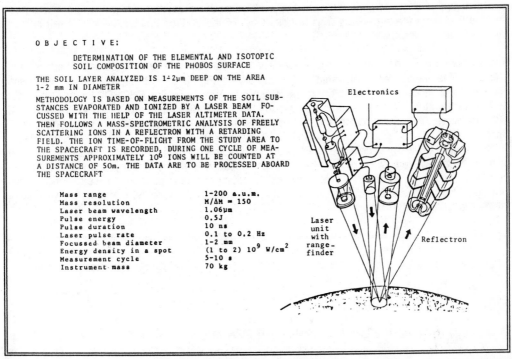

OBJECTIVE:

DETERMINATION OF THE ELEMENTAL AND ISOTOPIC
SOIL COMPOSITION OF THE PHOBOS SURFACE

THE SOIL LAYER ANALYZED IS 1÷2μm DEEP ON THE AREA
1-2 mm IN DIAMETER

METHODOLOGY IS BASED ON MEASUREMENTS OF THE SOIL SUB-
STANCES EVAPORATED AND IONIZED BY A LASER BEAM FO-
CUSSED WITH THE HELP OF THE LASER ALTIMETER DATA.
THEN FOLLOWS A MASS-SPECTROMETRIC ANALYSIS OF FREELY
SCATTERING IONS IN A REFLECTRON WITH A RETARDING
FIELD. THE ION TIME-OF-FLIGHT FROM THE STUDY AREA TO
THE SPACECRAFT IS RECORDED, DURING ONE CYCLE OF MEA-
SUREMENTS APPROXIMATELY 10^6 IONS WILL BE COUNTED AT
A DISTANCE OF 50m. THE DATA ARE TO BE PROCESSED ABOARD
THE SPACECRAFT

Mass range	1-200 a.u.m.
Mass resolution	M/ΔM = 150
Laser beam wavelength	1.06μm
Pulse energy	0.5J
Pulse duration	10 ns
Laser pulse rate	0.1 to 0.2 Hz
Focussed beam diameter	1-2 mm
Energy density in a spot	(1 to 2) 10^9 W/cm^2
Measurement cycle	5-10 s
Instrument mass	70 kg

Figure 2A-11: LIMA-D -- Remote Laser Mass-Spectrometric Analysis, Phobos Soil Composition
(Austria, Bulgaria, CSSR, GDR, Max-Planck Institute, FGR, USSR)

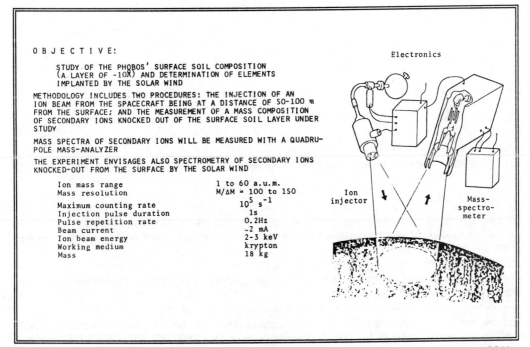

OBJECTIVE:

STUDY OF THE PHOBOS' SURFACE SOIL COMPOSITION
(A LAYER OF ~10Å) AND DETERMINATION OF ELEMENTS
IMPLANTED BY THE SOLAR WIND

METHODOLOGY INCLUDES TWO PROCEDURES: THE INJECTION OF AN
ION BEAM FROM THE SPACECRAFT BEING AT A DISTANCE OF 50-100 m
FROM THE SURFACE; AND THE MEASUREMENT OF A MASS COMPOSITION
OF SECONDARY IONS KNOCKED OUT OF THE SURFACE SOIL LAYER UNDER
STUDY

MASS SPECTRA OF SECONDARY IONS WILL BE MEASURED WITH A QUADRU-
POLE MASS-ANALYZER

THE EXPERIMENT ENVISAGES ALSO SPECTROMETRY OF SECONDARY IONS
KNOCKED-OUT FROM THE SURFACE BY THE SOLAR WIND

Ion mass range	1 to 60 a.u.m.
Mass resolution	M/ΔM = 100 to 150
Maximum counting rate	10^5 s^{-1}
Injection pulse duration	1s
Pulse repetition rate	0.2Hz
Beam current	~2 mA
Ion beam energy	2-3 keV
Working medium	krypton
Mass	18 kg

Figure 2A-12: DION -- Remote Mass-Analysis of Secondary Ions (Austria, France, USSR)

at a distance of 50 to 100 meters from the surface, and a measurement of the mass composition (derived from the secondary ions knocked out of the surface soil by the ion beam) is made by the mass spectrometer.

Radio Sounding -- A radar study of Phobos will make use of a special Soviet-designed radar package. The experiment, illustrated schematically in Figure 2A-13, represents an interesting aspect of the study of Phobos during the closest approach by the PHOBOS·88 orbiter, in that it involves the study of both surface topography and subsurface structure using the same instrument package. The investigation requires determining the electrophysical characteristics of the Phobos regolith, and utilizes radio sounding of the interior of Phobos. Sounding frequencies of 5, 130, and 500 MHz will be used, and sounding depths ranging from the surface layers down to perhaps 200 meters or more may be possible. Because both topography and substructure will be studied using this radar sounding technique, results of this experiment are expected to provide knowledge about the characteristics of the materials Phobos is made of as well as its internal layering and structure.

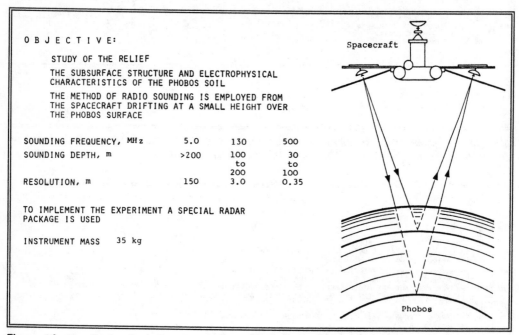

O B J E C T I V E:

    STUDY OF THE RELIEF

    THE SUBSURFACE STRUCTURE AND ELECTROPHYSICAL CHARACTERISTICS OF THE PHOBOS SOIL

    THE METHOD OF RADIO SOUNDING IS EMPLOYED FROM THE SPACECRAFT DRIFTING AT A SMALL HEIGHT OVER THE PHOBOS SURFACE

SOUNDING FREQUENCY, MHz	5.0	130	500
SOUNDING DEPTH, m	>200	100 to 200	30 to 100
RESOLUTION, m		150	3.0	0.35

TO IMPLEMENT THE EXPERIMENT A SPECIAL RADAR PACKAGE IS USED

INSTRUMENT MASS   35 kg

Spacecraft

Phobos

Figure 2A-13: Radar Sounding Experiment, Surface Relief and Subsurface Structure (USSR)

FREGAT -- The FREGAT experiment illustrated in Figure 2A-14 is particularly significant from a geologist's point of view. The objectives of the experiment are to obtain images and spectra of Phobos' surface, which will then be used primarily for position referencing to support measurements made by other experiments. However, it also will provide data to make a series of photometric maps, topographic maps, spectral maps, and maps representing specific textures and morphology.

The experiment includes three cameras, a spectrometer, and an on-board memory unit. The imaging system will make simultaneous measurements in three channels and will also allow multispectral images to be formed that can later be compiled. At the spacecraft's closest approach

O B J E C T I V E S:

- IMAGES AND SPECTROGRAMS OF PHOBOS' SURFACE
- POSITION REFERENCING FOR MEASUREMENTS MADE BY OTHER INSTRUMENTS
- THEMATIC MAPS OF PHOBOS (TOPOGRAPHIC, SPECTRAL RADIANCE, TEXTURE/ MORPHOLOGY MAPS, ETC.)

FREGAT FEATURES:

THREE CAMERAS, A SPECTROMETER AND AN ON-BOARD MEMORY UNIT

THE TV CAMERAS MAKE A SIMULTANEOUS SURVEYING IN THREE SPECTRAL CHANNELS AND THUS ALLOW FOR COMPOSITE IMAGES WITH THE FOLLOWING CHARACTERISTICS:

Distance to object to be surveyed,km	6300	800	50	0.05
Surface resolution /without blurring/,m	7000	900	60	0.06
One-frame coverage,km	3000x3000	400x300	25x19	25.10^{-3}x19.10^{-3}

THE SPECTROMETER IS SCANNING THE SURFACE UNDER STUDY BY A SLOT PARALLEL AND ABOUT EQUAL TO A TV FRAME SIDE, THE DISPERSIVE SPECTROMETER SYSTEM ENSURES SPECTROMETRY IN 14 SPECTRAL BANDS FROM 0.4 to 1.1 μm AT A RESOLUTION OF 50 nm.

CHANGES IN THE DIRECTION OF THE LINE OF SIGHT, USING A ROTATIONAL MIRROR, MAKE POSSIBLE VERTICAL AND WIDE-COVERAGE SURVEYS OF PHOBOS, NOT ONLY CAN IMAGES OF MARS AND PHOBOS BE OBTAINED BUT ALSO THOSE OF BRIGHTEST STARS WHICH ARE IMPORTANT FOR NAVIGATION.

THE ON-BOARD MEMORY UNIT PERMITS RECORDING 1100 FULL FRAMES (3 TV FRAMES AND SPECTROGRAM) AT A RATE OF 2Mbit.s^{-1} AND THEIR READING AND DUMPING - VIA THE TM CHANNEL TO EARTH AT A RATE OF 4Kbit.s^{-1}.

THE TOTAL MASS IS 50 kg.

Figure 2A-14: FREGAT -- Survey and Mapping Camera Package (Bulgaria, GDR, USSR)

during the 50-meter hovering activity, it will have a surface resolution of approximately 6 cm. Of course, imaging is conducted as the spacecraft moves in toward the satellite and then moves away, facilitating a wide range of resolutions and the opportunity to study a large area of the surface.

The spectrometer system has fourteen spectral bands ranging from 0.4 to 1.1 microns. It is also possible to make changes in the direction of the line of sight of the imaging system using rotational mirrors. This in turn makes possible the imaging of wider area, such as panoramas of the surface of phobos while at the closest approach range, for example. The on-board memory unit, a recorder, permits storage in excess of 1000 full frames.

KRFM/ISM/TERMOSKAN -- The KRFM, ISM and Termoskan experiments that are illustrated in Figure 2A-15 (next page) are intended to study the thermal physics and reflection properties of Phobos' surface in the infrared and visible spectral regions. Mineral compositions will be determined and mineralogical maps will be made using near-infrared spectra obtained over different wavelength ranges measured by each of the instruments: a radiometer, a photometer, and an infrared spectrometer. The photometer has 10 spectral subregions, the spectrometer 128, and the radiometer 6. The spectrometer can also operate in a scanning mode.

THE PHOBOS LANDERS

One of the most intriguing and interesting aspects of the entire PHOBOS·88 project is the concept reflected in its landers. There are two different kinds of landers used, and one of each kind will be deployed by each of the PHOBOS·88 orbiters.

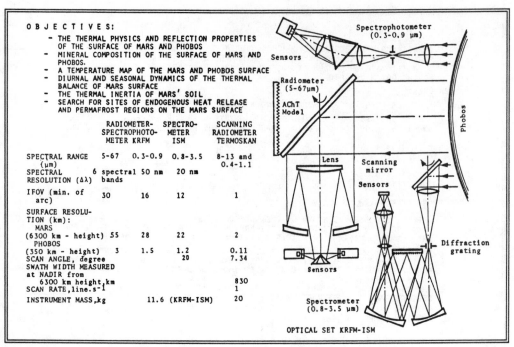

	RADIOMETER-SPECTROPHOTO-METER KRFM	SPECTRO-METER ISM	SCANNING RADIOMETER TERMOSKAN
SPECTRAL RANGE (µm)	5-67	0.3-0.9	0.8-3.5	8-13 and 0.4-1.1
SPECTRAL RESOLUTION (Δλ)	6 spectral bands	50 nm	20 nm	
IFOV (min. of arc)	30	16	12	1
SURFACE RESOLUTION (km): MARS (6300 km - height)	55	28	22	2
PHOBOS (350 km - height)	3	1.5	1.2	0.11
SCAN ANGLE, degree			20	7.34
SWATH WIDTH MEASURED at NADIR from 6300 km height,km				830
SCAN RATE,line.s⁻¹				1
INSTRUMENT MASS,kg		11.6 (KRFM-ISM)		20

Spectrophotometer (0.3-0.9 µm)

Sensors

Radiometer (5-67µm)

AChT Model

Phobos

Lens

Scanning mirror

Sensors

Sensors

Diffraction grating

Spectrometer (0.8-3.5 µm)

OPTICAL SET KRFM-ISM

Figure 2A-15: KRFM/ISM/TERMOSKAN -- Radiometric and Spectral Sensing (USSR, France)

Long-Term Automated Lander

When the orbiter spacecraft has achieved its hovering distance of fifty meters, the long-term automated lander (LAL) is separated; the Phobos approach and LAL separation events are depicted in Figure 2A-16. The LAL represents an excellent, comprehensive, long-term spacecraft for the investigation of Phobos' surface. It is deployed from the base of the spacecraft straight down to the surface of Phobos, while the hopper vehicle (second lander) I will describe shortly is deployed off the side. Once separated, the LAL begins descending toward Phobos' surface at several meters per second -- oriented for touchdown. To assure its orientation, the LAL will be "spun up" on its longitudinal axis after separation. In its landed configuration, as illustrated in Figure 2A-17, the LAL will essentially be "bolted" to the surface with an anchoring penetrator device. The main objective of the LAL is to conduct a variety of scientific experiments and observations on Phobos that require a longer period of time. A microprocessor is provided to control all of the LAL's systems. The science instruments/systems incorporated into the lander include:

- A spectrometer that uses alpha back-scattering and X·ray fluorescence to study the elemental composition of Phobos' surface;

- Optical sensors to determine the Sun's angular position (which will be useful in studying the Phobos libration and mass distribution;

- A spectrometer to study Phobos' internal structure and detect meteoritic impacts (it will also measure the deformation of Phobos in its orbit about Mars);

- A penetrometer that incorporates temperature sensors and an accelerometer;

- A small TV camera with which to study the microstructure of the soil;

- The on-board radio system will facilitate celestial mechanics studies of Phobos' libration.

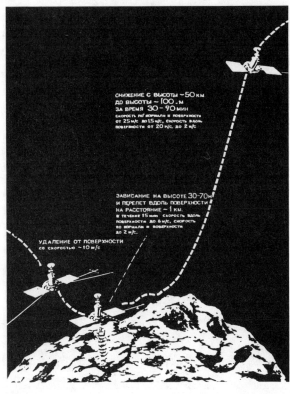

Figure 2A-16:

Soviet Depiction of Hovering PHOBOS·88 Orbiter and Deployment of Landers

< ■ *Orbiter approaches to hovering range (50 m), drifts at low-relative-velocity (2-5 m/s);*

■ *Stay time 15-20 minutes;*

■ *Deploys LAL from bottom of orbiter spacecraft and hopper from side.*

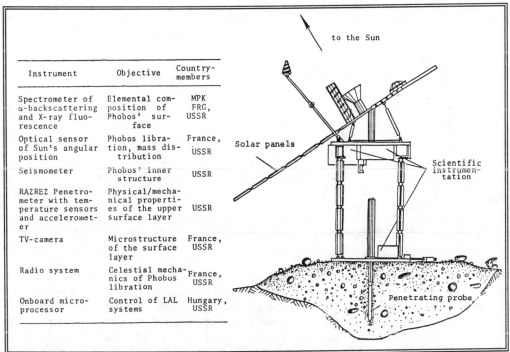

Instrument	Objective	Country-members
Spectrometer of α-backscattering and X-ray fluorescence	Elemental composition of Phobos' surface	MPK FRG, USSR
Optical sensor of Sun's angular position	Phobos libration, mass distribution	France, USSR
Seismometer	Phobos' inner structure	USSR
RAZREZ Penetrometer with temperature sensors and accelerometer	Physical/mechanical properties of the upper surface layer	USSR
TV-camera	Microstructure of the surface layer	France, USSR
Radio system	Celestial mechanics of Phobos libration	France, USSR
Onboard microprocessor	Control of LAL systems	Hungary, USSR

Figure 2A-17: Long-Term Automated (Phobos) Lander

The Phobos Hopper

The second lander, called a "hopper" because of its unique means for achieving mobility in a low-surface-gravity environment, is illustrated in Figure 2A-18. The hopper is not very large, being perhaps somewhere in size between a basketball and a medicine ball. After deployment off the side of the orbiter, the hopper "falls" about 50 meters to the surface and impacts at very low velocity. It is expected to bounce and roll a bit before coming to rest, and it then jettisons a structural element incorporated to keep it from rolling too far after its initial landing.

Surface Mobility -- After the initial landing events are completed, the hopper will be making geochemical and other measurements as it moves about on Phobos. The hopper will be cycled through a series of operations that cause it to hop across the surface, righting itself after each hop, and to perform a variety of experiments to measure surface characteristics. The "hop" itself occurs when a spring mechanism is activated (once the hopper has been positioned in the correct upright orientation by rod devices), propelling the little lander grasshopper-like to its next location[7]. This is a very interesting mobility concept that is capable of moving a surprisingly comprehensive package of instruments around on the surface of Phobos. That package includes an X-ray fluorescence spectrometer, a magnetometer, a penetrometer, and a gravimeter.

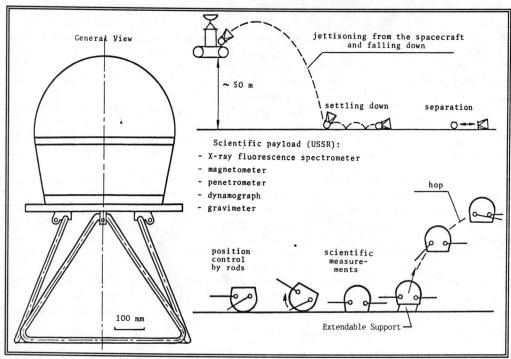

Figure 2A-18: PHOBOS·88 Hopping Lander (Hopper)

---

[7] The word "orientation," as used in the Soviet mission-description paper and by Dr. Head, appears to infer an upright rather than a directional orientation. The hopper's rounded shape (probably in association with weight distribution) is clearly designed to aid in the righting procedure, i.e., it helps the lander to roll into an orientation with the surface that is favorable to the operation of its orienting rods (if it isn't initially upright). The rods act to roll the hopper onto its only flat surface, which then provides a stable upright orientation. Considering the potential (perhaps probable) influence of topographic factors and the simplicity of the hopper's mobility mechanism, it appears that the directional aspect of its movement will involve considerable uncertainty. {Ed.}

For orienting itself into an upright configuration after a hop, the hopper is equipped with one stationary arm and one moveable arm which work together to complete the orienting operation. This mode of activity causes the hopper to roll (if necessary) and right itself for the hopping operation that actually moves it across the surface. The moveable orienting rod can be seen at different angles (Fig. 2A-18), relative to the body of the hopper and the stationary rod, as it first completes the orienting phase of its rotation while on the surface and then returns to its initial point while in flight during a hop[8]. I have to believe that the Bolshoi ballet was in on the design of the hopper because of the way it gets around; it represents quite a remarkable concept.

Hopper Mission and Science Events -- The experiments packaged within the hopper will measure surface composition and other characteristics of Phobos' surface. Taking its activities step by step, the first is basically that of stopping initial momentum after separation from the orbiter and impact on Phobos. It hits the surface and takes a bounce or two, initially covering perhaps two hundred meters. It then continues to tumble and roll, passing a point some 1.4 km from its first contact, and then continues to roll slowly out to a point that could be as much as 2.5 km from the point of its initial contact (assuming that it avoids topographic obstructions). Progressing in what would seem like slow motion, even though in real time, these events could take something on the order of 30 minutes before the hopper finally comes to rest.

At this point the protective device is jettisoned, probably causing the hopper to move slightly, and it then enters its orientation mode for the first time. Deploying and using its orienting rod devices, the hopper gradually manages to get itself upright. If it is stationary and properly oriented after coming to rest, the moveable orientation rod will simply rotate around in preparation for the jump. But, if the hopper is upside down, the rotation of the rod--which would then contact the surface while rotating--rights the hopper in preparation for its leap. Initially, the orientation mode conducted prior to the first hop event will probably move the hopper approximately one meter (assuming that it in fact is not upright and must be rolled to achieve correct orientation). The completion of the orientation activity takes about one minute.

This activity prepares the hopper for its first hop. The spring is activated, causing the hopper to "leap" approximately twenty meters in altitude and forty meters in distance. As this occurs, the moveable orientation rod rotates quickly back around to its starting position while the hopper is in flight. The small vehicle then impacts and rolls again, although not as forcefully or as far as when it first made contact on Phobos. This time it rolls only about ninety meters, if (as before) nothing impedes its rolling motion, and takes about eight minutes to come to rest. The hopper again executes its orientation procedure, using the rod devices to orient and right itself. The science instruments are then activated to make measurements, which takes six or seven minutes, and the science data are transmitted back to the PHOBOS·88 orbiter. The science data acquired in this manner represent the chemical composition and other surface properties of Phobos.

This series of activities is then repeated -- another hop, orientation, more science, another hop, orientation, more science, et·cetera, until the "Lone Ranger" mode occurs and the hopper disappears over the horizon into the sunset. If this concept is successful, there will be an incredible variety of measurements over the surface of Phobos to complement the long-term,

---

[8] The Soviet mission description in which this illustration was used does not explain the function of the extendible support element (identified lower-right). It may be associated with the mechanism that propels the hopper (spring-like) on its leaps, as is perhaps suggested by the sequence of events depicted in the illustration, but the nomenclature suggests the possibility of an elevating function that might otherwise be associated with science activity just prior to a hopping event. {Ed.}

two-lander mission activity. It will provide--in addition to the data collected by the Phobos-hovering orbiter--a very comprehensive study of the surface characteristics and chemistry of Phobos.

The total time of operation for the hopper is three hours and the total number of jumps currently planned is ten. My primary concern is that the arc of one of those first jumps will terminate at the bottom of a Phobos groove, although that might not be a bad event if the hopper survives the impact. One could then acquire analysis data within the groove along its length. In any event, there is plenty of time to think about such possibilities, and the hopper clearly represents a very interesting and imaginative concept when one considers the fact that--as in the Viking case--there are two spacecraft. With the potential science afforded by two LAL vehicles and two hoppers, I think the program will acquire a tremendous amount of information about Phobos that will bear on many of the scientific questions we've been asking and which the Soviets propose as their basic rationale.

THE SURFACE OF MARS

The significance of the surface of Mars, in terms of the planet's geological structure and evolution, is where it fits into what we know about comparative planetology. When one studies a geological map of the surface of Mars, one finds illustrated very amply the dichotomy (northern and southern hemispheres) discussed by others in these proceedings [in addition to Fig. 1C-7, pg 62 (NMC-1C); see geologic maps (NMC-1B): Figures 1B-1, 1B-2, and 1B-10 on pg A-2 in the color display section], particularly in the equatorial belt over which the PHOBOS·88 orbiter will be orbiting. It is generally along the equator that the edge of the young northern lowlands (discussed in detail by Mike Carr) encroaches on the older cratered highlands of the southern hemisphere. A tremendous amount of information can be gained from studying the dramatic features that define these two major hemispherical provinces, as well as from the volcanoes of Tharsis and the canyons, valleys, and channels found along or near Mars' equatorial belt.

After the encounter with Phobos, the spacecraft will return to Orbit 3, again circularized at 9700 km, for an additional 140 days. This will initiate an extensive Mars exploration phase wherein the spacecraft experiments will be oriented toward the surface of Mars. The martian surface will be studied by remote sensing methods as previously outlined, primarily in the visible, infrared, and gamma-ray spectral regions. These data will complement remote sensing data gathered earlier (prior to the Phobos encounter when the orbiter was in Orbit 1 with a pericenter very close to the martian surface). The Mars exploration by the orbiter spacecraft, then, will take place both in a highly elliptical orbit (Orbit 1) over some 25 days, and later from Orbit 3 just prior to and following the Phobos encounter, the latter phase spanning 140 days.

One of the most exciting investigations of the martian surface will employ radiometry and photometry in an attempt to compile a global temperature map. This map will, in turn, be used to study the diurnal and seasonal dynamics of the thermal budget and also to measure the thermal inertia of the regolith. This work will facilitate a search for areas that might be emitting indigenous heat and, conversely, regions where ground ice may be particularly concentrated. In addition, elemental abundances on the martian surface will be measured through gamma-ray spectrometry. Such data will help determine the nature of martian rocks in the entire equatorial region of the planet and provide a basis for measuring the abundances of the fundamental rock-forming elements and natural radioactive elements.

The Atmosphere, Ionosphere and Magnetosphere of Mars

As I indicated earlier, the PHOBOS·88 program plan also incorporates a series of experiments to study the martian atmosphere and ionosphere. In particular, these will acquire measurements of the vertical concentration distribution of ozone, water vapor, molecular oxygen, and dust, and of the vertical temperature and pressure profiles. An extensive amount of data will also be acquired regarding seasonal, local, and diurnal variations for a variety of atmospheric parameters. And finally, measurements of the deuterium/hydrogen ratios also will be made.

In particular, the AUGUST experiment illustrated in Figure 2A-19 was developed jointly by the Soviet Union and France to study--primarily--the chemical composition of the atmosphere. It utilizes solar radiation spectra acquired as the spacecraft first passes into and then emerges from the planet's solar shadow, and will make possible a variety of vertical distribution measurements--compositional, temperature, and pressure--within the atmosphere. As reflected in the illustration, the spectrometer utilizes Fabry-Perot interferometer technology.

The martian ionosphere will be studied using a pulse-radio-sounding technique. The objective is to learn more about the ionopause (structure) and to measure the electron density-height distribution under a variety of different solar illumination conditions. Indeed, the methodology

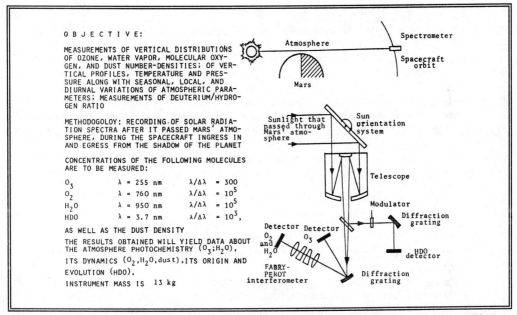

Figure 2A-19: AUGUST -- Spectrometric Studies of Martian Atmosphere (France, USSR)

associated with many of the atmospheric measurements is essentially that of recording the spectra of solar radiation detected as it passes through the atmosphere during spacecraft ingress and egress from the shadow of the planet.

The magnetosphere of Mars, which has proven difficult to detect and measure in the past, will be studied through the capabilities of two major experiments, APV-F and ASPERA, illustrated and defined in Figures 2A-20 and 2A-21. The APV-F experiment has been designed to study the fluctuation of specific electric and magnetic fields and a specific ion plasma component, as well as their spectral densities. It will also contribute to the understanding of plasma oscillations. The ASPERA experiment, mounted on a scanning platform, utilizes electrostatic analyzers (with magnetic deflectors) to determine ion mass in the magnetosphere and to detect magnetospheric boundaries. It will also be able to detect magnetospheric plasma interaction with the solar wind.

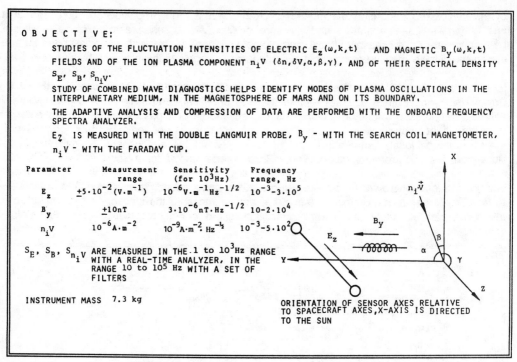

OBJECTIVE:

STUDIES OF THE FLUCTUATION INTENSITIES OF ELECTRIC $E_z(\omega,k,t)$ AND MAGNETIC $B_y(\omega,k,t)$ FIELDS AND OF THE ION PLASMA COMPONENT $n_i V$ $(\delta n, \delta V, \alpha, \beta, \gamma)$, AND OF THEIR SPECTRAL DENSITY S_E, S_B, $S_{n_i V}$.

STUDY OF COMBINED WAVE DIAGNOSTICS HELPS IDENTIFY MODES OF PLASMA OSCILLATIONS IN THE INTERPLANETARY MEDIUM, IN THE MAGNETOSPHERE OF MARS AND ON ITS BOUNDARY.

THE ADAPTIVE ANALYSIS AND COMPRESSION OF DATA ARE PERFORMED WITH THE ONBOARD FREQUENCY SPECTRA ANALYZER.

E_z IS MEASURED WITH THE DOUBLE LANGMUIR PROBE, B_y - WITH THE SEARCH COIL MAGNETOMETER, $n_i V$ - WITH THE FARADAY CUP.

Parameter	Measurement range	Sensitivity (for 10^3 Hz)	Frequency range, Hz
E_z	$\pm 5 \cdot 10^{-2} (V \cdot m^{-1})$	$10^{-6} V \cdot m^{-1} Hz^{-1/2}$	$10^{-3} - 3 \cdot 10^5$
B_y	$\pm 10 nT$	$3 \cdot 10^{-6} nT \cdot Hz^{-1/2}$	$10 - 2 \cdot 10^4$
$n_i V$	$10^{-6} A \cdot m^{-2}$	$10^{-9} A \cdot m^{-2} \cdot Hz^{-1/2}$	$10^{-3} - 5 \cdot 10^2$

S_E, S_B, $S_{n_i V}$ ARE MEASURED IN THE .1 to 10^3 Hz RANGE WITH A REAL-TIME ANALYZER, IN THE RANGE 10 to 10^5 Hz WITH A SET OF FILTERS

INSTRUMENT MASS 7.3 kg

ORIENTATION OF SENSOR AXES RELATIVE TO SPACECRAFT AXES, X-AXIS IS DIRECTED TO THE SUN

Figure 2A-20: APV-F -- Plasma Wave Fluctuations/Oscillations (CSSR, ESA, Poland, USSR)

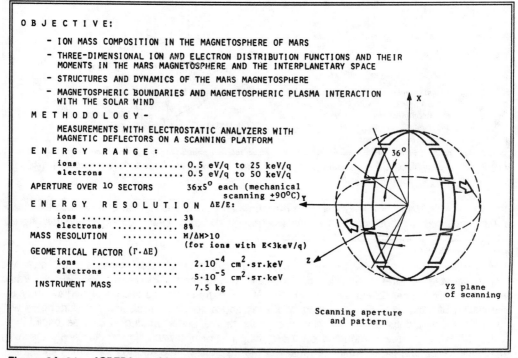

OBJECTIVE:

- ION MASS COMPOSITION IN THE MAGNETOSPHERE OF MARS
- THREE-DIMENSIONAL ION AND ELECTRON DISTRIBUTION FUNCTIONS AND THEIR MOMENTS IN THE MARS MAGNETOSPHERE AND THE INTERPLANETARY SPACE
- STRUCTURES AND DYNAMICS OF THE MARS MAGNETOSPHERE
- MAGNETOSPHERIC BOUNDARIES AND MAGNETOSPHERIC PLASMA INTERACTION WITH THE SOLAR WIND

METHODOLOGY-

MEASUREMENTS WITH ELECTROSTATIC ANALYZERS WITH MAGNETIC DEFLECTORS ON A SCANNING PLATFORM

ENERGY RANGE:

ions .... 0.5 eV/q to 25 keV/q
electrons .... 0.5 eV/q to 50 keV/q

APERTURE OVER 10 SECTORS $36 \times 5°$ each (mechanical scanning $\pm 90°$C)

ENERGY RESOLUTION $\Delta E/E$:

ions 3%
electrons 8%

MASS RESOLUTION $M/\Delta M > 10$ (for ions with $E < 3 keV/q$)

GEOMETRICAL FACTOR ($\Gamma \cdot \Delta E$)

ions .... $2 \cdot 10^{-4} cm^2 \cdot sr \cdot keV$
electrons .... $5 \cdot 10^{-5} cm^2 \cdot sr \cdot keV$

INSTRUMENT MASS ..... 7.5 kg

YZ plane of scanning

Scanning aperture and pattern

Figure 2A-21: ASPERA -- Magnetospheric Investigations

CONCLUSION: EXCITING POSSIBILITIES

With due respect for the remarkable things already discussed, there is one additional and very exciting possibility not yet mentioned -- although I hinted at it when I indicated that there would be spacecraft duplication. The PHOBOS·88 program operations, like those of the Viking and Voyager missions, will begin with two identically instrumented spacecraft. That is, there will be two orbiters (each functioning as primary spacecraft systems), and each will "mother" its own set of LAL and hopper vehicles. This is particularly valuable with respect to the use of the second set of landers.

If all goes well for both of the flight spacecraft during launch, flight, and arrival (MOI), and if the first Phobos encounter and surface investigations are successful with the first set of landers, there is a possibility that the second orbiter (and its set of landers) could be diverted to Deimos. This of course would represent a remarkable scientific (and navigational) accomplishment. The comparison of these two satellites, in light of what appears in Viking images to be unexplained differences in their surface character (regolith thicknesses and behavior), would be incredibly significant to our understanding of such bodies and their evolution. ∎

REFERENCES

This space is otherwise available for notes.

~~~~~~~~

(Dr. Head provided the following list of general references in preference to contextual references.)

Sagdeev, R.Z., Balebanov, V.M., Zakharov, A.V., Kovtunenko, V.M., Kremnev, R.S., Ksanfomality, L.V., and Rogovsky, G.N. 1986. *The Phobos Mission: Scientific Goals*. Academy of Sciences of the USSR, Space Research Institute. Moscow, USSR.[9]

Balebanov, V.M., Zakharov, A.V., Kovtunenko, V.M., Kremnev, R.S., Rogovskii, G.N., Sagdeev, R.Z., and Chugarinova, T.A. 1985. *Phobos Multidisciplinary Space Mission*. Academy of Sciences of the USSR, Space Research Institute. Moscow, USSR.

Cintala, M.J., Head, J.W., and Wilson, L. 1979. The nature and effects of impact cratering on small bodies. *Asteroids* (T. Gehrels, ed.). University of Arizona Press, pp 579-600.

Cintala, M.J., Head, J.W., and Veverka, J. 1978. Characteristics of the cratering process on small satellites and asteroids. Proceedings of the 9th Lunar and Planetary Science Conference. pp 3803-3830.

---

[9] In the two Soviet papers referenced on this page, the name G.N. Rogovsky appears in the first while G.N. Rogovskii appears in the second. It is highly probable that the same individual is named in each case, but we have been unsuccessful in attempts to learn which is the correct spelling. We offer our apologies. The translation of Russian names to English is sometimes based in part on phonetic pronunciation and may differ from one translation to another; for example, Academician Roald Sagdeev's last name is sometimes written in English as "Sagdeyev" in an attempt to achieve a more proper pronunciation by readers. {Ed.}

MARS OBSERVER: MISSION
[NMC-2B]

Mr. William I. Purdy
Project Manager: Mars Observer
Jet Propulsion Laboratory

The MARS OBSERVER program will utilize a design-to-cost spacecraft developed for a fixed amount of money. It is the first mission to reflect the PLANETARY OBSERVER concept, which dictates the use of a common, multi-mission spacecraft developed essentially from flight-proven systems and technologies. The flight system will be launched in 1992 from a Space Shuttle orbiter, and a high-energy Transfer Orbit Stage (TOS) will be used to inject the spacecraft into its interplanetary orbit. The scientific focus is specifically limited to Mars itself, and cruise science will not be performed. Initially, the spacecraft will be placed in a highly elliptical Mars orbit with a 4100-km periapsis. After about 55 days, the orbit's apoapsis will be lowered to 361 km (224 miles) above the martian surface, where the spacecraft's orbit also will be made circular and Sun-synchronous in preparation for the science mission (mapping) activity. For the most part, the mission is designed to be routine, repetitive and nonadaptive. However, the characteristics of the mapping orbit can be varied slightly to satisfy specific mission objectives. Altitude can be adjusted to achieve repeating ground tracks, such that the spacecraft either passes over the same point every 39 orbits or takes steps in a westward or eastward walk. The mission plan for Mars Observer science activity spans a full martian year, such that one complete cycle of seasonal variation can be observed, but mission extensions will be possible if science results and/or opportunities justify a continuation of data acquisition.

INTRODUCTION: OPENING COMMENTS

I am going to discuss Mars Observer engineering and management, and perhaps a little mission design, and Arden Albee [NMC-2C] will then discuss the Mars Observer science program. In doing my part, I will try to follow Jim Martin's advice to make it interesting[1]. For example,

---

[1] James S. Martin, Jr., is the former NASA project manager for the Viking Project and served as chairman of the organizing committee for the NASA Mars Conference. Mr. Martin, presently an aerospace consultant, retired from NASA following the conclusion of Viking primary mission activity at JPL. He had previously served as project manager for the highly successful Lunar Orbiter program (both Lunar Orbiter and the Viking Project were managed from NASA-LaRC). He returned to private industry where he held a series of executive positions with Martin Marietta Aerospace before retiring to devote full time to his consulting interests. {Ed.}

I will try not to get too deeply into the economic philosophy of the **Mars Observer** concept, which is to utilize a design-to-cost system for a fixed amount of money, and I will try not to give too much time to explaining the use of existing spacecraft designs and the minimization of new technology. However, one needs to understand that this kind of thinking makes up the foundation of the *Planetary Observer* philosophy that underlies several observer-based missions that will be described during this conference. I will also try not to be overly descriptive in discussing the two-year **Mars Observer** mission (687 days at Mars), which will be remarkably routine as we acquire the same data and perform essentially the same operations every day.

I would like to begin with a quote from Part 1 of the **Space Science Evaluation Committee** (SSEC) report[2] that serves as the basis for why we are going back to Mars and why we are continuing its exploration. I like this quote because it is a good initial response to the question of why we are going back to Mars. One must remember that the United States will not have sent a spacecraft to Mars (by the time **Mars Observer** arrives) for nearly fifteen years[3]. There is still a lot to learn, despite--and perhaps as one result of--the mass of **Viking** and **Mariner** data we have seen and heard about at this conference and which we still have on hand for advancing and continuing our studies. What we have learned has posed new questions, and I believe that **Mars Observer** is going to contribute to that data bank in a mighty way.

> *While knowledge of Mars is extensive, it contains significant gaps.*
> *More importantly, there are a number of first-order scientific ques-*
> *tions that can be best addressed from an orbital platform. The Mars*
> *geoscience/climatology orbiter [Mars Observer] will provide new*
> *observations not feasible from Earth or Earth-orbit, which extend*
> *and complement existing measurements and provide an improved basis*
> *for future intensive investigation.*
>
> *Space Science Evaluation Committee (SSEC)*

THE MARS OBSERVER MISSION

The essential aspects of the current launch and mission plan are presented in Figure 2B-1. We will launch **Mars Observer** in 1992[4] from a Space Transportation System (STS) Space Shuttle orbiter. The flight system occupies about half of the available space in the SSV's cargo bay

---

[2] Planetary Exploration through Year 2000. 1983. Part 1 of a report by the Solar System Exploration Committee (SSEC) of the NASA Advisory Council. U.S. GPO. [Ref. pg 141]. {Ed.}

[3] The Viking spacecraft were launched in late August and early September, 1975. {Ed.}

[4] As a result of major shifts in the STS Space Shuttle launch schedule, many programs have been delayed. At the time of the NASA Mars Conference, **Mars Observer** was on the STS manifest for an August 1990 launch. However, in April, 1987, NASA officially directed JPL to replan the **Mars Observer** program for a 1992 launch. As a result, specific facts related to mission scheduling have been changed in or deleted from this presentation to avoid misleading readers and to provide a more accurate representation of the **Mars Observer** mission operations time line. If figures reflect the former schedule (unchanged in material provided for these proceedings in spite of official changes), editing within the text will normally indicate the nature of the changes. {Ed.}

and does not stress the orbiter in any significant fashion[5]. **Mars Observer** is essentially a spacecraft whose major system components have been launched before and do not represent a problem for the STS system. We have planned for a standard 20-day planetary launch window, and we will deploy the flight system (spacecraft and Transfer Orbit Stage -- TOS) from the cargo bay in low Earth orbit (LEO). I will describe the launch configuration and preparations in greater detail later in this presentation.

LAUNCH, 1992

- 20-Day Launch Window, Deploy from STS/SSV Cargo Bay in Low Earth Orbit (LEO)
- Inject onto Trans-Mars Trajectory with Transfer Orbit Stage (TOS)

ONE-YEAR INTERPLANETARY TRANSIT

- Trajectory Correction Maneuvers, Instrument Calibrations

ORBIT INSERTION AT MARS

- 4100-km Apoapsis, 24-hr Orbit
- Insertion Orbit Drifts to Sun-Synchronous Orientation (60-80 days)

ATTAIN CIRCULAR, FROZEN-GEOMETRY MAPPING ORBIT (Propulsive)

- 361-km Average Altitude, 116-minute Period, 93° Inclination
- Sun-Synchronous with 2:00 am/pm Nodal (Equator) Crossings

SCIENCE MISSION

- Duration -- One Martian Year (687 days) in Mapping Orbit
- Repeating-Ground-Track Strategy Maps Entire Planet (59 days)

PLANETARY PROTECTION REQUIREMENT

- Orbit Altitude Raised at End of Mission Activity

Figure 2B-1: Outline of Mars Observer Mission Description

---

[5] The space vehicle commonly known as "Space Shuttle" is often referred to more specifically as the "Space Shuttle orbiter" (or less formally as "the shuttle orbiter" or simply "the orbiter"). In addition, some NASA and JPL illustrations refer to Space Shuttle as the SSV (Space Shuttle Vehicle). Because differing references cannot always be made consistent in a proceedings document of this type, readers are likely to encounter more than one name for the same vehicle or system, as in this presentation (text vs. illustration callouts). {Ed.}

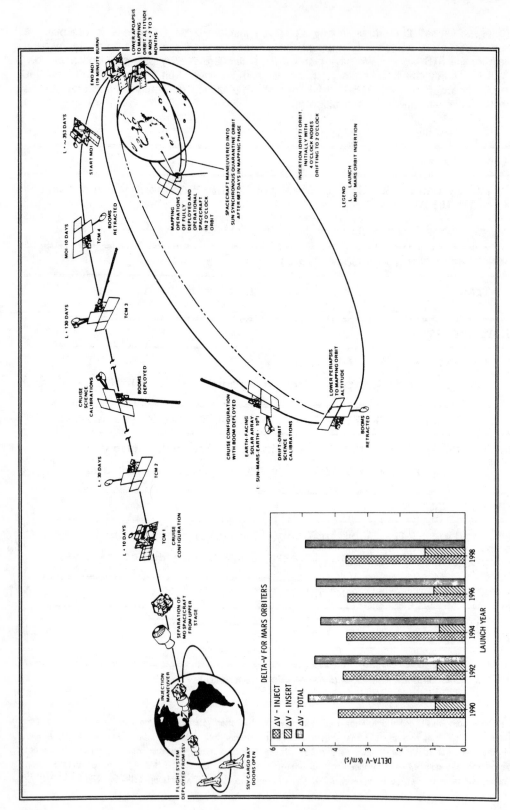

Figure 2B-2: *Launch and Mission Flight Scenario*

Interplanetary and Mars Orbits

The trans-Mars trajectory is achieved using a high-energy transfer orbit vehicle to inject **Mars Observer** into its interplanetary orbit. It will take about a year to reach Mars (359 days for the reference mission), and the only activities planned during that time are the necessary engineering operations. We will not perform cruise science; our scientific objectives are focused on Mars and Mars alone. This reflects another aspect of the **Planetary Observer** concept, wherein the scientific focus of an Observer mission is specifically limited in order to keep costs down and to make the mission as simple as possible.

The Mars orbit insertion (MOI) initially places the spacecraft in a highly elliptical orbit with a 4100-km periapsis. This initial orbit is in fact roughly the same orbit the Soviet **PHOBOS·88** (multidisciplinary Phobos/Mars investigation) spacecraft will use, described in these proceedings by Jim Head [NMC-2A], except that ours is utilized as a temporary 24-hour orbit. We will use it only to allow the orbital plane to move with respect to Mars in order to achieve a 2:00-pm nodal crossing point. That takes about 55 days. We will then lower the apoapsis and circularize the orbit at 361 km (224 miles) above Mars, which is significant considering that Mars has a radius of only 3400 km; an orbital height of 361 km is very close to the planet's surface.

The circular orbit is Sun-synchronous and serves as a mapping orbit. We stay in that orbit for a full martian year (687 days), which allows the observation of one complete cycle of seasonal variation. Extended missions are of course possible if finances and spacecraft health permit. At the end of mission operations, whether one martian year or an extended mission of perhaps two or more, the spacecraft orbit will be moved outward to assure that we satisfy the long-term planetary protection (quarantine) requirement not to impact Mars before the year 2038.

Figure 2B-2 (facing page) illustrates the trans-Mars and orbital scenario planned for **Mars Observer**, and reflects propulsion requirements (V) for Mars during the 1990's. To place approximately 770 kg of mass into Mars orbit with our SSV/TOS launch system, we will use the standard Type-II flight trajectory that **Viking** flew in 1975-76. It will take nearly a year to complete. On the other hand, a Type-I trajectory will be used for the Soviet **PHOBOS·88** spacecraft because they can achieve much better mass performance with the Proton launch vehicle they will be using; the **PHOBOS·88** spacecraft will reach Mars in about 200 days.

A heliocentric perspective of the trans-Mars flight is illustrated in Figure 2B-3 (note that the illustration reflects the previous 1990 launch date and mission schedule rather than that of the 1992-launch replan; the trajectory geometry, however, remains essentially the same). We are also studying trajectories that involve a large maneuver during the transit that would allow us to launch a little more mass. Using the preliminary reference mission trajectory illustrated for the 1990 launch date as an example, one can see that our arrival time is at about 4:00 pm; that is, the nodal crossing of the initial orbit at arrival occurs at about 4:00 pm. If one looks at the line to the Sun, it can be seen that this nodal crossing occurs at about 4 o'clock in the afternoon. That initial orbit is essentially fixed, but it drifts slowly with respect to the Sun (60-80 days) until we reach a

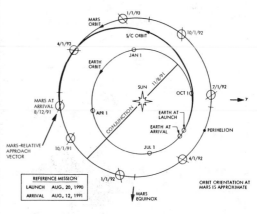

Figure 2B-3: Heliocentric Orbital Geometry

2:00 pm nodal crossing. At that point we inject the spacecraft into the circular mission orbit described earlier. When translated to the 1992 launch date, these orbital maneuvers and the closest approach to Mars will occur early in 1993[6].

Science Mission Plan

Figure 2B-4 presents a time line for the nominal science mission, although it should again be noted that the chronology reflects the 1990 rather than the 1992 launch date. Chronological specifics will of course be different for the new launch date, but the general relationship of mission events and activities over time will remain essentially the same (beginning early in 1993 rather than mid 1991). The illustration provides an overview of the mapping phase and associated strategy as well as the science data acquisition strategy. It should be understood that all of the instruments on the **Mars Observer** spacecraft are body-fixed instruments; we do not have a moveable platform as has been characteristic of **Mariner**-type spacecraft. However, we have two instruments mounted on booms, a magnetometer and a gamma-ray spectrometer. Arden Albee, **Mars Observer** Project Scientist, discusses the science mission objectives and instruments in greater detail in these proceedings.

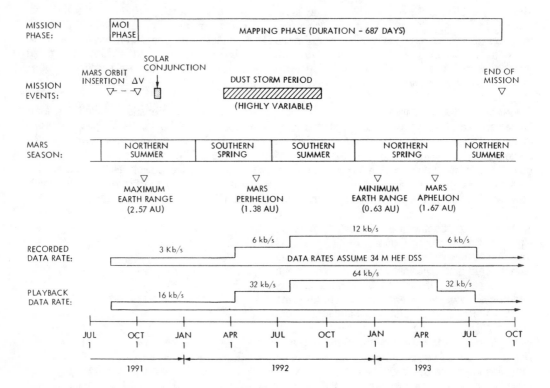

Figure 2B-4: Example Time Line for Mapping Science Phase (1990 launch)

---

[6] The propulsion requirements for **Mars Observer** improve somewhat as a result of the changes in launch planning. As noted in Fig. 2B-2, V requirements are lower for the mid opportunities during the 1990's. In this respect, 1992 is a better launch opportunity than was 1990. {Ed.}

A brief overview of the science data acquisition strategy is presented in Figure 2B-5. We won't perform cruise science, as I explained earlier, and will operate the science system enroute to Mars only to the extent necessary to calibrate the magnetometer and gamma-ray spectrometer. The acquisition of primary science data will not begin until the spacecraft is in its mapping orbit. The mapping mission itself will start following injection into the circular mission orbit following MOI early in 1993, and will carry into 1995.

INSTRUMENT MOUNTING AND POINTING

- ■ All Instruments Body-Fixed (No Scan Platform), Nadir or Limb Pointed
- ■ Some Instruments have Internal Articulating or Scanning Mirrors

---

NO CRUISE SCIENCE

- ■ Necessary Calibrations and Operations ONLY During Trans-Mars Flight
- ■ Necessary Calibrations and Operations in Insertion Orbit

SCIENCE MISSION

- ■ Occurs Only in Mapping Orbit (duration -- one Mars year)
- ■ Mapping Orbit Permits Repeating Ground Tracks (e.g., every 38 orbits)
- ■ Mapping Orbit Repeats Every 59 Days

SCIENCE DATA ACQUISITION/TRANSMISSION

- ■ Science Data Recorded Continuously at 1.5, 3, 6, or 12 kb/s Throughout Mission
- ■ Playback Data Rates Dependent on Telecom Link Performance
- ■ Single-Station Pass of Real-Time Science Data at 64 kb/s Every Third Day

Figure 2B-5: Overview Outline of Science Data Acquisition Strategy

Mapping Orbit Concept: Characteristics and Advantages -- The characteristics of the mapping orbit can be varied to satisfy specific mission objectives, as illustrated in Figure 2B-6. For example, we will adjust the altitude of the orbit to achieve repeating ground tracks just as one would with a polar orbiter at Earth, such that the spacecraft will pass over almost exactly the same point on Mars' surface every 39 orbits. If we adjust the altitude appropriately, we can move that point (at the equator) a number of kilometers in a westward or eastward walk (the reference mission design calls for steps of 10 km). Therefore, if we have an instrument with a 40-km swath width at the equator--the visual and infrared mapping spectrometer, for example--and then move that swath 10 km every 39 orbits, we can get a complete map of the planet in about 59 days. In fact, we can get about 12.5 complete maps of the planet over the course of the currently defined mission (one martian year).

This mapping strategy is demonstrated by the overlapping swaths at the equator (Fig. 2B-6); the scenario illustrated is for a day-side pass on a descending orbit. The ground track is offset by about 30 km every 39 orbits, which allows a 10-km overlap of two 40-km swathes. This produces about one complete equatorial walk around the planet every 59 days and generates a large amount of contiguous, repetitive data in much the same way Earth mapping missions are now conducted.

The **Mars Observer** orbital concept facilitates a number of mission advantages:

1) It permits relatively long integration times for gamma-ray spectrometers,

2) it allows repetitive observations of the same point for surface-observation instruments, and...

3) it produces a large amount of data.

It is important to note that the **Mars Observer** orbital concept allows us to adopt a philosophy that no single data pass is crucial or necessary, which reduces cost relative to mission operations (in commanding the spacecraft and in both the type of preparation required and the size of the science and engineering teams needed to run the mission). In this respect, the key words for the **Mars Observer** mission--as you have already heard and will hear again--are: *routine, repetitive* and *nonadaptive.*

Data Recording and Playback Strategies -- As previously noted (Fig. 2B-5), **Mars Observer** will record data continuously at either 1.5, 3, or 6 kilobits per second (kb/s), depending on range and on the performance of the telecom link, throughout the mission (24 hours a day. The spacecraft has plenty of data storage capacity, amounting to 1.5 billion bits in each of two recorders. Depending on link performance and on the availability of the large deep space network (DSN) antennas[7], we can get as much as 32 kb/s on the downlink, allowing us to record at 6 kb/s continuously. In addition, a 12-kb/s record rate is available for short periods; if there are some features the science team wants to record at higher data rates, it will be possible. Finally, on every third day we plan to conduct a real-time science data pass at 32 kb/s.

We will conduct the lower-rate data passes on a daily basis. We are planning on one eight-hour station pass per day, punctuated by occultation periods, and this allows about four-and-a-half hours of data playback a day. We start the mapping phase recording at 1.5 kb/s and then play it back at 8 kb/s. This occurs once a day. Because of Earth's range early in the mission, this is the highest rate we can achieve using a 34-meter (antenna) station. As conditions improve, however, we can increase the data rate until we have achieved the continuously-recorded data rate. The 6-kb/s budget is divided among seven instruments, and the radio science experiment uses the telecom link itself.

The playback strategy is illustrated in Figure 2B-7 using a three-day example. The spacecraft will record continuously and then play the data back for eight hours each day when the station is in view. The white areas in the playback bars represent occultation breaks when the spacecraft is behind the planet. Every third day a 32-kb/s real-time pass is conducted, and that downlink appears on the chart only on day 3.

---

[7] The three largest DSN antennas are the 64-meter (210-ft) dish antennas based at three different sites around the world. Their separation is an important factor in maintaining continuous communication links with deep space missions; they are approximately 120° (longitude) apart and are located near Madrid (Spain), Canberra (Australia), and at Goldstone in California. {Ed.}

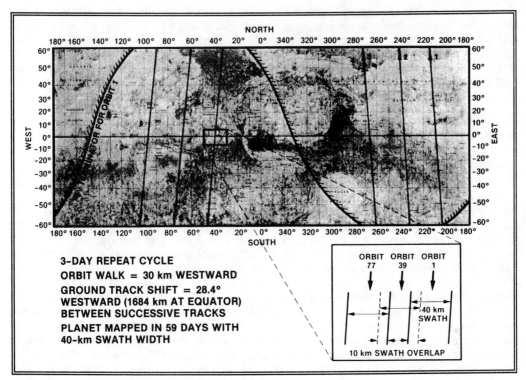

3-DAY REPEAT CYCLE

ORBIT WALK = 30 km WESTWARD

GROUND TRACK SHIFT = 28.4°
WESTWARD (1684 km AT EQUATOR)
BETWEEN SUCCESSIVE TRACKS

PLANET MAPPED IN 59 DAYS WITH
40-km SWATH WIDTH

Figure 2B-6: Mars Observer Mapping Strategy

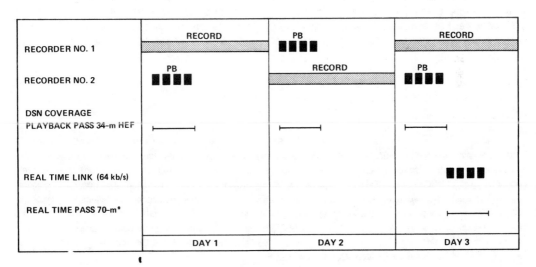

* FROM ~ 6-1-92 TO ~ 6-1-93 A 34-m STATION WILL SUPPORT 64 kb/s

Figure 2B-7: Mars Observer Data Return Strategy

We have been studying ways to improve the data return, and one possibility is that we could extend the coverage time. We are not using all of the coverage time available, but we sized our coverage at eight hours per day and there are longer passes available within the single-pass-per-day constraint. Among other possibilities for increasing the data return is the knowledge that we could use either an additional station pass or a seventy-meter (rather than a 35-meter) station. However, the current plan and the current budget is--for now at least--represented in the strategy I have defined.

MARS OBSERVER FLIGHT SYSTEM AND SPACECRAFT

The flight system is the vehicle we deploy from the Space Shuttle cargo bay. The recommendations of Tom Young's SSEC subcommittee[8] were to utilize to the maximum extent possible an existing spacecraft design. It was on that basis that the Mars Observer contract competition was conducted. An RFP (Request for Proposal) was issued in June, 1985, and RCA was selected as the winner in March, 1986. One of the fundamental tenets of the competition was that the selected spacecraft should be a known and proven spacecraft, and that premise was in fact honored. RCA's spacecraft is based on a SATCOM-K, Shuttle-qualified bus, and utilizes the TIROS/DMSP avionics which are well suited to this mission[9].

Only a small number of modifications had to be made to turn RCA's spacecraft into Mars Observer. One of those, as appropriate for the mission, was the addition of a Mars-horizon sensor for attitude control. Another, the bipropellant orbit-insertion subsystem, is essentially a new development and provides about 2.7 km/s V. The power subsystem and the way the power is switched during the trans-Mars journey is of course unique to Mars Observer, as is the telecommunications link. Relative to the latter, we will be using X-band up and X-band down, which represents a new development. The transponder is being developed independently and will be supplied to RCA.

The high-energy upper stage to be used to inject the spacecraft into its trans-Mars orbit is a new development. Orbital Sciences Corporation has privately financed the development of the Transfer Orbit Stage (TOS). But, while it is a new development, it represents well understood components. For example, it includes the 20,000-lb first stage motor of the Inertia Upper State (IUS) and some avionics from the Peacekeeper (MX) and other military missiles.

---

[8] Mr. A. Thomas Young was Chairman of the SSEC Spacecraft Technology Subcommittee. During a long and successful career with NASA, and following the completion of mission definition management responsibilities on the Lunar Orbiter program under Mr. James S. Martin, Jr., Mr. Young headed the development of Mars mission objectives at NASA-LaRC. He then became the Viking Project's Science Integration Manager, again working with J.S. Martin (who had been named Viking Project Manager). Mr. Young concluded his Viking association as Mission Director during primary mission operations at JPL. After serving in a variety of NASA executive posts (at NASA Headquarters, NASA-ARC, and NASA-GSFC), he left NASA and has since held a series of executive positions with Martin Marietta Aerospace. He was President, Martin Marietta Orlando Aerospace (Orlando, Florida) at the time of the NASA Mars Conference. {Ed.}

[9] TIROS and DMSP represent spacecraft technologies and programs developed to serve meteorological requirements. TIROS is an acronym for Television InfraRed Observation Satellite while DMSP is an abbreviation for Defense Meteorological Satellite Program. {Ed.}

The Flight System (spacecraft launch configuration)

Figure 2B-8 outlines the flight system's description and Figure 2B-9 presents an illustration of the Mars Observer flight system configuration in the SSV cargo bay. As an incidental note for comparison, the Soviet's PHOBOS·88 spacecraft weighs a great deal more than Mars Observer. The total on-orbit weight of our spacecraft is 769 kg and it carries only about 102 kg of science payload; the PHOBOS·88 spacecraft carries over 400 kg of science payload alone. However, the Soviet spacecraft will be launched by a Proton booster that has the power needed to help place the greater spacecraft mass of PHOBOS·88 in orbit at Mars.

FLIGHT SYSTEM

■ Consists of Spacecraft and Upper Stage

SPACECRAFT

■ Consists of Spacecraft Bus and Payload

+ Spacecraft Bus a Derivative of Earth Orbital Spacecraft (EOS)

RCA SATCOM-K Bus

RCA TIROS/DMSP Avionics

Mission-Unique Modifications/Developments:

- Mars Horizon Sensor
- Orbit Insertion Propulsion
- Power
- X-Band Telecommunications

+ Payload Consists of Science Instruments, Payload Data Subsystem

---

UPPER STAGE

■ Transfer Orbit Stage (TOS),
Orbital Science Corporation

■ TOS Motor is Derivative of IUS First Stage Motor

■ TOS Avionics are Derivative of Existing Missile and Spacecraft Avionics

---

FLIGHT SYSTEM AIRBORNE SUPPORT EQUIPMENT

■ Airborne Support Equipment,
Supplied by Orbital Science Corporation

Figure 2B-8: Outline of Mars Observer Flight System Characteristics

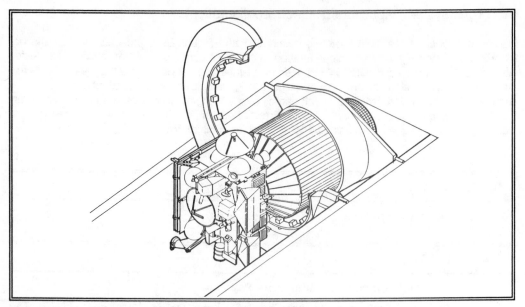

Figure 2B-9: Illustration of Mars Observer Flight System

The jaw-like clamp that holds the flight system in place during launch is illustrated in its open configuration (Fig. 2B-9), suggesting readiness for deployment. The spacecraft itself is seen as a complex array of instruments at the left in the illustration, while the TOS is the large object to the right. The whole flight system configuration rotates on an axis (visible across the cargo bay at the right) to erect itself, after which it is deployed. The TOS trans-Mars injection sequence is started about 23 minutes after deployment, and the spacecraft separates from the upper stage--and is on its own--about 45 minutes later.

Mars Observer Spacecraft

Figure 2B-10 illustrates, very generally, some of the spacecraft's physical characteristics (top) and includes an outline of information about its various systems (bottom). A color illustration of **Mars Observer** is presented on the title page of the color display section (pg A-1). The spacecraft is depicted as it might appear in orbit at Mars in both illustrations. The line drawing of the spacecraft (top, Fig. 2B-10) is from the RCA proposal, and it is not accurate relative to the present configuration. For example, it shows two instruments that were initially included in the proposal but which are not now on the payload: one of them is an ultraviolet photometer located just above the radar antenna, and the other is an ultraviolet spectrometer that appears to extend below the bus at the lower right corner.

The primary part of the spacecraft is the SATCOM-K bus. The high-gain antenna at the top of the bus (at the end of a deployment arm) is the one-meter-diameter telecommunications antenna. The large solar panel array is of course deployed on another deployment arm somewhat behind the bus from this perspective. There are, as I mentioned earlier, two six-meter booms at the ends of which are mounted the magnetometer (right), and the gamma-ray spectrometer (left). Radiators that look out into cold space are visible at the top of the bus, and they cool the VIMS (Visual and Infrared Mapping Spectrometer) detectors. The one-meter antenna on the bottom of the bus--pointed down-left--is the radar antenna, and it looks directly toward the planet. The PMIRR (Pressure-Modulated Infrared Radiometer) atmospheric instrument is contained in a long box mounted on the same face of the bus in the upper right corner.

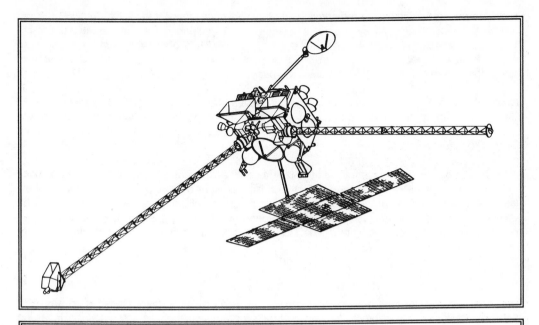

SPACECRAFT CONTRACTOR
■ RCA Astro-Electronics (RCA-Astro), Princeton, NJ

SPACECRAFT HERITAGE
■ SATCOM-K: Structure, Thermal Design, Solar Array/Solar Array Drive
■ TIROS/DMSP: Attitude Control, Command and Data Handling, Power Conditioning and Distribution, Telecommunications, Thermal Design
■ MO-Unique Development: High Gain Antenna (HGA) Pointing, Power Shunt Regulator, Bipropellant Propulsion, Mars Horizon Sensor

SPACECRAFT CAPABILITIES
■ 102-kg Science Payload: Includes Two 6-m Booms for Magnetometer and Gamma-Ray Sensor
■ 1.5 Billion Bits of Data Storage in Each of Two Tape Recorders
■ 40-W TWT and 1.4-m HGA Provide Downlink Data Rates of 8, 16, 32 and 64 kb/s
■ NADIR-Oriented Pointing Control is Better than 10 Milliradians Per Axis Using Mars Horizon Sensor/Reaction Wheels/Monopropellant Hydrazine RCS
■ 109 W Continuous Power Provided for Science Payload by Power Subsystem Utilizing 17.3-m² Sun-Pointed Solar Array; Two (2) 26.5 A-H Batteries

Figure 2B-10: Mars Observer Spacecraft (as on orbit) and Characteristics

Data Storage -- As previously indicated, the **Mars Observer** spacecraft will have two tape recorders for data storage. They are units designed and developed for the **Galileo** spacecraft. These recorders can store 1.5×10^9 bits per unit and facilitate about three times the capacity needed for a daily record rate of 6 kb/s for a Mars mission. Clearly, then, we have plenty of data storage capacity in the **Mars Observer** spacecraft.

Communications -- The telecommunications subsystem utilizes a 40-watt, X-band traveling wave tube (TWT) with a one-meter, high gain antenna, and it provides sufficient margin for the downlink rates. We won't have a problem transmitting the data rates planned, but we are not sizing it for any larger data rates.

Attitude Control -- The spacecraft will utilize Mars-horizon sensors for attitude control once in Mars orbit, comparable to the use of Earth-horizon sensors designed for Earth-orbiting satellites. For reference, the pointing control accuracy of 10 milliradians per axis is a little bit better than half a degree per axis. This is an autonomous control system with which the spacecraft is continuously nadir-oriented. That is, the spacecraft rotates at its orbital rate about Mars.

Conclusion

This concludes my description of project engineering relative to the **Mars Observer** mission and spacecraft design. A "companion" discussion of the program's science, as presented by Dr. Arden Albee, **Mars Observer** Project Scientist, is included in these proceedings of the *NASA Mars Conference* [NMC-2C]. ■

REFERENCES

This space is otherwise available for notes.

~~~~~~~~

MARS OBSERVER: SCIENCE[1]
[NMC-2C]

Dr. Arden L. Albee
California Institute of Technology
Mars Observer Project Scientist (at JPL)

MARS OBSERVER will be the first Planetary Observer mission. Mars was selected because it is readily accessible to the widest range of planetary sciences--geophysics of the interior, surface geology, and atmospheric science--and because it represents potential for future missions. By building on the strength of the existing database for the study of Mars (e.g., Viking), we enhance opportunities to compare and better understand the triad of terrestrial planets with atmospheres -- Venus, Earth and Mars. The selection rationale for the primary science objectives is that they are readily attainable using sets of mapping data obtained by instruments evolved from those already proven on Earth or during previous missions. Two will deal with surface and near-surface aspects, including surface/atmosphere interaction, and will produce maps associated with the two dimensions of the surface. The third is focused on establishing the nature of the elusive magnetic field. The fourth objective is to map in time and space and the last is to map in a profile through the atmosphere. The resulting data sets will be exploited in many ways. Some of the investigations are associated with specific instruments while some represent interdisciplinary studies. The most significant increase in understanding may result from the synergism between the instruments and the cross-working of their products through the interdisciplinary investigations. Soviet missions scheduled for 1988 and 1992 (or 1994) will afford opportunities for mutually beneficial cooperation.

INTRODUCTION: MISSION RATIONALE

Because Mars Observer is the first of the Planetary Observers, the question we must initially address is: Why is the first Planetary Observer going to Mars? Figure 2C-1 is one response to that issue, and I think this conference itself illustrates very well the reasons Mars was selected. Mars was the obvious choice in that it readily offers something for nearly everyone interested in planetary science -- geophysics of the interior, surface geology, and atmospheric science. Earth, Venus and Mars are closely related to one another. Through comparative planetary study,

---

[1] As in the case of the previous presentation by William I. Purdy [Mars Observer: Mission, NMC-2B], the content of this presentation (text and illustrations) has been updated to reflect NASA's replan for a 1992 Mars Observer launch. The illustrations are dated 7/15/87, reflecting the status of science planning, investigations, and instruments as of that date rather than that of the NASA Mars Conference held a year earlier (7/21-23/86). {Ed.}

then, we can begin to understand the climates of all three planets (relative to each other) as well as their evolution. It is this rationale that leads us to Mars as the natural choice for the first of the Planetary Observer missions.

- ■ Earth, Venus and Mars Form a Related Triad of Inner Solar System Planets with Atmospheres

- ■ Planetary "Intercomparison" Provides a Powerful Means of Testing Ideas about Origin and Evolution

- ■ First-Order Scientific Questions Remain to be Addressed

Figure 2C-1: Why Mars for First Planetary Observer Mission?

A key part of the Planetary Observer concept is to constrain and focus the choices of science, and that process involved--for example--looking at several possibilities for Mars missions. And, while **Mars Observer** was selected to be the first, another that is a strong candidate for a future mission--the **Mars Aeronomy Observer**--is also discussed in these Mars Conference proceedings [Hunten, NMC-2D]. Basically, the definition of the **Mars Observer** mission is that it represents science that can best be done from a near-circular polar orbit fixed at 2:00 nodal crossings in the afternoon and morning. So, the science objectives established for the **Mars Observer** mission represent those aspects of science that best utilize the perspective afforded by this orbit.

Science Objectives

The set of science objectives (five) listed in Figure 2C-2 represent those that are readily and reliably attainable. Each objective can be met using sets of mapping data that can be obtained by instruments already proven on Earth or during previous planetary missions. These objectives may at first seem a bit limiting, perhaps, in that they do not set out to specifically solve--for example--Mars' climate history or evolutionary origin. However, if we consider the broad basis for each of the objectives, we find that they are objectives to develop data sets that can be exploited in many ways.

---

(1)   • Determine the Global Elemental and Mineralogical Character of the Surface Material

(2)   • Define Globally the Topography and Gravitational Field

(3)   • Establish the Nature of the Magnetic Field

(4)   • Determine the Time and Space Distribution, Abundance, Sources, and Sinks of Volatile Material and Dust Over a Seasonal Cycle

(5)   • Explore the Structure and Aspects of the Atmosphere's Circulation

---

Figure 2C-2: Mars Observer Science Objectives

The first two objectives deal with the surface and near-surface realm, and result in maps associated with the two dimensions of the surface. The third objective (to establish the nature of the magnetic field) is perhaps somewhat unnatural for this near-circular orbit, but it represents a very important global question for Mars. Despite all of the missions that have gone to Mars, we *still* do not know whether Mars has a global magnetic field; if indeed one does exist, it must be very small. While the 1988-89 Soviet Phobos/Mars mission [discussed in these proceedings by James Head, NMC-2A] will also address this issue, the **Mars Observer** circular orbit has some advantages for trying to sound into the interior. This is achieved in part by using the third dimension of time to remove the time variation of the magnetic field at the planet.

Finally, the last two objectives are additional mapping objectives: mapping in time and space, and mapping in a profile through the atmosphere. As a result, the data sets for these two objectives clearly will be multidimensional. One might note that the fifth objective specifies "aspects of the circulation of the atmosphere." One of the **Mars Observer** constraints is that we are always near the 2:00 am/pm nodal point, and therefore will not be able to observe the diurnal variation. That is why we referred to this mission for some time as the *geoscience-climatology* mission. We will study the annual and seasonal changes in the atmosphere of Mars, but we will get much less information on the diurnal, daily variation.

Overview of Science Investigations and Teams

As one might expect, I have been asked: "How come there is only one atmosphere instrument?" I would like to make the point during this presentation, in leading toward an answer to this question, that the synergism between these instruments is *very* important to the nature of the investigations. Indeed, every science instrument on **Mars Observer** (except the magnetometer) also functions as an atmosphere instrument in some manner. Although the five prime objectives established for the instruments may not specifically state it, each will in fact contribute data for problems involving not only the surface, but the atmosphere and the atmosphere/surface interaction.

Mars Observer Science Investigations -- Figure 2C-3 lists all of the investigations, some of which represent efforts associated with specific instruments while some represent **interdisciplinary** tasks. The name of the scientist selected as principal investigator/team leader for each investigation is also given.

Though not readily apparent, the **Mars Observer** science investigations fall into several categories, largely as a result of their origin but sometimes on the basis of their interdisciplinary nature. In general, these categories are as follows:

(1)   Some are directly associated with instruments that were originally defined and worked upon as mission facilities,

(2)   some were developed by teams set up (in accord with Planetary Observer philosophy) to pre-develop certain kinds of instruments for such missions,

(3)   some were proposed by principal-investigator teams, and...

(4)   some reflect broadly-based primary scientific interests coordinated by five *interdisciplinary scientists* (IDS), each of whom will integrate, analyze and interpret data acquired by all of those instruments most relevant to their investigations.

INVESTIGATION	INVESTIGATOR/INSTITUTION
Gamma Ray Spectrometer	William Boynton University of Arizona
Magnetometer	Mario Acuna NASA-GSFC
Mars Observer Camera	Michael Malin Arizona State University
Pressure-Modulated Infrared Radiometer	Daniel McCleese Jet Propulsion Laboratory
Radar Altimeter/Microwave Radiometer	David Smith NASA-GSFC
Radio Science	G. Leonard Tyler Stanford University
Thermal Emission Spectrometer	Philip Christensen Arizona State University
Visual and Infrared Mapping Spectrometer	Larry Soderblom USGS-Flagstaff
Data Management and Archiving; Weathering and Volatile Cycling	Raymond Arvidson Washington University
Geoscience	Michael Carr USGS-Menlo Park
Polar Atmospheric Science	Andrew Ingersoll California Institute of Technology
Atmosphere/Surface Interaction	Bruce Jakosky University of Colorado
Climatology	James Pollack NASA-ARC

NOTE: Represents both instrument-specific and interdisciplinary investigations.

Figure 2C-3: Science Investigations and Investigators (Team Leaders)

Scientist Selection Process -- In addition to the scientists selected to lead team efforts associated with individual instruments or the interdisciplinary investigations, there also will be a later call for *participating scientists.* The science teams are quite limited, in terms of human resource, and do not incorporate a sufficient number of people to process and reduce the large amount of data **Mars Observer** will generate during the period of mission operations. The *participating scientists* will be invited into the activity close to launch time via a "Dear Colleague" letter from NASA headquarters. This will enable others, including those currently in graduate school, to look forward to the possibility of such an opportunity rather than to feel that the door has already been closed by the pre-mission selection process.

The scientists selected for interdisciplinary scientist (IDS) responsibilities, as previously defined according to the fourth category, are:

- Dr. Raymond E. Arvidson[2]   Data Management and Archiving; Weathering and Volatile Cycling

- Dr. Michael H. Carr[2]   Geosciences

- Dr. Andrew P. Ingersoll   Polar Atmospheric Science

- Dr. Bruce Jakosky   Surface/Atmosphere Interaction

- Dr. James A. Pollack[2]   Climatology

SCIENCE INSTRUMENTATION

Figure 2C-4 is a listing of the selected instruments, and appropriate measurement objectives are shown with each. I will discuss the instruments and their purpose one by one, and provide some background concerned with origin and heritage.

INSTRUMENT	PURPOSE
Gamma Ray Spectrometer	Elemental Composition of Surface
Magnetometer	Intrinsic and Local Magnetic Field
Imaging Camera	Global Synoptic Views, Selected Moderate and Very High Resolution Images of Surface and Atmosphere
Pressure Modulator Infrared Radiometer	Profiles of Temperature, Water, Dust; Radiation Budget Measurements
Radar Altimeter	Topography, Microwave Radiometry
Radio Science	Gravitational Field; Atmospheric Refractivity Profiles
Thermal Emission Spectrometer	Surface Mineralogy; Atmospheric Dust and clouds; Radiation Budget
Visual and Infrared Mapping Spectrometer	Surface Mineralogy, Volatiles, Dust; Surface/Atmosphere Interactions

Figure 2C-4: Mars Observer Science Instruments and Measurement Objectives

---

[2] Former Viking science team members:

Dr. Carr was team leader for the Orbiter Imaging Team [see NMC-1C]; Dr. Arvidson was multidisciplinary and supported several teams, principally the Lander Imaging Team and the Magnetic Properties Team, and was also involved in studies of global aeolian processes. Dr. Pollack, also a member of the Lander imaging Team, focused principally on the problems of measuring atmospheric properties (opacity, aerosols, dust) using lander imaging data, and more recently has developed general circulation models (with topography) of Mars' atmosphere. {Ed.}

Gamma Ray Spectrometer (GRS)

The gamma-ray spectrometer (GRS), illustrated in Figure 2C-5, is one of the facility instruments developed by a team (see Category 1, **Mars Observer Science Investigations** on previous page) chaired by James Arnold (University of California, San Diego). The flight team leader is William Boynton from the University of Arizona. The GRS is mounted at the end of a six-meter boom to cut down background contamination from the spacecraft. In the illustration, the instrument appears to be dominated by its radiator. In fact, the radiator does represent a very critical function; the GRS achieves its high spatial resolution using a cooled germanium crystal.

Objectives -- Mapping the elemental composition of the surface is a significant aspect of Mars Observer science. However, the precision and the spatial resolution with which these elements can be measured varies considerably. For some elements, the spatial resolution may be an entire hemisphere of the planet, while with others (when integrated over the lifetime of the mission) the resolution will be comparable to the spacecraft elevation -- 361 km (224 mi). The GRS will also produce information about the hydrogen-depth dependency and on the atmospheric column density, as well as information on gamma-ray bursts. One can see that the GRS is an instrument that will provide data about both the surface *and* the atmosphere.

Heritage -- It is important, based on Planetary Observer philosophy, to note the heritage of this instrument. Figure 2C-6 illustrates some of the GRS heritage with a spectrum (top) acquired during an Apollo mission. A sodium iodide (NaI) crystal was used in the Apollo instrument; note the lack of distinct peaks in its spectrum. The cooled germanium (Ge) crystal used in the Mars Observer GRS, on the other hand, will yield much better spectra, as illustrated in the lower spectrum (simulation -- same lunar composition) which has very distinct peaks.

OBJECTIVES
1. DETERMINE THE ELEMENTAL COMPOSITION OF THE SURFACE OF MARS WITH A SPATIAL RESOLUTION OF A FEW HUNDRED KILOMETERS THROUGH MEASUREMENTS OF INCIDENT GAMMA-RAYS AND ALBEDO NEUTRONS. (H, O, Mg, Aℓ, Si, S, Cℓ, K, Ca, Fe, Th, U)
2. DETERMINE HYDROGEN DEPTH DEPENDENCE IN THE TOP TENS OF CENTIMETERS
3. DETERMINE THE ATMOSPHERIC COLUMN DENSITY
4. DETERMINE THE ARRIVAL TIME AND SPECTRA OF GAMMA-RAY BURSTS

INSTRUMENTATION
A HIGH SPECTRAL RESOLUTION GERMANIUM DETECTOR COOLED BELOW 100K WILL MEASURE GAMMA-RAY FLUX IN THE ENERGY RANGE 0.20 TO 10 MeV. NEUTRONS WILL BE DETECTED WITH A PLASTIC SCINTILLATOR. INSTRUMENT IS MOUNTED AT THE END OF A SIX METER SPACECRAFT BOOM

TEAM MEMBERS
J. ARNOLD – UCSD
P. ENGLERT – SAN JOSE STATE UN.
W. FELDMAN – LANL
A. METZGER – JPL
R. REEDY – LANL
S. SQUYRES – CORNELL UN.
J. TROMBKA – NASA/GSFC
H. WANKE – MPI (FRG)

DETECTOR (MOUNTED ON REAR OF RADIATOR PLATE)

PASSIVE RADIATIVE COOLER

ANTI-COINCIDENCE SHIELD

PHOTOMULTIPLIERS

Figure 2C-5: Gamma Ray Spectrometer (GRS)

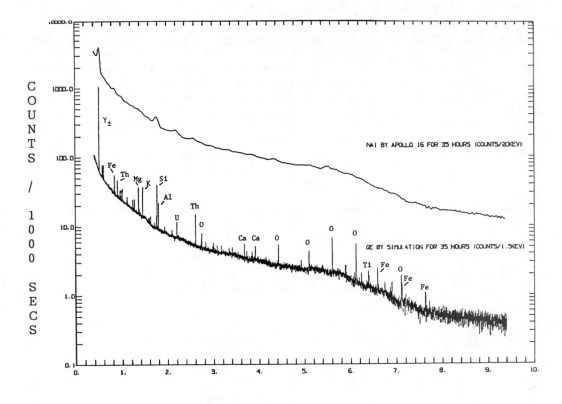

Figure 2C-6: *Spectra for Apollo NaI (TI) Detector (top) and Ge Detector (lower)*

(Comparison between a measured lunar spectrum with sodium iodide detector and a simulated lunar spectrum with a germanium detector responding to the same composition.)

Potential GRS Contributions -- The GRS instrument would make a valuable contribution even if it only generated one spectrum integrated over the entire planet. That would determine the uranium-thorium-potassium ratios, for example, for the whole planet, which would be a significant contribution. The fact that one can begin to get spatial resolution and map these elements over the surface of the planet is a bonus. The debate about the proposed martian origin of the SNC meteorites[3] has been discussed in these proceedings, and it may be that the only measurements **Mars Observer** will provide which could effectively contribute to this question are those reflected in the uranium-thorium-potassium ratios to be determined by the GRS.

---

3 As noted in other NMC presentations, SNC (sometimes orally expressed as "snick") is an abbreviation for three unusual types of meteorites: Shergottites, Nakhlites and Chassignites. These have been carbon-dated to be about 1.3 billion years old, which is remarkably young (in geological terms); most common meteorites have been carbon-dated to be about 4.6 billion years old. The SNC meteorites have igneous rock textures and are basaltic in composition, and gases detected within them are remarkably similar in composition to the martian atmospheric gases detected by the **Viking** landers. {Ed.}

Visual and Infrared Mapping Spectrometer (VIMS)

The visual-and-infrared-mapping spectrometer (VIMS) illustrated in Figure 2C-7 also has a significant heritage. The **Viking** orbiter camera system used a three-color camera[4], and so can be used as one example of VIMS heritage even though it did not incorporate a spectrometer. Since that time, however, a more significant heritage has evolved through the development of Earth-based aircraft instruments as well as the **Galileo** near-infrared mapping spectrometer (NIMS). In fact, the JPL group that is developing the **Mars Observer** instrument was responsible for the **Galileo** instrument, and the same group is also developing both SSV- (Space Shuttle Vehicle) and aircraft-based instruments to conduct multi-spectral mapping of Earth. Clearly, there is a significant background of development associated with the VIMS instrument. Both the development team and the flight team associated with the **Mars Observer** VIMS instrument are led by Larry Soderblom (USGS, Flagstaff).

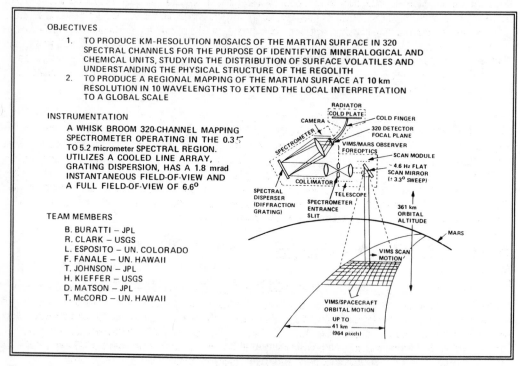

OBJECTIVES

1. TO PRODUCE KM-RESOLUTION MOSAICS OF THE MARTIAN SURFACE IN 320 SPECTRAL CHANNELS FOR THE PURPOSE OF IDENTIFYING MINERALOGICAL AND CHEMICAL UNITS, STUDYING THE DISTRIBUTION OF SURFACE VOLATILES AND UNDERSTANDING THE PHYSICAL STRUCTURE OF THE REGOLITH
2. TO PRODUCE A REGIONAL MAPPING OF THE MARTIAN SURFACE AT 10 km RESOLUTION IN 10 WAVELENGTHS TO EXTEND THE LOCAL INTERPRETATION TO A GLOBAL SCALE

INSTRUMENTATION

A WHISK BROOM 320-CHANNEL MAPPING SPECTROMETER OPERATING IN THE 0.35 TO 5.2 micrometer SPECTRAL REGION. UTILIZES A COOLED LINE ARRAY, GRATING DISPERSION, HAS A 1.8 mrad INSTANTANEOUS FIELD-OF-VIEW AND A FULL FIELD-OF-VIEW OF 6.6°

TEAM MEMBERS

B. BURATTI – JPL
R. CLARK – USGS
L. ESPOSITO – UN. COLORADO
F. FANALE – UN. HAWAII
T. JOHNSON – JPL
H. KIEFFER – USGS
D. MATSON – JPL
T. McCORD – UN. HAWAII

Figure 2C-7: Visual and Infrared Mapping Spectrometer (VIMS)

---

[4] Two identical cameras were used on each of the **Viking** orbiters. They were slow-scan vidicon units, both equipped with 475mm lenses. Each had a filter wheel with six filters: a clear filter to provide sensitivity across the near-UV and visible wavelengths, a filter restricted to the near-UV/violet for cloud and ice enhancement, a minus-blue filter to produce results the reverse of those of the violet filter, and three filters (red, blue and green) for the later-production of color images (at JPL or elsewhere). The cameras alternated during an imaging sequence, each recycling every 4.5 seconds. The vidicon could record up to 1056 lines by 1182 pixels per frame, with each pixel having a brightness value in the range of 0-127. Resolving power was dependent on the orbiter's altitude when an image was recorded (**Viking** orbits were highly elliptical, affording a wide range of resolutions on each orbit). Finally, both of the two frames required to produce a stereogram could be acquired by the same camera -- or both; the second frame was usually acquired during a different orbital pass to achieve the nominal binocular separation for the same target (preferably when the range and Sun angle were comparable). {Ed.}

Function -- The VIMS is, in effect, a "whisk broom." It incorporates linear sets of detectors, and it detects 320 channels with a spectral range from 0.3 to 5.2 microns. This covers the visual spectrum up into the near-infrared, detecting mainly electronic transitions. The range includes the sorts of things one can actually see, particularly the effects of iron and weathering as well as vibrational spectra of molecules like OH and CO_2.

Objectives -- There are essentially two mapping objectives. In one case, the VIMS instrument will provide very detailed spectra of selected 1-km points or profiles. These have the full spectral resolution that Earth-based observations of the surface of Mars have shown to be of value. In addition, a limited number of channels can be used to map at a resolution in the range of 1-10 km, depending on the data rate.

Basis -- It is helpful to examine some aspects of the VIMS science using **Viking** orbiter color images, such as those included with Larry Soderblom's presentation [see NMC-1B illustrations, pages A-2 and A-3 in the color display section]. In real or false color, one can see that there are differences across the martian surface. The question of course is: do these color differences reflect differences in bedrock, weathering, and particle size -- or something else? With only two or three channels to look at (in **Viking** images), there is already significant divergence in opinion as to how much bedrock information one can gain from such color data, as opposed to how much of it may simply represent surficial windblown sand, et·cetera.

Two teams--one at Washington University (St. Louis, MO) and another at the University of Washington (Seattle, WA)--have worked in very great detail to understand the spectra of individual rocks in areas viewed by the **Viking** landers. Examples of the nature of such features are visible in the **Viking** images presented in Figure 2C-8. Many of the differences are due to different slope angles or simply to the aspect angle associated with the line-of-sight to a particular rock. Other differences are due to a variation in weathering. Both groups have concluded that there are no distinctly different kinds of rocks but that there are distinct differences in the weathering layers on the surfaces of boulders.

Thermal Emission Spectrometer (TES)

The thermal emission spectrometer (TES) instrument illustrated in Figure 2C-9 is a new development. The instrument does have some relevant heritage, however, in the IR radiometers flown on **Mariner 9** to provide atmospheric data. The TES flight team is led by Philip Christensen from Arizona State University. Note that the TES team, like other **Mars Observer** science teams, represents what I think to be a healthy mixture of old hands from the **Mariner/Viking** programs and relatively new people coming into planetary mission activity for the first time.

Design and Function -- The TES instrument was designed predominantly to look at the martian surface and to study the nature of the surface materials. Instead of having a fixed set of channels, the instrument has a Michelson interferometer capable of producing a detailed spectra over the range of 6.25 to 50 micrometers. Its spectral range can detect variations in the silica-oxygen bonds and aluminum-oxygen bonds. The instrument can therefore provide information on differences in rock type that might not show up in the visual and near-IR data.

Mission Potential -- The primary TES objectives deal with understanding the surface of the planet, but the instrument also represents objectives that capitalize on the fact that it can identify water ice. In this respect, it will certainly be used to look at the growth and retreat of the icecap, as suggested in the **Viking** orbiter picture presented in Figure 2C-10. In addition, since it works with six pixels in a rectangular array, the TES instrument also can serve as an atmospheric instrument for looking at the planet's limb in profile.

Figure 2C-8: Martian Boulders and Surface as Imaged by Viking Landers
(VL-1 top, VL-2 bottom)

268

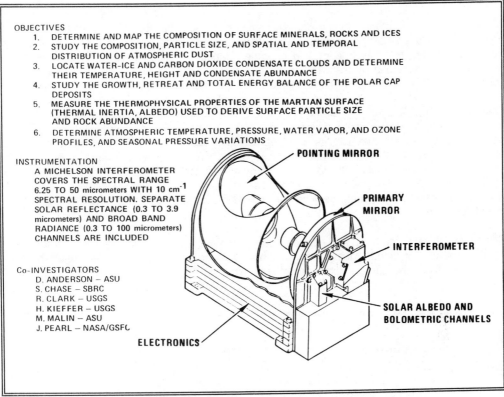

OBJECTIVES
1. DETERMINE AND MAP THE COMPOSITION OF SURFACE MINERALS, ROCKS AND ICES
2. STUDY THE COMPOSITION, PARTICLE SIZE, AND SPATIAL AND TEMPORAL DISTRIBUTION OF ATMOSPHERIC DUST
3. LOCATE WATER-ICE AND CARBON DIOXIDE CONDENSATE CLOUDS AND DETERMINE THEIR TEMPERATURE, HEIGHT AND CONDENSATE ABUNDANCE
4. STUDY THE GROWTH, RETREAT AND TOTAL ENERGY BALANCE OF THE POLAR CAP DEPOSITS
5. MEASURE THE THERMOPHYSICAL PROPERTIES OF THE MARTIAN SURFACE (THERMAL INERTIA, ALBEDO) USED TO DERIVE SURFACE PARTICLE SIZE AND ROCK ABUNDANCE
6. DETERMINE ATMOSPHERIC TEMPERATURE, PRESSURE, WATER VAPOR, AND OZONE PROFILES, AND SEASONAL PRESSURE VARIATIONS

INSTRUMENTATION
A MICHELSON INTERFEROMETER COVERS THE SPECTRAL RANGE 6.25 TO 50 micrometers WITH 10 cm^{-1} SPECTRAL RESOLUTION. SEPARATE SOLAR REFLECTANCE (0.3 TO 3.9 micrometers) AND BROAD BAND RADIANCE (0.3 TO 100 micrometers) CHANNELS ARE INCLUDED

Co-INVESTIGATORS
D. ANDERSON — ASU
S. CHASE — SBRC
R. CLARK — USGS
H. KIEFFER — USGS
M. MALIN — ASU
J. PEARL — NASA/GSFC

POINTING MIRROR

PRIMARY MIRROR

INTERFEROMETER

SOLAR ALBEDO AND BOLOMETRIC CHANNELS

ELECTRONICS

Figure 2C-9: Thermal Emission Spectrometer (TES)

Figure 2C-10: Recession of CO_2 Winter Cap Observed by Viking Orbiter (8/76)

Mid martian summer (northern hemisphere) VO-2 frames, imaged at a range of 4000 km, show probable maximum recession of north polar cap to residual, permanent ice cap of water ice. During winter, polar region is shrouded by much broader hood of frozen CO_2 frost and haze.

269

Imaging Camera

A camera was not included in the original payload, but one has been selected as a probable addition. The **Mars Observer** camera (MOC), illustrated in Figure 2C–11 is a significant change from the cameras that have been used on other planetary spacecraft. The MOC represents a unique design developed by team led by Michael Malin (Arizona State University, Tempe). Constructed of carbon–fiber epoxy composites, it will provide three different kinds of images.

Capability and Application –– For one application, the camera will view limb-to-limb (≈3,000 km) to observe clouds, dust storms, et·cetera, on a large scale. Figure 2C–12 presents **Viking** orbiter images which illustrate the benefits of being able to image the planet on a large scale over time. The picture on the left, taken in February of 1977 (early spring on Mars), is of a surprising early storm that quickly enveloped much of the southern hemisphere. The picture on the right (a mosaic) was imaged in June of that same year, and reveals a second massive storm that became essentially global in scale over only a few days, much as had the first. Both storms lasted for weeks.

The **Mars Observer** camera will also have an intermediate resolution that will facilitate more detailed tracking of clouds, dust storm development, and other ephemeral features. With these resolutions, then, the camera fulfills an atmospheric science function. However, it also has a

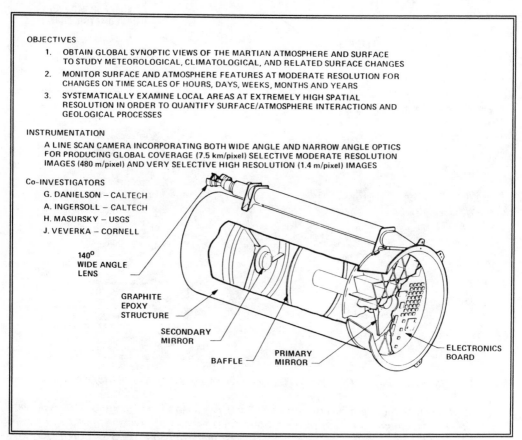

OBJECTIVES

1. OBTAIN GLOBAL SYNOPTIC VIEWS OF THE MARTIAN ATMOSPHERE AND SURFACE TO STUDY METEOROLOGICAL, CLIMATOLOGICAL, AND RELATED SURFACE CHANGES
2. MONITOR SURFACE AND ATMOSPHERE FEATURES AT MODERATE RESOLUTION FOR CHANGES ON TIME SCALES OF HOURS, DAYS, WEEKS, MONTHS AND YEARS
3. SYSTEMATICALLY EXAMINE LOCAL AREAS AT EXTREMELY HIGH SPATIAL RESOLUTION IN ORDER TO QUANTIFY SURFACE/ATMOSPHERE INTERACTIONS AND GEOLOGICAL PROCESSES

INSTRUMENTATION

A LINE SCAN CAMERA INCORPORATING BOTH WIDE ANGLE AND NARROW ANGLE OPTICS FOR PRODUCING GLOBAL COVERAGE (7.5 km/pixel) SELECTIVE MODERATE RESOLUTION IMAGES (480 m/pixel) AND VERY SELECTIVE HIGH RESOLUTION (1.4 m/pixel) IMAGES

Co-INVESTIGATORS

G. DANIELSON – CALTECH
A. INGERSOLL – CALTECH
H. MASURSKY – USGS
J. VEVERKA – CORNELL

140° WIDE ANGLE LENS

GRAPHITE EPOXY STRUCTURE

SECONDARY MIRROR

BAFFLE

PRIMARY MIRROR

ELECTRONICS BOARD

Figure 2C–11: Mars Observer Camera (MOC)

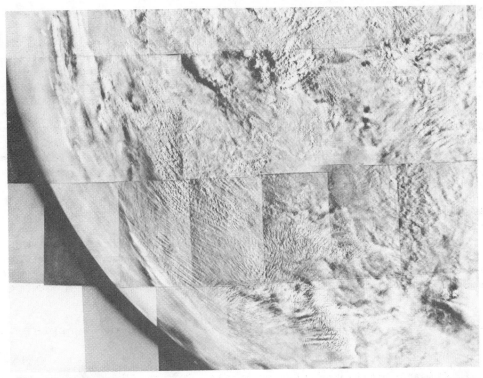

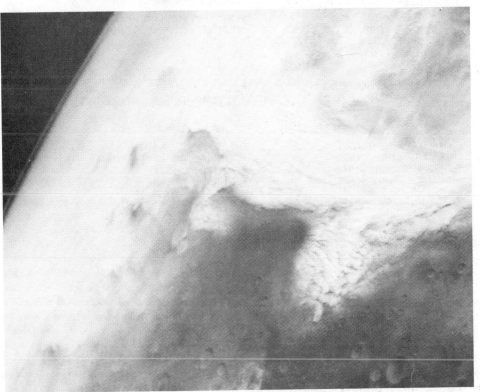

Figure 2C-12:   Wide-Field Viking Orbiter Views of Large Martian Dust Storms   (left, 2/77 – right 6/77)

very high resolution mode capable of 1.4-meter (surface area) pixels, which is almost an order of magnitude better than the very best orbital photos produced by the **Viking** orbiters[5].

Operational Constraints -- The camera operates within severe restrictions in terms of data rate, so it is characterized by a feature common to many of the **Mars Observer** instruments -- the need for much more memory than instruments have required in the past. It must be able to both condense and store data within the instrument itself. The high resolution mode, in particular, will require this capability. In that mode, "snapshots" will be taken at random to some extent, but their content can be located by reference to intermediate-resolution frames registered on both side. A snapshot at the higher resolution will be about two kilometers on a side, and we will also get both VIMS spectral data and TES (thermal emission) data for roughly the same area. In this way, the snapshot-area data can be correlated with other significant data sources to aid in interdisciplinary interpretation.

Radar Altimeter and Radiometer (RAR)

Figure 2C-13 summarizes the investigation related to the radar altimeter and radiometer instrument (RAR). This team, led by David Smith (NASA-GSFC), will emphasize both altimetry and

OBJECTIVES
1. TO PROVIDE TOPOGRAPHIC HEIGHT MEASUREMENTS WITH A VERTICAL RESOLUTION BETTER THAN 0.5% OF THE ELEVATION CHANGE WITHIN THE FOOTPRINT
2. TO PROVIDE RMS SLOPE INFORMATION OVER THE FOOTPRINT
3. TO PROVIDE SURFACE BRIGHTNESS TEMPERATURES AT 13.6 GHz WITH A PRECISION OF BETTER THAN 2.5K
4. TO PROVIDE WELL SAMPLED RADAR RETURN WAVEFORMS FOR PRECISE RANGE CORRECTIONS AND THE CHARACTERIZATION OF SURFACE PROPERTIES

INSTRUMENTATION
A Ku-BAND (13.6 GHz) RADAR ALTIMETER/RADIOMETER WITH AN ADAPTIVE RESOLUTION TRACKING SYSTEM WILL RETURN AMPLITUDE, SHAPE AND TIME DELAY OF ECHO ALONG WITH BRIGHTNESS TEMPERATURE USING A ONE METER NADIR-POINTED ANTENNA

Co-INVESTIGATORS
H. FREY — NASA/GSFC
J. HEAD — BROWN UN.
J. GARVIN — NASA/GSFC
J. MACARTHUR — APL
J. MARSH — NASA/GSFC
D. MUHLEMAN — CIT

G. PETTENGILL — MIT
R. PHILLIPS — SMU
S. SOLOMON — MIT
F. ULABY — UN. MI
H. ZWALLY — NASA/GSFC

Figure 2C-13: Radar Altimeter and Radiometer (RAR)

[5] Following **Viking** primary mission activity, the orbiters' periapsides were gradually lowered, making increased surface resolution possible. This process ultimately produced high or very high resolution images of a few selected areas on the planet. The best resolution possible early in the **Viking** orbital imaging activity was roughly 100 meters. However, when the orbiters were near their lowest periapsides late in their missions, resolutions were improved in a few instances to better than 10 meters. As an example of resolving power, the best **Viking** images could not have resolved one of the landers (roughly three meters across). The **Mars Observer** camera will be able to resolve "distinct" features of smaller size (if imaging and atmospheric conditions are extremely favorable), although it is believed unlikely that a **Viking** lander would be recognized as such. {Ed.}

information from the return shape of the wave to get roughness and slope (topography) characteristics, and it will also perform radiometry in one channel.

Some of the most important aspects of the topography information derived from the altimeter data are the ways in which it will supplement imagery. We will be able to learn, for example, whether the bigger stream channels run downhill or uphill. Such information would of course be very important relative to the history of the channels. If the apparent flow does not run downhill, for example, one would be led to suspect that there has been topographic deformation since the channels were formed. More importantly, the RAR instrument will produce global topography data that are compatible with gravimetry data, such that we can look at global gravity relative to global topography for Mars in comparison with the other terrestrial planets.

Radio Science

The radio science investigation is summarized in Figure 2C-14, which focuses on two primary areas of interest -- atmosphere studies and gravimetry. Led by Len Tyler[6], the radio science

---

OBJECTIVES

ATMOSPHERE
1. TO DETERMINE PROFILES OF REFRACTIVE INDEX, NUMBER DENSITY, TEMPERATURE, AND PRESSURE AT THE NATURAL EXPERIMENTAL RESOLUTION (APPROX. 200 m) FOR THE LOWEST FEW SCALE HEIGHTS AT HIGH LATITUDES IN BOTH HEMISPHERES ON A DAILY BASIS
2. TO MONITOR BOTH SHORT TERM AND SEASONAL VARIATION IN ATMOSPHERIC STRATIFICATION
3. TO CHARACTERIZE THE THERMAL RESPONSE OF THE ATMOSPHERE TO DUST LOADING
4. TO EXPLORE THE THERMAL STRUCTURE OF THE BOUNDRY LAYER AT HIGH VERTICAL RESOLUTION (APPROX. 10 m)
5. TO DETERMINE THE HEIGHT AND PEAK PLASMA DENSITY OF THE DAYTIME IONOSPHERE
6. TO CHARACTERIZE THE SMALL SCALE STRUCTURE OF THE ATMOSPHERE AND IONOSPHERE

GRAVITY
1. TO DEVELOP A GLOBAL, HIGH-RESOLUTION MODEL FOR THE GRAVITATIONAL FIELD
2. TO DETERMINE BOTH LOCAL AND BROAD SCALE DENSITY STRUCTURE AND STRESS STATE OF THE MARTIAN CRUST AND UPPER MANTLE
3. TO DETECT AND MEASURE TEMPORAL CHANGES IN LOW DEGREE HARMONICS OF THE GRAVITATIONAL FIELD

INSTRUMENTATION
SPACECRAFT RADIO SUBSYSTEM X-BAND UP AND DOWN-LINK SUPPLEMENTED WITH AN ULTRASTABLE OSCILLATOR DURING OCCULTATION EXITS

TEAM MEMBERS
G. BALMINO – BGI (FRANCE)   W. SJOGREN – JPL   R. WOO – JPL
D. HINSON – STANFORD   D. SMITH – NASA/GSFC

Figure 2C-14: Radio Science

---

[6] Dr. G. Leonard Tyler, representing Stanford University's Center for Radar Astronomy, played a crucial role during the very uncertain period preceding the first Viking landing, and previously during mission planning relative to the selection of candidate landing sites. Although very difficult to interpret, Earth-based radar data (when available for a specific area) helped identify potentially hazardous surface roughness well below the resolution of early Viking orbiter images. Following MOI, initial imagery clearly revealed that the planned VL-1 landing area was much rougher than anticipated, a fact that led to the decision not to land there and to delay landing until an alternate site could be found. New Aricebo radar data (which included scans of potential landing areas) was added to older radar data, and Len Tyler's team had the task of interpreting this composite of information very quickly. Dr. Tyler's contribution to the landing site evaluation/decision process is credited with having been a vital factor in replanning the landing for VL-1's safe touchdown in Chryse Planitia. {Ed.}

team is always finding new ways to milk another piece of information from the characteristics of radio waves that pass through an atmosphere, and I suspect that our current list of six objectives will be larger by the time we actually fly the mission.

We particularly want to develop a high-resolution global gravity-field map that is comparable in resolution to the topography map. In addition, however, there are a whole set of atmospheric observations planned for the radio science. By utilizing two occultations per orbit, we will acquire two profiles through the atmosphere on each orbit.

Magnetometer

It is very important to establish the magnetic field of Mars for comparison with the other terrestrial planets. The magnetometer is illustrated in Figure 2C-15, and it is the one **Mars Observer** instrument that I can truly say is well developed. Mario Acuna (NASA-GSFC) is the team leader for the magnetometer. It is important to note that since there is only one space-craft and one magnetometer, the time-variant nature of the magnetic field in which Mars is located (within the solar wind) must be subtracted out in order to understand the relatively small residual fields.

Pressure Modulated Infrared Radiometer (PMIRR)

The characteristics of the pressure-modulated infrared radiometer (PMIRR) instrument are summarized in Figure 2C-16. The PMIRR was derived under the leadership of Daniel McCleese (JPL) from instruments used at Venus by Fred Taylor (Oxford, UK), but also represents newer technology currently being prepared to fly on UARS (Upper Atmosphere Research Satellite).

The instrument uses nine channels in the infrared. It has two pressure-modulated cells, one containing CO_2 and one containing H_2O, to provide filtering on two of the channels. It acquires direct information on pressure, temperature, CO_2, H_2O, atmospheric dust, and--indirectly--on an array of other atmospheric features. As one of its prime modes, the PMIRR will record a profile of the limb of the planet roughly every 30 seconds in its nine channels. In addition, it facilitates nadir and off-nadir viewing, which provides information about column density, for example. The instrument also will provide information on low-lying fogs and clouds like those visible in many **Viking** images.

INTERDISCIPLINARY INVESTIGATIONS

The *interdisciplinary scientists* (IDS) represent a very important element of the mission. The interdisciplinary objectives may in fact be the most essential objectives of the mission. Although the individual instruments will enable a significant increase in the understanding of Mars, through their specific capabilities, it is highly probable that the most important increase in understanding will come about as a result of the synergism between the instruments and the cross-working of their data products. A summary of this interdisciplinary effort, along with the names of the scientists responsible for the five interdisciplinary tasks, is presented in Figure 2C-17. These scientists are concerned with fields that cut across the specifically focused objectives associated with each instrument.

Basis -- This effort requires that we very carefully think through the mission as well as the science plan. The **Mars Observer** mission is, as Bill Purdy indicates in his presentation for these proceedings [NMC-2B], a very repetitive mapping mission. Because the work will be so repetitive, there will be a tremendous amount of data generated (representing two years of mission activity) that must be compared in considerable detail.

OBJECTIVES
1. ESTABLISH THE NATURE OF THE MAGNETIC FIELD OF MARS
2. DEVELOP MODELS FOR ITS REPRESENTATION, WHICH TAKE INTO ACCOUNT THE INTERNAL SOURCES OF MAGNETISM AND THE EFFECTS OF THE INTERACTION WITH THE SOLAR WIND
3. MAP THE MARTIAN CRUSTAL REMANENT FIELD USING THE FLUXGATE SENSORS AND EXTEND THESE IN-SITU MEASUREMENTS WITH THE REMOTE CAPABILITY OF THE ELECTRON-REFLECTOMETER SENSOR
4. CHARACTERIZE THE SOLAR WIND/MARS PLASMA INTERACTION
5. REMOTELY SENSE THE MARTIAN IONOSPHERE

INSTRUMENTATION
TWO TRIAXIAL FLUXGATE MAGNETOMETERS AND AN ELECTRON-REFLECTOMETER WILL BE MOUNTED ON A SIX METER SPACECRAFT BOOM. TWO TO SIXTEEN VECTOR SAMPLES PER SECOND WILL BE ACQUIRED

Co-INVESTIGATORS
K. ANDERSON – UCB
S. BAUER – UN. GRAZ (AUSTRIA)
C. CARLSON – UCB
P. CLOUTIER – RICE UN.
J. CONNERNEY – NASA/GSFC
D. CURTIS – UCB
R. LIN – UCB
M. MAYHEW – NSF
N. NESS – NASA/GSFC
H. REME – UN. TOULOUSE (FRANCE)
P. WASILEWSKI – NASA/GSFC

Figure 2C-15: Magnetometer (MAG)

OBJECTIVES
1. MAP THE THREE-DIMENSIONAL AND TIME-VARYING THERMAL STRUCTURE OF THE ATMOSPHERE FROM THE SURFACE TO 80 km ALTITUDE
2. MAP THE ATMOSPHERIC DUST LOADING AND ITS GLOBAL, VERTICAL AND TEMPORAL VARIATION
3. MAP THE SEASONAL AND SPATIAL VARIATION OF THE VERTICAL DISTRIBUTION OF ATMOSPHERIC WATER VAPOR TO AN ALTITUDE OF AT LEAST 35 km
4. DISTINGUISH BETWEEN ATMOSPHERIC CONDENSATES AND MAP THEIR SPATIAL AND TEMPORAL VARIATION
5. MAP THE SEASONAL AND SPATIAL VARIABILITY OF ATMOSPHERIC PRESSURE
6. MONITOR THE POLAR RADIATION BALANCE

INSTRUMENTATION
A LIMB, OFF-NADIR AND NADIR SCANNING RADIOMETER. MEASUREMENTS ARE MADE IN NINE SPECTRAL BANDS WITH FIVE FILTER CHANNELS AND TWO PRESSURE MODULATOR CELLS (ONE CONTAINING CARBON DIOXIDE, THE OTHER WATER VAPOR)

Co-INVESTIGATORS
R. HASKINS – JPL
C. LEOVY – UN. WASHINGTON
D. PAIGE – UCLA
J. SCHOFIELD – JPL
F. TAYLOR – OXFORD (ENGLAND)
R. ZUREK – JPL

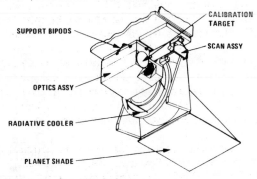

SUPPORT BIPODS

CALIBRATION TARGET

SCAN ASSY

OPTICS ASSY

RADIATIVE COOLER

PLANET SHADE

Figure 2C-16: Pressure Modulated Infrared Radiometer (PMIRR)

IDS/INSTITUTION	INVESTIGATION
R. Arvidson, Washington U.	Data Management and Archiving Weathering and Volatiles Understanding weathering mechanisms and their temporal variation, and the cycling of volatiles through the sedimentary system.
M. Carr, USGS-Menlo Park	Geosciences Better understanding of the role of water in surface evolution, characterization of the planet's volcanic history and determination of the nature and cause of the uplands/plains dichotomy.
A. Ingersoll, Caltech	Polar Atmospheric Science Definition of atmospheric circulation at all seasons to specify the poleward transport of CO_2, water, dust, and energy, as well as the radiative and surface fluxes in polar regions.
B. Jakosky, U. of Colorado	Surface/Atmosphere Interaction Determine nature of the interaction between the surface and atmosphere to better understand the processes involved in the formation and evolution of the martian surface and atmosphere.
J. Pollack, NASA-ARC	Climatology Assess the influence of dust on the atmospheric circulation, the transport of dust, the factors which control the life cycle of dust storms, the role of dynamics in the seasonal water cycle, the constraints on a dense (early) CO_2 atmosphere, and the modulation of the atmospheric circulation that is a result of astronomical variations.

Figure 2C-17: Interdisciplinary Scientists (IDS) and Investigations

In order to optimize the science return from the mission, we must be very careful about how we share the downlink time--between the instruments, between day and night, and between the martian seasons--over the course of at least one martian year. If we successfully achieve this planned sharing of data, its integrated total value will be much greater than the combined product of the individual instruments alone.

Data Management and Archiving -- Of the five interdisciplinary tasks, that of data management and archiving, may be the most important to the rest of the planetary science community. We are fortunate that Ray Arvidson (Washington University, St. Louis, MO) has accepted responsibility for data management and archiving in addition to work he will be doing relative to understanding processes associated with weathering and volatile cycling. This will bring his considerable Viking expertise to the task. His selection again emphasizes the importance of the cross-instrument character of the mission and of bringing the data together such that it can be readily correlated and understood, not only by the investigators on the project but by the rest of the planetary community as well. The community for these kinds of data sets is, after all, much larger than those few people whose names are associated with Mars Observer.

In order to do that, we *must* have an exceptional data distribution and archiving system that is consistent both with recommendations and with the planetary data system. In addition, the Mars Observer system will also include the ability to distribute digital data rather than hardcopy, such that the planetary science community can efficiently process and work on the mission products. Those are the plans now being carried forward by Ray Arvidson's working group on data management and archiving.

CONCLUSIONS: OPPORTUNITY FOR OVERLAP WITH SOVIET MISSIONS

In concluding my presentation, I would like to address one other aspect as a follow-up to Jim Head's presentation [NMC-2A]. The 1988-89 Soviet multidisciplinary Phobos-Mars investigative mission, PHOBOS·88, will open a dramatic period of Mars exploration with its 1988 launch, and it will then be followed by Mars Observer in 1992 and still another Soviet mission (Vesta) in 1992 or (more likely) 1994. Finally, having been well studied by both NASA and the Soviets, rover/sample-return missions might be moving toward reality by the end of the period (prior to the year 2000). Mars rover and sample-return missions, as studied by NASA, are discussed at length in these NASA Mars Conference proceedings [see presentations NMC-2F and NMC-2G].

It is therefore appropriate to suggest that we must look at this composite of international missions as an attack on a common problem. All of the investigator groups, both American and Soviet (as well as those from other nations), share common interests. From the point of view of climatology, for example, the mere fact that we and the Soviets will jointly be investigating Mars over a long time span, extending our base for understanding the martian climate, is very important. The Viking lander meteorology data [see Leovy, NMC-1G] represents a very significant climatology database, but if we can extend it over the period of time covered by these three missions, that base will prove even more important.

In addition, the orbits utilized by the three missions are very complimentary. Some of them are polar while some are essentially equatorial (e.g., the PHOBOS·88 mission orbit is nearly equatorial). Mars Observer is Sun-synchronous and therefore won't see any of the daily variation, whereas the other missions are not Sun-synchronous and will record daily variation. So, the synergism between these orbits and the various mission plans is very important. There are, of course, a variety of other measurement approaches that are important as well, and it is possible that there will be sufficient mission overlap to permit direct comparison of simultaneous measurements.

All of these missions share some common implementation needs. We need better models and descriptions of the atmosphere, the gravity, the topography, and the surface properties. We are contemplating, as one example of how we might help with this process, having a period of time at the very start of the **Mars Observer** mission devoted to understanding the gravity field better, simply as a device designed to reduce the mission costs relative to navigation. One can see that it would make sense to share some of this work with the Soviets. In turn, they certainly will have data that would help us. They have a gamma-ray instrument on both of the **PHOBOS·88** orbiters, for example, and information produced by those instruments might be very valuable to the **Mars Observer GRS team** relative to developing an understanding of the kind of background they will experience.

In conclusion, then, I think it is very important that we continue to look at our own program as a component of a series of missions that together will lead to a better understanding of the planet. I think that we should continue to improve communications and the sharing of data with our Soviet colleagues, so that we and they can indeed take advantage of this beneficial opportunity to significantly advance our mutual understanding of Mars. ∎

REFERENCES

This space is otherwise available for notes.

~~~~~~~~

MARS AERONOMY OBSERVER
[NMC-2D]

Dr. Donald M. Hunten
Professor, Planetary Science
University of Arizona

AERONOMY is a relatively new term now being used to represent the science of upper atmospheres, and it is principally concerned with complex processes like photochemistry and ionization. However, it considers lower atmospheres as well, taking into account phenomena that affect atmospheric structure and circulation. As a result, the MARS AERONOMY OBSERVER ranks well among the missions proposed to follow Mars Observer in NASA's Planetary Observer mission queue. The program would utilize the same spacecraft bus selected for Mars Observer, and the spacecraft/payload mass factors would be essentially the same. Broadly considered, the mission would: (1) determine the diurnal and seasonal variations of the upper atmosphere and ionosphere, (2) determine the nature of the solar wind interaction, (3) verify whether Mars has an intrinsic magnetic field, and (4) measure the present thermal and nonthermal escape rates of atmospheric constituents and determine what these rates imply for the history and evolution of the martian atmosphere. To maximize mission effectiveness, the spacecraft's elliptical orbit would actually dip well into the upper atmosphere to make direct measurements. The science working team has recommended a core payload of essential instruments and a secondary payload that would enhance such a mission. The recommended instruments are effectively compatible with the Planetary Observer concept of investigative synergism.

OPENING COMMENTS AND INTRODUCTION

Let me start by first explaining what this mysterious word "aeronomy" means. I suspect that many may have been unfamiliar with the term before encountering it in the NASA Mars Conference agenda or in these proceedings. It was originally coined--within living memory--by Sydney Chapman, a distinguished physicist and geophysicist. The word's usage is supposed to be analogous to that of "astronomy," and, in simple meaning, it focuses on knowledge of the air. More specifically, aeronomy is often thought of as the science of upper atmospheres, and it is particularly concerned with things like photochemistry and ionization. However, it also considers lower atmospheres as well. For example, Mike McElroy discussed lower atmospheric aeronomy on Mars in his Mars Conference presentation [NMC-1F]. Here on Earth, scientific problems concerned with things like ozone and air pollution, for example, would be in the realm of aeronomy.

Mars Aeronomy

The subject of aeronomy as it relates to Mars in perhaps best introduced using a couple of familiar **Viking** pictures. Figure 2D-1 is a black-and-white mosaic comparable to the color limb picture that appeared on the cover of the **Viking** tenth anniversary edition of the Planetary Society's *Planetary Report*[1], although a piece of the mosaic is missing from the upper left corner in this early **Viking** release (7/76). While craters tend to dominate one's general perception of the region, upon closer examination one might note that there appears to be a thin veil of either atmospheric dust or ground haze diffusing surface detail on the floors of some of the craters. One can also see atmospheric hazes at the horizon, illustrating another aspect of the martian atmosphere[2].

I would also like to refer readers to Figure 1A-8 (page 36), in the presentation given by my University of Arizona colleague, John Lewis [NMC-1A], which is another familiar **Viking** orbiter picture showing the great volcanic mass of Olympus Mons shrouded in clouds. It reveals still another aspect of the atmosphere. Again, forget about the volcano itself for a moment and instead note the clouds and the wave phenomenon. These features are also examples of the kinds of atmospheric processes and phenomena with which my presentation--and the **Mars Aeronomy Observer** mission--will be concerned.

Figure 2D-1:

Horizon Perspective Across Argyre Impact Basin

(Viking Orbiter Mosaic)

---

[1] The Planetary Report. 1986 (Jul/Aug). Vol. VI, No. 4.

[2] The horizon in Fig. 2D-1 is at a range of ≈19,000 km; the view is generally northward, with north at the upper left. The large impact basin, Argyre Planitia, dominates this particular perspective. With a diameter of 700 km, it is one of the largest clearly defined impact basins on the planet -- though still much smaller than Hellas Planitia. Argyre is located deep in the southern hemisphere, extending from ≈43° S to ≈57° S, and its floor exhibits bright frost during the winter and turbulent dust storms during the late spring and summer. {Ed.}

Mars Aeronomy Observer Mission and Spacecraft

At about the time that **Mars Observer** was evolving nicely, Geoff Briggs (Director, Solar System Exploration Division, NASA Headquarters) asked JPL to study several other options. One of these was the **Mars Aeronomy Observer** mission, which was one of four missions recommended for the first phase of the core program for exploring the inner planets (as set forth by the Solar System Exploration Committee -- SSEC).

The **Mars Aeronomy Observer** program will of course use the same spacecraft bus selected for the **Mars Observer** spacecraft, making only those changes peculiar to the **Mars Aeronomy Observer** mission and instruments. The **Mars Observer** payload capacity is just over 100 kg in total weight, which is representative of the Planetary Observer model, and the entire recommended **Mars Aeronomy Observer** payload is well within that constraint. Our power and telemetry are also well within the limitations. This is the basis of our feelings, then, that the **Mars Observer** bus is also an excellent platform for the **Mars Aeronomy Observer** experiments.

Bill Purdy [NMC-2B] and Arden Albee [NMC-2C] discussed the SSEC recommendations for Planetary Observer missions in some detail, relative to how they pertain to the **Mars Observer** mission, and the same characteristics and constraints will apply to the **Mars Aeronomy Observer** program such that they need not be repeated in this presentation. I will therefore focus primarily on the aspects of our own spacecraft and mission, and suggest that the **Mars Observer** presentations be reviewed for information about the Planetary Observer concept that generally defines both programs.

MISSION DEFINITION

During the course of the **Mars Aeronomy Observer** mission, we will--in effect--be peeling away the upper atmosphere through a variety of investigations, while also observing the interaction of the solar wind with Mars' ionosphere. Figure 2D-2 summarizes the mission goals and objectives according to definition work performed by the *Mars Aeronomy Observer Science Working Team* (M/AO S·W·T), which I chaired[3]. The team was formed in mid-1985 by NASA and met at JPL. Jim Slavin, a JPL scientist, was vice chairman and did much of the work. The objectives represent the kinds of things the mission must do, and they were developed specifically to see if the Planetary Observer concept was valid for this particular kind of mission.

The M/AO S·W·T is listed in Figure 2D-3. There are a few **Viking** veterans on the list, notably Bill Hanson[4], but we also made a deliberate attempt to include a number of fresh new people on the team. For this reason many of the names may not be familiar, but I can attest to the fact that they certainly know their stuff. This blend of experienced and fresh participation made the M/AO S·W·T an excellent team for its task.

---

3 The M/AO S·W·T published a report of its work (which also reflects the results of two workshops held at JPL) as NASA Technical Memorandum 89202. The report was titled: Mars Aeronomy Observer: Report of the Science Working Team (October 1, 1986, pp 1-73). All of the illustrations and lists presented as figures in this presentation are contained in and were taken from that report. {Ed.}

4 Dr. William B. Hanson, University of Texas, was a member of the **Viking** entry science team led by Dr. A.O.C. Nier (U. of Minnesota). Dr. Hanson was principal investigator and instrument team leader for the **Viking** retarding potential analyzer (RPA) instruments (one RPA was mounted in each of the two **Viking** descent capsules). The RPA instruments were manufactured by Bendix. {Ed.}

GOALS

Almost nothing is known about the local-time, latitudinal, or seasonal behavior of the thermosphere or ionosphere, or of solar wind interactions on these regions. The goals are therefore essentially unchanged from those previously defined by the SSEC, and are:

- Determine the diurnal and seasonal variations of the upper atmosphere and ionosphere,
- Determine the nature of the solar wind interaction,
- Verify whether Mars has an intrinsic magnetic field, and...
- Measure the present thermal and nonthermal escape rates of atmospheric constituents and determine what these rates imply for the history and evolution of the martian atmosphere.

OBJECTIVES

To better understand the processes and characteristics of the upper atmosphere and ionosphere, as well as the interaction of the solar wind, the objectives of the **Mars Aeronomy Observer** mission are to determine:

- Spatial variations of the upper atmosphere and ionosphere (altitude, local time, latitude and longitude);
- Energy budget for the ionosphere plasma populations (heat sources and energy-loss processes);
- Dynamics of the atmosphere (energy and momentum transport processes);
- Temporal variations (seasonal, solar EUV variability, solar wind events); and...
- Sources of the nightside ionosphere (transport from dayside versus local production).

Figure 2D-2: Mars Aeronomy Observer Goals and Objectives

NAME	INSTITUTION
Donald M. Hunten (Chairperson)	University of Arizona
James A. Slavin (Study Scientist)	Caltech/JPL
Lawrence H. Brace.	NASA-GSFC
Drake Deming	NASA-GSFC
Louis A. Frank	University of Iowa
Joseph M. Grebowsky	NASA-GSFC
Robert M. Haberle	NASA-ARC
William B. Hanson.	University of Texas (Dallas)
Devrie S. Intriligator	Carmel Research Center
Timothy L. Killeen	University of Michigan
Arvydas J. Kliore	Caltech/JPL
William S. Kurth	University of Iowa
Andrew F. Nagy	University of Michigan
Christopher T. Russell	U. of Calif., Los Angeles (UCLA)
Bill R. Sandel	University of Arizona
John T. Schofield	Caltech/JPL
Edward J. Smith	Caltech/JPL
Yuk L. Yung	Caltech
Ulf von Zahn	University of Bonn
Richard W. Zurek	Caltech/JPL

Figure 2D-3: Mars Aeronomy Observer Science Working Team

The schematic diagram in Figure 2D-4 illustrates the interaction of the solar wind with Mars. It suggests how our concept of the **Mars Aeronomy Observer** mission might improve understanding of the rich plasma processes believed to result, even for an "unmagnetized" Mars (which was the case for this model[5]). The Sun impacts the upper atmosphere with both radiation and solar wind particles, and a major objective of an aeronomy mission would be to study the linking among the various atmospheric components (the ionosphere and the upper atmosphere of which the ionosphere is a part) as well as the solar wind interaction with both. We are also interested in aspects of the magnetic field that may be involved in that interaction process.

---

5 The term "Flux Rope" used in Figure 2D-4 is based on knowledge of Venus, and such features are defined in the Science Working Team's report as "bundles of magnetic field about 20 km across, [which] are created at the ionopause and slip into the planetary ionosphere. The ionosphere sweeps these "flux ropes" to the nightside [of Venus] where they may combine to form high-field, low-plasma-density **magnetic holes**." {Ed.}

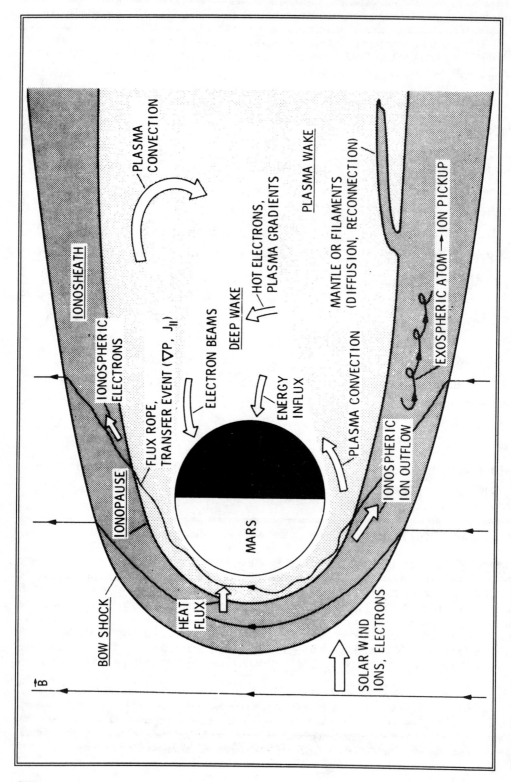

Figure 2D-4: Schematic Model Illustrating Solar Wind Interaction With "Unmagnetized" Mars

A summary of our recommendations for instrument selection is presented in Figure 2D-5. We endorse the idea of conducting this mission according to the Planetary Observer rationale, and we have recommended (1) a **core payload** that more or less fits within the envelope suggested by the SSEC and (2) a **secondary payload** that represents some extras we think should be considered. The M/AO S·W·T feels that the sooner these decisions are made, the better. Why wait!? Venus Pioneer has already performed this mission very successfully and it doesn't make sense not to be doing it at Mars.

PAYLOAD	MASS (kg)	POWER (W)	TELEMETRY[a] (b/s)
CORE			
Neutral Mass Spectrometer[b] (NMS)	10.0	8.5	180
Fabry-Perot Interferometer (FPI)	13.5	5.5	30
UV+IR Spectrometer (UV+IRS)	5.0	7.0	130
Ion Mass Spectrometer (IMS)	2.5	1.5	60
Retarding Potential Analyzer & Ion Driftmeter (RPA+IDM)	4.5	4.0	80
Langmuir Probe (ETP)	2.0	4.0	30
Plasma & Energetic-Particle Analyzer (PEPA)	10.0	9.0	320
Magnetometer (MAG)	3.0	3.5	200
Plasma Wave Analyzer (PWA)	5.5	3.5	130
Radio Science[c] (RS)	4.5	12.5[d]	---
	----	----	----
Core Payload Totals:	60.5	59.0	1160
SECONDARY			
Infrared Atmospheric Sounder (IAS)	8.0	7.5	260
UV+Visual Synoptic Imager (UV+VSI)	9.0	8.0	1000
Neutral Winds/Temperature Spectrometer (NWTS)	10.0	9.0	180
	----	----	----
Secondary Payload Totals:	27.0	24.5	1440
	====	====	====
CORE+SECONDARY TOTALS:	87.5	83.5	2600

[a]   Individual instrument rates can be highly variable and will depend upon the final payload and orbit selection. The rates listed are based upon typical duty cycles for each experiment, and they have been averaged over the orbit (i.e., 6000 x 150 km orbit has been assumed).

[b]   Includes limited wind measuring capability.

[c]   Consists of S-band transponder and stable oscillator.

[d]   10 W (continuous) for the stable oscillator and 25 W (10% duty cycle) for the S-band transponder.

Figure 2D-5: M/AO S·W·T Recommended Science Payload Instrumentation

Principal Science Questions

Figure 2D-6 presents the M/AO S·W·T list of principal science questions (below) addressed by the mission, and Figure 2D-7 then presents a synopsis of the measurements needed to provide the answers. The science investigation begins at the outer most part of the atmosphere, starting with the solar wind. We want to learn more about how the solar wind interacts with Mars.

QUESTIONS

[1] How does the temperature and density of the upper atmosphere respond to seasonal variations caused by the large eccentricity of the Mars orbit combined with the tilt of its pole?

[2] Do the current photochemistry models adequately explain the composition of the Mars ionosphere and upper atmosphere?

[3] Is the plasma pressure of the dayside ionosphere sufficiently large to exclude the solar wind and the interplanetary magnetic field (IMF), or does the solar wind penetrate the ionosphere and interact directly with the thermospheric neutrals?

[4] What are the primary processes for atmospheric escape from Mars? Does oxygen escape by photochemical energization exceed the loss rate due to ion scavenging by the solar wind at the top of the ionosphere?

[5] What are the sources and sinks for the nightside ionosphere? Are the intrinsic or induced magnetic fields large enough to inhibit the nightward flow of ions produced on the dayside by solar EUV? Are there important ion acceleration processes that cause planetary ion escape from the nightside ionosphere, as observed at Venus?

[6] Does Mars have a super-rotating thermosphere and mesosphere as do Earth and Venus; i.e., is super-rotation a universal feature of planetary atmospheres, even those having very rapid and very slow rotation rates?

[7] Does Mars have a cryosphere of type observed in the lower thermosphere on nightside of Venus?

[8] Do lightning-like electrical discharges take place in the martian atmosphere which generate "whistler" mode waves as are thought to occur at Venus, Jupiter, and Saturn?

[9] Do the large amplitude thermal tides and gravity waves that are believed to be generated in the lower atmosphere by solar heating, dust storm effects, and high winds flowing over high relief surfaces propagate into the upper atmosphere?

[10] What is the global mean circulation of the martian upper atmosphere? What are the altitudinal, latitudinal, diurnal, and seasonal dependencies of the global dynamic state?

[11] What is the global mean thermal structure of the upper atmosphere? What are the mesopause and exospheric temperatures? Does a temperature minimum exist near 100 km on the nightside?

[12] What are the altitude, spatial, and temporal variations of the homopause?

[13] What are the main forcing processes for the dynamics of Mars' atmosphere? What is the relative importance of in·situ thermal insulation, eddy diffusion, tidal forcing, et·cetera?

[14] What are the main perturbations from the global mean dynamic and thermal state? What are the causal mechanisms for such perturbations?

[15] What is the deuterium-to-hydrogen ratio in the martian atmosphere? What is the helium abundance?

Figure 2D-6: Principal Science Questions for Mars Aeronomy Observer Mission (M/AO S·W·T)

MEASUREMENTS

[1]   Measure the characteristics (i.e., composition, density, temperature and winds) of the neutral upper atmosphere and ionosphere, as a function of altitude, to the lowest possible altitudes...

$$(h \geq 130 \text{ km in·situ}, h \leq 70 \text{ km remote});$$

[2]   Measure the diurnal variation of the upper atmosphere throughout the entire range of altitudes...

$$(h \geq 70 \text{ km});$$

[3]   Measure these variations over the widest possible range of latitudes consistent with obtaining good diurnal coverage and resolution;

[4]   Measure the effects of changing season and solar activity upon the atmosphere and ionosphere.

Figure 2D-7: Mars Aeronomy Observer Instrument Measurements (to answer Fig. 2D-6 questions)

Comparative planetology is often at the heart of these interests. For example, we already know quite a lot about how the solar wind interacts with Venus, which has no intrinsic magnetic field. There is some indication that Mars has enough of a magnetic field (although we really do not yet know that it does) to make a small difference in how that interaction works. We will of course know more about Mars' magnetic field after the **Mars Observer** mission, which will utilize a magnetometer, but that mission will not be making the correlative measurements one really needs to understand this particular interaction. It is certainly necessary, therefore, to continue to make magnetic field measurements on the **Mars Aeronomy Observer** mission. We have also seen on Venus some pickup of planetary ions from the ionosphere, and we would like to know how important this process is on Mars. Similarly, there are a variety of atmospheric processes on Earth that are not very prominent on Venus, and we would now like to know how important they might be on Mars.

Ionosphere -- We have, relative to the ionosphere, a lot of measurements of electron density which have been acquired via the radio occultation method during prior missions. We will acquire still more of this kind of data from **Mars Observer**, but the only positive ion measurements we presently have were produced by instruments mounted in the **Viking** lander capsules (VLC) during the two landings. Each set (VLC-1 and VLC-2) of **Viking** data, then, was acquired at about the same local time during the very brief period associated with one of the landing events[6]. Therefore, a lot of questions remain about the physics of the ionosphere itself.

---

6 The ion data acquired during the **Viking** mission were measured by two retarding potential analyzer (RPA) instruments. One RPA was mounted in each of the two entry-descent capsules. The bottom half of each capsule was an aeroshell-heatshield upon which was mounted a variety of atmospheric instruments, including two upper atmosphere instruments -- the RPA and an upper atmosphere mass spectrometer (UAMS). The RPA began its ion measurements during the deorbit-coast phase prior to the entry and landing events. The landings occurred approximately three hours after capsule separation from the orbiters, but the primary descent--once entry began (at $\approx 800,000$ ft when 0.05g was sensed)--took only about ten minutes. The rest of that period of time was spent in a coast-to-entry mode initiated by a deorbit propulsive maneuver shortly after separation from the orbiter, during which the RPA made direct measurements in the very upper reaches of the atmosphere. RPA cycles (nominal mission plans called for eighteen) were designed to gather data over 5.5 minutes in each 6.67-minute period, and then to become continuous ≈ 40 minutes prior to the initiation of entry events; the instruments were designed to operate until an atmospheric pressure of 3×10^{-4} mb was reached. {Ed.}

The question of whether the ionosphere or the magnetic field is more important in standing off the solar wind is also represented in our objectives. We would also like to know if the solar wind induces motions in the ionosphere. We see strong motions of that sort on Venus and of a different sort on Earth. The night-side ionosphere also is of interest because there has only been one marginal detection of night-side ionospheric electrons -- that one by a Soviet space-craft. This phenomenon has not been detected by U.S. spacecraft because we tend to use higher frequencies for our radio links.

Thermosphere -- The neutral upper atmosphere--or thermosphere region--is supposed to be hot, but the martian thermosphere was not found to be so by **Viking**. Atmospheric escape was dis-cussed at this conference by Mike McElroy, but while many excellent theories have been put forth, there is very little real data available in the form of needed measurements. The atmos-phere itself is probably influenced by the solar wind in some manner we still do not understand.

The instruments we have defined can also make a number of measurements relative to the lower atmosphere, too. Indeed, there are a number of questions to be addressed in this respect; e.g., the third question, as I will show you in a moment, is a particularly important one. Finally, because the upper atmosphere sits on top of the lower atmosphere, and they are coupled, we can't really study one without studying the other.

Review of Knowledge

At this point, I would like to review our knowledge of Mars as established by prior missions (the Mariners and **Viking**). First, Figure 2D-8 (left) presents some indications of temperature profiles predicted by a model by Gierasch and Goody [1968], in which they assume the atmo-sphere is clear. Figure 2D-8 (right) reflects the same data base, but the theory has been mod-ified to take into account the presence of dust in the atmosphere and the heating effect of that dust as it is illuminated by the Sun. The resulting profiles are compared in both cases to a similar envelope of temperatures that were actually observed by radio occultation experiments on the **Mariner 6 and 7** missions (shaded areas). One can immediately see that--under relatively dusty conditions, at least--the presence of dust and its variability really must be taken into account. This produces some very different concepts of the thermal structure. Figure 2D-9 is based on temperatures observed by the **Mariner 9** IR radiometer instrument. The chart portrays surface temperatures as a function of time, showing the well known and very large diurnal variation of temperature that we believe should be reflected in the upper atmosphere.

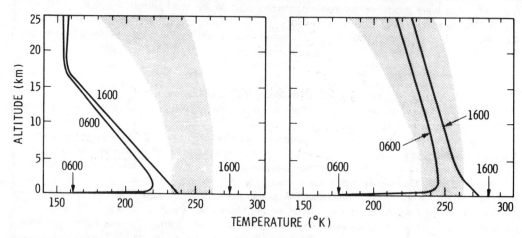

Figure 2D-8: *Temperature Profiles for Martian Atmosphere*

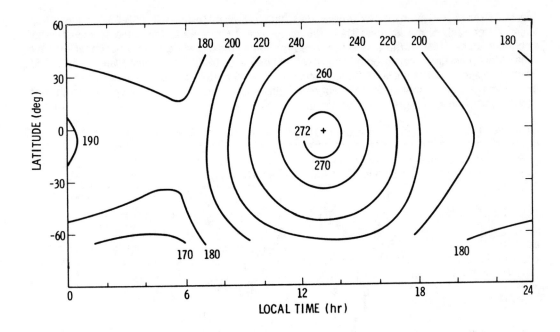

Figure 2D-9: Model of Mars Surface Temperatures (based on clear atmosphere)

Figure 2D-10 is an attempt to indicate all of the interactions that we need to consider. As previously indicated, a list would include the effect of dust (in the lower atmosphere) on the temperature profile, the motions in the upper atmosphere, and the related effects of the solar wind. Note the depiction of solar photons producing an ionosphere and associating the atmosphere, and also the escape processes discussed by Mike McElroy.

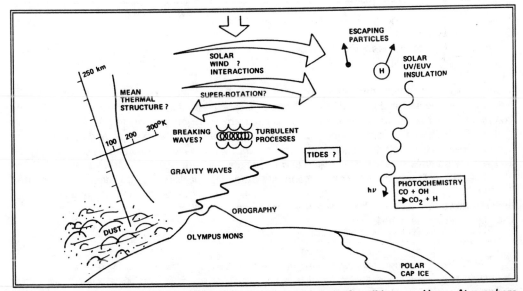

Figure 2D-10: Influences of Solar Wind and Lower Atmosphere Conditions on Upper Atmosphere

Figure 2D-11 portrays the now-familiar data that represent one year of Viking atmospheric pressure for each of the landers. The strips at top and bottom show the standard deviation for each martian day (sol). It must be emphasized that these pressures are somewhat larger than the mean Mars pressure because both landers were in depressed regions (Chryse Planitia and Utopia Planitia, respectively) relative to Mars datum[7]. One can see not only the formation of the polar caps, a process that sucks CO_2 out of the atmosphere, but also many waves due to passing storm systems and a significant "bump" due to the presence of a fairly large dust storm.

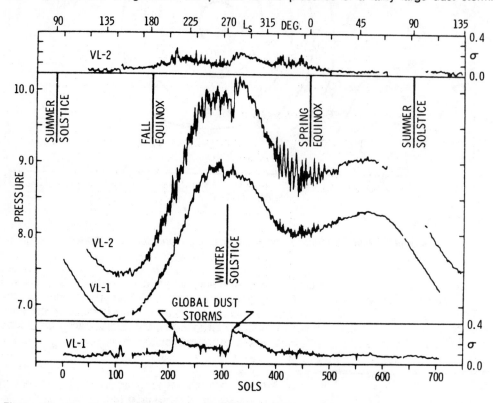

Figure 2D-11: Atmospheric Pressure Cycles (1 year at surface) Recorded by VL-1 and VL-2

Viking upper atmosphere entry data, which have not been discussed to any extent at this conference, are reflected in Figure 2D-12. These data are presented in the form of models of the composition of the thermosphere, starting at ≈300 km and going down to 120 km. At left is a model of the martian thermosphere, which is consistent with the neutral and ion data from the two Viking RPA instruments; at right, the neutral, ion, and electron temperatures are given. The various neutrals measured by the Viking upper atmosphere mass spectrometers (UAMS) are also included in the data. CO_2 can be seen, for example, but O had to be theoretically inferred. The neutral temperature goes only to 200°K at this time, although the ion and electron temperatures are much higher. These data were measured by Bill Hanson's RPA instruments on the two Viking entry aeroshells.

---

[7] Mars datum (zero) elevation, used in lieu of mean-sea-level, is the elevation on Mars at which the mean atmospheric surface pressure is that of the triple-point pressure of water; the pressure at which it can exist in equilibrium as a solid, liquid or gas at 273°K -- 6.1 mb [Hord 1972, Christensen 1975]. {Ed.}

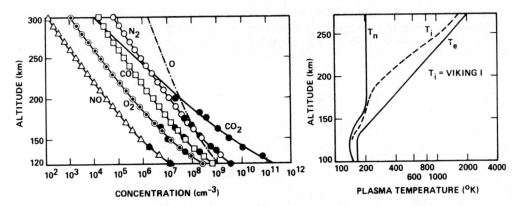

Figure 2D-12: Model of Martian Thermosphere Based on Neutral and Ion Viking Entry Data

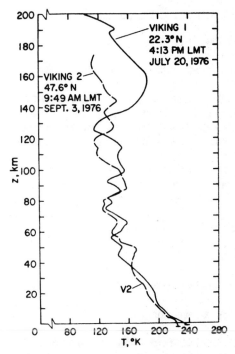

Figure 2D-13: Temperature Profile

Figure 2D-13 presents the entire temperature profile inferred from data acquired not only at very high altitudes, as previously mentioned, but also by measurements of recovery temperature and stagnation pressure (relative to drag during the descent) made on both of the aeroshells[8]. The data sensing process switched over to direct barometric measurements near and on the surface[9]. This is quite a remarkable temperature profile, in that we would have expected a rather sharp rise in temperature starting at about 300-400 km and going down to 120 km. This is one of the things that we really need to learn more about. The Viking landings were conducted during solar minimum, and perhaps this is a major reason the Viking temperature profiles are as cold as they are.

Figure 2D-14 indicates inferences of the atmospheric temperature determined from both Mariner and Viking radio occultation data (the temperature scale is valid only for pure CO_2 ions). The solid curve represents solar activity while the other data (represented by "X" plots) are measurements of the upper atmospheric temperature taken by the different missions. Note that, when comparing the inferred

---

[8] The descent-capsule aeroshells were equipped with sensors for high-altitude measurements of recovery temperature and stagnation pressure during the brief entry phase, and the landers themselves had pressure and temperature sensors which completed the profile of the atmosphere's structure at low altitude and on the surface. Lander accelerometer data also supported the atmospheric structure investigation. {Ed.}

[9] The aeroshells were separated from the Viking landers approximately 1200 meters above the surface, and data were then recorded with instruments on the landers themselves during the final moments before landing (involving, in quick succession, parachute and then propulsive deceleration technologies). Prior to that, the instruments on the aeroshells were used to record data during atmospheric braking and as the parachutes initially slowed and stabilized the descent capsules. {Ed.}

temperatures with those provided in the **Viking** data, the inferred temperatures are considerably higher than those from the direct measurements. There is a strong suspicion that these inferred numbers should be scaled down by quite a big factor, using more sophisticated models of the ionosphere.

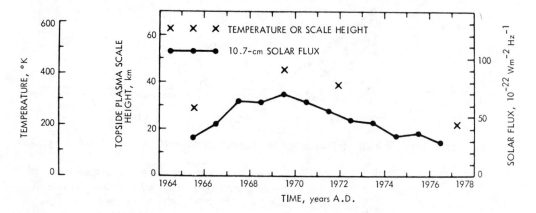

Figure 2B-14: Ionospheric Scale Heights Inferred from Mariner/Viking Radio Occultation Data

The flow diagram presented in Figure 2D-15 reflects interests associated with the complex and still controversial issues surrounding the biology question, as evidenced by discussions at this conference. This involves the lowest component of the atmosphere, and looks at the same physics that Mike McElroy, Norman Horowitz [NMC-1I] and Gil Levin [NMC-1J] have discussed. The odd-hydrogen catalytic cycle is depicted in the center, and the arrow at the inside-top of that box depicts the oxidation of CO into CO_2 with an OH molecule. The box at the bottom represents the precipitation into the soil of the various oxidants, which include a number of things other than just the hydrogen peroxide that may very well have had--as I believe--a major influence on the biology experiments.

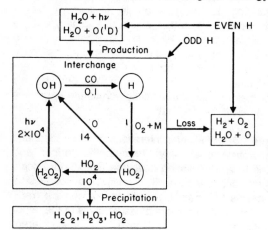

Figure 2D-15: Odd-Hydrogen Near Surface

Figure 2D-16 (a and b) illustrates rather directly the effect of dust storms on the lower atmosphere, using data acquired by **Mariner 9**. At left (Fig. 2D-16a), the altitude of electron density maximum is plotted against the spacecraft's orbit number (Rev.); the solid line is a model that allows for atmospheric contraction due to the dissipation of the massive dust storm that welcomed **Mariner 9** to Mars. Depicted at right (Fig. 2D-16b) are various temperatures, first during the dust storm and then as the dust cleared away (specific orbital revolutions are identified for cross-referencing). One can see an enormous drop in the stratospheric temperature at the same time during the **Mariner 9** mission. Measurements of the height of the ionospheric peak are represented, indicating very directly the inflation of the upper atmosphere (by a matter of 25 km) due to the high temperatures in the lower atmosphere during the first couple of months. This clearly suggests that it doesn't make sense to measure the upper atmosphere without also having some knowledge of the lower atmosphere.

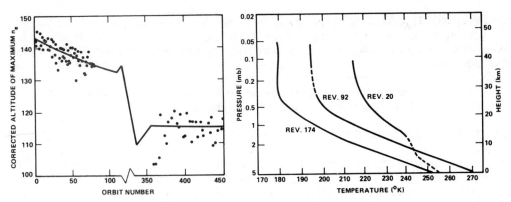

*Figure 2D-16: Left, a) Altitude of Electron Density vs Mariner 9 Revs;
Right, b) Temperature Profiles for Specific Mariner 9 Revs*

Solar Wind Interaction with Mars

The question of just how the solar wind interacts with Mars is reflected in Figure 2D-17. The sketch at right, which represents a purely ionospheric interaction, reflects the nature of what has been seen at Venus (with the **Pioneer Venus** spacecraft and during a number of Soviet **Venera** missions -- the latter in less detail) as it might be represented at Mars. The left sketch represents assumptions based on the existence of a significant intrinsic magnetic field. The nature of what we would expect to see--that is, the bow shock and the ionopause--is not terribly different in the case where there is negligible or no magnetic field or in the case of a magnetic field that is propagated by electric currents induced in the ionosphere. There are suspicions that Mars is somewhat like that, with just enough internal magnetic field to make some difference, and it would be very interesting to be able to investigate this possibility. On the other hand, if Mars is like Venus and essentially has no magnetic field, it would give us another example of a field-free planet to contemplate.

In another case, measurements made by the **Dynamics Explorer** at Earth have produced an ultraviolet image of Earth's auroral zone. Its perspective is essentially that of a polar view, including the day side and visible daylight airglow. With this kind of data, wind directions in the polar cap of the thermosphere can be readily discerned at altitudes well into the ionosphere. Indeed, the precipitation of energy in the aurora profoundly affects not only the thermal behavior but also the motions in the upper atmosphere. This is quite well understood for Earth and we feel that similar measurements at Mars are quite feasible. Indeed, this kind of product demonstrates what we should be able to do with comparable instrumentation on the **Mars Aeronomy Observer**, and the taking of such measurements is therefore part of our recommended payload.

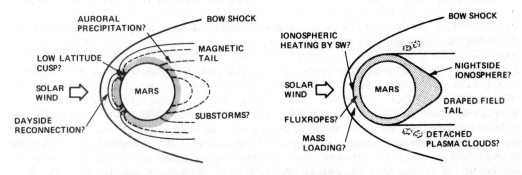

*Figure 2D-17: Solar Wind Interaction Scenarios -- **Magnetospheric** (left), **Ionospheric** (right)*

RECOMMENDED PAYLOAD

The nature of the measurements to be taken by the core and secondary payload instruments (listed previously with their technical characteristics in Fig. 2D-4) are summarized in Figure 2D-18. Let me again emphasize that these payloads and instruments do not have official status and simply represent the recommendations of the M/AO S·W·T. The actual payload would be selected by NASA on the basis of competitive proposals.

Core Payload Instruments

Most of core payload instruments have flown on **Pioneer Venus** or Earth-orbiting missions. We have, for example, included a neutral mass spectrometer and a Fabry-Perot interferometer (the latter measures doppler shifts of spectral lines). An ultraviolet+infrared spectrometer also is included to perform a variety of tasks, such as measuring ozone (O_3) or detecting water vapor near the surface (using the IR channel). It can also do more essential things, like measure the CO_2 or develop altitude profiles of H, O, C, and N using limb observations.

We have also focused some attention on obtaining ionospheric measurements. The instruments tentatively identified for this purpose include an ion mass spectrometer and a retarding potential analyzer with an ion driftmeter. The RPA/driftmeter measures a different aspect of the ions and also measures their velocities. A Langmuir probe, which is an electron temperature probe, has been included for the purpose of measuring electron temperature and density in the ionosphere; it would also serve as a solar EUV flux monitor.

Among the instruments concerned with the interplanetary and near-Mars plasma and fields, as well as the solar wind, are: a plasma and energetic particles analyzer and a magnetometer (the word "energetic" implies energies up to perhaps a few kiloelectron volts). And, while there will be a magnetometer on **Mars Observer**, that type of instrument is also a necessary part of the **Mars Aeronomy Observer** plasma package. Finally, our recommended core payload includes a plasma wave analyzer and the radio science systems already discussed.

Secondary Payload Instruments

In addition, we have recommended several instruments as secondary payload candidates. The first one is an infrared atmospheric sounder, similar perhaps to one of the instruments planned for **Mars Observer**. And, while a visual synoptic imager--comparable to the **Mars Observer** live-scan camera--has been included on our secondary payload list as well, because we too feel that imaging is a good idea, we have added a far-UV channel that we think could be integrated into such an instrument. It would facilitate auroral images similar to those made by the Dynamics Explorer during Earth observations, which I previously described as an example of this kind of science product. The last of these secondary instruments is the neutral-winds-and-temperature spectrometer, which would be used to measure doppler shifts at lower altitudes to determine thermospheric winds and temperatures.

Figure 2D-19 represents an attempt to correlate the various instruments with the realms of investigation I have discussed. The information relates specific instruments to atmospheric regions and processes. For example, roughly seven instruments bear on the issue of neutral atmospheric structure. The same premise is represented for ionospheric structure and interaction with the solar wind, and then--moving from regions to processes--for photochemistry and escape rates. Clearly, we can put together a selection of instruments whose correlated interdisciplinary results would tell us a great deal about each of these various regions and phenomena, reflecting the synergism philosophy expressed so well by Arden Albee in his discussion of the **Mars Observer** science strategy.

INSTRUMENT TYPE	MEASUREMENTS
Core Payload	
■ Neutral Mass Spectrometer	Number densities of neutral species, isotopic abundances, temperatures, and two components of cross-track wind velocity.
■ Fabry-Perot Interferometer	Atmospheric vector wind and temperature altitude profiles in the lower thermosphere (h < 200 km), metastable densities, volume emission rate profiles, rotational temperatures, and velocity distributions for escaping atomic oxygen.
■ UV+IR Spectrometer	H, O, C, and N altitude profiles from limb observations and nadir column densities of O_3 and CO_2 which allow modeling of escape rates and surface densities; IR channel measures water vapor.
■ Ion Mass Spectrometer	Number densities of ion species and their isotopic abundances.
■ Retarding Potential Analyzer and Ion Driftmeter	Ion temperatures, densities, and velocities in the ionosphere.
■ Langmuir Probe	Electron temperatures and density; solar EUV flux monitor.
■ Plasma & Energetic Particle Analyzer	Solar wind and magnetospheric particle velocity, density, temperature, and composition.
■ Magnetometer	Magnetic field properties in the solar wind, magnetosheath, magnetosphere, and ionosphere.
■ Plasma Wave Analyzer	Plasma wave properties in the solar wind, magnetosheath, magnetosphere, and ionosphere.
■ Radio Science	Atmospheric and ionospheric density and temperature altitude profiles.
SECONDARY PAYLOAD	
■ Infrared Atmospheric Sounder	CO_2, H_2O, aerosols, thermal structure, and winds (h $\leq$ 70 km).
■ UV+Visual Synoptic Imager	Global observations of stimulated emissions due to charged particle precipitation and NO, O_3, and dust in lower atmosphere.
■ Neutral Winds and Temperature Spectrometer	Thermospheric winds and temperature.

Figure 2D-18: Mars Aeronomy Observer Science Instrument Capabilities

REGIONS		
Neutral Atmospheric Structure	Ionospheric Structure	Solar Wind Interaction
NMS	IMS	PEPA
FPI	RPA+IDM	MAG
RS	ETP	PWA
UV+IRS	RS	IMS
IAS*	MAG	RPA+IDM
NWTS*	PWA	ETP
UV+VSI*	PEPA	RS

PROCESSES					
Photo-chemistry + Escape Rates	Global Energy Balance (Thermal Winds)	Nightside Ionosphere	Dayside Solar Wind Interaction	Pickup of Planetary Ions	Magneto-tail Dynamics
NMS	NMS	IMS	PEPA	PEPA	PEPA
IMS	FPI	RPA+IDM	MAG	MAG	MAG
UV+IRS	NWTS	ETP	PWA	PWA	PWA
ETP	ETP	PEPA	IMS	---	---
IAS*	IMS	MAG	ETP	---	---
NWTS*	RPA+IDM	PWA	RPA+IDM	---	---
---	PEPA	UV+VSI*	---	---	---
---	PWA	---	---	---	---
---	IAS	---	---	---	---
---	UV+VSI*	---	---	---	---

* Secondary Payload Instrument

Figure 2D-19:   Mars Aeronomy Observer Science Instrument Applications

Science Operations Rationale

One major way in which the **Mars Aeronomy Observer** mission differs with that of the **Mars Observer** is that the M/AO S·W·T has recommended a highly eccentric orbit during most of the mission, and one that dips into the upper atmosphere (the **Mars Observer** mission calls for a circular, fixed-altitude orbit). Although remote sensing is certainly a necessary part of their capability, many of the **Mars Aeronomy Observer** science instruments are intended to directly sense their ambient medium much as did the **Viking** upper-atmosphere instruments (RPA and UAMS). The primary interest in using them, therefore, is to actually sample the upper atmosphere to make in·situ measurements. A fairly elliptical orbit is required to facilitate these brief dips into the atmosphere, and it would probably have an apoapsis at a distance of several planetary radii. The periapsis, on the other hand, would be well below 150 km and perhaps as low as 120 km. The **Viking** orbiters effectively utilized highly elliptical orbits that were modified during the latter stages of their missions to achieve even lower periapsides for high resolution imagery[10].

Mission Options

There have been some interesting alternative mission concepts proposed that could significantly enhance the data product by incorporating comparative, data-correlation opportunities for Earth and Mars. One of these would use a single spacecraft, first for an Earth-orbiting mission and then for a similar Mars mission; the second would use twin spacecraft, one at Earth and the other at Mars.

The first concept, called the **Earth/Mars Aeronomy Observer (EMAO)**, is an idea proposed by Larry H. Brace (NASA-GSFC), who happens also to be a member of the M/AO S·W·T. His idea, which reflects a time when we had a more optimistic perspective of SSV (Space Shuttle Vehicle) capabilities, is to launch an **Explorer**-type spacecraft into orbit around Earth that would do the kinds of things we want to do at Mars. After a year or two we would recapture it using an SSV, mount it on an upper stage while still in orbit, and then send it off to Mars. We could also, of course, bring it back to Earth for some refurbishment before re-launching it to Mars. The beauty of this concept is that the same spacecraft and instruments would perform both missions, and the scientific objectives--as well as the nature of the data product--would be essentially the same no matter which way it was performed.

Another option would be to build twin spacecraft that might very well be launched by a single SSV, placing one in orbit at Earth and sending its twin to orbit Mars. The idea is to tap and utilize resources of both the **Earth Orbiting Explorer (EOE)** program and the Planetary Observer program, such that each would get more for its money.

---

[10] Following MOI events for both **Viking** flight spacecraft, the orbiters were established in orbits which had apoapsides slightly greater than 30,000 km and periapsides of roughly 1520 km. Slight differences were generally a result of periapsis latitude requirements for a given orbiter. These orbits were gradually modified to achieve increased imaging resolutions once primary mission objectives had been met. The apoapsides increased by more than several thousand kilometers as the periapsides were lowered to a little below 300 km. The problem of pointing accuracy became significant at lower periapsides, where the high resolution afforded made targeting a critical matter -- due largely to the effect of variability in the orbit periods as a result of Mars' uneven gravity field (particularly near Tharsis). Attempts to correct for orbital perturbations were only modestly successful in avoiding targeting errors, but intuitive/adaptive adjustments made by experienced **Viking** mission planners saved many of the high resolution imaging sequences. {Ed.}

Scientifically, there isn't much difference between these two missions, but the comparative aspect facilitated by EMAO (using the same spacecraft and instruments for both) could obviously be quite beneficial. It would afford significant advantages for the interpretation, correlation, and comparison of data from the two missions.

CONCLUSIONS

We have found that many of the options we have studied are in fact severely limited by the planetary quarantine factor -- the so-called "planetary protection" requirements[11]. This is an aspect of the program that is still to be negotiated. For the JPL version of this mission, the mission profile would look very much like the one Bill Purdy presented for the **Mars Observer** mission.

Nobody knows, of course, when any of this will happen. **Mars Aeronomy Observer** is somewhere in the queue among the next three Planetary Observers, although probably not next in line. We are doing our best to keep it near the front as the need to learn more about Mars continues to grow. ■

---

[11] Due to concern that lunar and planetary surfaces could suffer biological contamination if inadequately sterilized spacecraft were to land or impact on them, international cooperation was sought to avoid that possibility through the International Council of Scientific Unions. The resulting standards, established by the council's Committee on Space Research (COSPAR), were adopted in October, 1958. As implemented by NASA, these standards required the agency to adopt "the general policy of sterilizing, to the extent technically feasible, all space probes intended to pass in the near-vicinity of--or impact upon--the moon or planets." It is this policy that is generally known as the "planetary protection" policy -- sometimes called Planetary Quarantine. Concerns about spacecraft sterilization were compounded by the emerging desire to search for life elsewhere in the solar system, because an essential problem facing direct exobiology is that hard-impact or lander spacecraft instruments might simply detect "hitchhiking" Earth organisms, and it would be extremely difficult (if not impossible) to distinguish the "Earthlings" from alien organisms. The exobiology issue led to the creation of NASA's Office of Life Sciences in 1960, only two years after the creation of the "planetary protection" standards. In a specific application of the policy to the case of Mars, it has been agreed upon internationally that the surface of Mars is not to be contaminated before the year 2018. This means that even orbiting spacecraft must not be left in orbits that, due to decay, could cause the spacecraft to impact the surface prior to that year. The policy makes the development of hard-impact or lander systems, in particular, significantly more difficult and costly, first because the various sterilization technologies (e.g., heat) can be quite detrimental to sensitive components. The components must therefore be designed for the sterilization environment as well as those they will encounter after departing Earth. Secondly, a variety of special designs and methods must be employed to assure that sterilization is maintained until the spacecraft is free of Earth's environment. In the **Viking** case, some of the instruments (e.g., biology and molecular analysis) were chemically cleaned and decontaminated as a part of the flight-readiness procedure. Following a lander's final assembly, including stowage in its entry capsule, the entry capsule was sealed in a soft-skinned outer bioshield (filled and pressurized with inert gas) and carefully heat sterilized in a large, unique oven-like chamber ($\approx 233\,°$ F for 40 hours). Once sterilization was completed, the enclosed hardware was sterile and the bioshield served as a barrier against outside contamination until part of it was jettisoned following launch (once the spacecraft was safely in sterile space). {Ed.}

REFERENCES

This space is otherwise available for notes.

~~~~~~~~~

Blume, W.H., Low, G.D., and Tsou, P. 9/1983. *Planetary Observer mission descriptions.* JPL D-1248 (internal document).

Gierasch, P., and Goody, R.M. 1968. The effect of dust on the temperature of martian atmosphere. *J. Atmos. Sci.* 29:400-02.

Glickman, R.E. 2/6/1982. *Mars polar orbiter mission-orbit studies.* Ball Aerospace Systems Division, Report B6240-81.010.

KEPLER, an interdisciplinary Mars orbiter mission. 12/1982. European Space Agency. Report on Phase-A Study, SCI(82)5.

KEPLER Mars orbiter. 12/1985. European Space Agency. SCI(85)6.

Mars orbiting water mission. 8/1981. NASA-ARC. Final report.

Mars orbiter study. 9/1982. Hughes Aircraft Company. Vol. 2, Final Report, Contract NAS-2-11224.

Mars orbiter conceptual systems design study. 9/1983. TRW. Final report (for NASA-ARC), Contract NAS-2-11223.

Planetary Exploration Through Year 2000: A Core Program. 1983. Part 1 of a report by the Solar System Exploration Committee (SSEC) of the NASA Advisory Council. GPO.

Planetary Observer planning: FY84 Final Report. 9/1984. JPL D-1846 (internal document, J.E. Randolph, ed.).

Planetary Observer planning: FY85 Final Report. 11/1985. JPL 642-901 (internal document, J.E. Randolph, ed.).

Planetary Observer Planning: FY86 Mission Studies Report. 9/1986. JPL D-3649 (internal document, R.A. Wallace, ed.).

Ryan, R.E. 6/5/1980. *Low cost exploration of the solar system mission descriptions for Venus and Mars aeronomy.* JPL Interoffice Memorandum 374-80-43.

Ryan, R.E. 11/20/1980. *Limited-scope mission studies, Mars and Venus aeronomy missions.* JPL Interoffice Memorandum 379-80.

Study of Mars orbiter spacecraft derived from existing designs. 7/14/1981. Ball Aerospace Systems Division. Report, Contract NAS-2-11023.

MARS SAMPLE RETURN: MISSION
[NMC-2F]

Mr. James R. French, Jr.
Vice President, Engineering
American Rocket Company

There are four fundamental mission options available for the accomplishment of Mars sample return missions: (1) direct entry, direct return; (2) direct entry, Mars orbit rendezvous; (3) out-of-orbit entry, direct return; and (4) out-of-orbit entry, Mars orbit rendezvous. Numerous variations of these options are dictated by the nature of the mission, the types of mission vehicles involved, and launch mass relative to launch vehicle capability. Among the important technologies involved are certain aspects of how the orbit and entry at Mars are accomplished and how the return vehicle is dealt with at Earth. In both cases, the use of propulsive technology must be weighed against the value of using aerodynamic capture, braking, and maneuvering, which contribute significant savings in launch mass. Not only can aerocapture and aeromanuevering be used in place of propulsion, they could also be combined with propulsion to achieve a productive blend of these technologies. Rover technology is also evolving. Concepts with varying degrees of machine intelligence have been studied, ranging from small, very simple lander-tethered vehicles to autonomous rovers capable of making their way about with the help of interactive control from Earth. In addition, in·situ propellant production could provide some of the needed fuel mass at Mars. Depending on the nature of the mission and the spacecraft elements involved, sample return may require multiple launches with expendable launch vehicles in the absence of capacity previously afforded by the SSV/Centaur-G'. While difficult, however, sample return missions are feasible and within reach of our capability with continuing study.

INTRODUCTION: OPENING COMMENTS

Sample return missions have been discussed and considered since before the beginning of the space age as one of the primary goals of planetary exploration. The scientific aspect, which defines the advantages of sample return in very clear terms, are discussed by Doug Blanchard in his presentation for these **NASA Mars Conference** proceedings [NMC-2G], so I am not going to try to explain why we should pursue such missions. Instead, I will simply explain why they are complex and demanding missions, and why--by the standards we have come to accept for unmanned planetary missions--they can be rather costly.

MARS SAMPLE RETURN MISSION CONCEPTS

Sample return represents a very complicated mission objective that is both a blessing and a curse. There are a variety of different ways to accomplish such a mission, for example, and I could spend the entire allotted time on that topic alone. The nature of the four primary concepts currently defined for sample return missions are outlined in Figure 2F-1.

[1] DIRECT ENTRY / DIRECT RETURN ■ Single Launch ■ Dual Launch w/On-Orbit Assembly	
[2] DIRECT ENTRY / MARS ORBIT RENDEZVOUS ■ Single Launch ■ Dual Launch	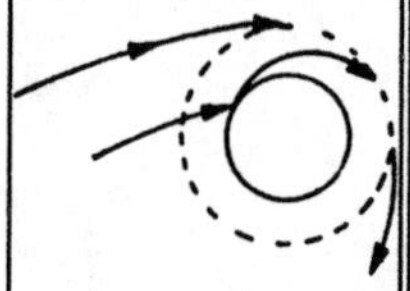
[3] OUT-OF-ORBIT ENTRY / DIRECT RETURN ■ Single Launch ■ Dual Launch w/On-Orbit Assembly	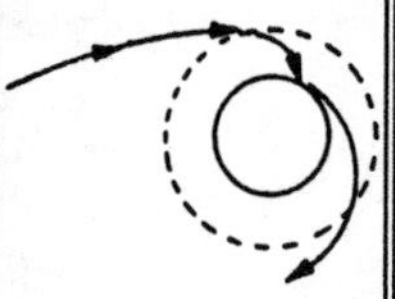
[4] OUT-OF-ORBIT ENTRY / MARS ORBIT RENDEZVOUS ■ Single Launch ■ Dual Launch	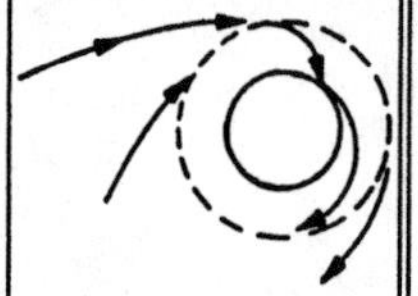

Figure 2F-1: Primary MSR Mission Options

I will briefly introduce them at this point, and then add detail as I progress through variations that have evolved out of a series of studies. First, however, it should be noted that these concepts involve some spacecraft configurations that are broadly defined on the basis of their primary function. They are: (1) Mars Sample Return (MSR), which is most often a generic reference related more to mission type or to a nonspecific--perhaps composite--flight spacecraft, (2) Earth Return Vehicle (ERV), and (3) Lander/Ascent Vehicle (LAV[1]). The four MSR mission options make use of these spacecraft systems in different ways, as suggested in the following:

[1] *Direct Entry and Return* -- The most straightforward way to do a MSR mission is to simply plunge in directly (without orbiting) and land. After collecting a sample for return, the lander's ascent vehicle would then serve as an ERV and take off for a direct return to Earth, pretty much as Flash Gordon might have done it.

---

[1] The term "lander/ascent vehicle" (LAV) refers to an integrated vehicle that actually lands on Mars, after which its ascent element transports the samples collected back to Mars orbit or to Earth. In part, this is analogous to the Apollo lunar landing module, which left its lander element behind when it was time to ascend to lunar orbit for rendezvous with the command module. However, the MSR LAV ascent vehicle could also function as an Earth-return vehicle (ERV) itself, negating the need for a rendezvous and an independent ERV. {Ed.}

[2]   *Orbiter/Earth-Return and Direct-Entry Lander/Ascent Spacecraft (one of each)* -- The second possibility is to send two spacecraft, landing one of them (the LAV) directly while the other (the ERV) waits in orbit to bring the sample home following rendezvous with the sample-bearing ascent vehicle launched from the lander.

[3]   *Composite Spacecraft with Ascent/Earth-Return Vehicle* -- This concept would place a single integrated MSR spacecraft in orbit, separate and land the LAV, and then send the sample directly back to Earth from the lander. In this case, the ascent vehicle also serves as the ERV.

[4]   *Earth-Return and Lander-Ascent Vehicles (one of each) with Rendezvous* -- The final concept first places both ERV and LAV spacecraft into orbit. The LAV lands from orbit and collects its sample, and an ascent vehicle then carries the sample back into orbit to rendezvous with the ERV for the flight home. This final concept is somewhat analogous with the way the Apollo missions were conducted, except that the LAV and ERV arrive at Mars separately.

There are, of course, many variations on these concepts based on how certain aspects of the mission are performed, e.g., how the orbit at Mars is achieved in the first place and how the MSR flight vehicle is to be dealt with at Earth. The MOI (Mars Orbit Insertion) could be achieved with propulsion, or aerodynamic braking could be used to achieve capture. It would also be possible to use aerodynamic braking just to circularize the orbit following a propulsive MOI. Similarly, the mission design would have to consider what is to be done when the ERV spacecraft gets back to Earth. The sample-carrying vehicle could enter directly or it could be placed in an orbit compatible with a space station, and again both propulsive and aerodynamic braking are potential elements of the equation.

HISTORIC PERSPECTIVE

Briefly, the history of Mars sample return can be thought of in terms of pre-Viking work, contemporary Viking work (mid-to-late 1970's), and post-Viking (current) work. Prior to Viking mission operations in the 1970's, there was at least one sample return study performed that involved NASA's Langley Research Center (LaRC), the Jet Propulsion Laboratory (JPL), and Martin Marietta's Denver Aerospace (MMDA) company[2]. I think it was optimistic in terms of anticipated performance, but it did represent a serious engineering look at MSR mission concepts.

As a result of the interest generated by Viking, a significant study of the MSR idea was mounted in the late 1970's. It involved JPL and NASA-JSC in particular, and many of the other NASA centers to a lesser extent. It was based on a Space Shuttle Vehicle (SSV) launch capability and used the twin-stage version of the inertial upper stage (IUS). It might be of interest to note that some of these older studies may become more important because of the cancellation of the *Centaur-G'* (-G prime), at least in terms of SSV applications, as well as changes in SSV launch constraints for the STS (space transportation system). These 1970's studies are outlined

---

2 The participants involved in this study were the principal team members on the Viking Project. NASA's Viking Project Office was based at NASA-LaRC, the orbiter program was managed at JPL, and Martin Marietta was prime contractor for the landers and integrating contractor for the flight spacecraft. A number of the early sample return studies involved follow-on Viking technology applications, wherein the lander was turned into a rover by using small drive tracks (achieving surface mobility with continuous-loop elastic titanium belts) in place of the original footpads (3). Later, sample return was added by mounting an ascent stage atop the lander. {Ed.}

* MARS SAMPLE RETURN STUDIES *

NASA–LaRC/JPL/MARTIN MARIETTA (Denver)
- Based on Viking Lander/Orbiter
- Dual Titan/Centaur Launch

JPL/NASA–JSC (post-Viking)
- Evaluated Variety of Mission Concepts
- Candidate Concepts:
  - Direct Entry/Direct Return
  - Mars Orbit Rendezvous
- Based on SSV/Twin-Stage IUS
- Introduced Aerocapture and In·situ Propellant Production (ISPP)
- Serious Study of Autonomous Rendezvous in Mars Orbit
- Dual Launch from Earth Orbit or On-Orbit Assembly Required
- Proceeded as far as **Pre-Project** Prior to Cancellation (1980)
- Aerocapture and ISPP Research Continued After Cancellation

* MARS SAMPLE RETURN CONCEPTS *

DIRECT ENTRY/DIRECT RETURN
- Semi-Ballistic (low lift/drag)
- Three-Stage Sterilized Solid Propulsion System
- Viking Lander Engines
- Requires On-Orbit Assembly at Earth

MARS ORBIT RENDEZVOUS
- Out-of-Orbit Mars Entry
- Propulsive Insertion Into Elliptical Mars Orbit
- Viking-Derived Lander
- Three-Stage Sterilized Solid Ascent Propulsion
- Orbiter Performs Circularization and Rendezvous w/Ascent Stage

Figure 2F–2: Mars Sample Return Studies and Concepts (1970's)

Within the realm of this work, we introduced the first serious studies of aerodynamic capture into Mars orbit and also of in·situ propellant production, i.e., using martian resources to produce propellants at Mars for the flight home. We also looked at direct-entry and direct-return concepts, as well as in-orbit rendezvous at Mars, et·cetera. We concluded that it really wasn't possible to do these kinds of missions with the two-stage IUS upper stage in a single launch; the rendezvous missions had to be performed with two spacecraft launched separately, or a couple of launches had to be made into Earth orbit to facilitate the assembly of the extra large spacecraft before sending it on its way. In either case, we could not avoid a multiple-launch requirement.

In fact, this concept proceeded as far as a pre-project study intended to lead to a hoped for project start, but planning was cancelled in the downturn of planetary exploration activities during the late 1970's. Indeed, it was horribly expensive; it would have cost--over a program period of eight years--perhaps ten percent (10%) of the annual national marijuana expenditure[3], making it quite unaffordable (speaking in frustrated jest). We considered many of the possible options, including the low lift-over-drag (L/D) semi-ballistic entry. In other words, the concept was analogous to an Apollo capsule with what seemed like the reasonable technology at the time. Relative to propulsion, we looked at the in-orbit Mars rendezvous concept and performed some of the first serious studies of autonomous rendezvous in Mars orbit.

However, that work was followed by a fairly lengthy hiatus, and studies concerned with MSR missions were always just beyond the planning horizon. Sample return was the mission of the future, and it looked as if it would *always* be the mission of the future. However, there was a new beginning of sorts in 1984, with some serious studies on the subject. This work again involved JPL and JSC, with some participation of other NASA centers, and it generally picked up where the activity of the late 1970's had left off -- but with an infusion of new ideas. Among its considerations, the work added a large rover to the system and developed new knowledge of ascent and rendezvous technologies. Of particular importance was the availability of work we had done concerning aerocapture, including some that had subsequently grown out of studies involving that technology during the 1970's, and it had advanced quite significantly.

Also, the idea of adding a rover to the MSR concept was introduced to facilitate a higher quality mission through the acquisition of a more selective set of samples. However, that enlargement made the spacecraft even bigger than it was before, and therefore more difficult to launch. We moved ahead on it because at that time we believed we would have the Centaur-G'[4] upper stage available, which we felt might make a single-launch program possible. One may wonder why I continue to favor the idea of a single launch. The Space Shuttle orbiter vehicle (SSV) is in fact fairly expensive to launch, particularly when it includes a Centaur-G' upper stage. Even in the case of a large mission for which major launch costs are justified, the launch costs are not inconsiderable. Therefore, if we can--for reasonable expenditure of effort--constrain program requirements to a single launch, it is certainly worth the effort in terms of reduced launch costs.

---

[3] Mr. French's comment was based on the fact that in 1984 it was estimated that the American population spent approximately $16 billion on marijuana. {Ed.}

[4] The Centaur-G' was a planned high-energy stage designed to serve as an alternative to the IUS, and as such was to be deployed from a SSV to support deep space planetary missions like Galileo. It was, in fact, a new version of the Centaur upper stages used to successfully launch the Viking, Voyager and Helios spacecraft on Titans (T3E vehicles). As such, it was a cryogenically-fueled stage (using liquid hydrogen and oxygen, LH/LO_2). Although it has been removed from STS planning with respect to manned missions, it is expected to be available for use with a new generation of ELV's (expendable launch vehicles). {Ed.}

Some good work was also performed--and in fact continues--in the area of realistic development associated with the guidance capability needed for aerocapture and another capability called *"aeromaneuvering."* Aeromaneuvering is the descent maneuvering needed after a spacecraft capsule enters an atmosphere to achieve a pinpoint landing, as opposed to having to consider the large "footprints" we had to deal with just prior to the **Viking** landings[5].

Because the 1984 study picked up at the point where we had stopped in the 1970's, but with the understanding of **aeromaneuvering** technology available for consideration, it applied this new concept to the same mission options in different modes -- including semi-ballistic entry (low L/D) with various propulsion modes, et·cetera. The concept proved to be too much mission for a single launch, however, even with the combined capabilities of the Centaur-G' and the SSV, and our conclusion once again was that we could not perform a MSR mission with one launch. We did find, however, that if we launched with the Centaur only partly loaded with propellant, and then finished the loading in orbit at a space station or via a second launch, some of the options were feasible -- even comfortably so.

The sequence for getting the MSR flight vehicle into orbit at Mars and a lander on the surface, and then transporting the sample back to Earth, is essentially that of placing a propulsive orbiter (the MSR spacecraft) in orbit at Mars, using a semi-ballistic direct entry for the LAV, and then achieving a Mars-orbit rendezvous with the MSR orbiter (by the lander's sample-return ascent vehicle) or returning directly to Earth. One must, of course, decide among the optional ways to get the lander safely and accurately on Mars, i.e., aeromaneuvering via direct entry as opposed to going into orbit first. One must also determine the best way to return the sample to Earth, i.e., by means of an orbital rendezvous or direct return.

Workable Options

The results of the 1985 study are outlined in Figure 2F-3. We essentially eliminated the direct-return missions because of the large mass requirement reflected in the propellant the lander would have to carry to the surface of Mars. That is, the amount of propellant needed to generate sufficient launch energy for a direct return to Earth is simply too great. It is interesting to note that the same concern drove the Apollo program to its ultimate mission profile. We also gave up on the semi-ballistic options, i.e., the low L/D-configured vehicles. We did not feel that we had sufficient control to insure aerocapture with good reliability when attempting to slow down to orbital speed, such that it could be achieved without either skipping out of

---

5 The process of selecting new **Viking** landing sites was agonizing, partly because of uncertainty about where--within a probability footprint--the lander might actually touch down. Several sizes of model footprints were used, the largest representing the greatest chance for entry perturbation (least accuracy). However, because of large-scale risk at the surface, as dramatically illustrated in the early **Viking** orbiter images (crater densities, ejecta debris, erosion etching, et·cetera), the smaller the footprint the better. The footprint itself was an ellipse placed on a line projected by the lander capsule's entry trajectory after the landing sequence was initiated by separation and retro propulsion. Topography within the ellipse had to suggest minimal risk to the lander. At best, surface resolution at the time was ≈100 meters (although improved by Earth-based radar data), meaning that risk to the lander had to be inferred from very large features. The ellipses were generally between 50 and 200 km in length, depending on the probability of landing within them. However, both **Viking** landers touched down remarkably close to the crosshairs within their respective footprints, and were apparently only minimally perturbed by unanticipated entry dynamics. Even so, the need to achieve pinpoint landing accuracy (with great certainty) for MSR sites has driven related studies toward those technologies that will make it possible. One must remember that all such landings on Mars must be fully automated, because such events occur very quickly and communication time between the planets makes interactive Earth-based human control impossible. {Ed.}

the atmosphere like a flat rock on a pond or plunging through the atmosphere and crashing, extremes that nevertheless reflect concerns associated with these kinds of techniques. Instead, this option was found to dictate a high L/D vehicle, which in turn presents still another set of problems. I have discussed the aerocapture and aeromanuevering concepts in greater detail in the second of my two presentations for these proceedings [NMC-3E]: *Key Technologies for Expeditions to Mars.*

The option requiring that everything be placed in orbit by propulsion (No. 3, Fig. 2F-3) results in a mass that is much too heavy, such that a single launch for such a mission is simply not viable -- not even when the Centaur is partially fueled in orbit (as described earlier). Indeed, such a mission would require a minimum of **three SSV/Centaur-G'** launches.

Similarly, the fourth option, which places the MSR orbiter in Mars orbit using its own propulsion system but utilizes direct entry for the LAV, is also too heavy. However, this kind of mission could be made light enough by invoking another use of the atmosphere. If a small propulsive maneuver is performed to get the MSR orbiter captured in a very loose elliptical orbit at Mars, we could then allow a controlled orbital decay to occur. A large drag shield installed on the spacecraft could--over the course of many orbits (a period of several weeks to a few months)--circularize the orbit using drag produced during multiple passes through the upper atmosphere. This might allow us to reduce propellant mass enough to keep the mission in a single-launch regime. The aspect one must be careful about, of course, is knowing when to quit using aerodynamic braking before suffering the "**Skylab syndrome**." Nevertheless, this does appear to be a viable option, and it is a relatively simple option to implement. That is, even though it is somewhat complex operationally, it is relatively simple with respect to hardware.

SEMI-BALLISTIC OPTIONS ELIMINATED

DIRECT RETURN ELIMINATED

CONTINUED STUDY OF OPTIONS:

[1] Aerocapture All · Aeromaneuver Entry · Orbit Rendezvous

[2] Aerocapture Orbiter · Aeromaneuver Direct-Entry ·
    Mars Orbit Rendezvous

[3] Propulsive All · Aeromaneuver Entry · Mars Orbit Rendezvous

[4] Propulsive Orbiter · Aeromaneuver Direct-Entry ·
    Mars Orbit Rendezvous

VEHICLE CONCEPTS STREAMLINED -- RESULTS

■ First Too Heavy (integration into single-aeroshell complex)

■ Second May Allow Single-Launch/Dual-Aeroshell Integration

■ Third Option Much Too Heavy for Single Launch

■ Fourth Option Too Heavy as Presented -- but Promising

   · Orbiter Inserted Into Elliptical Orbit
   · Circularize Slowly Using Aerodynamic Drag (aerobraking)

Figure 2F-3: Mission Options, 1985 Study

Alternatives for Future Consideration

Among the alternatives we didn't consider, and that probably deserves some serious study, is *in-situ propellant production* (ISPP). The use of martian resources to manufacture propellant for the return journey, for example, is an important possibility. There are a variety of ways ISPP could be employed, but a relatively simple way it might be applied to a small mission like sample return (compared to a manned mission) is to make only oxygen (O) from the martian carbon dioxide (CO_2) atmosphere. The oxygen could be used as an oxidizer with a fuel component transported from Earth, a relatively simple concept that might be viable within the scope of MSR mission requirements. Alternatively, both the oxidizer and a fuel like carbon monoxide (CO) or methane (CH_4) could be produced, although that would be somewhat more complicated.

In addition, we did not study closely the low-thrust, high-energy propulsion options, specifically electric and solar sail. We tended to neglect them because their flight times tend to be extremely long. While they are fine technologies for many kinds of missions in which flight time is not an important factor, they are not readily adaptable to a MSR mission.

Favored Concepts

We essentially wound up our 1984-85 studies with two aerocapture versions that utilize aeroshells to produce aerodynamic drag as the vehicles encounter the atmosphere[6]. These are the first two items outlined in Figure 2F-4. The third option reflects the use of a propulsive MSR orbiter and a direct-entry, aeromaneuvering LAV.

AEROCAPTURE ORBITER · AEROMANEUVER DIRECT ENTRY

■ Dual Aeroshells

■ Launch Vehicle Integration May Increase Mass

AEROCAPTURE ALL · AEROMANEUVER ENTRY

■ Integration in Single Aeroshell Causes CG Problems

■ Dual Use of Entry Heatshield

PROPULSIVE ORBITER · AEROMANEUVER DIRECT ENTRY

■ Each Vehicle Optimized for Its Task

■ Simplified Integration

■ Drag Brake Incorporation on Orbiter is Straight Forward

Figure 2F-4: Vehicle System Concepts -- 1984-85

---

[6] Those familiar with the nomenclature of the Viking entry/descent capsule may recall that the term "aeroshell" applied only to the bottom half of the capsule. It served as a mounting platform for science instruments and deorbit propulsive components internally and as a heatshield externally. In the context of this presentation, the term applies to the entire configuration utilized with each space vehicle, although design characteristics differ relative to the application. [Ed]

Aerocapture Orbiter · Aeromaneuver Direct Entry -- The first of the aerocapture concepts involves aerocapture with an orbiting MSR spacecraft while independently conducting a direct entry with a LAV, the latter aeromaneuvering to a precise landing. That requires each of the vehicles to be packaged in its own aeroshell. And, since we require high L/D, these are rather slim, streamlined aeroshells. It looked as if that combination was quite good with respect to mass, in terms of being within the launch capability of a single, SSV/Centaur-G'. Unfortunately, integration of these two lifting bodies in the same package presents significant problems. We did not get far enough into the study during the 1984-85 tasks to look in detail at the matter, but this kind of integration appears to be quite difficult and is a fairly serious problem. One must remember that this kind of problem usually translates itself into a launch mass penalty, so it may be that the advantage of this particular scheme is more apparent than real when one gets into the actual engineering of it.

Another problem associated with the dual use of atmospheric aeroshells (although probably less of an issue than the one just discussed), is concerned with multiple exposures to aerodynamic environments and heating. This is related to why we have reusable heatshields on SSV's -- the various materials used (tiles, et·cetera) must stand up to multiple re-entries. One of the great advantages of doing aerocapture at Mars is that, even when nearing Mars at maximum approach velocity, which any spacecraft is likely to do on a normal trans-Mars trajectory, spacecraft velocity relative to the upper atmosphere is still only about seventy-five percent (75%) of the relative airspeed an SSV (or any other vehicle) re-entering Earth's atmosphere would encounter. Clearly, then, the martian atmosphere is a fairly benign environment in terms of the kinds of entry effects we have long learned to deal with at Earth. The fact that the martian atmosphere is CO_2 makes it a little bit worse, but that is a relatively minor aspect compared to the effect of the velocity. Therefore, the dual-use of aeroshells is not a serious problem even though the payload integration matter is a major challenge.

Aerocapture All · Aeromaneuver Entry -- The second aerocapture concept initially looked quite good. For this concept, we aerocapture a single, fully integrated MSR spacecraft into an orbit at Mars, and then divide the original flight spacecraft and use only its forward element (with an aeroshell) to aeromaneuver to a landing. The problem is that one again encounters some serious integration difficulties, particularly in how the center-of-gravity (CG) is located within the aeroshell. As one can imagine, locating the center of mass within the aeroshell is quite critical to maintaining the proper angle-of-attack trim during entry. During the course of our work, we have never successfully solved the problem of how one maintains CG control in both configurations. That is not to say "it isn't doable," but it will not be easy.

Propulsive Orbiter · Aeromaneuver Direct Entry -- The scheme of putting the MSR orbiter into orbit propulsively, and using aeromaneuvering for the direct entry of the LAV separately, appeared to be quite feasible if we invoked aerodynamic braking later for orbit circularization. This idea has not been studied in detail, but it appears to be a relatively simple thing to do. Mars is a fairly benign place to do slow circularization of an orbit, as well. Generally speaking, a bent biconic shape is dictated by the L/D requirements. While it is not the only shape that will deliver the kind of L/D we need, it is among the better candidates for this purpose.

Integration-packaging presents something of a problem in this type of vehicle. The sample canister, which looks a little like a "cookie jar" with a handle on top, must be located toward the rear of the vehicle. The front end is basically reserved for spacecraft systems that keep everything in order during the journey home. The canister, which is discussed in more detail by Doug Blanchard, departmentalizes a variety of sample types for the return trip. It can carry about 5 kg of carefully selected Mars samples to Earth.

Finally, as its title implies, a small propulsion motor is included in the configuration to circularize the orbit after successfully achieving aerocapture. Since aerobraking is performed going into orbit by approaching the planet at a relatively low passing altitude through the atmosphere, the spacecraft can "capture" into orbit as desired. But, unless something is done quickly to adjust the periapsis upward, the spacecraft will come back into the atmosphere the next time around -- and that could be terminal. It is therefore crucial that we circularize the orbit right away, investing a reasonable amount of propulsive energy in doing so, even though aerocapture technology is productively represented in this option.

THE ROVER EVOLUTION

One of the ideas now being coupled with that of the sample return mission is the use of a rover. That aspect has been summarized in Figure 2F-5, which reflects a relatively current state of understanding on rover technology (1984-85). We had studied rovers quite extensively back in the 1970's, and a rover mission was in fact the leading candidate for a major Mars mission that actually got as far as a NASA cost review before biting the dust. Indeed, we had even studied the idea of sending two small rovers at the same time. The rationalization was that one could help the other get out of trouble, e.g., push it out of a hole if it got stuck, and there is always (of course) greater safety in redundancy should one fail or make a wrong turn. However, I suspect that another reason is that it was interesting--if not very efficient--to have two of them trucking around on the surface, where they could take pictures of each other as they conducted their activities (for whatever purpose that might serve).

NUMEROUS MISSION STUDIES (mid/late 1970's, mid 1980's)

- Dual Rover
- Large Autonomous Rover for 1984 Launch
- Mini-Rover for Sample Return Missions
- CNES Inflatable Mars Ball (wind propulsion)
- U·of·A* Inflatable Sectored-Tire Mobility

RESULTS:

- Large Autonomous Rover Study Overly Optimistic
- Modest-Size Rover Requires High Degree of Machine Intelligence or Extensive Human Control Interaction/Intervention
- Large Inflatable Rovers Substitute Size for Intelligence

*Advanced Mars Ball concept and model developed as student research project at University of Arizona.

Figure 2F-5: Rover Studies/Concepts -- 1970's/80's

The dual-rover idea developed into the large autonomous rover program that was proposed for a 1984 launch. It didn't make it, as I said, and in retrospect it is probably just as well. I think we were highly (overly) optimistic about the degree of autonomy we thought we could build into the rover at that time. Many people were infected by their perception Star Wars (the motion picture) of robotic capability, and there was a feeling that if R2D2 and C3PO could do all of those remarkable things, we certainly ought to be able to build a rover that could find its way around Mars. It didn't turn out to be quite so easy (perhaps **THE FORCE** hasn't been with us).

We also considered a mini-rover, which was--quite accurately--the size of a bread box. In support of a sample return mission, this small rover would simply truck around in the immediate area of the immobile lander and collect samples. This kind of solution wasn't generally considered to be good enough, however, because what "they" really wanted was a rover capable of substantially more range and sophistication[7].

Test Models: Large Inflatables vs Conventional

The French space agency suggested the use of a large inflatable Mars ball as a mobility system. The idea was that the ball, known as the CNES rover, would be blown around by the wind. Along with other problems, the idea didn't work very well with our improved knowledge of Mars. However, that original idea led to still another inflatable-rover mobility concept, a model of which has been built by students at the University of Arizona[8]. This concept, somewhat inaccurately known as the **Mars Ball** because of its distant CNES rover heritage, is discussed and illustrated in detail by Doug Hilton in his Mars Conference presentation [NMC-2H]. It is an intriguing and fascinating idea. Its premise is to substitute large size for intelligence, one that seems to work well for many sporting events and so might work well on Mars, too.

A full-scale rover model of the more conventional type was built for the purpose of conducting some operational research in 1984-85. I would not dignify it by calling it a prototype, but it was at least an operating model of a Mars rover concept as we envisioned such a vehicle at that time. It didn't move very fast, but it did move along at a pace that--had we been able to guide and power it properly--would have allowed it to cover a few hundred kilometers in the course of a Mars year. The model is still in storage someplace at JPL, and it is put through its paces once in awhile for visiting dignitaries. Its configuration is not particularly important, but it does give one an idea of the general size and magnitude of the kind of vehicle needed to cover long distances. It is roughly the size of an office desk and weighs approximately 450 kg (Earth weight). One of its main problems, other than the obvious one (guidance), is to be able to deliver enough power to proceed at a reasonable pace over a long period of time.

Rover Control Concept

The current state of rover-control technology is a blend of common-sense machine intelligence and human interaction/intervention. One often hears statements to the effect that rover-control

---

7 At its earliest stage, the concept of using a restricted-range rover took two forms. In the simplest case, a small rover was tethered to the lander and operated at the end of its umbilical cable essentially as an extension of its "parent" system. Another took the form of a simple autonomous rover capable of executing commands transmitted to it from the lander or an orbiter, which received its own command loads from Earth based on rover or lander images. These scenarios were closely related "next step" ideas designed to utilize--to the extent possible--the Viking technology base still in place at that time, such that a reasonable cost savings could be realized. They were reflected in both MSR and in·situ sample analysis concepts. As Viking mission activity progressed, evidence of site modification (both physically and chemically) by the plumes of the landers' terminal descent engines was suggested, and this in turn suggested that autonomous rovers with greater range than these relatively limited early concepts afforded would be necessary to find pristine, uncontaminated samples. {Ed.}

8 Both concepts, the original CNES Mars (wind) ball and the current U·of·A **Mars Ball**, were originated by Jacques Blamont, who is with the National Center for Space Studies (CNES -- Centre National d'Etudes Spatiales) in France [extracted from NMC-2H, Douglas A. Hilton]. {Ed.}

technology is too complicated and that we can't possibly do a rover mission anytime soon. That is certainly true if one postulates a highly intelligent rover that could, for example, do about everything that each Mars Conference attendee did as he or she came in the front door for the first time and had to navigate through the Academy of Science auditorium in search of an appropriate seat. That is essentially what one is asking a rover to do if it is required to navigate autonomously over the surface of an unknown planet. That takes a fair amount of artificial machine intelligence even if it doesn't necessarily have to be human-level intelligence.

For now, at least, that level of machine intelligence seems to be beyond anything that we can reasonably anticipate anytime soon. With human intervention (Earth-based interaction and guidance input), however, it appears that we can move a reasonable distance in a reasonable period of time without highly advanced robotic intelligence. One scenario, for example, starting with a rover at spot X, might be as follows:

■ The rover transmits stereo (panorama) images to Earth;

■ The mission controllers review the image data (which incorporates very precise three-dimensional range and elevation measurements based on the stereo images) and, knowing roughly which way he or she wants the rover to go, determines a reasonable path and picks out the best series of perhaps three path segments in the direction of the rover's objective (bypassing obstacles the rover can't negotiate);

■ A command sequence is then prepared (and perhaps tested on a simulation model) that represents the path segments desired, out to a range where reliable imaging data are no longer available;

■ The command sequence is transmitted to the rover to update its sequencing computer, and essentially says "go to this point, then stop and take more pictures!"

■ The rover then proceeds along the appropriate path segments (at what would be a glacially slow rate), using precise guidance instruments (gyrocompass and odometer functions) until it reaches the point established by the command sequence -- unless it first encounters an unanticipated obstacle or hazard). In either case, it stops and takes another set of stereo images which will essentially initiate a repetition of the procedure (or correctional movements to get it out of trouble).

With this type of rover mission plan, there need be only enough machine intelligence on board the rover to detect unseen holes that it might fall into or obstacles that it couldn't get over, et·cetera. When it encountered one of these, it would simply stop (the familiar "no-go" in Viking language), take pictures, and yell for help[9]. This does not require a high level of robotic intellect, and it does appear to be something that we could do. Admittedly, the concept is quite operations-intensive, but smart programming in this manner could facilitate the ability to cover a few tens (perhaps even a few hundreds) of kilometers in a year.

---

9 The Viking landers contributed valuable if fundamental experience with hazard avoidance logic. The lander surface samplers experienced stoppages due to hardware or topographically-induced stress loads that exceeded maximum specifications, and would there upon cease operation in a manner that came to be known as a "no-go." Controllers would then study engineering data and take pictures to determine the extent of the problem, and, on the basis of those evaluations, new command sequences would be initiated to correct, avoid, or work around the cause of the problem. {Ed.}

For these reasons, then, one should not write off rovers as being something totally beyond our technology, anymore than one should be as wildly optimistic as we were about the 1984 rover -- thinking that we could simply turn it loose to do its thing without our interaction. When approached sensibly and in the light of technical reality, rovers *are* realistic, and I think they represent a technology we should seriously contemplate for a near-term Mars mission.

Sample Handling and Quarantine

Sample handling can be a relatively difficult problem. As has been suggested by others, it is increasingly clear that samples must be uncontaminated and as pristine as possible. The implications of this are that samples must experience minimal disturbance during acquisition, they must be maintained to some extent in a martian environment (e.g., pressure and temperature), and they must be isolated from possible outside contamination.

From the viewpoint of the hardware designer, these requirements pose some challenging problems. Clearly, the scientists who will be analyzing the sample would like it to be in the best possible condition, i.e., as much like it was while part of the surface composition of Mars as can be reasonably assured. There are, however, limitations on that. One can't dig a core sample, for example, without disturbing it to some degree. So, while we want minimum disturbance, we must be realistic in our expectations. We also want to maintain, as nearly as we can, the martian temperature and pressure to which the sample is accustomed, and we would like to isolate it from any sort of outside contamination during its handling (particularly during the return at Earth).

Though perhaps not as critical, there is also a problem when recapturing into Earth orbit rather than simply taking the bull by the horns and bringing the ERV (or some part of it) directly into Earth's atmosphere. A direct-entry return is certainly quite feasible; indeed, it is probably the easiest approach from the engineering/spacecraft design viewpoint. However, it creates problems for maintaining the sample in its pristine condition. While there are a number of ways it could be done, at least to an extent that would satisfy most of the requirements, it would not be easy. If, however, all one wants to do is determine elemental composition and other fundamental sample characteristics that are not severely affected by environmental conditions, the requirements would not be too difficult to satisfy using direct entry. Only when one wants to bring the sample back in a pristine, clean, unmodified condition does it get to be a bit of a challenge.

Another problem that is distantly related to the subject of pristine samples is that of *planetary quarantine*. Even though majority opinion now appears to support the belief that Mars is not a viable place for life, we cannot absolutely rule it out. We certainly, in any case, wish to avoid contaminating either the place or the data. Thus, there are definite reasons for cleanliness in the fabrication process and for the sterilization of at least some elements of the hardware that will be in direct contact with the surface environment and the sample[10].

Similarly, in the unlikely event that there are Mars organisms, we **must** be concerned about back-contamination, as well. This is now of less concern than it was before Viking, of course, and it is a little hard to imagine something that grew up in a benign place like Mars being able to survive in the horrible, oxidizing atmosphere we have here on Earth. However, no one can guarantee that, so we have to be a little careful -- if for no other reason than to avoid all the criticism that might arise if we aren't. To achieve this, sample processing and storage hardware that will be returned from the surface of Mars via the ascent vehicle must have minimal contact with the planet while closely exposed to it, and samples will have to be kept in total isolation.

---

[10] **Planetary quarantine** (planetary protection) issue is explained on page 300 [NMC-2D]. {Ed.}

RECENT LAUNCH VEHICLE CONSIDERATIONS

Figures 2F-6 through 2F-9 demonstrate the work of Gail Klein (with R. Frisbee, M. Nakamura, and D. Stetson) at JPL, which she has been carrying on since my departure. Not surprisingly, one of my recommendations is definitely being pursued -- that of finding out what can be done with other launch vehicles that are or may be available. Of particular interest are the latest variations on the Titan family of launch vehicles, Titan IV[11], originally T34D7, and the SSV with an IUS. *Editor note:* Because the Titan name change is a fairly recent development (late 1987), T4's are still identified as T34D7's in these figures.

Figure 2F-6 is a graph reflecting the kinds of things one can do with these launch vehicles. The circled numbers identify the vehicles, and one can see, for example, that a T34D7 (T4) with a Centaur-G' has quite a large capability. Note, too, that the SSV is depicted with a twin-stage IUS. The other Titan configurations included (T34D) use the smaller solid rocket motors

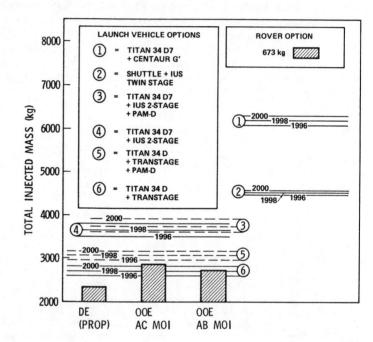

Figure 2F-6:

Total Injected Mass for Possible Launch Vehicle Configurations

(Includes rover option + 100 kg of science; lies within T34D7 + Transtage + PAM-D capability.)

---

[11] The Titan IV (T4) is essentially a new generation (by name) of Titan launch vehicles that nonetheless reflect, in particular, the configuration heritage of the most familiar Titan III (TIII or T3) vehicles: those with two large solid rocket motors (SRM's) strapped to a taller two-stage core (with or without an upper stage for additional propulsive boost and/or orbital maneuvering). Indeed, the new T4 was originally designated the T34D7, a complicated T3 designation identifying it as an enhanced Titan IIID equipped with new 7-segment SRM's (other SRM-equipped T3 vehicles used 5-segment SRM's). Although the T3D designation was for an autonomous vehicle without an upper stage (in support of launch requirements at the USAF launch complex in California), T3E vehicles with 5-segment SRM's and a Centaur upper stage were used very successfully to launch the two Viking spacecraft as well as both Voyagers and two Helios spacecraft from the Cape Canaveral launch complex. The T4 can be configured with any of several high-energy upper stages, including Centaur G'. The first T4 was delivered to the Canaveral AFS Titan integration/launch facility for final assembly early in 1988. [Ed]

(5-segment rather than 7-segment) and are coupled with either IUS, PAM-D, or Transtage upper stages[12]. The other numbers refer to launch year opportunities -- 1996, 1998, 2000. One can also see the mass required to perform a rover mission, although it is not a sample return mission in this case. This is a rover with about 100 kg of science, and it was included because it opened up the options somewhat for consideration. If one considers a direct entry with it, the mass requirements are well within the capabilities of just about all the launch vehicles represented. If one places it in orbit first and then enters, there is some mass penalty; but, if aerocapture and aerobraking are utilized for MOI, the mass penalty is lower.

Looking at it another way, this additional mass is the mass margin over and above that of the rover mission I have already discussed. There has been some consideration of including some avant-garde features, such as a very large balloon that could transport a rover from place to place. I don't know how it would be controlled (in terms of directional propulsion) relative to a specific target site, but I will leave that to the current-day visionaries. Figures 2F-7 and 2F-8 provide an idea of what these different vehicle options allow in the way of ability to perform such a mission. I believe the mass of the balloon system defined requires a fairly powerful launch vehicle, most probably something on the order of a Titan IV with Centaur-G'.

As stated in the brief report provided by the JPL team:

In addition to the transport of a rover carrying 100 kg of science to the surface of Mars, substantial mass margin is available for transport of additional science or balloon(s). For example, a balloon (sized to transport a rover carrying a 100-kg science payload) will require an additional launch mass of 1550 kg. This additional launch mass could be easily accommodated by a Titan 34D7 [Titan 4] plus Centaur-G'.

A number of mission options have been proposed as an alternative to the Shuttle-launched Mars Sample Return mission outlined in the JPL **Mars Sample Return Mission Study**, (1984). One possible option would be to use an expendable launch vehicle [ELV] to launch a spacecraft (containing both the rover plus ERV) on a trajectory to Mars. The spacecraft would utilize aerocapture technology to place it into a Mars orbit, whereupon the orbiter containing the ERV would be deployed. A lander could either separate from the spacecraft at this point, or it could separate prior to the Mars atmospheric entry. Following either a direct or an on-orbit entry, the lander would then aeromaneuver to a previously selected landing site. A parachute and terminal descent engine would be utilized by the vehicle during the last phase of its trajectory.

In a separate launch, either coincident with the present launch or at a later date, another ELV would be used to launch the ascent vehicle on a Mars trajectory. This spacecraft could utilize either a direct entry or an on-orbit entry followed by an aeromaneuvering landing to reach the surface of Mars. Because of the difficulty associated with targeting the ascent vehicle to the rover site, an on-orbit entry may be required.

An alternative to the above scenario would include the ERV in the same launch with the ascent vehicle. Studies indicate, however, that this option could satisfy the launch vehicle injection mass constraints ONLY if the ascent vehicle utilizes a direct entry (with its associated targeting problems) followed by aeromaneuvering to land on the surface of Mars.

JPL -- Examination of Alternative Mars Mission Options

---

[12] **Transtage**, not currently in production, is an upper stage that is closely related to Titan core-propulsion technology (hypergolic). The heritage of this upper stage is extensive, having flown very successfully over a long period of time on T3C vehicles at the USAF Cape Canaveral complex. Though less powerful than a Centaur (cryogenic propulsion), the T3C **Transtage** has a reputation for great reliability in performing complex propulsive maneuvers (achieving multiple restarts/deliveries) in near-Earth space out to geosynchronous orbital range. {Ed.}

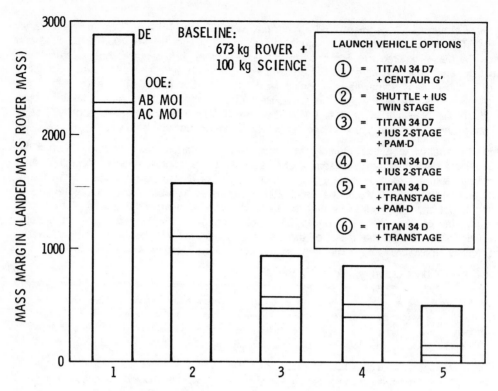

Figure 2F-7: Availability of Mass Margin for Additional Science -- e.g., Balloon(s)

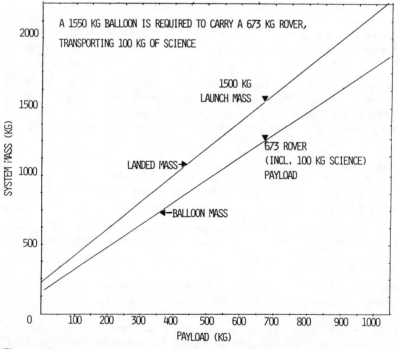

Figure 2F-8: Balloon Mission System Mass vs Payload

The point is that there are a number of launch vehicles and launch concepts that will allow us to carry out a MSR mission for the opportunities we will have in the mid to late 1990's. Therefore, we're certainly not out of options relative to the bigger surface missions on Mars.

Dual-Launch, Two-Lander Concept

When considering sample return in the complicated context of a dual launch, involving two separate lander vehicles (rover and LAV) and a surface rendezvous (as reflected in Fig. 2F-9), it does appear to be within the capability of at least the big Titan and Centaur (T4/Centaur-G'). Indeed, dual launch appears to be the name of the game, and it may always have been that way even when we had the SSV/Centaur-G'. That is, we were marginal for a few of our concepts even with the STS configuration. This particular MSR approach involves incorporating the rover and the ERV for the first launch. At Mars, the rover vehicle would land while the ERV waited in orbit for an ascent vehicle with surface samples. The second launch would be the LAV. At Mars, it would proceed to enter and land according to whatever landing option was selected, and it would then rendezvous with the rover -- busy gathering samples since its own landing. The samples would then be transported up to the ERV aboard the ascent vehicle for the flight back to Earth.

This is clearly not a simple mission, but none of the missions I've discussed could be called "simple." The easiest would employ the direct-entry, direct-return option, but that option is only simple in the sense that it minimizes the number of vehicles. It is otherwise still quite difficult to develop and perform. The surface rendezvous mission would be no more difficult than many of the other possibilities we've contemplated. The point one should arrive at from all of this is that, even with the STS/SSV problems and the demise of the SSV/Centaur-G' option, we have a stable of expendable vehicles that are capable of performing MSR missions if we really want to do so.

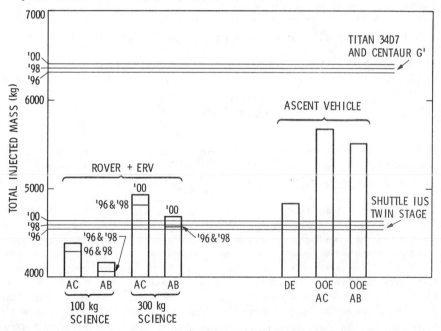

Figure 2F-9: Comparison of Launch Requirements (T4/Centaur-G' vs SSV/IUS) for a Specific Mission; (ERV and separate landings of rover and LAV, with on-surface rendezvous of rover and LAV)

SUMMARY OF RECOMMENDATIONS

I developed some recommendations based on my work with sample return studies up to the time I left JPL, but I will supplement that basis with some more recent work performed since my departure. The primary recommendations I will discuss are:

- ■ Evaluate Impact of Current Launch Vehicle Situation
  - · Expendables (ELV's)
  - · SSV w/Twin-Stage or Two-Stage IUS

- ■ Further Investigation of Low-Thrust Options

- ■ Continue Aerocapture/Aeromaneuver Work

- ■ Reintroduce In-situ Propellant Production (ISPP) Study

- ■ Continue Rover Work

- ■ Consider International Cooperation

One must remember that the launch vehicle situation has pushed a lot of what I have talked about into a realm of great uncertainty, and it may be a very long time before we again have something comparable to a "space tug" type of vehicle with significant power[13]. Even with the STS again operational, we will not have the Centaur-G' to use with the SSV. And, even when we had the Centaur-G', we were not entirely out of the woods in terms of being able to integrate what we needed into the vehicle configurations dictated by some of our concepts, i.e., squeeze them into single-launch missions.

Launch Vehicle Situation, Multiple-Launch Options -- We certainly need to look at what other alternatives may be available in the way of expendable vehicles and other upper stages suitable for STS applications. We may very well find it more appropriate to pay the tariff associated with multiple launches in support of MSR missions, and to be satisfied with the dual-launch concepts discussed earlier. The use of two or three ELV launches, wherein the LAV is launched on one and the ERV on another, or to facilitate final assembly in Earth orbit, are certainly quite feasible and may now represent the most reasonable course to follow.

Low Thrust Options -- The low-thrust options, i.e., electrical and perhaps solar sail, probably deserve another look. To be honest, they are not what I would like to have because of the long

---

13 "Space Tug" is a term that came into use when several upper stage technologies were being considered as a basis for the development of a high-energy, on-orbit STS propulsive vehicle that might--in some cases--be recovered and reused. The Space Tug was to provide transfer propulsion for a variety of spacecraft and missions, thus functioning somewhat like a tug at a shipping port. Although the original concept has not been realized, solid rocket technology was selected over that of cryogenic (e.g., Centaur) and hypergolic (e.g., Transtage) systems in the interim (as reflected in the IUS). Centaur-G' was developed as an IUS alternative to provide the increased degree of propulsive energy required by large, deep-space payloads, but the **Challenger** tragedy emphasized the risks associated with cryogenic hydrogen/oxygen systems, even when they are not at fault. While Centaur technology has historically shown itself to be operationally safe, it is now clearly inappropriate to consider flying one on a manned SSV. It should be noted, however, that Centaur-G' represents a significant advancement of the highly successful Centaur technology, and that it is still expected to play a major role in partnership with ELV's (like Titan IV) and perhaps with unmanned STS vehicles now under consideration. {Ed.}

flight times involved. They are not easily adapted to operating in a planetary gravity field, but they may, nevertheless, be worth additional consideration -- given the current uncertainties.

Aerocapture/Aeromanuevering -- Aerocapture and aeromaneuvering afford a couple of important advantages. Aerocapture allows one to get into orbit around either planet (Mars or Earth) with a minimal expenditure of propulsion and at a mass penalty much less than would be required if it was done totally by propulsion. Aeromaneuvering allows us to land with reasonably good accuracy. While I would not call it "pinpoint" accuracy, I believe it would be better than what improving the **Viking** methodology could produce[14].

ISPP -- In·situ propellant production (ISPP) is one of my pets, and I would like to see it continued for two reasons. First, I think it offers the possibility of performing sample return missions with a smaller investment of total spacecraft launch mass. Second, I view it as an enabling technology for the future exploration of more difficult targets, whether the missions be manned or unmanned[15].

International Participation -- The complexity of these missions, because of the multiplicity of vehicles and spacecraft systems involved, lends itself to international cooperation. In fact, they may be better suited to international participation than the more conventional kinds of missions, in the sense that one nation can build one kind of vehicle while another country builds the other. Interfacing can be relatively simple and a lot easier to manage at that level of activity than in the case, for example, where one country is building a sophisticated instrument that involves a very technical component-level interface in another country's spacecraft. It is hard enough to coordinate the latter kind of activity when we are all in the same city, let alone on the other side of the Earth. Finally, the fact that such missions are quite expensive certainly makes international participation attractive, presumably saving money by sharing the cost of the effort with others. ■

---

14 The **Viking** landings were reasonably accurate, and it is uncertain how much the **Viking** methodology could be improved -- based on the same technologies. The essential challenge would be to make the descent more manageable, as Mr. French has alluded. It should perhaps be noted that the most critical aspect of a nominal landing is experienced during the last few meters of the terminal descent, when large boulders or sharply defined topographic features (small vertical elevations) that cannot be seen in orbital images may be encountered (e.g., **Big Joe** at the VL-1 site or the face of a rock outcrop). A landing encounter with boulders (perhaps in clusters) could certainly damage a landing vehicle or immobilize (trap) a rover, for example, and our knowledge of Mars suggests that such blocks are probably quite common over much of the global surface -- certainly in some of the most interesting areas being considered for MSR missions. Reliable, fully automated terminal landing control, perhaps comparable to that achieved with the lunar landing module design (but without an astronaut), will, in some way, be necessary to avoid such hazards and assure a safe landing and mission opportunity. {Ed.}

15 In·situ propellant production (ISPP) is just one aspect of the exciting challenge associated with extra-terrestrial resource exploitation to support deep space missions. It is widely believed that advanced unmanned and manned planetary exploration can only be sustained in a cost effective manner--and technologically--through the utilization of resources available in space and/or on the bodies being explored. At the most fundamental level, this involves the making of materials like oxygen and water, perhaps even propellants, and using surface material to provide insulation and radiation shielding (for manned missions on the surface of the moon or Mars). These topics are discussed at greater length in those papers grouped in the third section of these proceedings, including the second of Mr. French's two presentations [NMC-3E]. {Ed.}

REFERENCES

This space is otherwise available for notes.

~~~~~~~~

MARS SAMPLE RETURN: SCIENCE
[NMC-2G]

Dr. Douglas P. Blanchard
NASA Johnson Space Center

The study of the SNC meteorites, and of lunar samples returned by Apollo manned missions and a Soviet automated mission, demonstrates the importance of having even very small amounts of actual material to study. Laboratory equipment affords better resolution and precision than flight instruments constrained by severe size, weight, and power limitations; samples remain accessible in the future for rapidly improving analytical technology; and the variety and complexity of laboratory instrumentation available is essentially unlimited. Our strategy in determining what kinds of Mars samples to collect involved asking fundamental questions in terms of how samples might help us answer them. This suggests what opportunities should be favored and what we have to do to pursue them. Once we understand the questions in these terms, we can relate that understanding to areas on the planet that can best provide samples suited to a specific issue. For example, one would like to have samples of crustal rocks to learn something about the planet's formation: such samples could be found in heavily cratered terrain and perhaps in some volcanic deposits. Some kinds of sampling should be done at virtually any site, while others, e.g., sampling concerned with frozen volatiles in the cryosphere, would favor sites near or perhaps within a polar region. Sample acquisition technologies--including rovers, tools, and storage devices--need to be developed. Apollo experience has proven to be a good basis for modeling rover missions, but the process of selecting, acquiring, storing, and returning samples presents many challenging problems.

INTRODUCTION

For my NASA Mars Conference presentation, I will discuss the science concept associated with sample return from Mars. We were given an eloquent discussion of Viking-style in-situ science on Mars early in these proceedings, which included sample processing and analysis in each of the landers. My interest is in samples and what they can tell us about Mars, as well, but in the context of a Mars sample return (MSR) mission. That is, how to get samples back to Earth in a condition that will allow them to tell us about Mars in a more specific and precise manner. That problem has preoccupied much of our thinking relative to MSR planning for quite some time. I'm not going to discuss the mission technology options available to potential MSR programs, because Jim French discussed these thoroughly in his Mars Conference presentation concerned with MSR mission concepts [NMC-2F]. Instead, I will discuss some of the science goals for Mars exploration.

MSR science goals have been expounded several times by various Mars science working groups. When we worked on the study conducted in 1984-85, we decided we were not going to create a new science rationale. There was ample science rationale already available, and I will review some of that shortly. I will also contribute some explanation about landing sites as well as the mobility questions associated with them. And, finally, I will discuss the samples themselves: How much sample material is enough? How do we sample a planet that is many light minutes away relative to our ability to control what is happening? How should the system pack and protect its samples so that we get back what we paid for? And how do we receive the samples at Earth?

Figure 2G-1 is an artist's conception by Mark Dowman of Eagle Engineering, and it gives one a sense of how a MSR mission might get under way. It illustrates the staging of the biconic-shaped, aerocapture-designed MSR vehicle at the NASA Space Station prior to launch. An orbital transfer vehicle (OTV) is seen latching up with the MSR vehicle. The space station affords the possibility of doing this assembly in orbit, and it creates some additional possibilities relative to weight and propulsion that are not available to us now.

[The sketch presented here as Fig. 2G-1 is a legend for a color illustration in the Color Display Section appendix on page A-4.]

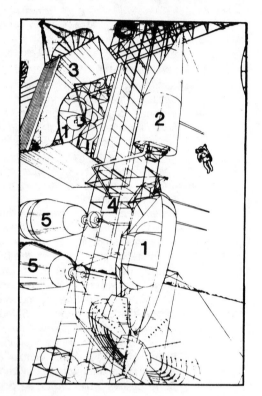

Figure 2G-1:

MSR Spacecraft
Preflight Preparations
At U.S. NASA Space Station

1) Orbital Transfer Vehicle
2) MSR Spacecraft (inside aeroshell)
3) OTV Hangar • 4) Mobile RMS
5) Propellant Storage Modules

SAMPLE CURATING AND VALUE OF SAMPLES

I will start by telling a detective story. The meteorite pictured in Figure 2G-2 [Color Display Section, page A-4] is the Shergottite meteorite collected in Antarctica (sample EETA-79001) that has been referred to a number of times during these proceedings. The detective story goes like this. The meteorite was picked up during a program sponsored by the National Science Foundation to collect meteorites on the Antarctic ice, and its composition happened to place it in a rare class of basaltic meteorites (SNC) in which there are only five or six other examples[1].

---

[1] The Shergottite meteorite is one of the SNC class of meteorites, sometimes referred to phonetically as "Snick" meteorites. SNC is an abbreviation for three unusual types of meteorites: Shergottites, Nakhlites, and Chassignites, all of which have been carbon-dated to be about 1.3 b.y. old (most common meteorites have been dated to be about 4.6 b.y. old). The SNC meteorites generally have igneous rock textures and are basaltic in composition. Gases detected within them are remarkably similar to the atmospheric gases detected by the Viking landers on Mars. {Ed.}

Meteorites Linked to Mars

When it was dated, the Shergottite meteorite had a particularly young age that made it quite anomalous. In an attempt to verify the age using another method, Don Bogard (NASA-JSC) irradiated part of this rock and performed an $^{39}Ar/^{40}Ar$ age procedure. When he put it in his mass spectrometer, the read-out results did not make sense. Being a good scientist, he reconfigured and decided he would try to perform the process on this meteorite without irradiating it, using only part of the technique. When he did that, he found that the natural gases in the meteorite bore an amazing resemblance to the martian atmosphere as determined by the Viking landers on the surface of Mars[2]. A minor part of this rock produced the same kind of signature (he had to take the rock apart very carefully to pick out those parts that gave him this data). There have been other studies since: e.g., nitrogen isotopes have been analyzed, as well, and they corroborate Bogard's data. Figure 2G-3 is an oxygen-isotope diagram, and it suggests what this data might mean.

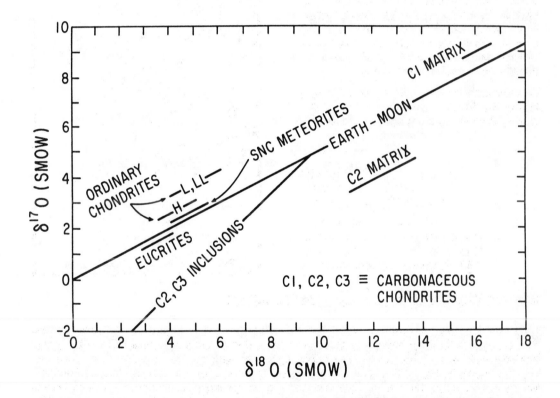

Figure 2G-3: Oxygen Isotope Diagram -- Comparison of SNC Meteorites to Chondrites

---

2 Precise atmospheric analyses were conducted using the Gas Chromatograph/Mass Spectrometer (GCMS) instruments (one in each lander). The primary purpose of the GCMS was to search for evidence of organics in the soil, but it could also function as an atmosphere instrument. Along with precise measurements of major and minor atmospheric constituents, an enrichment procedure was used to increase trace elements to detectable amounts so that their isotopic ratios could be determined -- a bonus product of the instrument's capability. {Ed.}

Let's assume, for a moment, that the meteorite did come from Mars. When we look for parent bodies of meteorites on the oxygen-isotope diagram, then, results associated with the SNC meteorites are tantalizing. The slight difference between the line for SNC meteorites and the Earth-moon line may or may not indicate a different parent body. There is, of course, a little controversy, in that not everyone believes this sample is from Mars. However, we have a lot more of the Shergottite sample locked up in a vault in Houston (not with the lunar samples), and researchers can come to test their own ideas about how and from where this meteorite came to Earth.

One point of this detective story is to explain how samples are now--or would be--studied. The salient features are that we could do multiple analyses of different kinds on a single sample. We have the ability to meticulously subdivide the sample, i.e., to pick out those parts that are of greatest interest or which we feel can tell us the most. We also have the opportunity to learn from unexpected results, and there is generally more material available for the next investigator with a better idea.

BASIS FOR SAMPLE RETURN MISSION

We are often asked why it is desirable to bring samples back from Mars, and the following thoughts summarize our rationale:

- Laboratory equipment will always have better resolution and precision than flight instruments,

- Returned samples become resources that are accessible in the future for rapidly improving analytical technology,

- Unforeseen key discoveries can lead to new experimental designs,

- There are no weight/power limitations on laboratory instruments,

- The variety and complexity of laboratory instrumentation for sample studies is unlimited.

First, of course, we can't load a spacecraft with all of the magnets and power supplies that one can have available in a laboratory. Also, laboratory equipment will always be state-of-the-art, assuming one has access to the right laboratories. Analytical technology is advancing and we may make unforeseen discoveries that will change the experiment design. In short, then, there are no weight and power problems, as one would have on a spacecraft, and there would always be an incredible array of techniques available (along with the quality of work made possible by modern equipment in a flexible laboratory environment).

In terms of what we want to learn about Mars on the basis of sample analysis, we have adopted the attitude that everything we *can* do on Earth we *should* do on Earth. However, that is not to say that we should do *every*thing here on Earth, and there are some things that we must do on Mars that are dependent on the local conditions. For example, we can't measure martian weather on Earth. In other words, we need to be able to measure those things on Mars that can't be transported to or recreated on Earth.

328

In summary, then, the kind of science we need to do in situ on Mars can be summarized as:

- Measurements dependent on "local conditions" that are difficult to reproduce in the laboratory (e.g., determination of ambient radiation environment or identification of complex unstable molecular species at martian surface);

- First-order characterizations of the martian materials that will alter the mission plan;

- Time-dependent or fluctuating conditions, such as local weather conditions or the diurnal exchange of volatiles between atmosphere and soil;

- Measurements of properties (pH, low temperature phase changes, et cetera) that may not survive the return trip to Earth (e.g., due to accelerations or radiation of container effects).

We should especially try to measure those things that impact operational decisions during the mission, such that they can determine the course of the mission. For example, "Do we have one of those rocks already?" To make that decision, we must be able to make some determinations about a rock being considered before actually acquiring it as a sample. In addition, we should acquire measurements representative of those things that are transient or difficult to maintain, or whose properties (like pH and phase changes) might be lost on the way home.

Fundamental Science Questions

In Figure 2G-4 I have divided the Mars problems more or less arbitrarily into several bins, and then considered some of the questions we ask when pursuing these problems relative to the sampling strategies we need in order to answer them. The content of this process represents a kind of shopping list, suggesting what sorts of opportunities are favored and what we have to do to pursue and satisfy them.

Formation -- The first science item on the shopping list, and one of the major things we want to know, is how Mars formed. And, associated with this, we want to understand the bulk composition of the planet. It would also be useful to know the state and nature of any primordial gases retained on the planet. To learn these things, we first want to look for crustal rocks (we can find them in heavily cratered terrains). We particularly want to look for deep-crust rocks and perhaps even upper-mantle rocks. We might look for them in volcanic material, especially if it contains xenoliths (little bits and chunks of the upper mantle that were carried along with the volcanic eruption outflow). And, finally, we might find the samples of primordial atmosphere in unweathered igneous rocks.

Differentiation -- Another question to consider is that of how the planet differentiated. Did it form a core? What kind of processes were involved? To find the answers, one would want to study some of the oldest rocks in the heavily cratered terrain, representing pieces of the early crust of Mars. One would also want to know where the oldest rocks are located and to what they might relate (geologically).

One would especially want to know if there is a magnetic field trapped in any of the rocks, because that would contribute a great deal of information about core formation. Again, one would look for old rocks in the heavily cratered terrain to answer such questions, and one should also consider the older volcanic regions to study the sequences of old rocks. In this way, one can begin to establish time significance for the study.

Mars Formation

* What is the bulk composition of the planet?

* What primordial gases are retained in the atmosphere?

Sampling Strategy

* Crustal Rocks . . . . Heavily Cratered Terrain

* Deep Crust/Mantle Rocks . . . . Volcanics With Xenoliths[*]

* Primordial Atmosphere . . . . Unweathered Igneous Rocks

Primary Differentiation of Atmosphere, Crust, Mantle and Core

* What are the compositions of the rocks and soils in the old, heavily cratered Mars terrains?

* What are the oldest ages of rocks on Mars?

* Do the crustal rocks have remnant magnetic fields?

Sampling Strategy

* Old Crustal Rocks . . . . Heavily Cratered Terrain

* Sequence of Old Rocks . . . . Old Volcanic Regions

Volcanism and Igneous Petrogenesis

* What are the ranges of ages and compositions of martian igneous rocks?

* Do martian igneous rocks contain xenoliths and what do they tell us about the deep structure of the martian crust?

* Are the rocks of the northern lowlands igneous or sedimentary?

* How old are the rocks of the Tharsis-Syria rise?

Sampling Strategy

* Rocks With a Range of Ages . . . . Young and Old Volcanics

* Deep Crustal Rocks . . . . Volcanics (containing xenoliths)

* Sequence of Units . . . . Layered Terrains

[*] Xenoliths are foreign rock fragments within mass of other rock.

continued...

Figure 2G-4: Fundamental Science Questions for Mars (1 of 2)

Weathering, Erosion, Transport and Sedimentary Processes

- What are the physical and chemical characteristics of the martian regolith?

- Can properties of the regolith account for the apparent source and sink regions of valleys sculpted by flow processes?

- Can volatile and ice contents of the regolith help explain the cratering morphologies and flow lobes observed in the martian landscape?

- What are the ages and chemical and isotopic compositions of the materials in the equatorial and polar layered terrains?

Sampling Strategy

- Eolian Dust . . . . Polar Layers
- Variety of Soils . . . . All Sites
- Permafrost Samples . . . . Nonequatorial Sites
- Transported Sediments . . . . Fluvial Outwashes
- Layered Sediments . . . . Layered Regions

Martian Biology

- Are there organisms or organic compounds in the martian soils?

- Are there fossil records of life forms in the martian rocks and soils?

Sampling Strategy

- Volatile–Rich Materials . . . . Polar Sites
- Sedimentary Sequence . . . . Layered Regions

Evolution of the Atmosphere and Cryosphere

- What are the volatile contents of crustal rocks?

- What is the composition of the present martian atmosphere?

- What evidence can be found in the geological record for the past history of the martian climate?

Sampling Strategy

- Permafrost . . . . Polar or Near–Polar Sites
- Sedimentary Sequence . . . . Layered Regions
- Present Atmosphere . . . . All Sites
- Polar Ices . . . . Polar Sites

Figure 2G-4: Fundamental Science Questions for Mars (2 of 2)

Volcanism and Petrogenesis -- Volcanology and igneous petrogenesis reflect questions about more recent geological events, and help us determine what has occurred more recently on the planet. To work on these problems, one would want to look at the range of ages and compositions over the planet and try to determine how heterogeneous the planet is with respect to volcanic and igneous processes. One would want to study those things representing ranges of ages in order to get some sense of the timing and the sequence of happenings, by looking for deep crustal rocks and the time sequencing of such units. These units might perhaps be found in the layered terrains, which could be volcanic or sedimentary; each type produces its own kind of information.

Processes -- In a more contemporary sense, we would want to conduct studies of the processes or products associated with weathering and erosion (e.g., meteorological transport and sedimentary processes). To do so, we would want to collect a selective variety of dust and soil samples. Some of the questions we might ask about such processes focus on the search for different kinds of samples. For example, we'd like to know the physical and compositional character of the wind-carried dust; we might look in the polar regions where dust is periodically deposited between layers of ice. A variety of soils would be useful and should be acquired at any and every site as standard procedure.

In the case of permafrost samples, it is probable that we'd be more able to acquire them at latitudes close to the poles. When looking for transported sediments, we'd want to go into some of the channels or out-washes, again searching for layered units to establish an understanding of their sequencing. And, of course, we must keep the martian biology issue in mind, despite the prevailing opinion that life on Mars does not now exist. All I would venture to add, if we want to find out if life may have existed at some point in the past, is that wisdom suggests we should go to polar sites or possibly the layered regions. Those are the places where samples might reveal evidence of the past history of life, should it have once been there.

Finally, we are interested in the atmosphere and the frozen volatiles of the cryosphere. This would definitely require taking samples of the atmosphere that could be brought back, but some permafrost and polar ice samples would also be very helpful. The sedimentary sequence would also provide some historical information about atmospheric development.

SELECTION/ACQUISITION: SITES, SAMPLES, AND ROVER SCENARIOS

When their funding was cut during the mid 1970's, the Mars science working group had the good grace to write up their work. The product was a ten-volume set of documentation that summarized much of the thinking about sampling and site selection on Mars. The work continues even today in various places, USGS-Flagstaff among them, utilizing some of the late-mission, high-resolution Viking orbiter imaging data[3]. Using this high resolution data, they are able to see more detail at some of the sites. It is important to understand that while some of the candidate sites have various kinds of terrains available to sample, none of them have all of the terrains in which we are interested. Therefore, the nature of the candidate landing sites is

---

3 Once their pre-defined primary mission objectives had been met, the orbits of both Viking orbiters were modified to allow high-resolution imaging. Each orbiter was equipped with two identical cameras, and orbital range rather than differences in optical magnification was responsible for the variations is resolving power. Viking periapsis resolution was improved in this way from approximately 100 meters to roughly 5 meters. However, the number of frames imaged at the very highest resolutions is quite small, due to the highly elliptical nature of the orbits and because targeting became very difficult, but a significantly larger volume of frames was imaged at only slightly lower resolutions as each spacecraft was approaching or departing its periapsis. {Ed.}

clearly important, and Figure 2G-5 defines ten MSR sites (each of the bottom two lines on the chart represents two sites) in a manner that reflects the understanding I have described in the selection process.

SITE	SITE/SAMPLE CHARACTER
North Pole, Site A	Perennial Polar Ices
North Pole, Site B	Perennial Polar Ices Soil From Perennial Ice-Free Trough
Arsia Mons West	Young Volcanic Rocks
Apollinaris Patera Northwest	Young Volcanic Rocks Eolian Sediments
Chryse Planitia (VL-1 Site)	Impact Crater Ejecta Fluvial Sediments
Schiaparelli Basin Southwest . .	Ancient Heavily Cratered Terrain Oldest Martian Crustal Rocks
Terra Tyrrhena (Iapygia, MC·21) . . .	Oldest Martian Crustal Rocks Ancient Heavily Cratered Terrain Old Volcanic Rocks that Mantle Ancient Crust
Candor Chasma & Hebes Chasma . . .	Layered Rocks From a Canyon

Figure 2G-5: Mars Sample Return -- Candidate Landing Sites, Nature of Samples for Each

Apollo-Based Scenario for a Mars Sample Return Mission

Sample selection and acquisition are also constrained by the degree of mobility possible once a sample return vehicle has landed, such that one must also understand the mobility questions involved. Deciding that we had to come to a better first-hand understanding of what is involved in sampling a planet with an automated rover, we elected to use our best analogy of rover experience -- the Apollo rover. We chose as our model (somewhat arbitrarily) the Apollo 15 extra vehicular activity (EVA), which included three rover traverses. In particular, we studied that mission's EVA-2 rover traverse (i.e., all of the work performed on that traverse), and used that work load in the simulation scenario defined in Figure 2G-6.

333

OBJECTIVE

■ Construct operational scenario for Mars rover based on EVA experience with the Apollo 15 lunar rover.

MODEL ASSUMPTIONS

[1] One uplink and one downlink to rover per day;

[2] Rover can accept multiple tasks and can return multiple sets of data and images;

[3] Rover will **NOT** be directed out of the field of view of last image;

[4] Sets of samples (3-4) require one day to collect and characterize;

[5] Earth decisions will require no more than one day;

[6] Rover can travel OR sample, but not both at once.

Figure 2G-6:   Mars Rover Scenario Based on Apollo 15 EVA

Note that we instituted some simple mission rules for our simulation: only one uplink and one downlink with the rover was to be possible each day, during which the rover could accept multiple tasks or transmit a number of images and some sets of data. In this respect, our model rover is fairly smart in terms of keeping its operations team informed. In addition, we decided that the rover would never be sent beyond visible safe limits determined at its previous vantage point. That is, we would not unknowingly send it over the crest of a hill until we could see what was on the other side, which somewhat limits where or perhaps how far one can send it. For our simulation model we collected one sample per day, and the "committee-style" operations decisions made on Earth were also allocated that amount of time. Finally, the rover could either travel or sample, but not both, within the time span of one day.

Figure 2G-7 is a map of the Apollo 15 EVA traverse activity, showing circuit traces for EVA-1, EVA-2, and EVA-3[4]. The Lunar Module is near top-center (north end of area), and the EVA-2 circuit trace (used to model our simulation) is down and somewhat toward the right (southeast). These traverses seem apropos because they actually total a little less than 30 kilometers (km) in surface travel. This reflects the kind of rationale that we will probably adopt for a Mars rover. That is, our Mars traverse model begins with a short traverse during which high-priority samples are collected and returned to the ascent vehicle for storage. Then the rover would be programmed to venture farther and perhaps take progressively longer excursions. Indeed, once most or all of the priority samples are collected and secured, we can send the rover on riskier and/or longer traverses as may be considered appropriate to the enhancement of the mission.

---

[4] NOTE: The illustration provided for use as Fig. 2G-7 did not include a legend to indicate what the alphabetic and numeric identifiers (capital letters and numbers along the three EVA traverse traces) represent, but these elements do not appear to correlate directly with information provided in this presentation. {Ed.}

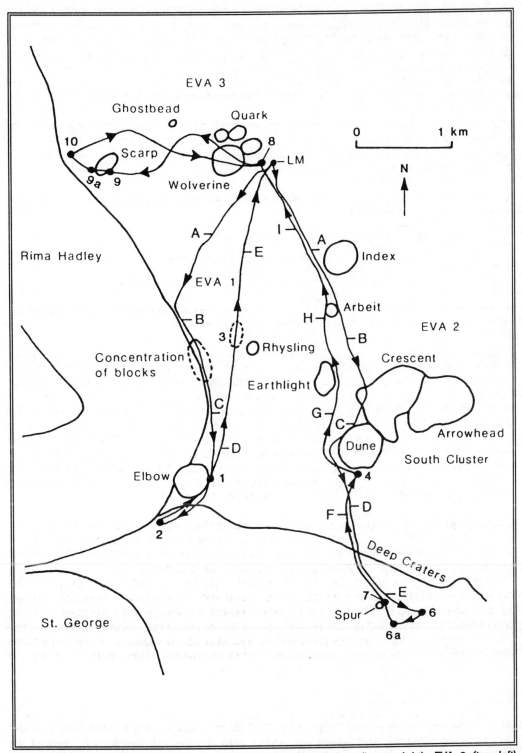

Figure 2G-7: Apollo 15 Excursions -- EVA-1 (center), EVA-2 (lower right), EVA-3 (top left)

335

Considering our modeling requirements, the Apollo 15 EVA-2 rover excursion produced a number of factors relating to our mission concept. It traveled 11 km, sampled at 5 distinct sites, and collected 45 rocks, 17 soil samples, and 8 cores. In order to accomplish the same things in our Mars-rover simulation, as reflected in Figure 2G-8, it took us 155 days[5]. Surprisingly, however, we spent only 15 days traveling. We invested 70 days in decision making and spent 70 days doing the work, and the rover was in motion only 31 hours. Still, while the speed of the rover is "glacial," as Jim French suggests, its rate of movement doesn't seem to be a problem.

What **does** seem to be a problem, however, is how sampling is accomplished on the rover relative to the amount of time given to the process. How can the rover be instructed to be clever about how it samples while we're concurrently trying to speed up the whole process? This problem is at least as significant as the mobility problem.

APOLLO 15 EVA-2 DATA

- Distance Travelled . . . . 11.2 km
- Sampling Stations. 5
- Samples Taken . . . . 45 Rocks
  17 Soils
  8 Soil Cores

RESULTS OF SIMULATION

- Days to Complete . . . . 155 Days
- Days Used to Travel . . . . 15 Days
- Actual Traveling Time . . . . 31 Hours
- Days Used for Sampling. . . . . 70 Days
- Days Used for Decisions (at Earth) . . . . .70 Days

Figure 2G-8:   Results of Mars Rover Simulation Based on Apollo 15 EVA-2

SAMPLING SYSTEMS, PROCEDURES AND PHILOSOPHY

The primary conclusion we have drawn from the simulation exercise is that we want to sample the widest diversity of units possible and that we should acquire one from every kind of unit that we can find. Therefore, mobility is essential to assure that a rover can get around in the area of the landing site to sample more than one vicinity. As an example, Figure 2G-9 is an artist's conception of a generic rover sampling a rock that has perhaps fallen down from a canyon wall (lander/ascent vehicle detail shown in Fig. 2G-13).

---

5 As in the case of Viking, although not reflected here, the term "days" is most probably intended as a reference to martian "sols," due to the difference in the length of a martian day (sol) and the need to make mobility decisions on the basis of local-time (daylight hours) Mars imagery, assuming that such decisions will be based primarily on visual rather than nonvisual (e.g., IR) imagery data. {Ed.}

Figure 2G-9: Depiction of Rock Coring Operation by Rover (lander/ascent vehicle in distance)

Quite obviously, one sample cannot satisfy *all* of the appropriate Mars-characterization objectives afforded at a given landing site, so mobility is needed to assure that we have access to more than one kind of sample. But it also seems clear, at least at this point in our thinking, that mobility speed is not particularly necessary. The greater issue is that of how to contend with potential obstacles (seen and unseen) and keep the rover on the move, albeit at a glacial pace, rather than having to stop and resolve a mobility problem at the slightest provocation.

Sample Collection and Analysis

The first problem in sample collection is deciding how much is enough. To study this problem, we again turned to the Apollo program. The Apollo samples are reviewed in the table presented in Figure 2G-10. As a product of the Apollo program, 382 kg of sample material (totaling over

APOLLO MISSION* >	11	12	14	15 (Rover)	16 (Rover)	17 (Rover)	TOTAL
Range and Time:							
Traverse Range (km)	0.25	2.0	3.3	27.9	27.0	35.0	95.5
EVA Time (hrs)	2.4	7.5	9.4	18.5	20.2	22.1	80.1
Samples:	58	69	227	370	731	741	2196
Sample Mass (kg):	21	35	42	77	96	111	382
Terrain:	Mare	Mare	Ejecta	Mare	Hilands	Mixed	

* Apollo 13 mission was aborted in flight due to fuel cell explosion.

Figure 2G-10: Apollo Mission Statistics

2,100 pieces) was collected. The Apollo rovers traveled almost 100 km on the moon, representing approximately eighty hours of EVA time on the lunar surface. We have reviewed the Apollo 15 tapes (at NASA-JSC) in an attempt to thoroughly understand the nature of rover activity, and one thing that is very clear is that the Apollo rover traverses were absolutely time-limited.

Application of Apollo-Based Experience -- If one translates (by calculation) the Apollo experience to our Mars rover scenario, 80 hours on the moon is equivalent to nearly 10,000 hours on Mars. This comparison clearly puts us in a whole new realm of thinking with respect to how we do MSR science, and the time can be used to considerable advantage. If a potential sample looked interesting on the moon, for example, we simply "grabbed it and bagged it" in the interest of time. With much more time available on Mars, we can and must be more selective. We can undertake some in-situ characterization and then decide whether to keep a sample on the basis of what we learn about it. Also, we can elect to acquire sub-samples, i.e., bring back small fragments of interesting rocks.

If we assume that two to three percent of the collection has been destroyed for science, which is close to correct, then only about 10 kg of the 382 kg (original mass produced by six missions) has been used. If we are allowed to send six missions over the course of a MSR program, duplicating the Apollo program, Mars rovers could acquire a total of 30 kg of the most interesting sample units. And, by doing so very selectively, we could achieve a scientific return comparable to that of the Apollo missions.

Luna 24 Sample Analysis Experience -- Another good example of how much can be accomplished with a limited amount of sample material is the story associated with the Soviet Luna 24 mission (1976). Luna 24 was an unmanned, automated spacecraft that landed on the moon and returned 160 cm of core sample. Results associated with sample analysis for the Luna 24 mission are summarized in Figure 2G-11. The U.S. was given only three grams to study. An incredible amount of research data was produced from those three grams of sample: major-element analysis was conducted, trace elements were identified, the petrology was determined, chronology was determined (by four different methods), a noble-gas analysis was conducted, the exposure indices were determined, physical properties were established, magnetic properties were studied, carbon chemistry was conducted, radioactive elements were detected and measured, and solar flare tracks were sought. We learned that the rock sample provided by Luna 24 contained more magnesium than expected and we found a basalt type we had not recognized before. The real bottom line, however, is that we still have--after all that work--1.3 grams of the original 3 grams!

Sampling Philosophy

To develop a fundamental sampling philosophy, we brainstormed with some specialists who have been involved in sample systems research about how best to sample a planet[6]. We tried to imagine what tools and techniques might be needed, and we thought about how much "machine intelligence" would be required to achieve the degree of control we wanted. Figure 2G-12 is a summary of some of our brainstorming ideas. I do not know the data rate required for any of them, nor do I know their mass or power-requirement factors; they are simply "idea tools." We may ultimately use some of them, all of them, or none of them. Still, they are worth a quick review to give one an idea of the kinds of things we are thinking about for use as sampling tools on the surface of another planet.

---

[6] Benton C. Clark and Ruth Amundsen, Martin Marietta Denver Aerospace: Study of sample systems for comets and Mars: final report, 3/87, NASA-JSC Contract NAS9-17511. Dr. Clark was deputy team leader of the Viking inorganic chemistry team led by Dr. Priestly Toulmin, III. He was extensively involved in the analysis of XRFS (X-ray fluorescence spectrometer) data and participated in the instrument's development. His expertise in sample analysis technology (for both Earth applications and space science) is widely acknowledged. {Ed.}

```
┌─────────────────────────────────────────────────────────────────────┐
│  Luna 24 Sample Material:                                           │
│                                                                     │
│  ■   160 cm of Core Material From Mare Crisium                      │
│                                                                     │
│  ■   Sample Mass Allocated to U.S. for Analysis -- 3 gm            │
│                                                                     │
│  ■   Amount of U.S. Allocation Remaining -- 1.3 gm                  │
│      (after studies below)                                          │
├─────────────────────────────────────────────────────────────────────┤
│  Studies Conducted (by U.S. scientists on allocated sample):       │
│  [1]   Major Element Analysis                                       │
│  [2]   Trace Element Analysis                                       │
│  [3]   Petrology                                                    │
│  [4]   Chronology by Rb-Sr Ar-Ar Nd-Sm U-Pb                        │
│  [5]   Noble Gas Analysis                                           │
│  [6]   Exposure Indices                                             │
│  [7]   Physical Properties -- Grain Size Analysis                  │
│  [8]   Magnetic Properties                                          │
│  [9]   Carbon Chemistry                                             │
│  [10]  Radioactive Element Tracks                                   │
│  [11]  Solar Flare Tracks                                           │
├─────────────────────────────────────────────────────────────────────┤
│  Significant Scientific Findings:                                   │
│                                                                     │
│  ■   New, Unanticipated Source for High Magnesium Lunar Material   │
│                                                                     │
│  ■   Evidence for a Third Family of Low-Titanium Basalts           │
└─────────────────────────────────────────────────────────────────────┘
```

Figure 2G-11:   Soviet Luna 24 Sample Productivity

In addition, the brainstorming list also reflects the kinds of things we want to sample, e.g., small rocks or large rocks (considering whether they can be picked up or must be "mined" for fragments). We would also want surface soils from several sampling realms, as well as permafrost material. Figure 2G-13 is an illustration of a lander concept in which the lander itself remains stationary and serves as a launch platform for the ascent vehicle; a rover is deployed from the lander to acquire the samples (both are also seen in Fig. 2G-9 in reverse prominence). The first objective, once on the surface, is to acquire a contingency sample at the site. If all of the other sampling strategies then fail, we will at least have that one sample[7].

_____

[7] A similar strategy was planned for the Viking landers, in that their initial computer load (ICL) of automated initialization commands included a sequence to acquire and distribute a surface sample "in the blind" (without a command uplink from Earth) as a contingency. Both landers landed safely and were operational, however, and the sample acquisition commands were updated on the basis of imaging data before the soil-acquisition ICL commands could be executed. As an example of the risk involved in such blind commands, it was later determined that these ICL sequences (for BOTH landers) would have resulted in the surface samplers encountering rocks during their extension. This probably would have resulted in a "no-go" and mission stoppage in both cases (another system response designed as a self-protection mechanism, and which successfully occurred a number of times on both landers). An attempt was ultimately made to push "ICL Rock" at the VL-2 site, and "the rock wouldn't budge!" [Reiber, D. B. 10/31/78. Viking Mission Status Bulletin (No. 46, final edition), Viking-Mars: Anatomy of Success (Gluttons for Punishment, pg 13). GPO 1978--684-463.] {Ed.}

MARS SAMPLE TYPES	SAMPLING METHOD
Rocks:	
■ Small Rocks	• Rake • Sieve
■ Large Rocks	• Chipper • Crusher • Slabbing Saw • Rock Holding Device
Soil:	
■ Wind-Blown	Wind Sock, Cup Arrays
■ Surface Material	• Scoop, Contact Samples • Magnet
■ Deep Soil	• Coring Drill • Drive Tube • Trencher
Ice-Laden Permafrost:	• Coring Drill • Drive Tube

Figure 2G-12: Summary of Possible Sample Acquisition Devices (relative to sample type)

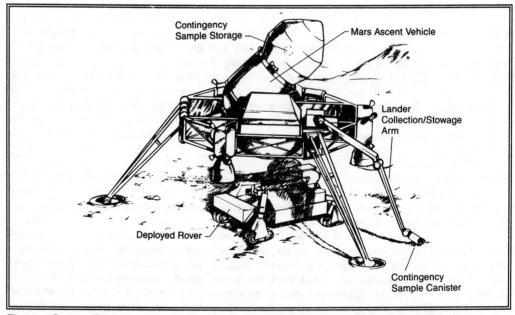

Figure 2G-13: Stationary Lander/Ascent Vehicle Equipped for Contingency Sampling

Figure 2G-14 is an illustration of a generic rover, showing the kinds of sampling systems we expect it would need. This particular rover has both a "strong arm" and a "smart arm," and it has a variety of tools. For example, it has several sampling devices that allow it to contend with differing sample acquisition problems, and it has two sample storage canisters.

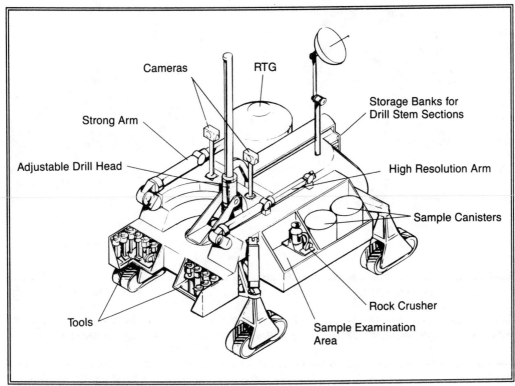

Figure 2G-14: MSR Generic Rover

Drilling Techniques and Problems

When drilling into the surface for buried samples, a rover might encounter unanticipated units. One scenario is a situation in which there could be an overburden of dry soil, within which the rover's drill might encounter a boulder. On the other hand, if one wants to penetrate the surface in search of permafrost as depicted in Figure 2G-15, which represents an entirely different drilling situation, one might run into a variety of materials above or associated with the permafrost. In addition, we might want to sample standing rocks, requiring the rover to tip its drill up and drill into the rock (as illustrated previously in Fig. 2G-9). Special techniques would have to be employed to recover samples properly once the drilling is accomplished, to be sure that the sample comes out with the drill.

We have looked at possible damage to the samples as a result of the drilling process. Basically, heating depends on how much energy is required to cut a certain kind of material. This is illustrated in Figure 2G-16. An important aspect of this is the "specific heat" generated by the drill, in terms of how much can be absorbed by the sample without significantly changing the sample's temperature. In turn, this seems to be primarily a function of how much water is in the sample, which is associated with the conductivity properties of the material.

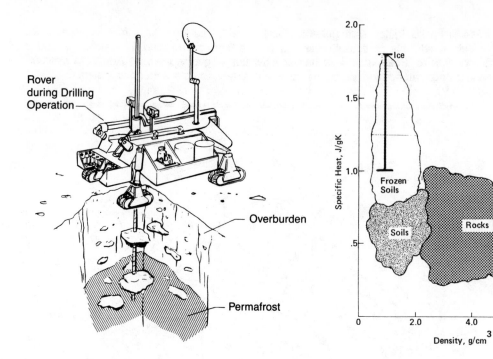

Figure 2G-15: Permafrost Sampling by Rover

Figure 2G-16: Variation of Specific Heat (with density)

Our conclusion was that the amount of heating depends on how rapidly one can disperse the heat produced by the drilling process. We are at a very preliminary stage in these studies, but it appears that one can do an appropriate amount drilling in the anticipated materials with very little heat-related damage. The units most likely to experience heat transfer are the basalts, but they can stand the greatest amount of heating without losing their essential qualities.

The soil sampling device illustrated in Figure 2G-17 is known as a "straight drive tube." In simple terms, it is a sand drill. If sample material is too loose and noncohesive to be retrieved with a standard core-tube device, something like this closed-tube device could instead be used. It is driven into the soil material, and small windows are then opened to let a sample fall inside. The windows are then closed to retain the sample when the tube is withdrawn.

Figure 2G-17:

Sand Drill for Sampling Noncohesive (loose) Soils

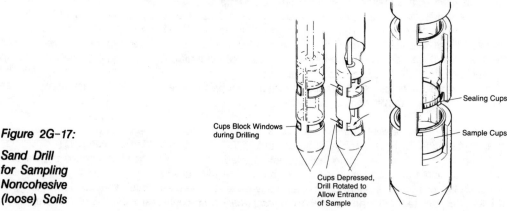

342

Other Rover Tools and Techniques

Figure 2G-18 illustrates some of the sampling tools that might be attached to either the rover's strong arm or its high-resolution arm. We have also been sensible enough to create a tool that could reach under the rover and snip off the drill stem if it gets stuck. The mini-core drill indicated in the illustration is important for performing the sub-sampling activity. A reliable

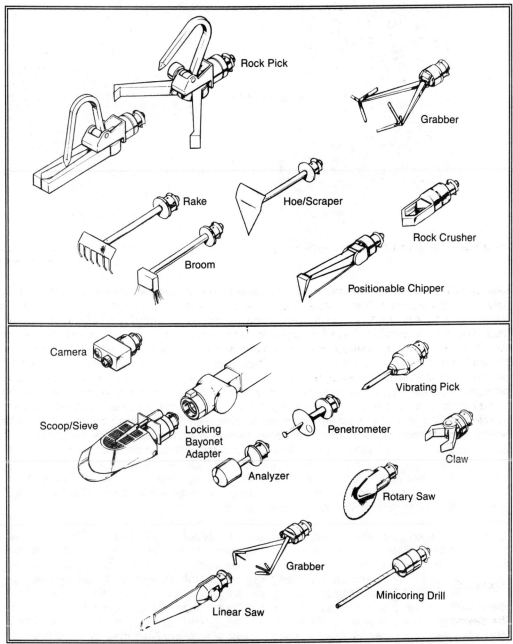

Figure 2G-18: Tools for Rover's Strong Arm (top) and High-Resolution Arm (bottom)

rock crusher would be needed, and one concept is illustrated in Figure 2G-19. It is based on the principle that stress is cumulative, i.e., if a rock is tapped long enough, it will break! Finally, there are several devices included to collect airborne dust.

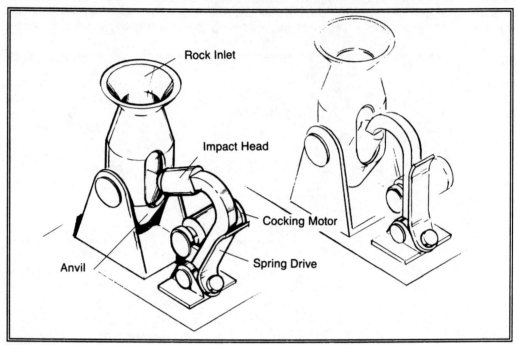

Figure 2G-19: Rock Crusher

Processing, Transporting, and Handling Samples

It is also important to know how best to process, pack, and unpack the samples in order to get them safely back to Earth. Figure 2G-20 illustrates one way to handle many samples. In one case, the sample (a larger rock) is restrained to facilitate the removal of a smaller sample (fragments or small chips) from it for storage and return. The same procedure could be used on core samples, or a saw might be employed. As I suggested earlier, this supports the mission philosophy concerned with collecting interesting samples without bringing back all of the associated mass. This kind of sample preparation would also facilitate in-situ analysis of a rock to see whether one like it had already been collected. Once smaller samples have been acquired in this manner, they could then be poured into storage tubes and capped to keep them separated. There are various modes possible for sealing the samples that prevent intermixing and maintain--to a large extent--the relative environment.

Figure 2G-21 is an interesting concept for a sample canister -- the "little cookie jar" referred to by Jim French. Basically, it is made up of hexagonal units, either as single units that are big enough for one drill core or several of the mini cores, or of open hexagonal units that are big enough for rocks and soils. Extra canisters could be included in the payload so that storage space could be reconfigured during the flight to Earth or while still on Mars. If we have an efficient sample storage system like the "little cookie jar," and bring back the kinds of rock and soil loads reflected in Figure 2G-22, calculations can then be made to determine how much of each sample type one can bring back on a 1-, 2- or 5-kg mission -- as indicated in that data.

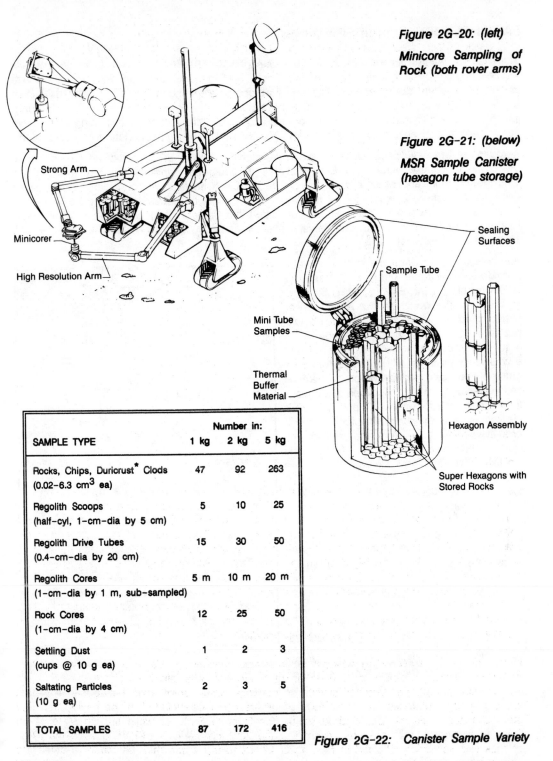

Figure 2G-20: (left)

Minicore Sampling of Rock (both rover arms)

Figure 2G-21: (below)

MSR Sample Canister (hexagon tube storage)

Strong Arm

Minicorer

High Resolution Arm

Sealing Surfaces

Sample Tube

Mini Tube Samples

Thermal Buffer Material

Hexagon Assembly

Super Hexagons with Stored Rocks

SAMPLE TYPE	Number in:		
	1 kg	2 kg	5 kg
Rocks, Chips, Duricrust* Clods (0.02-6.3 cm³ ea)	47	92	263
Regolith Scoops (half-cyl, 1-cm-dia by 5 cm)	5	10	25
Regolith Drive Tubes (0.4-cm-dia by 20 cm)	15	30	50
Regolith Cores (1-cm-dia by 1 m, sub-sampled)	5 m	10 m	20 m
Rock Cores (1-cm-dia by 4 cm)	12	25	50
Settling Dust (cups @ 10 g ea)	1	2	3
Saltating Particles (10 g ea)	2	3	5
TOTAL SAMPLES	87	172	416

Figure 2G-22: Canister Sample Variety

* "Duricrust" is the name given to a cohesive caliche-like soil layer discovered at both Viking landing sites. It exists at the surface (exposed by engine exhaust plumes) or just under a veneer of loose surface material. {Ed.}

EARTH-RETURN CONSIDERATIONS

There are a variety of problems associated with how the Earth-return vehicle (ERV) ends its journey at Earth. The primary concern is sample heating generated both by encounters with Earth's atmosphere during any application of aerocapture technology and by radiated IR from Earth itself (once the ERV is in orbit). The nature of the orbit (elliptical vs circular) into which the ERV is captured is quite important, and it essentially dictates how long the vehicle can remain in a given orbit before recovery. The availability of a space station is also a factor.

Aerocapture Out of Elliptical Orbit -- When we plan to capture ERV's out of an elliptical orbit, the threat of sample heating by Earth-radiated IR could be significant. This means that we must get up there and capture a vehicle before the Earth heats it. We considered how quickly this must be done and what kind of recovery system might be necessary. In a case where the vehicle is placed in an elliptical orbit, we concluded that Centaur-class performance would allow rendezvous and capture of the sample vehicle within an acceptable period of time -- about one 12-to-24-hour elliptical orbit.

Aerocapture out of Circular Orbit -- We also considered this problem from the perspective of having placed the ERV in a circular orbit. In this case, we would have a much shorter period of time in which to capture the vehicle, but it would actually take a couple of 3-to-4-hour circular orbits to capture it. However, it appears to be within the performance range of that postulated for the orbital maneuvering vehicle (OMV) associated with plans for the space station system. Figure 2G-23 is an illustration of the quarantine facility, which essentially stands off from the rest of the space station to achieve a degree of quarantine isolation. The ERV and its canister can be seen, and the OMV--having just returned from the mission to capture the ERV--is visible in the illustration.

[The sketch presented here as Fig. 2G-23 is a legend for a color illustration in the Color Display Section appendix on page A-4.]

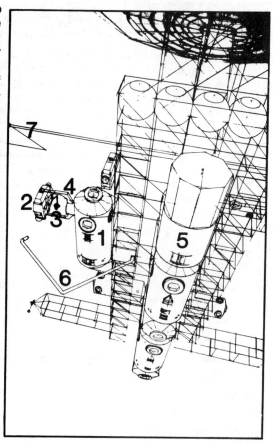

Figure 2G-23:

Delivery of Mars Sample
to the NASA Space Station
Quarantine Module

1) *Quarantine Module*
2) *Orbital Maneuvering Vehicle (OMV)*
3) *Returned Mars Sample*
4) *Airlock/OMV Hard-Dock*
5) *Other Modules of Growth Space Station*
6) *Mobile RMS* · 7) *Radiators*

CONCLUSIONS

Sample return missions enable very significant science opportunities to study Mars. With mobility, several science problems can be attacked at any site, but at no site will it be possible to attack all of the science problems in which we are interested. We have a distinct time advantage at Mars, with plenty of time to understand things as we progress, and we will be able to sub-sample and maximize the quality of our return payload while still on the martian surface. *Selectivity* is the crucial factor; we essentially begin curating the samples on the surface of Mars by collecting ONLY the ones we want -- leaving the ones we don't want behind! However, we need to do considerably more work concerning the kinds of systems and tools we must have to accomplish sample acquisition.

Benefit of Calibrated Rover -- At the end of the mission on the surface of Mars, when we launch the sample-return system (whether to an orbiter or back to Earth), we will leave behind a calibrated rover. That is, we will be able to compare new analysis data from the rover with that of samples it has previously analyzed and returned, and then correlate it with laboratory analyses of the same samples -- thereby calibrating the rover. Then, during an extended mission, we could send the rover off to explore, not only without the previous requirement to return to the lander but with a specific, highly detailed understanding of its data product.

Benefit in Use of Space Station -- While a space station is not essential to these missions, it does offer some launch and recovery opportunities that might not otherwise be available. A space station might also may play an important role in how the samples are returned, in terms of how the vehicle is captured and of how the samples are quarantined and curated prior to bringing them down to Earth. ■

REFERENCES

This space is otherwise available for notes.

~~~~~~~~

MARS BALL ROVER MOBILITY:
Inflatable Sectored Tire Concept[1]
[NMC-2H]

Dr. Douglas A. Hilton
Student Administrator, Mars Ball Project
University of Arizona

The University of Arizona MARS BALL project is named for a mobility concept originally envisioned as a large inflatable ball, and the idea of using inflatable components evolved from that idea. In its new form, the concept involves two wheel hubs, with large inflatable tires, mounted on a single axle. The tires are not continuous air chambers in the conventional sense, but instead are comprised of air bags called "sectors" arranged radially about the hub. These can be deflated and inflated individually in a sequence that results in slow, controlled movement. The large size of the tires and the deflation/inflation capability help the vehicle to move easily over obstacles that would frequently cause a conventional rover to stop and either work out an avoidance maneuver or wait for corrective action determined on Earth. Such operations are costly, making the ability to minimize the number of obstacles that can deter progress quite important. The Mars Ball can virtually ignore most of the rocks visible in Viking lander pictures of the martian surface, which means that it does not have to be extremely smart and that very few hazards will force it to stop and wait for new instructions. While the design of the Mars Ball sectored tires is still evolving, the technology is working well in the project's research model. The concept is therefore believed to be a viable candidate for future consideration.

INTRODUCTION

The two Mars Sample Return (MSR) presentations [see French and Blanchard, NMC-2F/2G], as well as others given in these Mars Conference proceedings, have repeatedly made the point that there is a scientific advantage to be gained from having mobility on the martian surface. I am going to discuss one possible type of roving vehicle which we think should be considered for any future manned or unmanned missions. Clearly, however, much further study of the various rover candidates is required before a final selection can be made.

---

[1] The NASA Mars Conference organization committee wishes to express its appreciation to the University of Arizona and its Lunar and Planetary Laboratory, as well as to Donald Hunten and Douglas Hilton, for bringing the Mars Ball research rover to the grounds of the National Academy of Science for demonstration during conference proceedings. The demonstration clearly reflected the hard work that has been invested in the project as well as the support it has been given, and, in turn, the appreciative audience it drew demonstrated the intense interest that has evolved with respect to rover technologies for use on Mars. {Ed.}

THE MARS BALL PROJECT (University of Arizona)

Figure 2H-1 outlines the *Mars Ball Project* that has produced the work I am about to discuss. The work of the project is being performed by a group of graduate students in the Department of Planetary Sciences at the University of Arizona in Tucson, under the advice of Dr. Donald Hunten. We have been working for three years on the design and construction of a model of an unmanned roving vehicle for possible use on Mars.

FACULTY ADVISOR

Donald M. Hunten, Lunar & Planetary Laboratory, U. of Arizona

---

STUDENT PARTICIPANTS

Douglas A. Hilton (Student Administrator)

Daniel M. Janes
Robert E. Eplee
David H. Grinspoon
Alan R. Hildebrand
Thomas D. Jones
Robert L. Marcialis
Mark S. Marley
Elisabeth A. McFarlane
Shelly K. Pope
Bashar Rizk
Nicholas M. Schneider
Ann L. Tyler

Funded by NASA Grant NAGW-546
with additional support from The Planetary Society.

Figure 2H-1: U·of·A Mars Ball (Rover Concept) Study Team

Mobility on Mars

While those attending the Mars Conference probably do not need to be reminded of what the surface of Mars is like, Figure 2H-2 is photographic evidence provided by a **Viking** lander. The picture was acquired by VL-2 at its Utopia Planitia site. One can see sample trenches dug by the landers surface sampler in the center-foreground. The object to the right of the trenches is the surface sampler's protective shroud, which was ejected during the lander's automatic initialization procedure shortly after landing.

One can see many rocks on the surface as far as the resolution of such objects is possible. It is believed that VL-2 landed near (perhaps within) an outer lobe of ejecta associated with a fairly large impact crater -- Mie[2]. The placement and sizes of the rocks near the landers

---

[2] **Mie** (pronounced **Me**), roughly centered at 48°N, 220°W, about 200 km northeast of the VL-2 landing site, is 100 km in diameter. Like many martian craters, its ejecta pattern exhibits expansive flow-like lobes, although subdued and more difficult to distinguish. Because Mie is the product of a relatively large impact, its ejecta attained significant range. VL-2 is thought to have landed on the thin outer margin of a remnant lobe. {Ed.}

Figure 2H-2: Rocky View at VL-2 Site in Utopia Planitia

have been well measured as a product of the **Viking** missions. It has been determined that there is a rock measuring one foot (or larger) in diameter roughly every ten feet. Further, one must remember that both of the **Viking** landing sites were chosen for their presumed lack of hazards[3]. Therefore, a realistic rover candidate should presumably be able to negotiate terrain at least as rough as that found in the vicinity of the **Viking** landers as a minimum requirement.

There are two basic land-based methods that can be employed for getting a rover around on the martian terrain, and these are shown schematically in Figure 2H-3. In this illustration, one

---

[3] As explained elsewhere in these proceedings, the "presumed lack of hazards" relative to the first **Viking** landing was determined on the basis of visual image resolving power no greater than 100 meters and Earth-based radar data that could only suggest apparent relative smoothness at a scale somewhat closer to that of the lander. None of the surface detail (rocks) visible in the lander pictures at either site was large enough to be detected by any means available at that time, such that rock hazards could not have been detected or, on the basis of experience, presumed present in advance of either landing. It should also be noted that the **Mars Observer** camera (at its highest possible resolution) may be able to resolve only the very largest boulders present on the surface in isolated areas, and will still be unable to see rocks like those that dominate the **Viking** lander images and might be a threat to future landers or rovers. These images, however, now provide a basis for extrapolation, based on the relativity of local topography at those sites to larger regional topographic features that can be seen in **Viking** orbiter pictures and which undoubtedly will be even more clearly resolved by **Mars Observer**. One would also hope that advanced landing technology will improve the landing vehicle's ability to "see" and avoid hazards (at the local scale) during the final moments of terminal descent (when threatening features could be sensed and while small increments of trajectory correction would still be possible). {Ed.}

is labeled "Conventional" (which may not be the best term in this context) and the other represents the U·of·A **Mars Ball** concept[4].

Conventional Mobility Method -- The "Conventional" method (Fig. 2H-3, left) represents a mission philosophy in which the rover zig-zags along a path to an objective, along which course changes are dictated by obstacles as they are encountered. When it detects an obstacle, such as a large rock, a conventional rover performs the necessary operations that result in charting a new course, and it essentially wanders along in this manner until it finally reaches its intended destination. These relatively frequent course corrections might be performed by on-board machine intelligence (highly advanced computer), or a picture can be transmitted to Earth at each corner of the zig-zag as the basis for a decision to take a new course -- the commands for which are then transmitted back to the rover. Since two-way communication time between Earth and Mars can be as much as 40 minutes, the amount of time involved in these operations could become extensive. While time itself may not be a major problem, this would involve intense Earth-control interaction with the rover and could impose a significant cost factor.

The Mars Ball Method -- The method employed by the U·of·A **Mars Ball**, the results of which are illustrated in the right half of Figure 2H-3, also involves sending a picture to Earth at each obstacle large enough to impede its progress. However, the design and large size of the **Mars Ball** vehicle is such that it can overcome many obstacles on its own, and fewer surface features will be obstacles to it. Course changes will therefore be farther apart, resulting in a straighter, less complicated path to the objective. Many--if not most--of the rocks that would be a problem for the smaller "conventional" vehicle are all but ignored by the **Mars Ball**, and the need for on-board image processing and/or course revision is all but eliminated. Its large size allows the **Mars Ball** to continue moving until it encounters a major obstacle not identified when the path segment was designed (such as a cliff, sharp depression, or large boulder), at which time a hazard detector would cause the vehicle to stop and transmit a picture for analysis. One will quickly note that it isn't really a ball, as its name implies. The name evolved out of an earlier, less successful concept for a large sphere that would roll around in the martian wind -- not unlike a beach ball. Both concepts were originated by Jacques Blamont, who is with the National Center for Space Studies (Centre National d'Etudes Spatiales -- CNES) in France[5].

---

[4] Rover research is an on-going process, and the term "conventional" at the time of the Mars Conference typically implied a rover that would move about on small, continuous-loop drive belt tracks (looking a little like belt sanders) rather than wheels. Some of the early rover concepts studied were follow-on modifications of the **Viking** lander itself (with and without sample return), exhibiting three-point drive track surface mobility, but these have since evolved into vehicles with four-point drive track mobility [see NMC-2F/2G]. Some of the concepts studied employed landers with their own roving capability while others utilized stationary landers that deployed either tethered or limited-range autonomous rovers (which returned samples to a lander for processing and/or ascent delivery). In a more exotic sense, robotic vehicles that could slowly and carefully step over large rocks, functioning not unlike long-legged spiders (with four or more highly articulated legs), have been modestly considered, and studies have also been done on the possibility of using flight vehicles (e.g., small, automated, airplane-like winged craft that could fly in thin atmospheres and land periodically to conduct surface investigations). Balloons have also been studied as a means for moving landers or rovers (as well as balloon-unique instrument packages) about on the surface. Some of these may perhaps become increasingly worthy of serious consideration as their mass constraints or need for considerable machine intelligence are overcome by advancing technologies, but their present states (including certain vulnerabilities to the martian environment peculiar to their designs) currently support the practicality of tracked or wheeled rovers over such concepts. {Ed.}

[5] Jacques Blamont is a consultant to CNES. In addition to the original **Mars Ball** concept, he is credited with a variety of balloon-based exploration concepts, including those utilized by the Soviet **Vegas 1** and **2** missions at Venus (enroute to Halley's Comet). He has proposed similar concepts for Mars [see **Exploring Mars by Balloon**. The Planetary Report, Vol. VII, No. 3. May/June, 1987.] {Ed.}

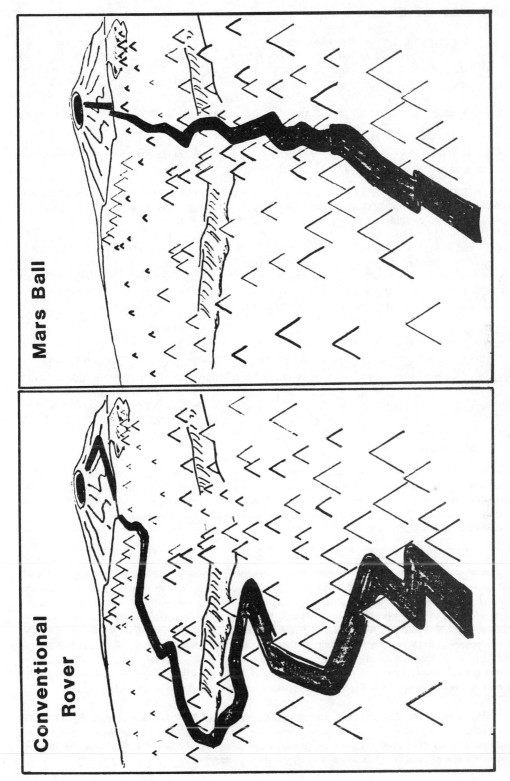

Mars Ball

Conventional Rover

Figure 2H–3: Comparison of Path Selection Based on Obstacle Tolerance

355

MARS BALL VEHICLE

Figure 2H-4 presents a side-view photographic perspective of the U·of·A **Mars Ball** test model developed and constructed at the University of Arizona (a color photo presented as Fig. 2H-4 in the Color Display Section, page A-4, provides a front view). The vehicle is 12 feet high, 16 feet wide, and weighs 1100 pounds (499 kg). The vehicle is clearly dominated by its two large wheels, which are inflatable fabric tires attached to hubs located at opposite ends of a single axle. The payload section is simply slung under the center of the axle, where it is free to rotate and/or swing like a pendulum. Because of this configuration, the **Mars Ball** essentially has no up or down orientation in the conventional sense, and therefore can't be turned upside down. This solves one of the problems that is somewhat of a consideration for the more conventional wheeled or track-driven vehicles.

There is a rigid hub at the center of each wheel, and these hubs are four feet in diameter. Making up the tire itself, and arranged around each hub in a radial orientation (like pieces of a pie), are eight airtight bags that we refer to as *sectors*. For purposes of referencing, they have been made more visible through the use of alternating red and white colors (dark and light in black-and-white reproductions). These airbag-like sectors receive their inflation air from a blower located inside the center of each hub. The numbered hardware objects arranged around the exterior face of the hub (painted white in some of the pictures but black in others) are valves that control the inflation and deflation of the sectors. It is the sequencing of this inflation/deflation process that causes the vehicle to roll in the desired direction.

Operation

Our test model is powered by ordinary 115v house current, which is wired either from a nearby building outlet or from a portable generator. Because this information tends to produce an image of a spacecraft operating at the end of a very long extension cord, I hasten to explain that the way we power our research model does not invalidate our concept. As in the case of the **Viking** landers[6], the amount of power the vehicle would need on Mars can be supplied by on-board radioisotope thermoelectric generators (RTG's) and batteries. We have calculated that a **Mars Ball** could operate properly on 170 watts[7].

I want to emphasize that our research model is not a prototype of an actual Mars rover vehicle. It was built to help develop an understanding of the sector-tire propulsion concept, and we have made no attempt to include all of the systems that would be needed on an actual Mars rover, as demonstrated by the nature of our current power supply. I will elaborate a bit more on what a real **Mars Ball** would have to be like shortly.

---

[6] Each of the **Viking** landers was equipped with two AEC radioisotope thermoelectric generators (RTG's) rated for a combined power output of 90 watts (69 W continuous). On Mars, their performance was essentially nominal and reflected a slight decay against time, as predicted. Two wet-cell NiCd battery packs, each containing two 24-cell batteries, were available in each lander to handle loads in excess of RTG capacity. They were rated at 27 vdc (total), 8 A-h per battery. Total battery storage capacity per lander was 1060 watt-hours. Peak loads were experienced during the entry phase when brief requirements reached in excess of 400 W. A maximum peak occurred briefly during terminal descent when a load of approximately 700 W was experienced. On the surface, demand did not significantly exceed RTG capacity during peak science activity. RTG's have been used on other planetary spacecraft, as well, including the two **Voyagers**, and will be used on Galileo. {Ed.}

[7] The figure of 170 W is a revised determination relative to the 300-W estimate given at the time of the Mars Conference. Dr. Hilton and his colleagues have since made more accurate calculations relative to their experience with **Mars Ball** power requirements. Though large (and potentially conservative), it is within the realm of power-generation capability represented in existing RTG technology. {Ed.}

Figure 2H–4: U-of-A Mars Ball Sectored-Tire Mobility Model (side view); [see also Fig. 2H–4 in Color Display Section, Pg A–4, for front view]

357

Operational Control -- The vehicle is controlled by the computer system shown schematically in Figure 2H-5. The top three boxes represent the three major components of the vehicle itself, the payload (in the center) and two wheels (one at each end of the connecting axle), each of which has a computer in it. Our research model also has a remote computer with a keyboard for human input, and commands are sent to the vehicle via a hard-wired link. However, this was done only for reasons of practicality, since telemetry is already well understood and there is no point in including such an expensive refinement in a vehicle designed essentially for the study of a mobility concept. The payload computer receives the signal, determines which wheel the particular command is for, and relays it to that wheel's hub computer. The hub computer then controls the inflation or deflation of the sector bags by operating the appropriate valves.

The photos of the model presented in Figure 2H-6 (*a*, *b* and *c*) on the facing page represent a sequence to give one a feeling for how the U·of·A **Mars Ball** moves. In this case, the vehicle is moving from left to right with its wheels rotating in a clockwise motion, and four sectors are labelled for reference. In the first photo (Fig. 2H-6a), note that sector 2 is just beginning to touch the ground. At this point, sector 4 is almost fully reinflated and sector 3 is essentially at minimum inflation. The weight of the vehicle will help deflate sector 2 as it is pushed forward, first by the reinflation of sector 4 and then sector 3 (a sequence that rotates sector 2 to the bottom of the tire). As sector 2 deflates, the tire sinks downward and forward, allowing sector 3 (the sector seen fully deflated in Fig. 2H-6a) to rise rearward into position for reinflation (Fig. 2H-6b). Its inflation lifts the rear of the tire even more and causes further forward rotation to the right (Fig. 2H-6c), such that sector 1 rotates into position for deflation on the forward side of the tire. This sequence is repeated over and over for both wheels, propelling the vehicle slowly along its directed course.

The vehicle has three computers because signals must pass between rotating and non-rotating parts through slip-ring connections. In order to eliminate the need for a separate rotating

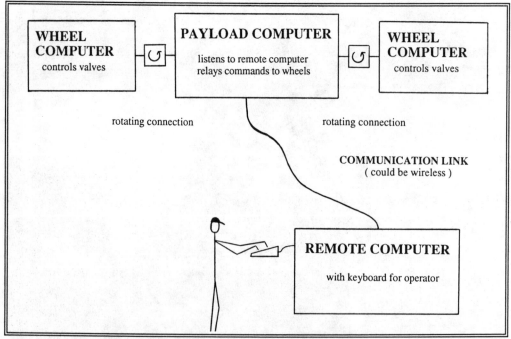

Figure 2H-5: Mars Ball Tri-Computer System -- Primary (payload) and Wheel (valve control)

358

Figure 2H-6: Mars Ball Inflation/Deflation Mobility Sequence (a, b, c -- top to bottom)

connection between the payload and each sector, a slave computer was installed in each of the hubs. These slave computers have the sole function of controlling valve electronics according to instructions received from the payload computer. The software requirements are very simple, as well, since the only variables are the inflation/deflation state of each sector. The control software for an actual Mars version could be just as simple, and guidance software requirements would be much less severe (than for some other types of vehicles) because no on-board course determination would be necessary.

Test Program

We have been testing the vehicle since May, 1986, shortly after completing its construction. Thus far in the program we have performed a number of successful tests, including speed measurements. We have determined, for example, that the vehicle can move forward at about 1 m/minute over a 23-minute average. Its speed is limited by how fast the air is squeezed out of a sector bag during the deflation process. This could obviously be improved by a better gas plumbing (inflation/deflation) system. We also found that it is relatively easy to turn the vehicle; it operates much like a tank in this respect, in that it is turned by rolling one wheel forward and the other backward.

Figure 2H-7 illustrates some of our test work, wherein a gentle slope is being negotiated by the vehicle. The slope pictured is a six-degree incline which the vehicle was able to ascend and descend with no problem. A six-degree slope is not the steepest slope that the **Mars Ball** can climb, it just happened to be the only slope we had been able to try at that point in time. Before we built the large, more complete vehicle we are now using, we first built a very crude model that incorporated a four-foot-diameter version of the wheel with roughly the same proportions, sector sizes and numbers, et-cetera. That model was able to climb a 24-degree slope. While we do not yet know if the current vehicle will be able to climb as well as the smaller one did, we expect its capability to be generally similar.

Figure 2H-7: U-of-A Mars Ball Model Moving Up Incline (6 degree slope)

Maneuverability over Obstacles -- The photographs of the model presented in Figure 2H-8 (*a, b,* and *c*, top to bottom) demonstrate sequentially the way in which a **Mars Ball** would contend with obstacles on Mars. Our simulation obstacle, measuring 20 inches high and 52 inches across the top, was made up of boxes. The time-lapse sequence shows the U·of·A **Mars Ball** essentially rolling slowly over the obstacle. Note that the lower edge of the vehicle's rigid hub remains at roughly the same height with respect to the ground as the wheel passes over the top of the simulated obstacle, such that the wheel's tire appears to "absorb" rather than climb the box.

Figure 2H-8:

Mars Ball Model Moving Over Simulated Obstacle

Direction of travel is from right to left:

a) Ascent at Obstacle

b) On Top of Obstacle

c) Descent from Obstacle

Other Advantages of Deflation Capability -- Figure 2H-9 demonstrates another advantage of the vehicle -- the fact that it is collapsible. The photograph was taken as the U·of·A **Mars Ball** was being assembled. With the tire sectors fully deflated and packed onto the hubs, the wheels are only a little over four feet in diameter (only slightly larger than the diameter of a hub by itself). It is in this configuration, presumably, that such a vehicle would be transported to Mars. In its collapsed configuration the **Mars Ball** constitutes a relatively small payload, with savings in both weight and cost. Note, too, that when the vehicle tires are deflated, the payload is sitting almost on the ground. This   uld be an advantage for geochemical experiments, for example, by making it possible to use in·situ detectors or sampling devices that must be very close to--or in contact with--the soil or a rock. On Mars, the vehicle would simply squat down on a sample site to facilitate such experiments.

Figure 2H-9: Deflated/Packed U·of·A Mars Ball (16-sector tires)

Note, too, (by counting the alternating light and dark sectors packed on the hub in Fig. 2H-9) that the vehicle pictured has more than eight sectors in each of its tires. We originally designed the vehicle to have sixteen sectors per tire, each being six feet in length compared to four feet for the current vehicle's eight-sector tires. The six-foot sectors proved to be too long, in that they were spindly and made the vehicle unstable; the whole vehicle would simply flop over to one side at times because there wasn't enough rigidity in the sector bags to hold the tires vertical. We solved the problem by shortening the sectors and making them fatter. And, because of their radial fit around the hub, we also had to reduce the number of sectors to eight. It was recently suggested that wind was the cause of our instability, but that simply is not correct; the wind was only occasionally an aggravating factor. Wind would only rarely be a problem on Mars, and the ability to deflate the tire sectors would facilitate both the mechanism and an adequate procedure for "securing" the vehicle in the event of a martian dust storm.

CONSIDERATIONS FOR A MARS CONFIGURATION

Figure 2H-10 outlines the features that would have to be represented on a Mars version of the vehicle. It is probable that the tires would again be made up of 16 sectors. Our experience suggests that 16 sectors may be more appropriate for coping with steeper slopes and numerous blocky obstacles. A Mars vehicle would probably be equipped with redundant sectors (uninflated

362

```
┌─────────────────────────────────────────────────────────────────┐
│                                                                   │
│   ■   More Sectors, Probably Redundant;                           │
│                                                                   │
│   ■   Endcaps (outer face of wheels/hubs) to Prevent              │
│         Being Upended;                                            │
│                                                                   │
│   ■   Compressors Instead of Blowers;                             │
│                                                                   │
│   ■   Recycling of Compressed Atmosphere;                         │
│                                                                   │
│   ■   On-board Power Source (≈170 W);                             │
│                                                                   │
│   ■   Tough Bag Materials to Withstand Abrasion,                  │
│         Low Temperatures, UV Radiation;                           │
│                                                                   │
│   ■   Possibly Larger Size.                                       │
│                                                                   │
└─────────────────────────────────────────────────────────────────┘
```

Figure 2H-10:   Features a Flight-Mission-Configured Mars Ball Would Have

until needed); if a primary sector is punctured, another can then take its place. Durable bag materials will be required to withstand abrasion, puncture, low ambient temperatures, and ultraviolet radiation. This set of requirements for the material appears to need no new technology, however, and the best current candidate is a Kevlar-based laminate.

Although a **Mars Ball** vehicle has the inherent advantage that it can't be turned on its back (because it has no up/down orientation), there is a possibility that it could be turned on its side with the axle in a vertical orientation. This might happen, for example, if one of the tires encountered a relatively high, ramp-like obstacle while the other remained on level ground. The problem could be prevented by putting large, hemispherical endcaps (some might think of them as "hubcaps") on the ends of the axle outside each hub. With these in place, axle orientation could not stabilize vertically if the vehicle got tipped up on one side, and would instead cause it to simply roll over and flop back down on both tires.

The ambient pressure on Mars is roughly 6-8 mb, depending on the time of year and where one is relative to datum elevation[8]. To support the vehicle in this low-pressure environment, we believe the working pressure within the sectors would have to be 12-18 mb. For this reason, compressors would be needed in place of the blowers we are currently using. The compression of inflation gas provides a means of saving energy, as a payback, by efficiently cycling the compressed gas between sector bags (i.e., pumping it to the next sector being inflated rather than exhausting it into the atmosphere and recompressing, as we do now). Our calculations suggest that the recycling of compressed inflation gas could save as much as fifty percent (50%) of the power consumption now anticipated for the recompression process.

The vehicle would need an on-board power source, probably comprised of RTG's and batteries as already discussed. It might also utilize solar cells in some capacity, remembering that another power source would be needed at night or during longer periods of severe atmospheric opacity. The rate of power consumption depends to some extent on how fast the vehicle must

---

8 As more fully explained elsewhere in these proceedings, the 0-km datum elevation has been established in lieu of a martian mean-sea-level, and is represented by an ambient atmospheric pressure of 6.1 mb. Both of the Viking landers were below datum elevation and therefore experienced slightly higher ambient pressures than would have been the case at 0 elevation. References and additional explanation are provided in Atlas of Mars by Batson, Bridges and Inge, NASA SP-438 (Appen C, Contour Mapping, by Sherman S.C. Wu). {Ed.}

move. The point was made by Dr. Blanchard during his Mars Conference presentation [NMC-2G] that slow speed is not a negative factor. Thus, power consumption could be reduced by cutting back on the rate of mobility. We originally calculated power requirements for a **Mars Ball** moving at about three times our current speed, and the results were excessive. Now, however, calculations suggest that a 250-kg vehicle would consume ≈170 W on Mars, assuming: 1) a speed of 2 m/minute (nearly 1 km in 8 hours), 2) a maximum tire pressure that is 9 mb more than the ambient atmospheric pressure, and 3) a conservative compressor efficiency of 10 percent.

It may be appropriate to make the Mars version of the vehicle larger than our current research model in order to reduce the number of hazards at which it must stop. However, a larger vehicle consumes more power, weighs more, and requires more payload space, and the tradeoffs necessitated by larger size would require very careful study.

OPERATIONAL ASPECTS AND FUTURE WORK

Figure 2H-11 outlines the major advantages and disadvantages we have been able to identify for a **Mars Ball** vehicle, and I will summarize some of the important ones here. These, in turn, guide our thinking relative to future ns.

ADVANTAGES

- Large -- Can Ignore Most Surface Features,
- Simple to Control,
- Less Contact with Earth Necessary,
- Collapsible -- Saves Spacecraft Weight and Cost,
- Essentially Immune to Upset.

DISADVANTAGES

- Requires Durable Bags and Reliable Gas Plumbing,
- Possible Large Power Consumption.

Figure 2H-11: Advantages/Disadvantages of Mars Ball for Mars Mission

Advantages, Disadvantages and Future Plans

Advantages -- The vehicle is large and can ignore many of the rock sizes we have seen in the **Viking** pictures. It is fairly easy to control, so the software is relatively simple compared to the degree of artificial intelligence needed in some of the other rover designs being considered. Finally, the fact that the vehicle is collapsible saves spacecraft weight and cost, and it will be essentially immune to upset once the "hubcaps" are in place.

The first two of these points imply that required contact with Earth can be minimized. In operation, we envision a scenario that would work in this manner: take a picture and transmit it to Earth; evaluate the picture and transmit a command load that would send the vehicle

364

perhaps halfway to the horizon (assuming no major surface obstacles arise to impede progress along the way); and then stop, take another picture, and repeat the procedure. The vehicle would have hazard sensors to detect cliffs, very large boulders, or other major obstacles along its direction of travel.

Disadvantages -- Some of the disadvantages we have identified include: a requirement for durable sector bags (fabric) to assure the ability to reliably travel long distances over rough, hazardous terrain under very difficult mission and environmental conditions[9]. Also, the inflation/deflation process requires a reliable gas plumbing system with which to control the sectors.

Planned Enhancements

Figure 2H-12 presents a summary of future plans our group has established for continuing work on the U·of·A **Mars Ball** concept. For example, we want to do a more thorough study of the vehicle's ability to cope with slopes and obstacles, to determine what it can and can't handle and how the concept can be improved in this respect. We want to develop increased automatic mobility, i.e., give the vehicle the ability to run by itself without an operator at the terminal controlling its routine functions. An improved computer control system has already been put into use that facilitates a degree of autonomous operation, such that the vehicle can now sense the positional state of a tire and then inflate or deflate its sectors automatically. We have already used the automated system on level terrain.

- Slope and Obstacle Limits,
- Automatic Rolling on Various Terrains,
- 16-Sector Tires,
- Control by Means of On-board Camera Only,
- Testing in Desert as Simulation of Mars.

Figure 2H-12:   Nature of Continued U·of·A Mars Ball Research Studies

We are hoping to successfully redesign our sectored tire so that we again have sixteen sectors rather than eight. We feel that sixteen-sector tires will perform better on both obstacles and slopes. We also want to develope a control system that utilizes a TV camera, such that the vehicle can be operated solely on the basis of transmitted pictures (which would obviously better simulate an actual Mars mission). To achieve and test these objectives, we are intending to conduct some research with the vehicle out on the Arizona desert as well as elsewhere in the southwestern United States. ∎

---

9 Both **Viking** landers consistently recorded temperatures at least as cold as $-85\,°C$ ($-120\,°F$), and VL-2 recorded winter temperatures of about $-123\,°C$ ($-190\,°F$) at its northern latitude site. Because future missions might include objectives to look for, analyze, and perhaps acquire samples associated with the cryosphere (e.g., surface/near-surface materials like ice or permafrost near or in a polar region), the ability to develop systems, tools, and materials capable of working reliably in a very cold environment is important. [See NMC-2G] {Ed.}

REFERENCES

This space is otherwise available for notes.

~~~~~~~~

NASA Mars Conference
Session 3, July 23, 1986

Issues & Options for Manned Exploration

Morning Session Chairman, Dr. Michael B. Duke

PRESENTATION   PRESENTER

---

[1] At the time of the NASA Mars Conference, Dr. Paine was chairman of the National Commission on Space and had just presented its report to the United States Congress and to President Reagan.

NASA MARS CONFERENCE
Session 2: Speaker Profiles

NMC-3A · DR. MICHAEL B. DUKE is chief of the Solar System Exploration Division at NASA's Johnson Space Center (JSC) in Houston. He has been with NASA for many years, extending back well into the Apollo era, and has principally been associated with manned programs. Dr. Duke was the original curator of the lunar sample archive activity at JSC. However, he has also exercised an active interest in study associated with planetary exploration, and participated in some of the conceptual planning conducted during the mid and late 1970's which it was hoped would result in a sample return mission by 1990. His knowledge of both manned mission technology and advanced mission planning established him as one of the chairpersons for the third and final session of the *NASA Mars Conference*, which addresses the issues and options challenging the future human exploration of Mars.

~~~~~

NMC-3B · MR. JOHN C. NIEHOFF is the Space Sciences department manager and a senior research engineer for Science Applications International Corporation (SAIC), located in Schaumburg, Illinois. Well published on topics associated with solar system exploration and NASA long range planning, Mr. Niehoff reflects 25 years of experience in celestial mechanics, lunar and planetary mission research, and program strategy planning -- including cost estimation and control. He has served on many NASA workshop teams and committees, and has at times been responsible directly to the NASA Administrator. Among his continuing responsibilities are advanced studies in support of NASA's Solar System Exploration Division and Shuttle Payload Engineering Division (OSSA/NASA), and he is also participating in advanced concept and assessment studies with a number of NASA research centers.

~~~~~

NMC-3C · MR. WILLIAM C. SNODDY has been deputy director of the Program Development Directorate at NASA's George C. Marshall Space Flight Center, Huntsville, Alabama since 1983, and he has been with NASA since MSFC (formerly an Army research facility) became part of the agency in 1960. His directorate is responsible for the planning and conceptual development of future projects involving manned and unmanned space facilities, launch vehicles, scientific spacecraft, and orbital transfer vehicles. He earned both a BS and MS in physics at the University of Alabama (Tuscaloosa), and also has a MS in administrative sciences (U. of Al., Huntsville). Mr. Snoddy often has had key responsibility for important science programs. He was project scientist for the **Skylab** Kohoutek (comet study) project and was manager of the **Skylab Apollo Telescope Mount** data analysis program. He spent a year at NASA Headquarters (1978-79) as acting manager of advanced programs and technology for the Solar-Terrestrial Division, and chaired the Science and Applications Space Platform Study Team. Of particular interest among his credits are studies of tethered satellites, the thermal control of satellites and spacecraft, and microgravity processing of materials. He is a Fellow of the AIAA and chaired the AIAA 16th Aerospace Sciences Meeting, and he has been awarded the Hermann Oberth Award of the Alabama section of the AIAA. Mr. Snoddy is a recipient of NASA's Exceptional Scientific Achievement Medal in recognition of his Skylab contributions.

~~~~~

NMC-3D · MR. BARNEY B. ROBERTS is the mission analysis and systems engineering agent for NASA's Office of Exploration, which is charged with defining options for human planetary exploration. As such, he leads an engineering team that will define the details of a long range space exploration plan for the agency. He earned his BS degree in physics at Georgia Tech, MS degrees in both physics and public management at the University of Houston, and is currently in pursuit of a Ph.D in public administration from the University of Colorado (Denver). Mr. Roberts has been with NASA-JSC for 25 years. His work has involved crew training, aerodynamics, and subsystem management (space shuttle). He began work in advanced programs at NASA-JSC in 1982 and was responsible for the Manned Mars Mission study sponsored jointly by NASA and the Los Alamos National Laboratories. He then became involved in strategic planning activities intended to define program options and concepts for new space initiatives, and this effort became the basis for his present assignment.

~~~~~

NMC-3E · MR. JAMES R. FRENCH, JR. is chief engineer and vice president of engineering at American Rocket Company in Camarillo, California. His expertise relative to planetary technology evolved out of many years of related studies at JPL. He was previously profiled for his Session·2 presentation [NMC-2F] which focused on Mars sample return mission concepts.

~~~~~

NMC-3F · DR. ARNAULD E. NICOGOSSIAN, M.D., is director of NASA's Life Sciences Division in Washington, D.C. He is both a physician and a scientist, having first earned a Ph.D in medicine at Teheran University in 1964 and then a MS in aerospace medicine after moving to the United States. Dr. Nicogossian established his medical credentials in the United States at Mt. Sinai's School of Medicine and Mt. Sinai Hospital between 1966 and 1970 (focusing on internal medicine and pulmonary diseases). He then developed his present specialties in aerospace medicine and nuclear medicine at Ohio State University, where he earned a MS (1972), and at the Baylor College of Medicine (1975). He has been associated with NASA since 1972 when he served as flight surgeon and worked in cardiovascular research at NASA-JSC. He has been at NASA Headquarters since 1976 when he served as manager of medical operations, and he became Director of the Life Sciences Division in 1983, succeeding Dr. Gerald Soffen (former Viking Project Scientist and Session·1 chairman for these proceedings of the NASA Mars Conference). Concurrent with his NASA responsibilities, Dr. Nicogossian has held or still holds faculty positions at USUHS (department of preventative medicine) in Bethesda, MD, and at Georgetown Hospital in Washington, D.C. He is a Fellow of several medical associations and a member of both the International Academy of Astronautics and the New York Academy of Sciences. His work has been recognized through NASA medals for exceptional service and leadership, the Bauer Space Medicine Award, the Strughold Space Medicine Award, the AIAA Jeffries Medical Research Award, the AAS Melbourne W. Boyton Award, and the Yuri Gagarin Medal (awarded by the USSR Astronautical Federation). He has authored or contributed to several books and over 30 scientific papers on topics related to aerospace and nuclear medicine.

~~~~~

NMC-3G · DR. PENELOPE J. BOSTON is a microbiologist and atmospheric chemist with the National Center for Atmospheric Research in Boulder, Colorado. She earned her Ph.D at the University of Colorado, where she was one of the founders of the *Case for Mars* conference series (which had its origin as a graduate project at the university) and served as editor for the proceedings of **The Case for Mars** conference (first of the series, now at three, published by AAS). Dr. Boston is a strong advocate of the human exploration of Mars, particularly with respect to life sciences issues. Among her research credits are studies of global biochemical processes and the interaction of the biosphere and atmosphere. Presently on a leave-of-absence from the Boulder center, she is working at NASA-LaRC where she is a National Research Council Associate. She is studying the emissions of trace gases from biomass burning, and the biogenic emission of gases from soil relevant to global climatic change. She has also been a consultant to NASA in planning life support for advanced manned missions, which is closely related to her topic for these Mars Conference proceedings.

~~~~~

NMC-3H · DR. CHRISTOPHER P. McKAY is a research scientist and exobiologist in the Solar System Exploration Branch of the Life Science Division at NASA's Ames Research Center (NASA-ARC). He earned his Ph.D in astrogeophysics at the University of Colorado in 1982. Dr. McKay's work at NASA-ARC has focused on understanding the relationship between the physical and chemical evolution of the solar systems and the appearance of life. He is a member of the Mars Sample Return Science Steering Group, serves as interdisciplinary scientist for exobiology on the Comet Rendezvous Asteroid Flyby (CRAF) mission, and is co-investigator for the Planetary Society's Mars Institute. He has augmented his work with extensive terrestrial research in the Antarctic dry valley, studying the region and microbial life found there as a possible analogy to Mars. He has had an active role in the "**Mars Underground**" since its inception and is an advocate of the potential of "terraforming" on Mars, the process of transforming the planet--through human engineering--into a more Earth-like body by altering the state of its own resources and environment. [see *Making an Earth of Mars*, The Planetary Report, Vol. 7, No. 6 (Nov/Dec, 1987), pp 26-27.] Dr. McKay served as editor for the proceedings of **The Case for Mars II** conference, published by AAS.

~~~~~

NMC-3I · DR. THOMAS O. PAINE was, at the time of the **NASA Mars Conference**, chairman of the **National Commission on Space**. Its report to Congress and President Reagan was later published under the title: *Pioneering the Space Frontier*. Dr. Paine earned both an MS and a Ph.D in physical metallurgy at Stanford University after serving as a submarine officer in the Pacific during WWII. He joined General Electric in 1949 and later was manager of GE's TEMPO Center for Advanced Studies (1963-68). President Johnson appointed him to the post of NASA Deputy Administrator early in 1968, and he became Acting Administrator later that same year -- on the eve of the first Apollo flight. As Administrator, he presided over five more Apollo flights, including the first lunar landing (Apollo 11), before returning to private life in 1970. Dr. Paine has served as president and CEO of Northrup Corporation, and has held executive posts with Eastern Airlines, Quotron, NIKE, RCA/NBC, Orbital Sciences, and General Electric. He's a leading advocate of the expansion of humankind into the solar system and of the establishment of permanent evolutionary outposts on Mars, and he has produced the first flag of Mars which supporters hope to have included in the cargo manifest for the first manned mission to the Red Planet.

■■■

WHY HUMANS SHOULD EXPLORE MARS
[NMC-3A]

Dr. Michael B. Duke
NASA Johnson Space Center

Among the explanations most often given as a general rationale for the manned exploration of Mars is the opportunity it represents for more intensive and adaptive science, the establishment of a technological and systematic gateway to the outer solar system, the opportunity to exploit the resources of Mars and its moons, and the encouragement of international cooperation. However, another fundamental reason an objective as significant as that of the human exploration of Mars is important is that it represents broad, exciting goals that could provide a unifying basis of challenge for all mankind. We need such goals to stimulate and motivate the minds of tomorrow's leadership and to help build its societies on both the national and international scenes. Not only do goals of this magnitude serve to focus minds on the positive aspects of the achievement toward which they strive, but they may in fact represent the one kind of human project that can bring about global cooperation and unity at a scale of real significance. Only by accepting the challenge reflected in goals of this kind can we hope to be ready for what lies ahead.

INTRODUCTION: WHY HUMANS SHOULD EXPLORE MARS

Krafft Ehricke[1] sat--because he was gravely ill at the time--at the podium of the **National Academy of Science** in October of 1984 and said, "...if God wanted man to become a space-faring species, he would have given man a moon." I think it is possible that if Krafft were alive and could contribute his great wisdom to these Mars Conference proceedings, he would say something like "...if God had intended for mankind to inhabit another planet, he would have given us a Mars." I would like to reflect my own views about why we should go to Mars by making a few comments perhaps more on the order of what was discussed during the Session·2 panel

---

[1] Krafft Ehricke was one of the great German rocket scientists who later came to the United States and helped build the foundation of America's space program. He previously worked with Werner von Braun's group at Peenemünde on the development of the V2, and he played a leading role in the development of the Atlas and Centaur rocket systems in this country. His education in aeronautical engineering at the University of Berlin included studies in celestial mechanics and nuclear physics. Beginning in 1965, he began to write about concepts for lunar settlements and the industrialization of space, which he introduced through a series of papers and a book still to be published. The book (being edited for publication) is: **The Seventh Continent: Industrialization and Settlement of the Moon**. {Ed.}

discussion chaired by Carl Sagan[2], but which echo the sentiment expressed with such wonderful simplicity by Krafft and reflect my own views about why we should go to Mars.

I have been involved in Mars programs for almost as long as I have been involved with the moon. I first encountered Dr. Gerald Soffen (former Project Scientist for the **Viking** program) in 1976 when we were working on conceptual sample return missions for Mars, and he might now be reminded that I will owe him a fine bottle of wine in 1990. I wagered that by 1990 we would have the Mars sample return mission under way, and I am clearly going to lose that bet. I'm a perpetual optimist. It is appropriate, then, that I'm leading off today's session concerned with manned Mars missions. However, I'm not about to make any bets.

Robotic vs Manned Exploration

During the Session·1 and Session·2 proceedings of this Mars Conference, we were shown the present-state and near-future development of the science of Mars based on our current understanding of the planet. These discussions provided clear evidence of what can be accomplished with robotic extensions of man's intelligence. The discussions presented in the proceedings of Session·3 turn to the issues that must be addressed in undertaking the human exploration of Mars, and they also contribute understanding about why this exploration should be accomplished by people even though so much is possible with robotic systems. This is certainly a crucial concern in view of the fact that we are having great difficulty finding the needed resources even for the scientifically exciting robotic missions described in the Session·2 proceedings.

Rationale for Manned Exploration

Among the rationales for the manned exploration of Mars people have put forth are: (1) the possibility for more intensive and adaptive science, (2) the establishment of a technological, systematic gateway to the exploration of the outer solar system, (3) the opportunity to exploit and utilize the resources of Mars and its moons, and (4) the encouragement of international cooperation. But the fundamental reason many of us are interested in an objective as significant as that of human exploration of Mars is that we are seeking exciting goals for all mankind. The goal of leaving human footprints on the surface of Mars--to stake humanity's claim--is that kind of exciting quest.

Purpose for Setting Goals

Why do we need such goals? I think we need them for a large number of reasons, which were essentially expressed during the Session·2 panel discussion. We need exciting goals to motivate people to devote the extra effort needed for the achievement of excellence in the interest of humanity, and we need to have these goals in front of us for extended periods of time to drive us onward. We need exciting goals to motivate our youth to excel scholastically and to participate in building our society. We need exciting goals to maintain our focus in an environment that is ever more diverting, ever more complicated. We need objectives that will focus us. The human exploration and settlement of Mars *is* that kind of exciting goal. Indeed, it can be one that motivates the whole world by helping to unite humanity in at least one common project. And when we have done it, we will be able to face each other across the continents and the oceans and say -- not *"they* did it", but *"we* did it ... we did it together!"

---

2 The panel discussions held during Session·1 and Session·2 of the **NASA Mars Conference** could not be included with the proceedings due to the preparation and review complexity associated with them, but they are being processed and it is hoped that they will be made available through NASA at a later date. {Ed.}

Awareness and Commitment Required to Make Mars Goal a Reality

It may be inevitable that humans will explore Mars. This is certainly an assumption that many of us hold. But it is not guaranteed. The value of the Session·3 discussions is their ability to make us more aware of the possibilities, but it will take a concerted effort--technically, scientifically, politically--to turn these conceptual possibilities into realities. The pathway to Mars will be complicated. It will reflect, in my view, the complexity of our own personal lives. None of us can predict our own futures, and yet all of us have goals we use to motivate our capacity for future achievement.

INSPIRATION FOR THE FUTURE

I would like to close my remarks by recognizing four people who have played a part in my personal journey, and who I think exemplify some of the qualities that will get us to Mars.

Harrison Schmitt -- First, I would like to recognize Harrison (Jack) Schmitt, the former astronaut and U.S. Senator who has been a leader in the campaign to send human expeditions to Mars. As a freshman and classmate of mine at Caltech, Jack encouraged me and convinced me to embark on a career in geology. Even then, Jack had his sights set much further ahead than I could have imagined at the time. Later, as a United States Senator, he argued that the U.S. should have a manned mission to Mars as a long-term goal. More recently, as an eloquent example of his vision for the future, he has advanced the concept that the human exploration of Mars should be *the* millennial project toward which we must all strive.

Krafft Ehricke -- I would also again like to mention Krafft Ehricke, who, to the very end of his life, held onto the vision of man's expansion into the solar system. He gave every ounce of his being to the work of convincing us to join him in his quest. Krafft Ehricke was very ill when he appeared at the **National Academy of Science** podium in 1984, and he died soon after. When he spoke here of the need for human exploration, it was to be his last major contribution to the concept he used to drive himself and motivate others for so long.

Thomas R. McGetchin -- I would also like to recognize and remember my good friend, Tom McGetchin[3], who, at the time of his death, was Director of the Lunar and Planetary Institute. Tom was completely devoted to--and had an infectious enthusiasm about--Mars. He was one of those who first laid the basis for our recognition that the Shergottite meteorites (one of the three meteorite types considered to be in the SNC class defined elsewhere in these proceedings) might be fragment samples of Mars. He would have been delighted to participate in this conference, and he would have been even more delighted to go on the first manned mission to the Red Planet.

Steven D. Howe -- Finally, I would like to recognize someone who is still very active in the concept of human planetary exploration, Dr. Steven Howe. Steve is a young scientist at the Los Alamos National Laboratory who has a single-minded purpose to qualify for the astronaut corps. He wants very much to become one of the explorers of space and perhaps to have the opportunity to be part of the first manned mission to Mars. His is an example of the type of enthusiasm we must be able to instill and sustain in the minds of our young people.

---

3 Thomas R. McGetchin was a research scientist in the fields of igneous petrology, volcanology, and planetary science. He earned his Ph.D in Geology at Caltech in 1967, where he was one of Bruce Murray's students in planetary science. He later held faculty positions at the Air Force Institute of Technology and then at Massachusetts Institute of Technology (MIT). He overcame his initial encounter with cancer in 1973 and went on to establish the geoscience group at Los Alamos National Laboratory. Dr. McGetchin became director of the Lunar and Planetary Institute (Houston, TX) in 1977, and served in that position until his death in 1979.

Conclusion: Concerted Effort Needed

It will be people like those I have mentioned, working together, who will achieve the human exploration of Mars. Perhaps, as John Logsdon said during the Session·2 panel discussion, we really cannot bring about the decision to undertake these explorations by ourselves. However, I think it is likely that the leaders of such a movement are probably attending this conference, and I hope that we at least can help motivate them to continue the long effort it is going to take to accomplish the task that lies ahead -- on Earth *and* on Mars. ∎

REFERENCES

This space is otherwise available for notes.

~~~~~~~~

PATHWAYS TO MARS:
New Trajectory Opportunities
[NMC-3B]

Mr. John C. Niehoff
Space Applications International Corporation

The purpose of discussing trajectory and staging scenarios associated with future manned Mars missions is to provide a sense of familiarity with both traditional and innovative concepts for getting to and from Mars. The characteristics and requirements of several alternative flight modes are briefly defined, discussed and compared, and the subject matter is addressed from three perspectives: Earth–Mars pathways, planet–centered pathways, and propulsion options. The concepts represent a variety of mission types and trajectories that are primarily defined as ballistic transfers, Cyclers, and Escalators, and they involve staging options for both Earth and Mars. Decisions regarding these missions and options will ultimately be affected by a number of evolving factors, such as technology development paths, and should be developed through an active process of study and debate under NASA leadership.

INTRODUCTION: TOPIC ORIENTATION

My purpose is to help others develop some familiarity with both traditional and innovative flight profiles to and from Mars. I will briefly discuss the characteristics of flight time and orbit mass performance associated with these pathway alternatives. This discussion complements and supports several other **NASA Mars Conference** presentations that address specific concepts associated with the technical aspects of getting to and from Mars [NMC-3C/3D/3E].

The scope of my presentation covers four subjects. First, I will discuss the pathways between Earth and Mars, and I will then review some planet-centered strategies. I will also discuss some of the propulsion options and finally touch on some of the mission design implications that these pathways suggest. This is a fairly comprehensive presentation, but with insufficient opportunity to provide a detailed explanation for many of the charts used. Nonetheless, most of the illustrations can be easily understood with a little supporting discussion as we proceed.

Mars' Influence on Mission Design

First, I will provide some orientation. In a broader sense, this subject matter involves more than just Earth and Mars, and must incorporate some information about Venus as well. Figure 3B-1 illustrates the orbital relationship of the solar system's three best known inner planets. Earth's orbit is, of course, at 1 AU (93 million miles) from the Sun, and Mars' somewhat more eccentric orbit is about half again as far (142 million miles, average of perihelion and aphelion distances). Venus, on the other hand, is about thirty percent (30%) closer to the Sun than Earth.

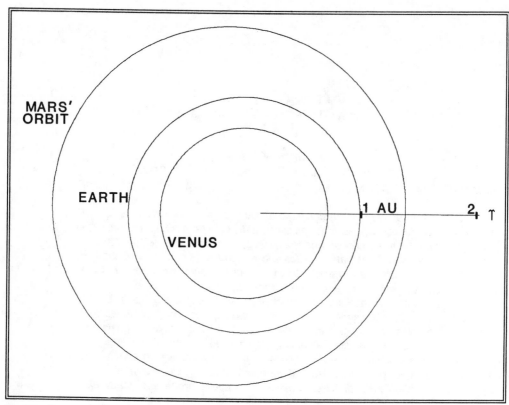

Figure 3B-1: Orbital Relationship of Earth and Mars

Mars' orbit is slightly elliptical, coming closest to the Sun at about 1.4 AU before moving out to about 1.6 AU. The motion of Mars, which is farther from the Sun than Earth, is slower. Consequently, Earth and Mars move in and out of favorable phasing for transfers to and from each other. If the correct phasing were to occur along the reference line indicated on the illustration, it would take approximately 26 months (roughly two years) before the phasing was again appropriate, and it would have advanced (counterclockwise) approximately 50 degrees.

Consequently, over the span of seven of these intervals, the phasing moves completely around the orbit. Every 15 years (two phase revolutions), a complete cycle of phasing characteristics is repeated. Hence, good launch opportunities to Mars repeat every 26 months, and they repeat nearly exactly in character and heliocentric orientation every 15 years.

The planets themselves have both similarities and differences, as indicated in Figure 3B-2. Mars is about half the size of Earth and its gravity is about one-third; that is, one would weigh about one-third as much on Mars as on Earth. Mars has two satellites rather than one, but they are quite small compared to our moon. These satellites are only about 14 (Phobos) and 7.5 (Deimos) miles in diameter. And, because Mars is about 1.5 times Earth's distance from the Sun and moving slower, the planet requires nearly two Earth years (687 days) to complete one orbit around the Sun (one martian year). A martian day (one sol), on the other hand, is only slightly longer (24 hours, 37 minutes) than our own day, and Mars' spin axis is tipped to the orbit plane by 25 degrees -- producing seasons on Mars just as Earth's obliquity does for us.

	EARTH	MARS
Diameter (miles)	7,925	4,125
Surface Gravity (g's)	1	1/3
Satellite (dia., miles)	Moon: 2,160	Phobos: 13.7 Deimos: 7.5
Average Solar Distance (AU)	1	1.5
Planet Year (Earth years)	1	1.9
Planet Day (hrs)	23.9	24.6
Seasons	Yes	Yes
Atmospheric Surface Pressure (bars, msl/datum)	1	1/200 (6.1 mb)
Highest Point	Mt Everest 29,028 ft	Olympus Mons 88,176 ft
Land Area (Earth = 1)	1	0.85
Surface Temperature Range (°F)*	−130° to 140°	−92° to −13°
Primary Atmospheric Gases	N and O	CO_2
Payload Mass Fraction -- Launch to Orbit (%)	≈3	≈20

* The surface temperature range shown for Mars reflects VL-1 data, i.e., northern hemisphere at only 23° N. Globally, as given in the case for Earth, the coldest winter temperatures on Mars are believed to be generally about 148° K (−125° C/−193° F) at the South Pole, and possibly as high as 295° K (22° C/71° F) at the southern midlatitudes during the summer. The winter polar temperatures are cold enough to condense CO_2, and VL-2 (located at 48° N) registered winter temperatures nearly as cold -- or about −190° F. Atypical cold extremes were observed by the orbiters when apparent super cooling (≤130° K/−143° C/−226° F) was briefly detected at each pole during its winter season. This was attributed to the possibility that CO_2 may have been freezing out of the atmosphere faster than it could be replaced.

Figure 3B-2: Comparison of Physical Properties for Earth and Mars

There are some really striking differences between the surfaces of Mars and Earth. For example, the huge martian shield volcano (Olympus Mons) is about three times the height of our tallest mountain (Mount Everest), while Valles Marineris--if placed on Earth--would stretch from New York to San Francisco and the Grand Canyon would be little more than a short ditch on its floor. Therefore, the fact that Mars is the smaller planet has not affected the enormity of some of its features.

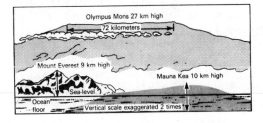

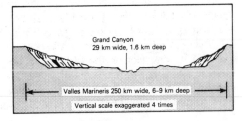

Perhaps surprisingly, however, the land area on Mars is about the same as on Earth. This is because a large portion of Earth's surface is covered by oceans of liquid water not found on Mars. The martian atmosphere is principally carbon dioxide and is much less dense (as little as 1/200th that of Earth's surface pressure for martian elevations at or above Mars datum). As we have learned from Viking data, temperatures on Mars are both much colder and change by a much greater range during the daily and seasonal periods [Leovy, NMC-1G].

Payload vs Gravity Well

For mission analysis, a critical parameter that the mission designer must often worries about is the amount of payload one can get out of the gravity well with a launch vehicle. On Earth, approximately three percent (3%) of what leaves the launch pad ends up as useful payload in low Earth orbit (LEO), while as much as twenty percent (20%) of what leaves Mars can be payload. Clearly, then, it is easier to get off Mars than it is to leave planet Earth, at least in terms of payload mass.

PATHWAYS: ORBIT TRANSFERS

Let's begin by first examining the more traditional set of orbit transfers to and from Mars. Two cases are illustrated in Figure 3B-3, representing what are generally referred to as *ballistic transfers*. This means that a spacecraft basically coasts from one planet to the other, using an impulse provided by a high-energy chemical rocket stage to get started. A spacecraft rocket motor would also be used to achieve Mars orbit insertion (MOI) at arrival and again at departure.

Ballistic Transfer Missions

Free-Return Flyby (Single Spacecraft) -- The first case that I want to discuss is the easiest to perform in terms of energy. It is called the **free-return flyby**. After being launched at Earth, the spacecraft follows a shorter arc to Mars and simply flies by the planet. This provides only a few hours to observe the planet from what can be a fairly significant distance, and then must continue on to complete 1.5 revolutions about the Sun (due to Earth-Mars phasing) before it returns to Earth. That amounts to about three years of flight time for approximately two hours of optimal viewing at Mars, a relatively unsatisfactory arrangement.

Short-Stay/Flyby-Rendezvous (Two Spacecraft) -- Mission productivity can be improved by including a landing on Mars. This Short-Stay/Flyby-Rendezvous mission, also depicted in Figure 3B-3, involves sending a second spacecraft (lander with ascent/rendezvous vehicle) timed to arrive and land on Mars thirty (30) days before the anticipated flyby of the interplanetary spacecraft. Doing so provides a month of surface operation time before the ascent vehicle must take off to rendezvous with the interplanetary spacecraft as it flies by, which proceeds to complete the 1.5 revolutions back to Earth in essentially the same manner as a free-return flyby mission spacecraft. In this case the primary interplanetary vehicle never actually stops at Mars, making it a fairly efficient concept, but it still provides only thirty days of mission time at Mars as a payback for the rather long three-year transfer.

Conjunction Class Mission

The Conjunction Class mission is perhaps the most traditional concept, as well as being the most frequently studied mission type. It is a low-energy mission that has a relatively long Mars

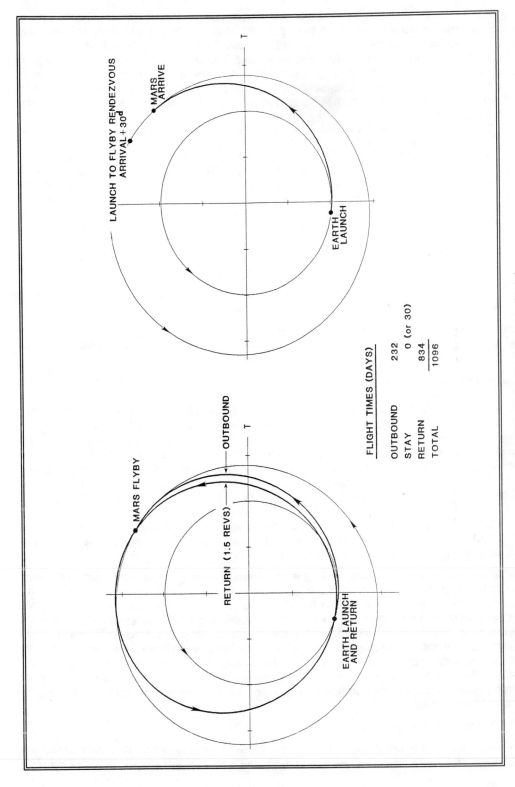

Figure 3B-3: Free-Return Flyby (left), Added Short-Stay With Flyby-Rendezvous (right)

LAUNCH TO FLYBY RENDEZVOUS

MARS ARRIVE

ARRIVAL + 30ᵈ

EARTH LAUNCH

MARS FLYBY

OUTBOUND

RETURN (1.5 REVS)

EARTH LAUNCH AND RETURN

FLIGHT TIMES (DAYS)

OUTBOUND	232
STAY	0 (or 30)
RETURN	834
TOTAL	1096

stay time. As Figure 3B-4 shows, the trajectory transfer path to Mars spans about 180 degrees. At Mars arrival, Earth is moving into conjunction with Mars, hence the name *Conjunction Class* for the mission type. The two planets are out of phase for the return, so a long stay time of more than 1.5 years is required. This might be quite appropriate if there are many things to do, as would be the case after an early exploration stage. When the planetary phasing is again correct, the Mars stay time is followed by a relatively short return flight. The long stay time and the shorter transfer times lead to about 2.8 years of total trip time for one of the lowest low-energy cases.

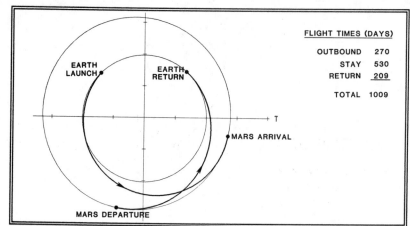

Figure 3B-4:

Conjunction Class Mission

Opposition Class Mission

If a significant improvement in the mission/trip time is desirable, we can move to the other extreme and use the *Opposition Class* mission illustrated in Figure 3B-5. This mission type is a very-high-energy mission. Now we arrive at Mars as Earth is leaving opposition with Mars, hence the identification as an **Opposition Class** mission. The stay time is relatively short, approximately 20 days, and then the spacecraft must rapidly get back onto a return trajectory to catch up with Earth (which is moving out of phase). To do so, the Earth-return vehicle must fall below Earth's orbit and closer to the Sun to achieve the higher velocity needed. This leads to a situation in which the spacecraft catches Earth at a very high approach velocity, which consequently requires higher energies to become recaptured. The advantage of a high-energy mission of this class is that total flight time is reduced to about 1.6 years, but stay time permits only about two or three weeks on the surface of Mars.

To improve on this type of mission, we have found that the addition of a Venus swingby makes shorter trip-time missions more favorable from an energy point of view, as reflected in Figure 3B-6. The same type of outbound transfer is used, and we now stay at Mars for roughly 60 days (two months instead of three weeks) before again beginning our catch-up maneuver on the return flight. But, on the inward leg for this mission case, we find Venus[1]. The swingby at Venus conveniently reshapes our orbit, giving us the velocity change required, and also provides a tangential return at Earth -- reducing the amount of energy needed to slow us down. Total trip time increases slightly to about 1.9 years (about 0.4 of a year greater than the non-swingby mission), but this mission has a much more favorable energy spectrum and provides a more reasonable stay time at Mars.

---

[1] Venus played an important role in reshaping the **Mariner 10** trajectory for its Mercury mission and will soon have a critical role in helping the replanned **Galileo** mission achieve its orbital objectives at Jupiter. {Ed.}

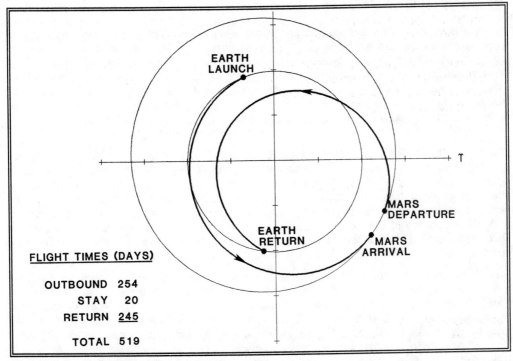

Figure 3B-5: Opposition Class Mission

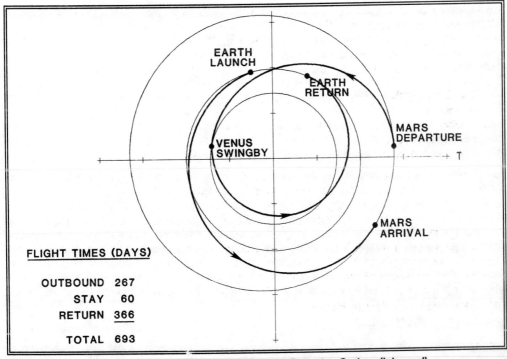

Figure 3B-6: Opposition Class Mission with Venus-Swingby Option (inbound)

Low-Thrust Transfer Missions

The next mission type I will discuss is comprised of those missions we call *low-thrust trajectories*, which are represented in Figure 3B-7. These use an entirely different propulsion system that continues to provide low thrust over a portion of the flight arc between Earth and Mars. The propulsion systems that can be used for this kind of mission are either ion engines or solar sails. Solar sails, as their name implies, utilize the energy found in solar wind; ion engines derive thrust from an electric power system, getting their power either from solar arrays or, in the case of manned missions, from a nuclear power plant. The low-thrust trajectories reflected (Fig. 3B-7) are typical for both of these propulsion concepts.

Low-thrust trajectories differ from the ballistic cases in that they begin with a slow, outward-spiraling orbit of the planet for a considerable period of time after the mission is initiated, before they finally escape Earth's gravity well. They continue to thrust for a portion of the interplanetary flight before the engine is shut down (in the case of ion propulsion) or before setting a solar sail so it no longer flies against the solar wind. The spacecraft then coasts for a significant portion of the trajectory before again beginning to thrust in order to reshape the spacecraft's orbit to that of Mars. As the spacecraft approaches Mars, it again slowly begins a spiraling flight path, this time inward toward its arrival in orbit about Mars.

Stay times range from 100 to 200 days; the example mission illustrated in Figure 3B-7 reflects a 100-day stay time. At the end of the "stop over," the whole process is repeated -- first spiraling outward from Mars to begin the return flight to Earth. The total mission time for a low-thrust mission is about 2.5 years. This is somewhat faster than can be achieved with the least energetic ballistic cases, and there are some specific performance advantages with these kinds of propulsion systems. Their current drawback is that they tend to be more advanced in technology requirements than a chemical rocket and are certainly beyond our present propulsive capabilities.

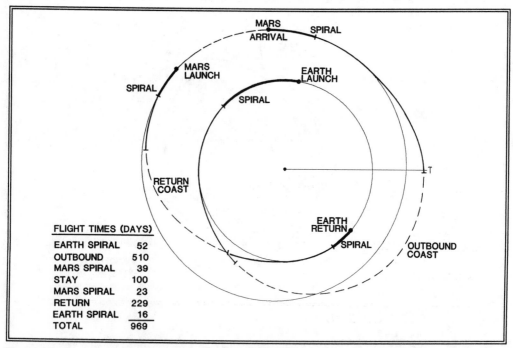

Figure 3B-7: Low-Thrust (transfer) Trajectory Mission

PATHWAYS: CYCLER-ORBIT MISSIONS

Recently, an old idea has been revived which uses a more innovative approach to flying to and from Mars. This promising concept involves what are called *Cycler orbits*. These are orbits that are designed to repeatedly re-encounter both Earth and Mars, such that their application is relevant to a potential Mars base which must be supported continuously with both supplies and people. It allows the establishment of an interplanetary infrastructure; a spaceport orbiting in interplanetary rather than Earth space, cycling back and forth like a shuttle between the two planets. As it passes each planet, a shuttlecraft *taxi* vehicle is launched to rendezvous with the spaceport, carrying supplies and new or replacement crew members. The spaceport itself simply continues to fly along its Cycler orbit.

VISIT (Cycler) Orbits

The first of the Cycler orbits I want to discuss is what we call *VISIT orbits*[2]. This concept was introduced last year at the *Steps to Mars Conference*. The most attractive of the two cases illustrated in Figure 3B-8 is the VISIT-1 orbit, which has a 1.25-year period. This orbit is designed to have a commensurability with Earth of four-to-five (4:5), which means that VISIT-1 goes around the Sun four times while Earth goes around five times. Consequently, the re-encounters at Earth occur once every five years. With Mars, the same orbit has a commensurability of three-to-two (3:2); consequently, VISIT-1 completes three orbits while Mars goes around twice, and re-encounters with Mars occur once every 3.75 years. Working through the Earth and Mars encounters reveals that the VISIT orbits then completely repeat their encounter sequences every fifteen years.

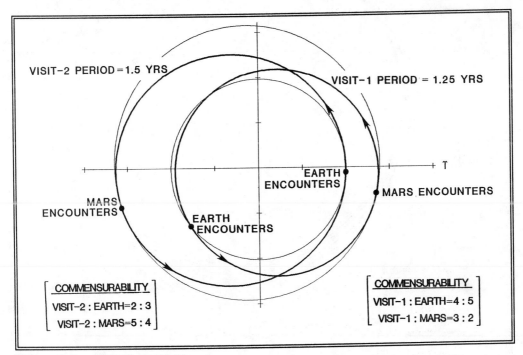

Figure 3B-8: VISIT Cycler Orbits

---

2 V·I·S·I·T = Versatile International Station for Interplanetary Transport.

The VISIT-1 orbit has now been verified through computer analysis. Figure 3B-9 represents a twenty-year propagation of the orbit, showing that it is quite stable and that the encounters have a slow, regressive-encounter location on both of the planetary orbits. Eventually, the orbit has to be re-tuned before coasting again for another twenty years. It does not use planetary swingby effects to perturb the station, other than for navigational purposes, being basically a natural orbit in space. With VISIT orbits, one must arrange two or three of the orbits at several different orientations in Earth-Mars space to create more encounters than one every five years at Earth and one every 3.75 years at Mars. Ultimately, then, a network of these orbits could be utilized to establish a total transportation system to support a base on Mars.

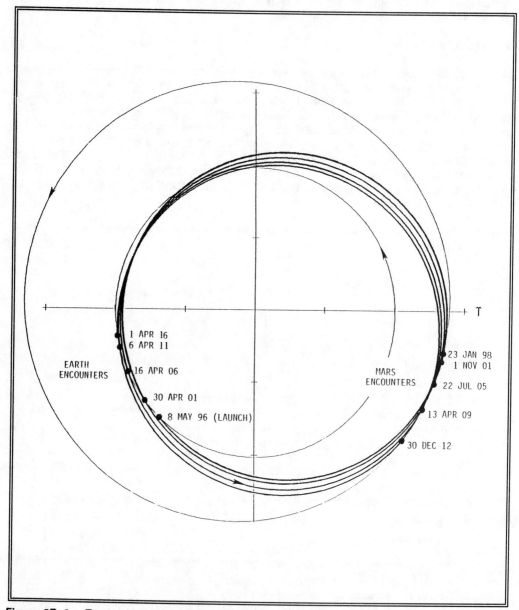

Figure 3B-9: Twenty-Year Propagation of VISIT-1 Orbit

Escalator (Cycler) Orbits

An alternative Cycler concept has been proposed by Dr. Buzz Aldrin (Apollo XI astronaut). His concept, illustrated in Figure 3B-10, has come to be known as the *UP/DOWN-Escalator orbits*. I will first describe how this orbit concept works and then contrast it with the VISIT orbits.

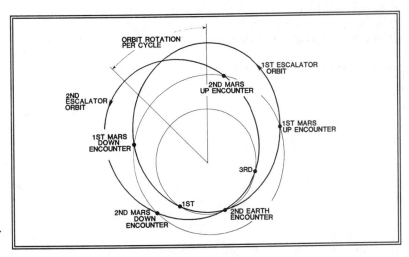

Figure 3B-10:

UP/DOWN Escalator Cycler Orbits

The UP component of an Escalator orbit, originating with the first Earth launch, proceeds slightly inside Earth's orbit at first and then outward toward the first Mars encounter. It is essentially a short transfer "up" to Mars. It continues on by Mars for a longer transfer back "down" through Earth's orbit to the second Earth encounter -- the DOWN component of the Escalator cycle. At this point, a critical gravity-assist maneuver (provided by Earth) is utilized, rotating the major axis of the orbit. This sets the orbit on a new path for the second Mars encounter and, subsequently, for another long return to the next Earth encounter before again repeating the gravity-assist at Earth. What is actually happening with this orbital process is that the orbiting station is using the Earth-swingby effect to precess itself and keep up with the progressive Earth-Mars phasing orientation. It continues to do so as this phasing geometry moves around the Sun.

One thing that can be done to enhance this orbital concept is to change the Mars encounter point. If one alters the approach (UP-Escalator trajectory) to re-time the encounter geometry, the Earth-return loop (DOWN-Escalator trajectory) can be significantly shortened. It then becomes, essentially, a mirrored image of the UP-Escalator. The disadvantage of the Escalator orbit is that it results in fairly high encounter velocities at Earth, as well as somewhat higher energies at the Mars encounters, compared with those of the VISIT orbit. On the other hand, it results in much more regular encounters, which occur once every two years. This suggests that only one or two of these orbits need be utilized rather than the network of three or more VISIT orbit stations.

The Escalator orbit propagation also has been computed, and Figure 3B-11 illustrates its fifteen-year propagation. It is not necessary to follow and fully understand the orbits illustrated, but it is important to note that they do need a little bit of nudging (propulsion) at certain times. Over a fifteen-year period, an accumulation of about two kilometers per second in additional impulse is needed to keep these Escalator stations in phase and on track with the Earth-Mars geometry. Figure 3B-11 simply reflects the characteristics of the DOWN-Escalator trajectory and demonstrates that it has been verified by computer analysis.

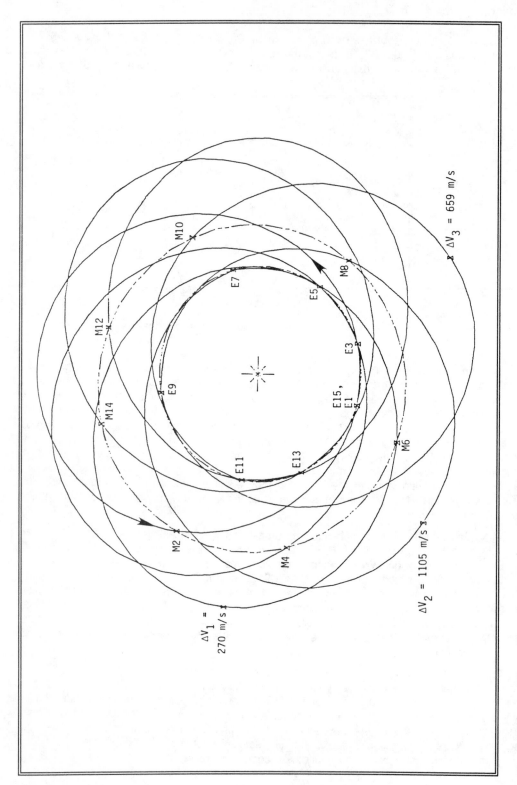

Figure 3B–11: Fifteen–Year Propagation of DOWN–Escalator Orbit

PLANET-CENTERED STRATEGIES

The planet-centered strategies I alluded to in my opening comments are important in developing a manned Mars mission design, and once I have reviewed that aspect of the work, I will touch on some of the performance comparisons.

Earth Staging

Figure 3B-12 depicts two straight-forward staging strategies for departing Earth, and the most traditional way is to begin, quite simply, in a circular LEO. At the appropriate time (at perigee), an impulse is performed that achieves both the correct direction and speed for transfer to Mars. On the other hand, since we may be supplying our staging base with propellants from both Earth and the moon (perhaps delivering oxygen manufactured on the moon, and hydrogen, other supplies, and people from Earth), the L1 libration point represents a particularly interesting point at which to stage. This is because it makes possible the opportunity to split the energy required to supply the staging base from both the moon and Earth.

This escape procedure begins at the libration point with a small, initial impulse, dropping the escape system into a transfer orbit prior to perigee. At perigee, a second impulse is provided to get the system back on the correct escape path. This libration point offers a somewhat lower energy requirement for escape than the conventional LEO, but reaching it from Earth requires a little more energy. The beneficial end result of using this orbit (and the procedures just described), rather than staging at LEO, is that it is more efficient to deliver materials produced on the moon.

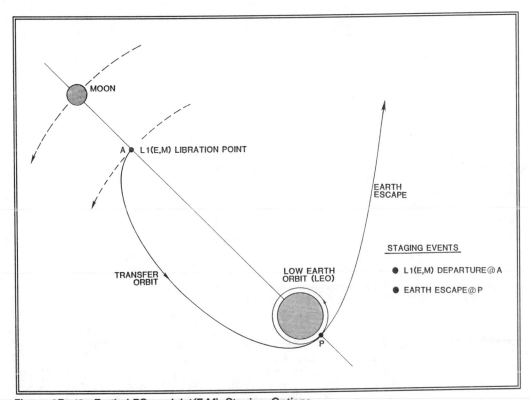

Figure 3B-12: Earth LEO and L1(E,M) Staging Options

Earth-Moon Cycler Orbit -- Figure 3B-13 is a rather complex diagram representing a staging base in an *Earth-Moon Cycler orbit*. This is another of the concepts proposed by Buzz Aldrin. The idea is to fashion an orbit that repeatedly goes back and forth between Earth and the moon, picking up at each encounter supplies and system components needed to build the Mars staging base. In this case, the initial flight orbit incorporates an approach to the moon that results first in a lunar swingby. It then falls back down to perigee where a very small impulse burn is made, placing the staging base in a phasing orbit. It completes two revolutions in the phasing orbit before another very small burn sends it toward another lunar swingby, such that the process can be repeated again and again.

With each lunar swingby, the orbit's major axis is rotated a number of degrees in a retrograde fashion, depending on how large the virtual orbit (dashed arcs in Fig. 3B-13) must be. This is an important characteristic of a Cycler orbit, because one needs to reorient this kind of orbit to match the correct conditions required to escape Earth and achieve a Mars transfer trajectory. This kind of orbit is also a minimum-energy orbit, in the sense that it minimizes the amount of combined energy needed to, for example, put oxygen from the moon on the Cycler in addition to supplies like hydrogen from Earth. However, it is a relatively high-energy orbit relative to escape, so that the escape maneuver is quite low. Clearly, this is a very advanced concept, and it could bear on longer range plans for sustaining a colony on Mars.

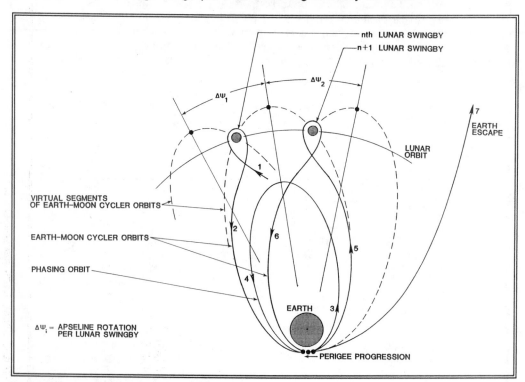

Figure 3B-13: Earth-Moon Cycler Staging Option

Low-Thrust Spiral Departure -- Figure 3B-14 illustrates the final Earth-departure staging option to be discussed, *low-thrust spiral orbits*. Not all of the revolutions of the spiral are shown, but it is clear that an awful lot of time is spent very close to the planet and much less time is spent at the end of the spiral. This has a problem associated with it, in that very long periods of time must be spent in the radiation belts.

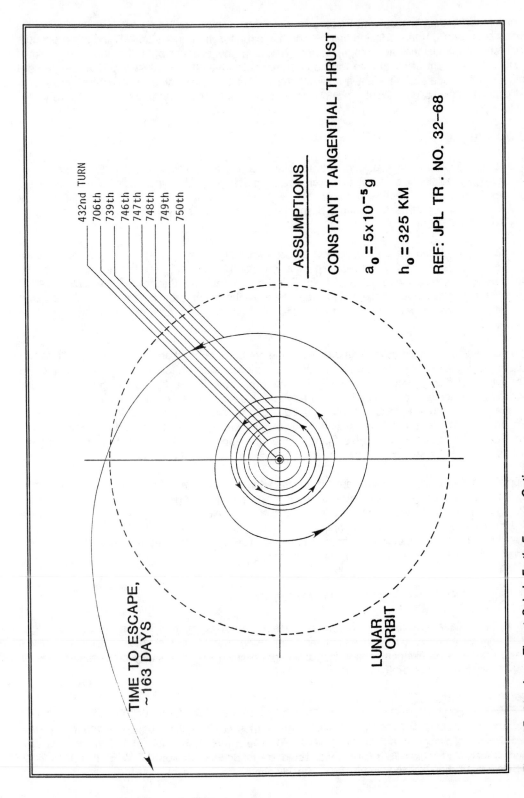

Figure 3B-14: Low–Thrust Spiral, Earth Escape Option

Consequently, people would not board a low-thrust vehicle departing for Mars until after its spiral emerged from the radiation belts -- perhaps at geosynchronous orbit. Using this procedure facilitates an additional advantage, in that boarding the spaceship at this later point would reduce the amount of spiral time for the passengers to only 60 of the 163 days that the process requires. This strategy, incidentally, allows the use of the same type of low-thrust propulsion at capture and escape from the planets as is used in interplanetary space, so it is a very efficient system.

Mars Staging

Figure 3B-15 outlines several concepts for Mars staging at arrival. The first I will discuss is the low-energy, traditional concept in which one impulse is used at periapsis to first place the spaceship in a loose, elliptical orbit. Another small maneuver is performed at the apoapsis point which allows the vehicle to begin a descent trajectory, with entry occurring approximately at Point E (Fig. 3B-15).

An alternative, based on the application of somewhat higher energy, uses a much larger burn at the periapsis, placing the spaceship in a low, circular orbit. This alternative concept then requires a somewhat longer burn at Point B in order to achieve the descent trajectory needed to enter the martian atmosphere. The advantage of the lower orbit is that it reduces phasing problems associated with both the landing site and with the departure, compared with an elliptical orbit.

For resource production as well as the establishment of a staging base to service a Mars colony (but also suggested as a candidate for the first manned Mars mission), the option to stage at the inner satellite of Mars (Phobos) has been proposed. Figure 3B-16 illustrates the orbital maneuvering needed to do so. First, capture is achieved at Mars periapsis, followed by a rise to the circular orbit of Phobos. A second burn circularizes the spaceship for rendezvous with Phobos. Phobos is also attractive as a staging base for an early mission not intended to land on Mars, as suggested by Dr. Fred Singer some time ago. In this sense, Phobos could serve as a base from which automated rovers could be sent to and controlled on the surface of Mars.

Supporting more advanced concepts is the fact that Phobos may have the composition of a carbonaceous chondrite. This suggests that it may contain hydrogen and oxygen that could be mined and manufactured on the satellite, thereafter to be used as propellant. This would help reduce the launch mass requirements for a manned Mars mission. In this sense, then, Phobos may be particularly important in shaping manned missions, at least as a staging base in Mars orbit and possibly as a propellant supply base, as well.

MISSION PERFORMANCE COMPARISONS

Now that I have reviewed the pathway concepts, I will compare the aspects of their flight performance potential (i.e., mass requirements) using the mission example illustrated in Figure 3B-17. The requirement posed in this sample problem is the delivery of 75 metric tons of useful payload to the orbit of Mars.

Setting the Stage

To accomplish this task for our ballistic round trips we will first require a 100-metric-ton spaceship. Following through on that, we would have a total flight mass of 175 metric tons departing Earth, of which 75 metric tons would be off-loaded at Mars. The Earth-return flight weight would then be 100 metric tons, the weight of the spaceship itself.

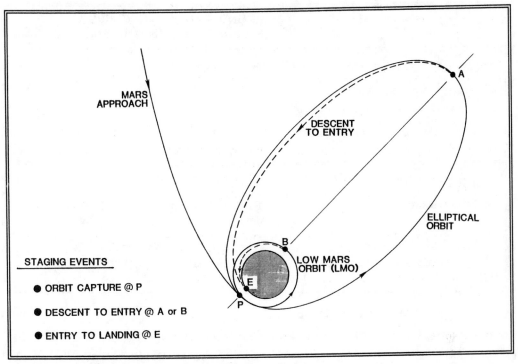

Figure 3B-15: Mars Elliptical and LMO Staging Options

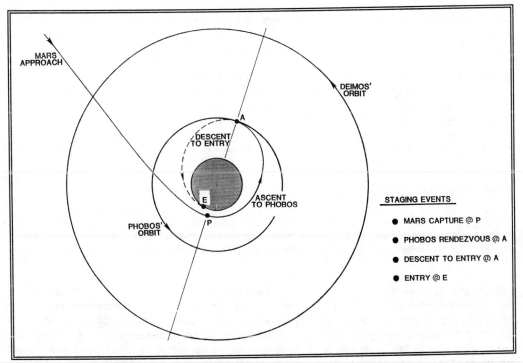

Figure 3B-16: Mars' Phobos Staging Option

	ROUND TRIP (Metric Tons)	CYCLERS (Metric Tons)
PAYLOAD ASSUMPTIONS		
■ Leaving Earth Orbit	175	85
■ Dropped in Mars Orbit	75	75
■ Returned to Earth Orbit	100	10
FLIGHT OPTIONS		
■ Ballistic Round Trips	Flyby-Rendezvous, Conjunction, Opposition, Venus-Swingby	
■ Ballistic Cyclers	VISIT-1, UP/DOWN-Escalators	
■ Low-thrust Round Trip	NEP, Solar Sail	
PLANET-CENTERED STAGING		
■ Earth (ballistic)	Earth-Moon Cycler	
■ Earth (low thrust)	Low Circular Orbit	
■ Mars	Phobos	
PROPULSION		
■ Chemical	H_2/O_2 Stages (460-s I_{sp})	
■ Nuclear-Electric (NEP)	8-MWe System (MPD thrusters, argon propellant)	
■ Solar Sail	25-km^2 (1 g/m^2)	
PROPELLANT OPTIONS		
■ Option 1	From Earth/Moon	
■ Option 2	From Earth/Moon and Mars/Phobos	

Figure 3B-17: Sample Problem for Performance Comparisons

However, if we have an interplanetary staging base operating in a Cycler orbit, we won't need that 100-metric-ton spaceship because much of what it represents is already embodied in the staging base. In fact, the most we might require is a 10-metric-ton shuttling vehicle to get to and from the Cycler base. The payload is still the same, however, so our total departure mass at Earth in this case will be 85 metric tons -- 10 metric tons for the transport and 75-metric tons of cargo).

The ballistic round trips, as previously defined, are represented in the example by three types of missions: Short-Stay/Flyby-Rendezvous, Conjunction Class, and Opposition Class (including the Venus-swingby case). For the ballistic Cyclers, one should consider both the VISIT-1 and the UP/DOWN-Escalator concepts. Also, as previously discussed, the low-thrust round trips are represented by nuclear-electric (ion) propulsion (NEP) systems and solar-sail propulsion systems.

For planet-centered staging, both the ballistic **Earth-Moon Cycler** and the low-thrust spiral cases are compared. Note that this puts low-thrust at a bit of a disadvantage because that system must carry itself up from LEO while the Cycler is already almost free of Earth's gravity well. At Mars, the Phobos staging base is chosen in this example. For chemical propulsion, cryo (hydrogen-oxygen) stages are assumed, and an eight-megawatt magnetic plasma dynamic thruster (with argon propellant) is the energy source selected for the NEP system. The solar sail considered has a five-kilometer (on a side) surface with a density of one gram per square meter.

Two additional options are worthy of special mention, one in which we produce propellant only at Earth and a second that involves the production of propellant at both Earth and Mars. It should be emphasized that these options apply **only** to the chemical propulsion cases.

Results

Figure 3B-18 presents the results. Concentrating first on the ballistic cases (with Earth-moon propellant only), we find that the Opposition Class mission (with a 1.5-year total flight time) requires an enormous amount of Earth-departure mass -- 3000 metric tons. However, the Venus-swingby case dramatically reduces that requirement by more than half, and at a penalty of only about six months additional trip time (increasing it to about two years).

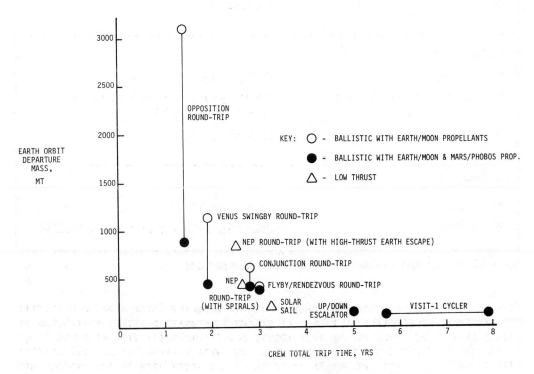

Figure 3B-18: Mass vs Trip Time Comparisons for Manned Mars Missions

When we consider the longer Conjunction Class mission, the flight mass is again halved but the trip time increases to three years. And, for a very simple Short-Stay/Flyby-Rendezvous mission with a stay time of only 30 days, we can again reduce the flight mass to about 500 metric tons. Production of propellant at Mars provides a dramatic reduction in the total mass, but still requires almost 1000 metric tons with the Opposition Class mission. Again, using a Venus swingby halves this requirement, but further improvements are not significant. In fact, relative to the Short-Stay/Flyby-Rendezvous case (in which we don't actually stop at Mars with our large mass), propellant production offers almost no improvement at all.

The NEP case is comparable to the best ballistic cases, even though the NEP case must start from LEO (deep in the gravity well) rather than from the more advantageous Cycler orbit. As an additional note, there is really no advantage to using high-energy thrust to escape Earth's gravity quickly, avoiding the spiral, before starting up the electric propulsion system, because the trip time really doesn't improve much and the mass is nearly doubled. The solar sail concept seems quite attractive. It imposes a slightly longer trip time but at only about half the metric tonnage otherwise required.

The UP/DOWN-Escalator case has a total trip time of five years and represents a very low tonnage when leaving Earth orbit. The VISIT Cyclers are similar but require longer periods of time. It should be remembered, however, that the Cycler orbits cases are (by intention) support-ing a Mars base, and a large portion of the associated mission time is spent at the base (by design). That is why these trip times are as long as they are. The UP/DOWN-Escalator actually spends about four years at Mars and only about a year going to and from the planet. With the VISIT-1 cases, time at Mars may be as much as six years while as much as two years may be spent on the transfer legs in transit.

SUSTAINED MARS BASE OPERATIONS

Figure 3B-19 provides a comparison of sortie (flight time plus stay time) trip times for sustained base operations on Mars. It also provides a little better idea of the total transport and expend-able-mass requirements for the Cyclers[3]. This illustration reflects a comparison of the two types of Cyclers I have discussed, the VISIT orbit and the UP/DOWN-Escalator, plus a simpler Escalator variant -- the DOWN-Escalator (just the DOWN component of the initial Escalator concept). We have also included the Conjunction Class mission, used in this context to support a Mars base, and a low-thrust NEP mission.

I am not going to review all the mission times given (Fig. 3B-19), but one critical aspect that should be noted is the amount of expendables (propellants and supplies) predicted to be con-sumed over a fifteen-year period. It can be seen, for example, that the UP/DOWN-Escalator option requires the highest amount of mass (expendables) at nearly 33,000 metric tons, but it is nearly matched by the Conjunction Class option (30,200 MT). The VISIT orbits reflect some-what longer mission times but require only a little more than 20,000 metric tons of expendables, a profile that is similar to that of the DOWN-Escalator. Interestingly, low-thrust NEP is quite impressive when compared with these large masses; its requisite expendable mass is less than 10,000 metric tons. However, this particular NEP concept requires a power plant capable of about twenty megawatts, which implies a requirement for extremely advanced technology.

---

[3] Previously given mass data represented only a shuttlecraft (taxi) vehicle. {Ed.}

FLIGHT MODE	FLIGHT TIME (YEARS)	STAY TIME (YEARS)	TOTAL TIME (YEARS)	NO. OF INTERPLANETARY TRANSFER STATIONS	TOTAL EXPENDABLES NEED FOR 15-YEAR OPERATION (MT)*
CONJUNCTION	1.6	3.2	4.8	2	30,200
VISIT-ORBIT CYCLER	1.2-6.3	1.6-5.9	5.7-7.9	3	21,500
UP/DOWN ESCALATORS	0.9	4.1	5	2	32,850
DOWN ESCALATOR	2.1	4.4	6.5	1	23,650
NUCLEAR-ELECTRIC LOW-THRUST	2.4	2.4	4.8	2	8,350

* Assumes 20-person average base size; L1(E,M) and Phobos staging nodes; in·situ propellant (H_2 and O_2) production at Earth's moon (O_2 only) and at Phobos and Mars.

Figure 3B-19: Flight Mode Comparisons for Sustained Mars Base Operations

SUMMARY CONCLUSIONS

In summary, it should be apparent that there are many pathways for human flight to Mars. Choices will be dictated by program maturity (exploratory vs sustained-base phase), technology readiness, risk tolerance, and the desired pace of exploration progress. These decisions will also be affected by space program activity in other areas, such as the Earth sciences, astrophysics, robotic solar system exploration, and human activity in Earth-moon space. The ultimate course and pace of a Mars exploration program should evolve from an active national/international process of studies, debate and policy development led by NASA. ■

REFERENCES

This space is otherwise available for notes.

~~~~~~~~

EARTH TO MARS:
Scenarios for Early Manned Missions
[NMC-3C]

Mr. William C. Snoddy
NASA Marshall Space Flight Center

Scenarios for manned missions have been developed as models for the human exploration of Mars, based on technology that is either currently available or is expected to be available in the near term. This discussion presents a description of the transportation and facility elements required, and includes scenarios involving a crew of six with site visits to the surface of Mars and/or Phobos. It is assumed that the space station would be used in the assembly of the Earth-Mars transit vehicle, which would weigh in excess of one million pounds when fueled. Cryogenic propulsion, together with aerobraking both at Mars and Earth, are used in the scenarios. Total mission durations range from 12 months (for a flyby) to 36 months. Artificial gravity technology, though representing a complex technical challenge, is reasonably possible if needed. Indeed, these scenarios suggest that there are no major technological impediments to the execution of missions allowing the exploration of Mars by humans.

INTRODUCTION

After a discussion of mission concepts like those described by John Niehoff in his **NASA Mars Conference** presentation, which defined pathways (trajectories and mission types) for manned missions to Mars [NMC-3B], one is naturally left with the question of what can realistically be undertaken relative to present and near-term technologies. I will begin to address that question by explaining our current thinking about how we might "get on" with the exploration of Mars. Figure 3C-1 presents an overview of what I will discuss during my presentation, the essential purpose of which is to generally define the more conventional ways to transport humans to Mars. I have pulled together a number of different concepts for the purpose of this presentation, and I will treat them as scenarios by examining each in the context of a series of charts that illustrate their technical parameters.

Mission and Science Rationales

Figure 3C-2 summarizes the mission and science objectives to provide a fairly complete picture of our rationale for manned missions to Mars. However, I won't dwell on many of the items outlined because the majority are reviewed in other **NASA Mars Conference** presentations. We are assuming, for the missions I will be discussing, that the martian satellites are desirable places to visit and that we will want to explore at least one of them (probably Phobos) very early in the manned exploration of Mars. Indeed, investigations of the satellites could represent a primary or even exclusive objective of several early missions, including the first.

OVERVIEW

- Objectives & Science Considerations

- Mission Types & Comparisons

- Space Station Implications

- Space Vehicle Configurations & Sizing Data

- Surface Systems Options

- Technology Implications

- Summary

Figure 3C-1: Overview of Scenarios for Early Manned Missions to Mars

MISSION OBJECTIVES:

SCIENTIFIC

- Understanding of Mars, Earth and Solar System
- Exploration of Mars and Vicinity
- Improved Understanding of Long-Duration Space Habitability, Effects on Humans (psychological, social, et·cetera)
- Search for Extraterrestrial Life (past or present)

POLITICAL

- Retention and Enhancement of U.S. Preeminence in Space
- Promotion of International Space Cooperation

HUMANISTIC

- Improvement of Quality of Life on Earth Through Advances in Science, Technology, and International Civilization
- Extension of Frontiers of Civilization

UTILITARIAN

- Development of Long-Life Space Transportation and Facility Infrastructure Systems
- Technology Advancement
- Commercialization of Space Activities
- Utilization of Space Resources

SCIENCE OBJECTIVES:

BASIC SCIENCE -- To Understand the Structure, Dynamics and Evolution of:

[1] Mars and Its Atmosphere

[2] The Martian Environment, Including Phobos and Deimos

[3] Martian Life Forms, Present and/or Past

[4] The Astronomical Universe

APPLIED SCIENCE -- To Provide Information on:

[1] Martian Resources (water, propellants, breathables, construction materials, power, fertilizer, et·cetera)

[2] Martian Weather (dust storms)

[3] Radiation Hazards (solar flares)

Figure 3C-2: Mission (top) and Science (bottom) Objectives for Scenarios

Mission

As expressed during the Session·2 panel discussion moderated by Carl Sagan[1], and as discussed at length by others during the conference, there are many reasons involved in the decision to develop manned missions to Mars. It is not clear at this time which of these will ultimately provide the basis for the first mission, and different groups will certainly use different reasons to support their rationale. There are, nevertheless, many excellent reasons for going to Mars, and I think it is clear that science will play an important and preeminent role in the initial manned exploration of the **Red Planet**.

Science

Figure 3C-3 gives one a sense of the vicinity of Mars and presents a quick review of the essential things we already know about Mars and its satellites. The numbered "X" points on the globe of Mars (upper right) are scenario landing sites that I will discuss in greater detail shortly.

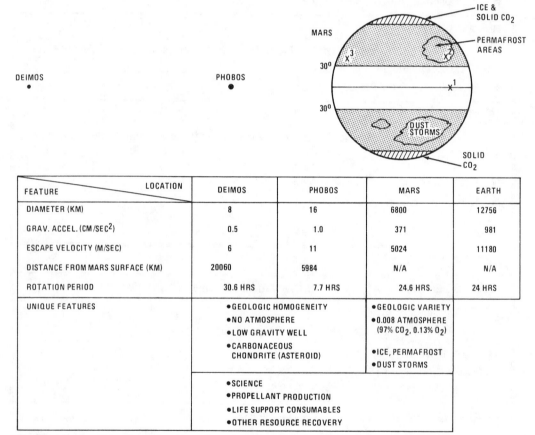

FEATURE	LOCATION	DEIMOS	PHOBOS	MARS	EARTH
DIAMETER (KM)		8	16	6800	12756
GRAV. ACCEL. (CM/SEC^2)		0.5	1.0	371	981
ESCAPE VELOCITY (M/SEC)		6	11	5024	11180
DISTANCE FROM MARS SURFACE (KM)		20060	5984	N/A	N/A
ROTATION PERIOD		30.6 HRS	7.7 HRS	24.6 HRS.	24 HRS
UNIQUE FEATURES		•GEOLOGIC HOMOGENEITY •NO ATMOSPHERE •LOW GRAVITY WELL •CARBONACEOUS CHONDRITE (ASTEROID)		•GEOLOGIC VARIETY •0.008 ATMOSPHERE (97% CO_2, 0.13% O_2) •ICE, PERMAFROST •DUST STORMS	
		•SCIENCE •PROPELLANT PRODUCTION •LIFE SUPPORT CONSUMABLES •OTHER RESOURCE RECOVERY			

Figure 3C-3: Mars Vicinity (Characteristics and Environments)

---

[1] The Session·1 and Session·2 panel discussions held during the **NASA Mars Conference** have not been included in the proceedings document due to the amount of transcription material involved and the anticipated time required to complete a satisfactory editorial review with the participants. It is not yet known whether or how they might be made available at a later date. {Ed.}

As previously noted (Fig. 3C-2), there are two types of sciences to be considered in developing a science rationale and plan for the human exploration of Mars. The first, *basic science*, represents the essential product of planetary exploration: to better understand the universe, our solar system, and Mars itself. The second is identified as *applied science*, although some refer to this classification as **operational science** because of its more specific nature. It involves the acquisition of information about Mars (its environment and resources) that will bear directly on mission planning, not only from the standpoint of making the first mission successful but to lay the groundwork for future missions, as well. If we are going to produce propellants and other materials on the surface of Mars or on its satellites (O_2 or building materials, for example), we need to know much more about the physical nature and composition of the potential resources in order to assure ourselves that they are actually available and accessible, and to determine how we might go about getting at them. In addition, this involves understanding the environment in which the work of developing those resources will have to be conducted (e.g., low gravity on the satellites or potentially reactive nature of surface materials when handled).

SCENARIO DEFINITION

Reaching and exploring the surface of Mars is of course the purpose of the whole program. In addition to the planetary characteristics given in Figure 3C-3, three landing sites used as models for calculating entry and landing trajectories associated with our scenarios are also illustrated. One of these (X^1) is on the equator and two more can be seen above it at about 35°N. One of the northern sites (X^2) was selected because of evidence of permafrost in that region, and the other (X^3) is of interest because of its geologic/topographic character. To provide a better sense of the regional topography associated with these sites, they are again identified on the topography maps of Mars presented in Figures 3C-4; the first two sites are characterized in the top chart, while the third is similarly illustrated in the bottom chart.

Mission Types

Figure 3C-5 essentially reviews several mission types in a summary fashion. First, of course, we have the possibility for performing flyby missions, with mission times as short as one year if we conduct our encounters appropriately and use a lot of propulsion to speed things up. Another possibility is based on going into orbit around Mars, a third alternative is to land on one of Mars' satellites, and a fourth alternative is to land on Mars itself. In all but the first of these alternatives, we have the option of traveling on either a *Conjunction* class or *Opposition* class trajectory, with typical travel times and stay times as shown in the figure.

I have highlighted the latter two mission types (dashed-line boxes in both segments of Fig. 3C-5), the Phobos or Deimos (PH or D) landing and the Mars landing, because they offer more science value and are the ones I will primarily focus on during my presentation. In fact, an evolving program will probably utilize a combination of these, so I will discuss a mix that might be accomplished in a coordinated fashion. Also, it is important to note that--in keeping with the technology available for early missions--the more traditional propulsion technology (cryogenic) is the only type considered in developing these scenarios, although others will be briefly discussed.

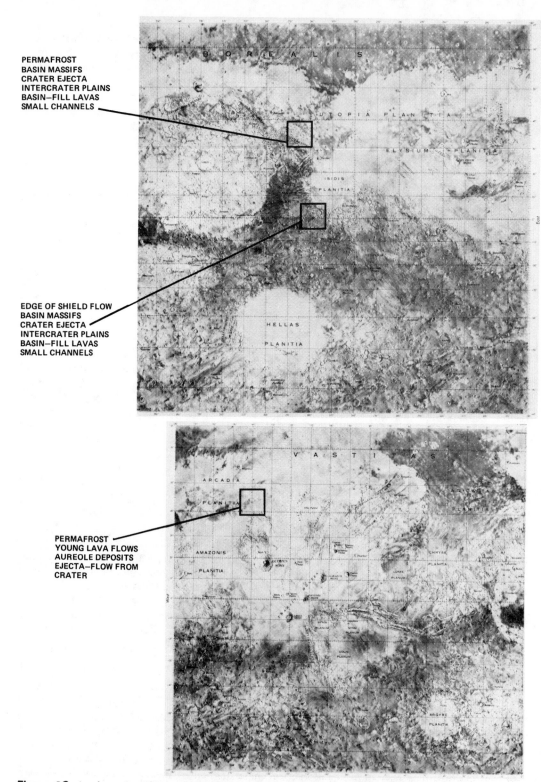

PERMAFROST
BASIN MASSIFS
CRATER EJECTA
INTERCRATER PLAINS
BASIN—FILL LAVAS
SMALL CHANNELS

EDGE OF SHIELD FLOW
BASIN MASSIFS
CRATER EJECTA
INTERCRATER PLAINS
BASIN—FILL LAVAS
SMALL CHANNELS

PERMAFROST
YOUNG LAVA FLOWS
AUREOLE DEPOSITS
EJECTA—FLOW FROM
CRATER

Figure 3C-4: Location/Characterization of Scenario Mars Landing Sites

410

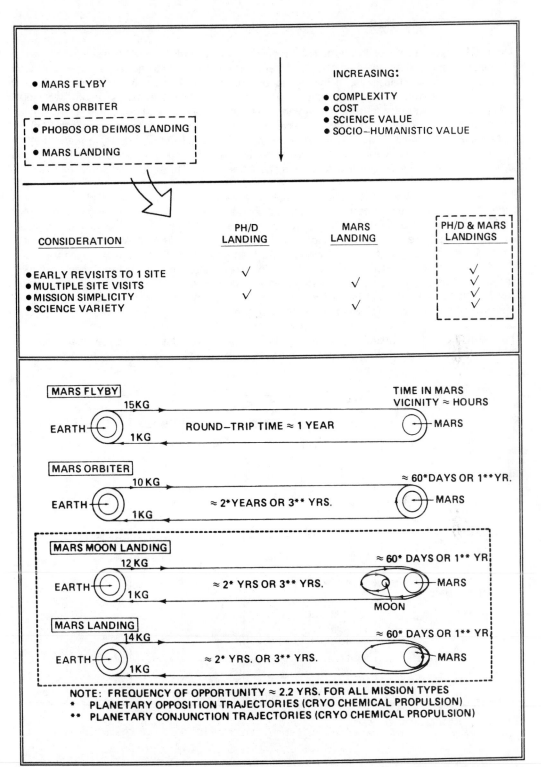

Figure 3C-5: Early Manned Mars Mission Types

411

Scenario Characteristics

Figure 3C-6 gives some of the essential aspects of the scenarios I will be discussing. Two of the scenarios are in fact demonstrated, the first represented by three missions and the second by two. For example, Mission-1 of the Scenario-1 group has a relatively short mission time of about two years, with a sixty-day stay time in the Mars vicinity. In this scenario, we would perhaps place a planetary transfer space vehicle (six-man crew) in a Mars parking orbit, where-after a smaller rendezvous vehicle would leave the main spaceship and rendezvous with Phobos, for example, to conduct some studies. That same vehicle would then go down to Mars to study a site on the planet's surface before returning its crew to the transfer vehicle for the trip home. This is a rather Apollo-like approach, as indeed all of these are to some extent.

The Scenario-1, Mission-2 concept represents a mission based on the idea of staying for a longer period of time (one year) -- a **Conjunction** class mission. We again have a crew of six, but it is assumed that all six will go down to the surface of Mars because of the long stay time, remaining there until it is time for Earth return. I don't believe they would want to stay in orbit for an entire year, having to spend more than enough time in the interplanetary space-ship during the transit periods. They might possibly spend their time at a site established during the earlier mission, or create a new site (perhaps one visited and studied on the first trip). The third mission of this Scenario-1 set would then essentially repeat the second, going to either of the two previous sites or again visiting a new one.

The scenario could be enhanced by including two excursion vehicles; in the cases just described we included only one. If we have two excursion vehicles and a crew of six, we would have

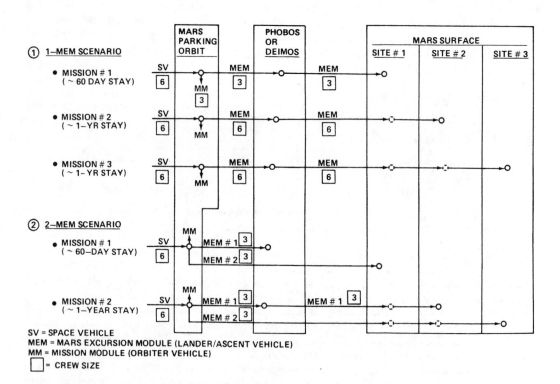

Figure 3C-6: Scenario Options for Early Manned Mars Missions

some options relative to how the vehicles are used. For example, one of them could visit one of the martian satellites while the second visited the surface of Mars, and that would give us certain synergistic capabilities. To perpetuate these benefits, we would again have two excursion vehicles during the second mission. They could then be used to visit two additional sites on the surface, for example, exercising their mission utility in another way. Scenario-2 is more "efficient" than Scenario-1, since three different Mars sites (as well as Phobos or Deimos sites) could be visited in two missions rather than three. I will describe these vehicles in greater detail later in the presentation, and will then provide a comparison of vehicle weights for the missions I have described.

SCENARIO SYSTEMS AND CONFIGURATIONS

With these kinds of scenarios in mind, Figure 3C-7 illustrates how such missions might unfold. Note that the Mars space vehicle is assembled with a high-energy stage needed to boost the vehicle out of Earth orbit and onto its Mars transfer orbit. We could perhaps recover the transfer stage, but that idea still needs further study to determine whether it would be cost effective to do so[2].

First, a lot of mass must be placed in Earth orbit no matter which of these missions is being considered or what kind of propulsion technology is applied (e.g., whether nuclear-electric or cryogenic). We very likely would assemble it at or in the vicinity of the space station. We will, of course, need launch vehicles capable of carrying that amount of mass into orbit, as discussed in greater detail by others at this conference. The mass of the kind of vehicle reflected in these scenarios amounts to more than a million pounds for the Mars space vehicle alone, and we *must* have launch vehicles that can carry this sort of mass into orbit if we are to conduct such missions as efficiently and economically as possible. The left portion of Figure 3C-7 depicts the trans-Mars vehicle arriving at and orbiting Mars, where the MEM (Mars excursion module) vehicle(s) would visit the martian satellites and/or the planet itself. Some portion of each MEM would serve as a launch and rendezvous (ascent) vehicle for returning to the orbiting spaceship prior to the Earth-return flight, and that departure would itself be facilitated by a departure rocket (right portion of Fig. 3C-7). This again is clearly a scenario that--at least in conceptual terms--duplicates the Apollo Lunar Module (LM) missions of the early 1970's.

---

2 As the concept for Space Shuttle and the Space Transportation System (STS) was maturing during the early 1970's, studies were conducted relative to the idea of developing a high-energy upper stage that would function as a "Space Tug" for STS deep space missions. It was hoped that such a vehicle, which is normally a very costly element of the launch system, could be recovered and reused. High-energy upper stages used to inject deep space probes like the Viking and Voyager spacecraft into their interplanetary trajectories (e.g., Centaur) typically trail along on deflected trajectories after separation and are therefore lost. Proven upper stages (advanced but unavailable growth configurations), representing the three primary propulsion technologies available--cryogenic (LH_2/LO_2), storable hypergolic liquid (hydrazine/nitrogen tetroxide), and solid--were studied. Space Tug (growth) configurations of Centaur and Transtage technologies were among those considered. Although the Space Tug concept has not been fully realized, the studies did lead to an intermediate concept based on solid rocket technology that ultimately resulted in the Inertial Upper Stage (IUS). The Transfer Orbit Stage (TOS) and Apogee and Maneuvering Stage (AMS) being developed for Orbital Sciences Corporation by Martin Marietta also reflect a partial response to some of the mission concerns that motivated the original Space Tug studies. These new stages can be used independently or together to satisfy a variety of mission requirements, and they are design-compatible with both STS and ELV (T4) systems -- as is IUS. {Ed.}

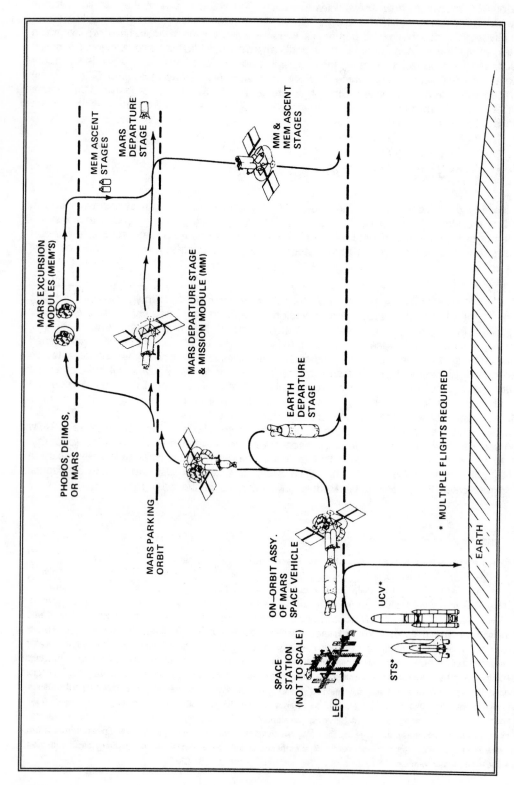

Figure 3C-7: *Typical Early Mars Mission Event Sequence*

Space Station Involvement

As outlined in the top portion of Figure 3C-8, our scenarios reflect implications for the NASA Space Station (Block·1 and beyond). Indeed, Mars missions will benefit significantly from space station technology, concepts, designs and experience, and all of these factors will feed directly into long-term program planning and design. In particular, space station experience in working with things like assembly, integration, logistics and operation--in a long-duration, zero-g environment--will first help us develop and build the interplanetary systems and then to prepare for and cope with the Mars missions. For example, space station experience would help us determine if we need to incorporate the capability to generate artificial gravity enroute or whether we can get along without it. The space station itself is, of course, very important as a staging base for departure and return, and the bottom portion of Figure 3C-9 is a conceptual depiction of these aspects of mission operations.

Interplanetary Manned Mars Space Vehicle

The two line illustrations presented in Figure 3C-9 (top and bottom) represent two very similar (except for differences primarily in the departure stages) configurations of the manned Mars space vehicle seen being built up at the space station in Figure 3C-8. It should be understood that these are zero-g configurations. Large cryogenic Earth and Mars departure stages, to be used for departing parking orbits at both planets, are identified. The Earth departure stage is jettisoned after its use, and the spaceship completes its transit to Mars with only the Mars departure stage still attached.

While the two configurations illustrated in Figure 3C-9 appear to be identical, the first (top) is for a year-2001 **Opposition** class mission while the second is for a **Conjunction** class mission in year 2003. There are differences in the departure stages, as previously noted, particularly in the Mars departure stage which must be significantly larger and more powerful for the 2003 mission.

Aerobrake -- Located at the nose of the vehicle, just ahead of a pair of large solar array panels provided for electric power generation, is the **aerobrake**. A large, bowl-like structure, it is used to help slow the vehicle (using atmospheric drag) at Mars, and again at Earth following the return flight. It is identified on both of the line illustrations (Fig. 3C-9), and it is seen being assembled in the conceptual illustration (Fig. 3C-8). By using aerobraking techniques at Mars and Earth[3], we can greatly reduce the mass of the whole system -- perhaps by as much as sixty percent for some trajectories. For this reason it is a particularly attractive technology.

The locations of three modules that would serve as crew habitats and laboratories during at least the transit flights to and from Mars are pointed out in both line illustrations. They are located behind the aerobrake and solar panels (two of them are also visible in the conceptual illustration at the bottom of Fig. 3C-8, numbered 1 and 2). These modules might, in fact, be derivations of the personnel modules used on the space station itself.

The MEM-1 and MEM-2 excursion vehicles are docked above and below the center line behind the aerobrake. MEM-2 is the lander module in this case, and would have its own aerobrake -- perhaps reflecting some Viking descent capsule and Apollo technology. MEM-1, assumed for the purpose of these illustrations to be only a rendezvous vehicle for exploring the martian satellites, will not encounter atmosphere and therefore does not need an aerobrake. It would serve as living quarters for its crew during the period spent in rendezvous with a satellite.

---

3 Aerobraking techniques are described in greater detail in these proceedings by James French [NMC-2F] and Doug Blanchard [NMC-2G] in association with discussions of unmanned sample return missions (second session of the NASA Mars Conference). {Ed.}

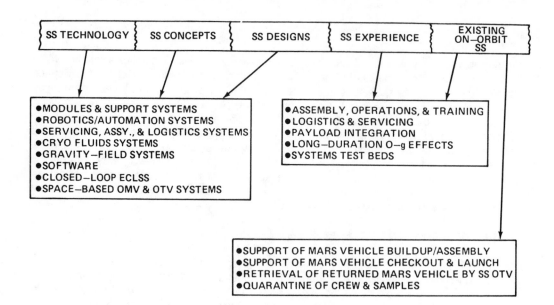

SS TECHNOLOGY	SS CONCEPTS	SS DESIGNS	SS EXPERIENCE	EXISTING ON—ORBIT SS

- MODULES & SUPPORT SYSTEMS
- ROBOTICS/AUTOMATION SYSTEMS
- SERVICING, ASSY., & LOGISTICS SYSTEMS
- CRYO FLUIDS SYSTEMS
- GRAVITY—FIELD SYSTEMS
- SOFTWARE
- CLOSED—LOOP ECLSS
- SPACE—BASED OMV & OTV SYSTEMS

- ASSEMBLY, OPERATIONS, & TRAINING
- LOGISTICS & SERVICING
- PAYLOAD INTEGRATION
- LONG—DURATION O—g EFFECTS
- SYSTEMS TEST BEDS

- SUPPORT OF MARS VEHICLE BUILDUP/ASSEMBLY
- SUPPORT OF MARS VEHICLE CHECKOUT & LAUNCH
- RETRIEVAL OF RETURNED MARS VEHICLE BY SS OTV
- QUARANTINE OF CREW & SAMPLES

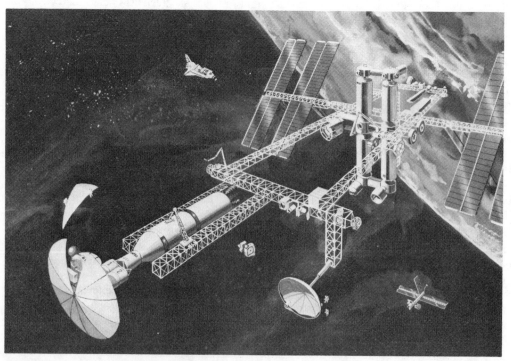

Figure 3C-8: Space Station Implications (top);
Conceptualization of Manned Mars Vehicle Assembly at Space Station (bottom)

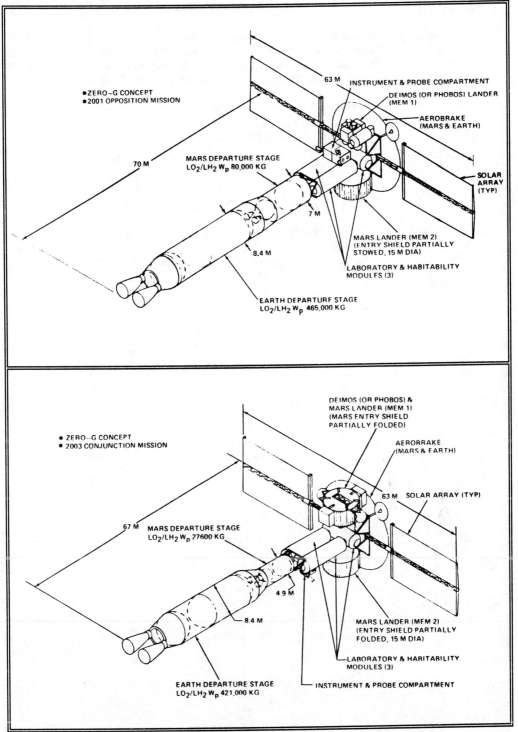

Figure 3C-9: 0-g Manned Mars Vehicle Concepts for Two-Lander Parallel Mission Scenario (top, Year-2001 Opposition Mission; bottom, Year-2003 Conjunction Mission)

417

Artificial Gravity

If it is determined that we must be able to generate artificial gravity enroute, it would probably be achieved using a rotational scheme of some kind -- as suggested by the configurations depicted in Figure 3C-10. In these concepts, the habitats would be extended outward in equilibrium from the center-line mass and would rotate about it. This is *NOT* something we would want to

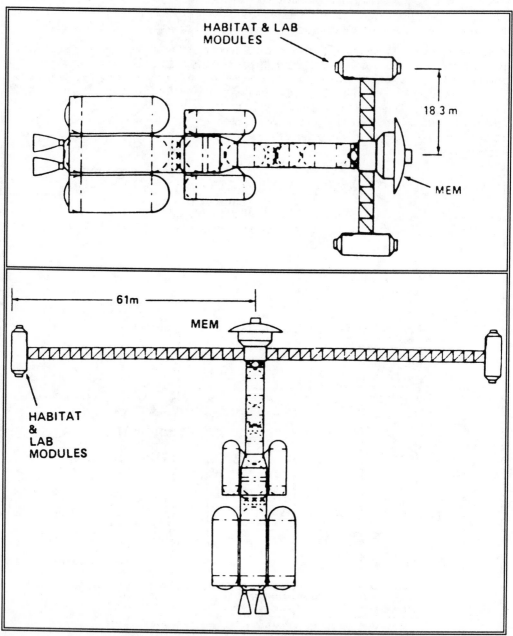

Figure 3C-10: Artificial-Gravity Configurations of Manned Mars Space Vehicle (top, 0.4-g option; bottom, 1-g option)

do unless it appears to be very necessary. For one thing, we would need to be able to pull the modules back in behind the aerobrake when approaching Mars, because the brush with the atmosphere would clearly be a destructive encounter for the unprotected modules. Also, personnel could not easily move between the modules because they are not connected environmentally; crew members could not, for example, live in a couple of them and work in another without taking on the risks and problems associated with frequent extra-vehicular activity. Rotation rates of 4 rpm were assumed for the illustrated configurations; later concepts have utilized a 2-rpm rate to further reduce coriolis effects. The lower rpm rate translates into much larger separation distances for the elements, and connecting structures tend toward being tethers.

There are a number of technical problems associated with the generation of artificial gravity for a system of the magnitude represented in our scenario, and they make one reluctant to take the route just described unless absolutely necessary. However, if some degree of artificial gravity *is* found to be necessary, it can be done and we will do it, but zero-g effects clearly must be better understood before we make significant decisions about how to achieve it. Because more work must be done on the gravity issue, both the space shuttle vehicle (SSV) and the space station will once again be important to the improvement of our knowledge relative to planning manned Mars missions.

MASS PERFORMANCE AND LAUNCH VEHICLE CONSIDERATIONS

Figure 3C-11 translates the missions I have discussed into space vehicle mass (millions of kilograms) for a year-2003 Conjunction class mission. One can see that the propellant requirement

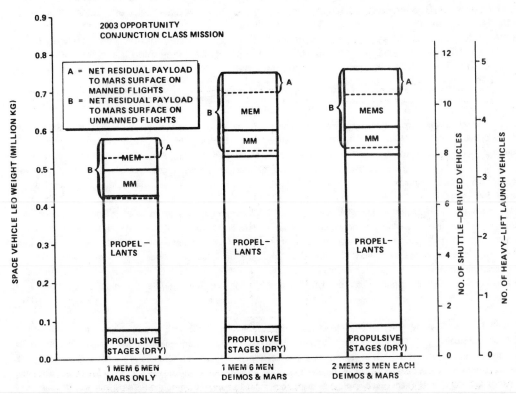

Figure 3C-11: Summary of LEO Mass for Manned Mars Vehicle and Projected Launches Required

is a very large part of the total mass, regardless of which mission type is being considered. A scale reflecting the number of flights needed to get all of the mission mass up into low Earth orbit (LEO) prior to departure for Mars is given at the right for conceptual SSV-derived and heavy-lift launch vehicles.

Putting this kind of mass into orbit is a major program in itself, and then many trips must be made just to fuel the Mars vehicles. If we were to use SSV as it now exists, it might take as many as thirty trips -- not a very efficient way to go about it. We will probably need a more powerful SSV-derived vehicle (SDV), or alternatively, we might want to develop a new heavy-lift (expendable) launch vehicle (HLLV) of the Saturn V class (or larger) to use for large-mass launching -- as the Soviets appear to be doing[4].

Launch Vehicles

Figure 3C-12 shows what some of the vehicles presently being considered in various studies might look like, and gives one a sense of their relative sizes. The new Titan IV (T4), referred to during its development as the T34D7 or T/CELV (Titan Complementary Expendable Launch Vehicle), is not represented here [for T4 performance data, see NMC-2F][5].

The launch vehicle on the far left (1) is the present STS SSV configuration. The SDV (2) is different from the current STS configuration in that the space shuttle orbiter is deleted and the external tank (ET) essentially becomes part of the core-stage booster with a cargo container on top. Another SDV concept being studied is one that attaches the cargo container to the ET much like the shuttle orbiter is currently attached; but, while it looks a little like a shuttle without wings or windows, its cargo bay is significantly larger. SDV configurations employ a high degree of STS element utilization, including the solid rocket boosters, most of the ET, and the SSV engines and avionics. The heavy-lift HLLV concept shown (3) has considerably more capability than a Saturn V. The Shuttle II (4) would be used primarily for carrying crews up into orbital space and to the space station, essentially functioning as a transfer or taxi vehicle. The final vehicle illustrated, (5) the Spaceplane, would also be a crew-transfer vehicle; it is based on advanced technology and represents a lower operational cost.

---

[4] Though developing a launch system and vehicle expected to be comparable to America's STS SSV, the Soviets have typically utilized what has been referred to conceptually in the United States as "big dumb boosters" [Arthur Schnitt, c1960-61]. The Soviet SL series of boosters, ranging from the SL4 (used to carry cosmonauts to the Soviet space station) to the largest and latest of the Proton rockets, the SL16 (believed comparable in mass performance to the Titan IV), have been "low-tech" systems. That is, these vehicles typically reflect significantly lower levels of propulsion and fabrication technologies in that they utilize common rather than exotic (costly) materials in their fabrication or as fuel. For example, steel is used in their fabrication and storable kerosene has been their primary fuel. Because of the simple nature of the fuel, propulsive systems and engines can be quite simple (dumb) compared to those that must contend with a difficult fuel like hydrogen. The Soviet philosophy has been well demonstrated through a busy launch schedule, with respectable reliability and cost performance. It is also represented to a lesser extent in the Soviet's new heavy-lift Energia (which also will provide launch propulsion for the USSR space shuttle). Although Energia (not quite as powerful as a Saturn V) is believed to burn hydrogen in its main engines, its booster engines continue to burn kerosene. {Ed.}

[5] The new Titan IV (T4) launch vehicle, designed to be compatible with upper stages developed for the STS SSV (including IUS, Centaur-G' and TOS/AMS), provides impulse performance increases in all three of its primary propulsive elements -- two seven-segment solid rocket boosters (SRB's) as one stage and two core stages. The core stages utilize storable liquid hydrazine and nitrogen tetroxide as a storable hypergolic fuel. Previous SRB-equipped Titans used slightly smaller core stages and five-segment SRB's. Martin Marietta also has studied a Titan LDC (large-diameter-core, 15 ft vs 10 ft) concept that could utilize two or four SRB's. {Ed.}

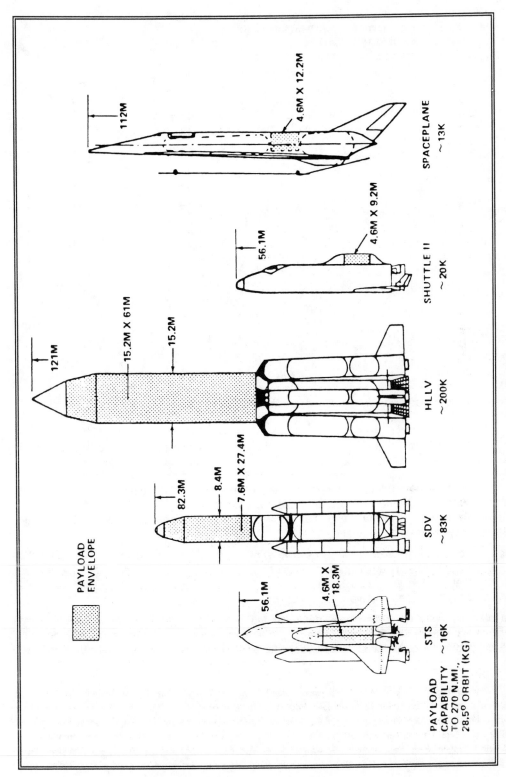

Figure 3C-12: Earth-to-Orbit Launch Vehicles (present and conceptual)

421

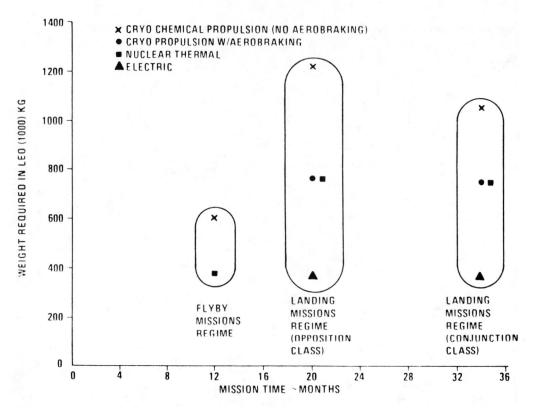

Figure 3C-13: Propulsion Performance Implication Comparison for Mars Flight Vehicle

Figure 3C-13 (above) is a very simplistic performance comparison of the different types of propulsion systems that could be used (departing from Earth orbit). If, for example, one considers the case of cryogenic-chemical propulsion (without aerobraking and designated by the "X" symbol in Fig. 3C-13) for the second class of missions, i.e., landing missions regime for the Opposition class, one can see that the total mass required in Earth orbit--to facilitate propulsive braking first at Mars and then back at Earth--is the greatest by far of any of the propulsion systems demonstrated.

If, on the other hand, aerobraking is used to slow down, first when arriving at Mars and then at Earth following the return flight for the same mission class (the "bullet" symbol), then one can see that the required initial mass is drastically reduced and proves to be comparable to that for a nuclear thermal system[6].

The best performer of all, as clearly demonstrated, is nuclear electric propulsion (NEP), which employs ion thrusters. NEP is quite attractive in terms of total mass for the same mission class, with specific impulses of 2,000-10,000 seconds. However, it represents a fairly advanced state

---

[6] Nuclear thermal rockets would use fission-energy reactors (operating at high temperatures) to develop high-thrust propulsive energy rather than low-thrust electric (ion) propulsion. Their specific impulses would be in the range of 800-2000 seconds (depending on design and propellant). Types of thermal nuclear rockets include: solid or liquid core, rotating fluidized-bed, and open- and closed-cycle gas core systems. The most common propellant is hydrogen, since its low molecular weight allows high exhaust-gas velocity. Principal operational issues are thermal cooling (chamber and nozzle), containment/control of the fission mass, and radiation shielding. {Ed.}

of technology, and it is for this reason that I have not discussed it for the purpose of this presentation. Conceptual configurations for space vehicles powered by both kinds of nuclear propulsion systems are shown in Figure 3C-14.

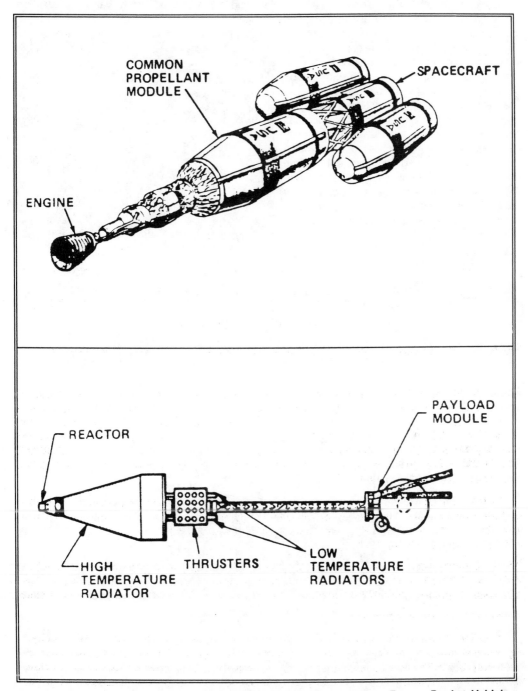

Figure 3C-14: Nuclear Propulsion Vehicle Concepts (top, Nuclear Engine Rocket Vehicle Application, NERVA; bottom, Nuclear Electric Propulsion, NEP)

SCENARIOS AT WORK

Figure 3C-15 is an artist's conception of our fully assembled space vehicle departing the space station for Mars on a two- or three-year mission, and again I would like to call your attention to the crew modules. Figure 3C-16 gives one a sense of what a laboratory module might be like. Representative of a space station module, it shows what might be a typical layout for a materials technology lab. I thought a laboratory depiction might be appropriate to use because one of the activities the crew might undertake during the long missions (assuming that some form of artificial gravity is not required) is research associated with a long-duration, zero-g environment. When traveling in the interplanetary medium, one will most certainly experience extremely low gravity. Indeed, except for the disturbances caused by the crew members themselves, it will perhaps be much lower than anything we could achieve in Earth orbit.

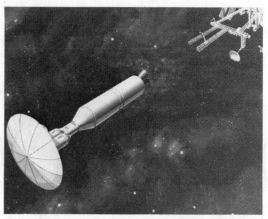

Figure 3C-15: Departing Space Station

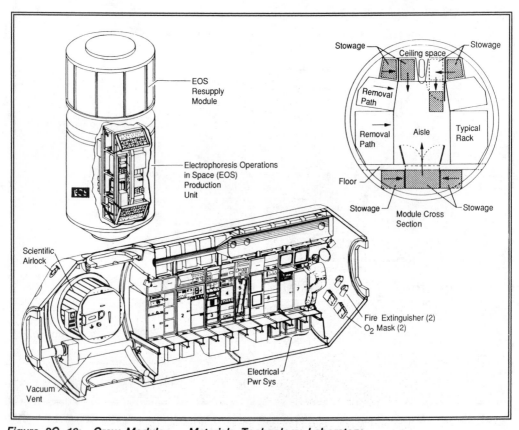

Figure 3C-16: Crew Module -- Materials Technology Laboratory

We think, for example, that 10^{-5} g to 10^{-6} g is probably about as low as will be possible at the space station. I would think, however, that the interplanetary medium will present a gravity environment that is even lower. We know that--for certain kinds of highly specific, low-gravity studies associated with convection, et·cetera--it is desirable to get the gravity factor down to as low as 10^{-8} g. Perhaps, then, low-gravity studies could occupy some of the crew's time quite productively during the long transition times to and from Mars, and the opportunity could produce some valuable new knowledge or processes. The equipment one might imagine to be represented in the materials technology laboratory (Fig 3C-16) might therefore be the instruments needed to do this kind of research.

Figure 3C-17 presents two conceptions that suggest some of the other studies that could be undertaken during a flight to Mars, depending somewhat on opportunities afforded by the transfer trajectory utilized. The crew could, for example, conduct solar studies enroute (left, Fig. 3C-17). They will have a somewhat different perspective and can therefore bring that view of the Sun--together with observations made simultaneously on Earth--to scientific advantage. Indeed, solar observation will be a necessary part of the crew's daily routine, because they will have to be alert to the possibility of dangerous solar flares and unusual radiation increases that would truly have an effect on them. In such instances, they might need to go into some sort of a "storm" shelter (a small area facilitized with greater radiation shielding as protection against an outburst capable of producing a hazard).

Assuming that we take advantage of Venus to provide a gravity-assist boost on these missions, we will have ample opportunity to learn more about Venus. We would swing by quite close to the planet on such orbits, possibly coming within 300 kilometers of the surface. Not only would we want to consider performing a series of Venus observation experiments (right, Fig. 3C-17), we might also want to drop off an instrument package as we went by -- an orbiter or even a small probe or lander.

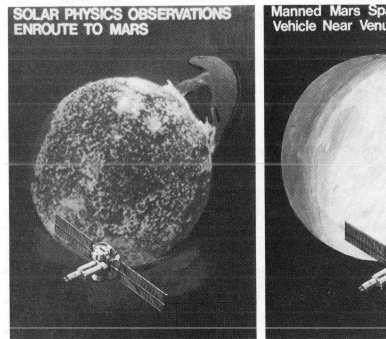

Figure 3C-17: Solar and Venus Investigations Possible During Trans-Mars Flight

Figure 3C-18 suggests that our space vehicle has finally arrived at Mars. Once in orbit around Mars, the excursion vehicle(s) can be used to visit the martian satellites and/or the planet's surface. Figure 3C-19 is a line illustration of a lander excursion module (MEM-2). In the case of MEM-2, a heatshield is necessary when entering the atmosphere to provide protection during the aerodynamic braking phase of the descent. Deorbit motors mounted on the heatshield are jettisoned before the terminal descent phase begins. Two additional propulsion systems are incorporated, one using three motors during the terminal descent phase (as did the Viking landers),

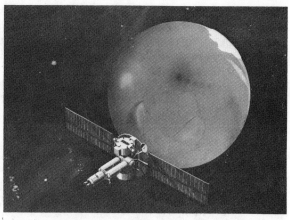

Figure 3C-19: Mars Vehicle Arrival at Mars

and a single-engine system to be used for the ascent phase following surface activity. A laboratory is incorporated in the descent vehicle for use while on the surface, and it is left behind when the ascent vehicle departs.

Only part of the landing vehicle can efficiently return to orbit, of course, and this smaller ascent stage essentially uses the framework of the descent vehicle as a launch platform -- much as the ascent stage of the Apollo LM did when returning to rendezvous with the command module in lunar orbit. Figure 3C-20 shows several additional lander MEM concepts, including one based on the familiar Apollo command module. In addition, two of the biconic configurations discussed briefly by Jim French in his Session·2 presentation are illustrated. These are essentially lifting-body vehicles, one of which is winged to facilitate better flight control. The large aeroshell concept offers potential advantages as a cargo lander as well as a manned lander. The large volume behind the aeroshell would accommodate the delivery of larger modules and other systems to the surface of Mars.

If one of the martian satellites (Phobos or Deimos) is to be visited using the MEM-1 excursion vehicle, it would certainly be desirable to have instruments and mobility in that environment, as suggested in Figure 3C-21. However, one must remember that the low escape velocity and gravity on these satellites could present some challenging problems. Perhaps, if the crew members move around *very* carefully, they will be able to stay on the surface (although tethers may be required for safety). In this sense, the martian satellites may be relatively easy to visit and observe, but quite difficult to physically explore and exploit.

Among the various pieces of apparatus illustrated (Fig. 3C-21) is a solar furnace, needed to process and analyze surface material to see what sort of oxygen content might be present (and which could then be retrieved and stored). A surface digging device can be seen remotely working on the surface in the foreground, dragging material up to where it can be transferred into a solar furnace for processing. A drilling device is being operated by a crew member who might, for example, drill holes in which devices could be placed in an attempt to retrieve materials using electrolysis or other techniques. We might also have to provide equipment for solar-electric power generation (visible in the distance) to provide energy for some of the activity. The round vehicle at the far left is the MEM-1 rendezvous vehicle, while the vehicle seen above the surface at the right (with an engine plume) has just taken off to return to the Mars space vehicle. Some of these techniques and devices might later be used on the surface of Mars itself, once tested and/or used on one of the satellites.

426

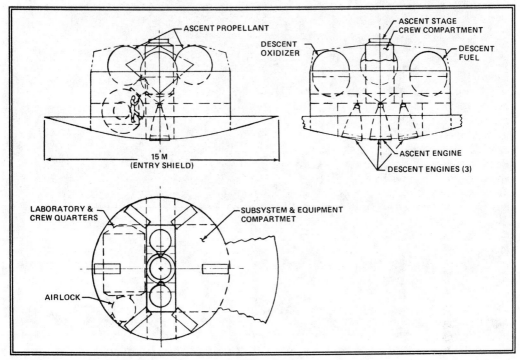

Figure 3C-19: Mars Excursion Module (MEM) Concept

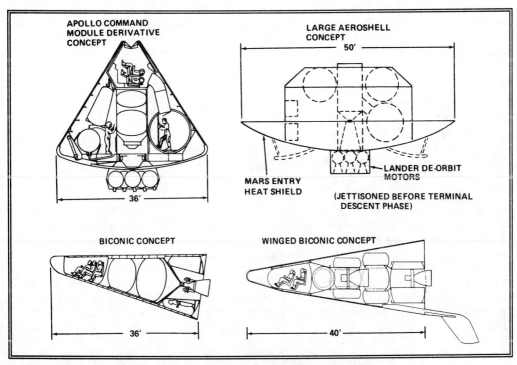

Figure 3C-20: Alternative Mars Excursion Module (MEM) Concepts

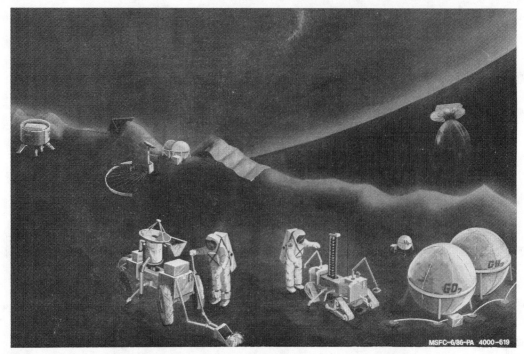

Figure 3C-21: Typical Early Mars Mission Activity on a Martian Satellite

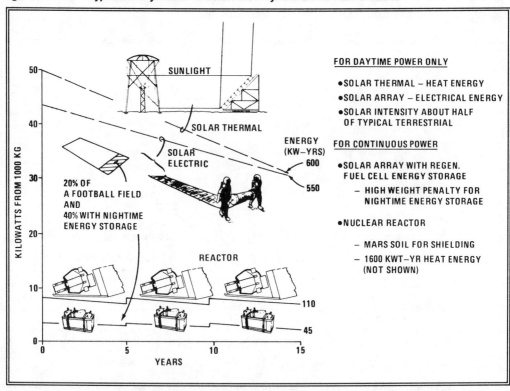

FOR DAYTIME POWER ONLY

- SOLAR THERMAL – HEAT ENERGY
- SOLAR ARRAY – ELECTRICAL ENERGY
- SOLAR INTENSITY ABOUT HALF OF TYPICAL TERRESTRIAL

FOR CONTINUOUS POWER

- SOLAR ARRAY WITH REGEN. FUEL CELL ENERGY STORAGE
  - HIGH WEIGHT PENALTY FOR NIGHTIME ENERGY STORAGE

- NUCLEAR REACTOR
  - MARS SOIL FOR SHIELDING
  - 1600 KWT–YR HEAT ENERGY (NOT SHOWN)

Figure 3C-22: Electric Power Generation Concept for Surface of Mars and/or Satellites

428

It would also be necessary to have electrical power on the surface of Mars, and Figure 3C-22 illustrates how the required energy might be generated and stored. These technologies are portrayed in a relatively simplistic manner to suggest different concepts: solar thermal, solar electric, fuel cell storage, and nuclear power. Solar electric is quite reliable and productive, but doesn't work very well at night; the Sun goes down on Mars just as it does on Earth! Power could also be provided by a nuclear generator scaled up from those frequently used on unmanned spacecraft, or by a fuel cell. In fact, it is probable that some combination of these will result in the best workable solution, but energy generation is a problem that needs further study if we are to sort out what makes the most sense.

CONCLUSION

The many technologies we need to consider for a manned mission to Mars are listed in Figure 3C-23. These are technologies that would be particularly applicable for the missions I have discussed. And, in turn, having such a mission as a target would beneficially focus and advance these technologies, which have other applications in space and perhaps even here on Earth. Therefore, there are many things we will need to start working on soon if indeed this country--perhaps with others--is ready to commit itself to such a monumental task.

- Long-Duration Space Environment Effects on Humans, and Preventive/Protective/Corrective Measures (low gravity, radiation, isolation, et·cetera)

- Advanced Medical Tools and Techniques

- Enhanced Earth-to-Orbit Transportation (heavier lift, lower cost)

- Aerobraking and Precision Recovery/Landing

- Long-Duration Cryo Fluid Storage and Handling

- On-orbit Assembly

- Long-Life, Maintainable, Light-Weight, Reliable, and Safe Systems

- Advanced Maintenance/Servicing Concepts

- Automation/Autonomy

- Advanced Closed-Environment Life Support System (CELSS)

- In·situ Resources Production (H_2O, O_2, propellants, materials, etc.)

- Tethers

- Advanced Power Systems (solar, nuclear, etc.)

- Advanced Propulsion Systems (chemical, electric, nuclear, etc.)

- Advanced Communication Systems (long-distance, high-rate)

- Advanced Data Management Systems

- Advanced EVA Equipment and Techniques

- Advanced Surface Transportation Systems

- Advanced Science Concepts and Equipment

Figure 3C-23: Key Technologies and Advanced Development Areas in Which Work is Needed

In summary, then, some of the important mission factors that must be considered in determining objectives for the scenario I have presented are:

- Early manned landings on both Mars and its satellites are desirable,

- A program incorporating landings on a satellite and Mars during each mission is both feasible and desirable,

- Weight and mission duration for a cryo-chemical propulsion, aerobraking manned space vehicle is comparable to that of nuclear vehicle performance,

- The potential benefits from the space station program (for a manned Mars program) are high.

We believe that sending human crews to land on and explore both Mars and its satellites is desireable for many reasons, beginning with the fundamental purpose of expanding our scientific knowledge of the solar system and Mars itself. We believe that both can be accomplished on a given mission, and that it is not only feasible but quite desirable to perform the early missions in the manner I have described.

More specifically, there are several recommendations worthy of consideration as a result of the studies that provided the context of my discussion. I would, for example, suggest that two excursion vehicles are preferable to one. While that conclusion might seem a simple statement to make, we in fact went through the numbers and performed complex calculations that led us to it. This work demonstrated, for example, that--for a variety of reasons--the use of a pair of MEM vehicles during one mission is more productive than two consecutive missions using one MEM on each. We also found that vehicle mass performance for cryogenic chemical propulsion, when coupled with aerobraking, is comparable to that of thermal nuclear propulsion, and is therefore probably still the propulsion concept of choice -- at least for the early missions. Finally, we *do* believe that the U.S. NASA Space Station will be extremely important to a Mars program, and the interface between the station and the Mars mission will have to be carefully developed as both programs evolve. ■

REFERENCES

This space is otherwise available for notes.

~~~~~~~~

A TRANSPORTATION SYSTEM
FOR ROUTINE VISITS TO MARS
[NMC-3D]

Mr. Barney B. Roberts
Mission Manager, Advanced Programs Office
NASA Johnson Space Center

A transportation system architecture will be needed to provide routine support for a base population of roughly 20 people on the surface of Mars. These people will be exploiting in-situ resources to reduce their own as well as the transportation system's dependence on Earth, although this will likely involve the acquisition of resources on the lunar surface and the martian satellites, as well. Our mission architecture for a routine transportation system would utilize in-situ resource production to support Mars missions via an Earth-Mars Cycler spaceport. Our Cycler concept would be able to generate artificial gravity while delivering additional consumables (such as food and propellant components not otherwise being produced on the surface of Mars or which can be produced more efficiently on the moons of both planets). Major secondary elements, including cargo and taxi vehicles, also would be needed. Among the issues that need to be addressed are key technologies, including propulsion and extraterrestrial material processing, and those associated with the human physiological/psychological (health) and social problems that will arise during long-duration missions. Such a system might evolve only after initial expeditions, but there are issues that must be addressed soon if we are to proceed. The specialized skills and tools needed to locate, extract, and process extraterrestrial resources are not well represented within NASA or its established aerospace/academic community, and new relationships must be developed with the technology centers best qualified to provide the expertise to develop these capabilities. The United States NASA Space Station now being planned will be a crucial element in the development of needed technologies and long-duration low/zero-g mission experience.

INTRODUCTION

First, I would like to put my presentation into context with other NASA Mars Conference presentations addressing similar or related topics. John Niehoff discussed various trajectory options (pathways) for access to Mars [NMC-3B], and Bill Snoddy described practical or conservative approaches for manned Mars missions that would be exploratory in nature or would serve as precursors to something more aggressive [NMC-3C], which perhaps is more on the order of what I will discuss. Finally, Jim French provides a more detailed perspective of some of the advanced technologies and techniques expected to play a role in future Mars missions [NMC-3E].

My presentation will provide an overview of a conceptual transportation system designed for routine visits to Mars. It builds on the *Visit Cycler* space station concept presented by John Niehoff, and it closely conforms to the infrastructure proposed by the **National Commission on Space** (reviewed by its chairman, Tom Paine, in these proceedings -- NMC-3I). While our concept tends to be somewhat modest compared to the more advanced proposals put forth, it still reflects and maintains the basic philosophy established by them. I have divided my presentation into four topical segments. First, I will discuss the theme that underlies the architecture of the routine Mars transportation system, which is that we assume there to be some sort of human imperative in space relative to Mars. Therefore, I will be describing a transportation system architecture that supports something on the order of 20 people on the surface of Mars, a base that relies substantially on the in·situ use of local resources to reduce its dependence on Earth. The second segment of my discussion will provide a degree of quantification relative to the potential value of in·situ resources needed to cut the umbilical cord to Earth. The third segment is probably the one that will generate the greatest interest, because it will provide a specific description of the transportation system proposed. In the final segment, I will make some closing remarks that reflect our desire to be realistic. That is, we should all be aware that we are not studying and discussing manned Mars missions today because we intend to "cut metal" tomorrow, but rather because there are some very important issues we need to face and resolve before we can proceed.

Some of these issues have been discussed by Bill Snoddy and others at this conference, and they are comprehensive in nature. Included are issues in the area of technology and issues with respect to human response and health (both physiological and psychological). We need to understand the impacts of these issues with respect to the planned first phase (Block·1) of the low Earth orbit United States NASA Space Station, such that we can plan the growth path of the station itself. To distinguish this space station from the advanced future space station concepts I will be discussing as well, I will refer to it as the NASA Block·1 Space Station or simply as NSS-1 (NASA Space Station, Block·1). The current system does not yet have a proper "personal" name and references to it may therefore vary, but we will have to do the best we can with the current nomenclature until something better is established[1].

Theme: Plotting a Strategy

Relative to our underlying theme, we have--within the agency--performed studies on concepts associated with manned geosynchronous orbit access, lunar bases, and manned Mars operations. We have found, perhaps not so surprisingly, that there is one common problem associated with living in the 1-g environment at Earth's surface -- how to get off of it! Launch operations, including the initial deployment and maintenance of the supply line, can actually represent half to two-thirds of the total cost for a given mission. Thus, the theme we have adopted, and which also was supported by the **National Commission on Space**, is that we want to plot a future space strategy that bootstraps itself through the solar system by the utilization of extraterrestrial resources along the way. Not only do we see this as being necessary to support the transportation system itself, but to relieve the dependency burden created as our reach broadens and extends outward into the solar system over time.

---

[1] References to the "U.S." NASA Space Station tend to imply that only the United States is involved in the undertaking when, in fact, Canada, the European Space Agency (ESA), and Japan will be developing components for the system, as well. The United States will shoulder the development of the primary architecture and operating systems/environments. Some of the advanced space stations discussed in these proceedings are located at Earth-moon Lagrange libration points (5) where Earth's gravity is balanced by lunar gravity. Spaceports located at these points would enjoy greater gravitational stability, requiring a minimam expenditure of energy to maintain their orbits relative to Earth and the moon. Points L1, L2 and L3 are on the Earth-moon line while L4 and L5 are on the lunar orbit. Similar libration points exist with planets relative to the Sun, and their L4/L5 points may "hold" groups of asteroids (as in the case of Jupiter). {Ed.}

434

"Hayburner" Analogy

By way of example, Figure 3D-1 (next page) reflects a "hayburner" transportation analogy created by Phil Garrison of JPL. He extracted some of his data from *The Expressmen*, published as one of the "Old West" series by Time–Life, which provided information on the performance and sustenance requirements for (in modern terms) various exploration systems in early America. For example, one of the more popular systems was the six–mule team and light wagon.

If, as illustrated, we load all of the needed consumables we can carry on the six–mule team wagon, we find--in doing the appropriate calculations--that we have a range of approximately 100 miles before the consumables are used up. This suggests that if we had explored the United States in the same way we explore space, our western boundary might well have been the Appalachian Mountains. Further, we discover when examining the data that the first 100 miles proved to be a near–suicide mission in itself. Having totally exhausted the consumables, we were apparently forced to eat a couple of the mules on the way back (suggested by the artist when he reduced the mule team from six to four; he must have assumed we were a third of the way back and therefore had eaten a third of the mules).

Lunar Resource Possibilities/Considerations

We have, of course, been considerably more serious in looking at how to support manned Mars missions than the "hayburner" analogy suggests. The type of manned Mars mission Bill Snoddy described, wherein one supports these missions from a lunar base using lunar propellants, represents a good example of that strategy. The cross–hatched bars presented in Figure 3D-2 reflect a comparison of Earth-to-LEO (low Earth orbit) launch mass in metric tons versus that for moon-to-LLO (low lunar orbit) or the L2 libration point (L2 is on the far side of the moon). The most interesting ordinate on the plot is the one that reflects "percent of baseline launch requirement." The baseline is 100 percent for Earth-to-LEO liftoff mass, and the required mass to LEO declines to approximately 50 percent as we consider the different options.

Some important tradeoffs must be made as one does this. For example, to deliver lunar oxygen to LEO to facilitate a departure from LEO (Case 4, Fig. 3D-2), we are able to substantially reduce the mass delivered to LEO. However, we have had to incorporate a fairly significant lunar logistical system to supply the oxygen. The analytical results reflected in Case 5 (Fig. 3D-2) are for missions departing from the Earth-moon L2 libration point. Such a mission is certainly conceivable and does provide some benefits, but we encounter some operational problems--as one might expect--for departure from the backside of the moon when our surface base is on the front side. Therefore, the mission architecture that I will describe provides a departure from L1 with only a few very minor performance penalties.

To achieve the factor-of-two savings, we had to establish a lunar logistical system to deliver the propellants. I want to compare this with the logistical system associated with lifting the same mass to LEO from Earth. The mass ratio to lift something from Earth's surface (to LEO) is 15-to-1 (15:1), whereas the mass ratio for the logistical system needed to lift something into these other positions can vary from about 1-to-1 (1:1) to maybe 2-to-1 (2:1). Therefore, we are exchanging a logistical system that is on the order of one-fifteenth of the logistical system needed to operate from Earth's surface.

It seems reasonably clear to us that it would not be cost effective to build a lunar base just to supply propellants for a Mars mission. That simply wouldn't pay. But, if we have a lunar base for other justifiable reasons, it would be beneficial--as demonstrated in Figure 3D-3--to supply our manned Mars missions from the lunar surface by operating on marginal costs made possible by that lunar operation. The bottom line, then, is that it appears as though utilizing lunar-produced propellants, made on the margin of a more comprehensive lunar base, could be a very productive alternative to Earth-launched propellants.

OUTFIT	DAILY DISTANCE	PAYLOAD
Pack Mule	50 Miles	300 lb
6-Mule Team and Light Wagon	20-25 Miles	2,000 lb
20-Mule Team and Heavy Wagon Tandem	16 Miles	45,000 lb
12-Ox Team* and Heavy Wagon Tandem	14-16 Miles	45,000 lb
6-Horse Team and Stagecoach	110 Miles Change Horses Every 12 Miles	9 Passengers (25 lb luggage ea) Plus Driver/Guard

* Best life-cycle cost.

REF: Data taken from **The Expressmen** (Time-Life "Old West" series).

OXEN FACTORS

- Ox Cost = 0.4 x Mule Cost
- Fewer Animals Required for Heavy Load
- No Operational Cost for Ox -- Only Grazing Required
- Indians Would Not Steal Oxen as Long as Buffalo Were Available

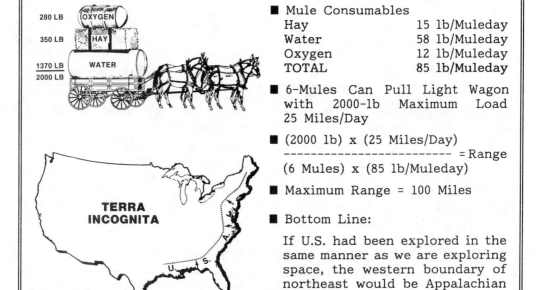

- Mule Consumables

Hay	15 lb/Muleday
Water	58 lb/Muleday
Oxygen	12 lb/Muleday
TOTAL	85 lb/Muleday

- 6-Mules Can Pull Light Wagon with 2000-lb Maximum Load 25 Miles/Day

- $$\frac{(2000 \text{ lb}) \times (25 \text{ Miles/Day})}{(6 \text{ Mules}) \times (85 \text{ lb/Muleday})} = \text{Range}$$

- Maximum Range = 100 Miles

- Bottom Line:

  If U.S. had been explored in the same manner as we are exploring space, the western boundary of northeast would be Appalachian Mountains!

Figure 3D-1: Hayburner Analogy

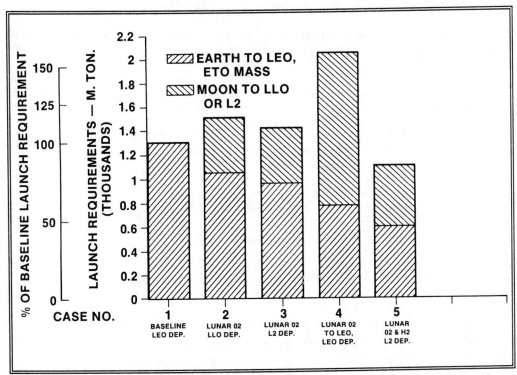

Figure 3D-2: Launch Requirements vs Scenario (oxygen and hydrogen production)

Baseline Manned Mars Case . . . .   ETO Mass = 1300 MT

LO_2 Delivered to Low Lunar Orbit (LLO) . . .   ETO Mass = 1040 MT

LO_2 Delivered to Earth-Moon L2 . . . .   ETO Mass = 930 MT

Lunar LO_2 Delivered to LEO . . . .   ETO Mass = 780 MT

LO_2 and LH_2 Delivered to Earth-Moon L2 . . .   ETO Mass = 600 MT

BOTTOM LINES:

■ Lunar-produced propellants may (on the margin) reduce the costs of manned Mars missions.

■ Earth-moon L2 appears to be best staging point.

LO_2 = Liquid Oxygen, LH_2 = Liquid Hydrogen, LLO = Low Lunar Orbit, L2 = Libration Point 2, MT = Metric Tons.

Figure 3D-3: Lunar- vs Earth-Produced Propellants

SELF RELIANCE: IN·SITU RESOURCE EXPLOITATION

Expectedly, as various mission options and alternatives are studied in a context of in·situ extraterrestrial resource utilization, mission sets evolve and are strongly coupled. I have indicated how this mission-set interaction can evolve in Figures 3D-4. What this implies, in simple terms, is that once we make the decision to utilize in·situ resources wherever we go, there are going to be some significant impacts on the way we do business. For example, our LEO missions will then become dependent on lunar-supplied propellants. We may have lunar products serving as shielding, or as feed-stock or raw material for finished products that we will ship from the moon to LEO. Geosynchronous missions also may interact with lunar bases, and all deep space missions--including the Mars missions--may become essentially dependent on lunar propellants and lunar launch services.

Mission Sets Become Coupled Through Use of Extraterrestrial Resources

■ LEO Missions May Depend On:

[1]Lunar Propellants
[2]Lunar Products (feedstock, shielding, finished)
[3]Mars Propellants from Phobos

■ Geosynchronous Missions May Depend on Lunar Resources:

· Large-Scale Manufactured Products
· Shielding
· Servicing

■ Mars Missions (and other deep space) May Depend On:

· Lunar Propellants
· Lunar Launch Services

■ Bottom Lines:

· Mission Sets No Longer Independent
· Individual Mission Analysis No Longer Valid

Utilization of Extraterrestrial Resources Requires Skills That are Scarce Within NASA

■ Resource Locating/Yield Estimation ⎫
■ Mining   WITH OPERATIONS
■ Beneficiating   IN NON-TERRESTRIAL
■ Transporting   ENVIRONMENT
■ Processing   WITH VERY LITTLE
■ Manufacturing   HUMAN INTERVENTION
■ Storage ⎭

■ Bottom Lines:

· Agency Must Aggressively Enter New Fields
· New Technologies Required
· New Relationships Required (private sector, academia)

Figure 3D-4: Sideline Implications of Extraterrestrial Resource Utilization

On the other hand, it is also possible to acquire propellants for utilization in LEO or at the moon from one of the martian satellites--Phobos, for example--if we find that we can more easily get hydrogen and oxygen from Phobos than from the lunar surface. The problem with the case of using Phobos-derived propellants, of course, is that the transport system needed to deliver the propellants could do so only as its orbital cycle permitted, which is roughly on 2.14-year periods. In any case, the bottom line is that once we commit to the use of in-situ resources, we can no longer do individual analyses of our mission sets; they *must* be considered as a unit.

An additional factor associated with a commitment to the utilization of in-situ resources is that it requires skills that are presently scarce within NASA. While we have a plentiful resource of rocket and space system designers, we do not have the skills associated with the problems of resource location, yield estimation, mining beneficiating, et-cetera. This expertise exists relative to our terrestrial environment, supported to some extent by our lunar experience, but we now need to begin to fold it into the agency in a manner that applies it to remote, non-terrestrial environments (wherein there is little opportunity for human intervention, and in which mass considerations are critical).

Once again, then, the bottom line is that we need to aggressively enter new fields if we are going to pursue these kinds of advanced, self-reliant missions. New technologies will be required and new relationships with institutions of specialized skills will have to be developed that differ from the more familiar relationships we have with our aerospace contractors, both in the private sector and in academia.

Benefits Derived from Extraterrestrial Resources

As an example of the kinds of benefits that might be derived from in-situ propellant production (ISPP), Figure 3D-5 is a plot of how benefits (in terms of mass reduction at LEO) could evolve over time through the production of ascent propellants on the surface of Mars. I will spend a little time explaining the plot shown in Figure 3D-6. I will then briefly review some similar examples that afford additional benefits, first using Figure 3D-6 which reflects propellant production on Phobos alone, and then Figure 3D-7 which reflects the combined benefits of propellant production on the martian surface and Phobos. I will also describe what is produced.

ISPP on Mars -- The horizontal scale in Figure 3D-5 is unusual because it represents opportunities to depart for Mars on **Opposition Class** missions, and the year indicated on the horizontal scale is the year in which a specific opportunity occurs. Consider, for example, a 20-year mission set; I have plotted the savings in total LEO mass for different assumptions made concerning the propellants and their sources. The direct-savings factor is plotted in metric tons and the left-most scale is a percentile reflection of the mass-saved-over-time relationship, serving as a relative measure of the benefit.

The factor that we have maintained as constant is the delivery manifest to the surface of Mars for the 20-year period reflected in these charts. However, we must account for the ISPP plant masses. For the Phobos plants, we deliver the plant masses concurrent with our "baseline" Mars surface manifest. Therefore, there is an "investment mass" cost that shows as a negative "savings-in-LEO-mass" in Figure 3D-6. For the surface ISPP plant, we accounted for the plant mass differently. When we take something down to the surface of Mars, we offset (in time) our mission delivery manifest for the delivery of plant components. Therefore, when delivering surface ISPP plant-facility components, we shift our productive mission set three or four opportunities to allow these first few opportunities to be used for plant delivery and set-up. This is why the plot curves in Figure 3D-5 are initially flat before we start seeing some payback on our investment.

ISPP on Phobos -- The bottom line for Figure 3D-5 is that we achieve a five percent (5%) reduction if we produce oxygen on the surface of Mars, and we can improve that benefit by two percent (2%) if we produce both oxygen and fuel on Mars. However, while the development of ISPP on the martian surface is clearly beneficial, it is not as dramatic as we might like. This drives us to begin looking at Phobos, the results of which are reflected in Figure 3D-6. The numbers improve noticeably in the case of Phobos because it becomes practical to refuel the larger stages with which we perform most of the ΔV and move the largest masses.

As mentioned earlier, the plot reflected in Figure 3D-6 differs from that presented in Figure 3D-5 in how we account for the "investment mass" of the ISPP plant. In Figure 3D-6, we transport the same productive mission manifest to the surface of Mars concurrently with delivery of the plant mass to Phobos. However, we have to penalize our initial departure from LEO to compensate for the weight of the ISPP production system units we must deliver with the first few missions. Once the ISPP plant becomes operational on Phobos, we see our benefits improve rather significantly, whether for just refueling the space vehicles or to access the propellants for descent stages, as well. As one can see, we achieve a mass reduction of as much as 25 percent over that period of time.

Combined Mars/Phobos ISPP -- Figure 3D-7 projects ISPP benefits on the basis of combining the product of both of the resources (Phobos and the martian surface). Note that using extraterrestrial propellants has allowed us to achieve a 31 percent reduction in mass that must be transported from Earth's surface to LEO. The bottom line of these considerations, then, which is reflected in the comparative summary presented as Figure 3D-8, is that there is considerable potential for greatly reducing mission costs through the production of extraterrestrial in·situ propellants.

DESCRIPTION: ROUTINE TRANSPORTATION SYSTEM

The mission architecture for our routine transportation system does utilize these "gas stations" in space, and it also follows some of the principles suggested by the **National Commission on Space**. It utilizes the cycling trajectories that I and others at this conference have discussed, and--in overview--it broadly resembles the mission illustrated in Figure 3D-9. The Earth-Mars *Cycler* orbit portrayed does not reflect a good representation of its orbital geometry because I can't get it all into a single, simplified illustration in a precise fashion. However, referring to John Niehoff's discussion of the **Cycler** concept will refresh one's memory on what these cycling trajectories really look like. Represented in Figure 3D-9, we have **Cycler** stations in the trajectories defined by John Niehoff and we have a similar space station positioned in the Earth-moon L1 libration point[2].

---

[2] Advanced interplanetary space stations, like the Cyclers discussed in these proceedings, are essentially continuous-operation transportation systems which function as transport support bases or spaceports. The latter is perhaps a more apt term for space systems that service other missions. However, they are often referred to as "space stations," which one should keep in mind when discussions also include references to the more current NASA Space Station (the first phase of which is now being planned). As previously noted, references to this system tend to be generic because it is as yet unnamed, such that one must be careful not to confuse it with advanced spaceports like the Cyclers. An unofficial acronym has been devised for this presentation only, to provide a more distinct identification for the station: NSS-1 (NASA Space Station, Block 1). Cyclers have large orbits that would bring them into proximity with planetary bodies other than Earth on a cyclic basis (e.g., the moon or Mars). Ultimately, in fact, conceptualists see a layered system of spaceports, beginning with those in Earth orbit (like NSS-1, the Soviet Mir, and ESA's proposed Columbus). These, in turn, might service spaceports at Earth-moon libration points or Earth-moon Cyclers, which could then service Earth-Mars Cyclers. {Ed.}

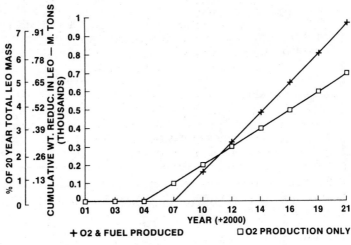

Figure 3D-5:

Mars Surface ISPP
(ascent stage propellant)

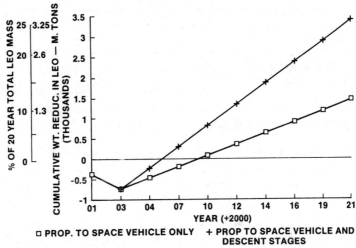

Figure 3D-6:

ISPP at Phobos
(O₂ & fuel production)

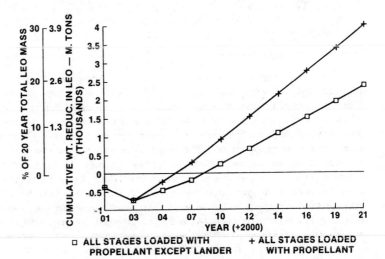

Figure 3D-7:

Mars/Phobos ISPP
(O₂ & fuel production)

	ETO MASS 20-YR PROGRAM	REDUCTION IN ETO MASS (%)
Baseline Case	13,000 MT	N/A
Surface ISPP		7%
■ 5%/2% Oxidizer/Fuel		
■ Plus 3 Mission Delay		
Phobos ISPP		25%
■ 12% for Space Vehicle Refueling		
■ 13% for Lander Refueling		
Both		13%
Bottom Line:		
■ ISPP at Mars has Potential to Significantly Reduce Mission Costs		

Figure 3D-8: ISPP at Mars

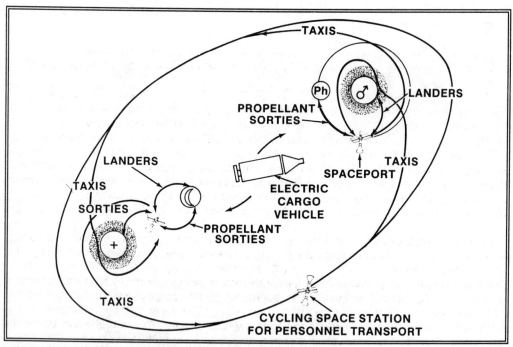

Figure 3D-9: Routine Transportation System Scenario Overview

ISPP Technology and Requirements for Phobos

We have conducted detailed analytical calculations for a scenario in which we must produce and deliver Phobos propellants. We also determined the requirements for each of our different elements, as reflected in Figure 3D-10.

Requirements:

- Process Surface Material of Phobos to Produce H_2 and O_2 Propellants (assumed to have 5% water by weight)
- Production: 600 MT/Year

Configuration

- Rock Melter Technology Developed by Los Alamos National Laboratory
- 6 MW Electrical Energy (nuclear) for Resistance Heaters to Remove Core Material
- 10 MW Thermal Energy to Heat Core Material (900°K)
- Evolved Gases Separated by Condensation and Adsorption
- H_2O Separated into Propellants by Electrolysis
- Separate 1-MW Reactor for Processing and Power for Single Habitat
- System Mass: 107 MT

Figure 3D-10: Phobos Propellant Plant Requirements and Configuration Attributes/Capabilities

Figure 3D-11 (page A-5, Color Display Section) is an artist's conception of a propellant production plant on Phobos. The device illustrated has a mass of 107 metric tons and processes Phobos material to produce hydrogen and oxygen propellants. As discussed by others, there are good indications that Phobos has the properties of a carbonaceous chondrite. If so, its composition would include from 1 to 20 percent water. For our case, we assumed the water component to be five percent and we designed a device to manufacture 600 metric tons of the propellant elements a year. These assumptions, when combined with reasonable estimates of production losses, indicate that we would consume only 0.001 percent of Phobos per year.

Our system configuration for propellant production on Phobos utilizes rock melter technology pioneered by the Los Alamos National Laboratory. Resistance heaters actually melt through the rock to facilitate the removal of core material, which will be accomplished by mechanical methods. That material is then mechanically transported into a ten-megawatt (10 MW) nuclear reactor to raise the core material temperature to approximately 900 degrees Kelvin (627°C/1160°F), which will drive off most of the water as water vapor. The water is then condensed and is processed by electrolysis to separate the hydrogen and oxygen. The vertical boom above the rock melter (Fig. 3D-11) is equipped with radiators, and it supports a separate 1-MW reactor to supply energy for the temporary crew habitat and the processing equipment.

Applicable Cycler Spaceports

The cycling space stations we have included in our system (referred to variously as **Cyclers**, **Earth-Mars Cyclers** and **Earth-Moon Cyclers**) reflect requirements determined by John Niehoff and his team at SAIC [Space Applications International Corporation]. An interplanetary cycling station's general orbital attributes are given in Figure 3D-12, and an artist's conception of a Cycler is presented in Figure 3D-13 (pg A-5, Color Display Section). It should be noted that the last item defined in Figure 3D-12, a system called a "**single dumbbell**," is itself a modest version of a Cycler designed to function as a spaceport.

Requirements:

- Crew: 17
- Facilities for Taxi Vehicle Servicing
- Gravity (g) Environment
- 1500 m/s in 15-Year Period

Configuration

- Length:

  Structure -- 224 m (arms)

  Tethers -- 3700 to 4300 m
    - Angular Momentum Management
    - Maintenance of Center of Rotation at Despun Axis

- Power: 300 kW, Nuclear (despun axis)

Propulsion: Ion Thrusters (despun axis)

- Transportation Systems on Despun Axis

  - Pressurized Hangar
  - Propellant Storage
  - Taxi Storage

- Eight (8) Space Station Modules

  - Habitation -- Closed-Environment Life Support
  - Control
  - Laboratories
  - Recreation

- System Mass: 460 MT

Spaceport Version -- Single Dumbbell

- Mass: 450 MT
- Crew: 10
- Location: Mars Orbit; Earth-Moon Libration Point

Figure 3D-12: Cycling Space Stations -- Characteristics, Configurations

Artificial Gravity -- It is important to understand that the primary **Cycler** spaceport concepts represented in our system rotate to produce artificial gravity. We did that not so much because we believe that an artificial gravity system is already a certainty, but because it is an engineering solution to uncertain requirements from the people in life sciences. I have found that there are bodies of opinion on either end of the continuum debate batting us back and forth; those on one end of the issue say "we probably won't need *g* (artificial gravity)" while those on the other end say "solve the g problem." All I know for certain is that if we design a system with artificial gravity to begin with, we will have a solution that satisfies the largest number of life scientists. In the final analysis, though, artificial gravity may not be necessary.

Earth-Mars Cycler -- The space station circulating in the Earth-Mars **Cycler** orbit is used for the major interplanetary transport of personnel. A similar spaceport is located in orbit about Mars, and it of course has access to Phobos such that we can conduct propellant sorties back and forth between the various elements of the system at Mars. To get to and from the **Cycler** station at either Mars or Earth, we use taxi vehicles. When departing the **Cycler** to access either planet, the taxi vehicle must aerobrake into the planet's atmosphere, and propulsive boost is required to depart either planet for return to the **Cycler** stations. We have also included an electric (ion propulsion) cargo vehicle in our mission architecture, and it is designed to carry unmanned equipment back and forth between Mars and Earth. The cargo vehicle utilizes the trajectories John Niehoff defined for low-thrust vehicles, and I will describe it in a little more detail shortly.

Cycler Characteristics

As indicated, we will need a **Cycler** crew of approximately 17 people to maintain a permanent base of about 20 people at Mars. The **Cycler** station has a mass of 460 metric tons and its power is provided by a 300-kW nuclear generator positioned on the despun axis. Our **Cycler** has facilities for **taxi** vehicle services and also provides--as I explained earlier--an artificial gravity environment. The term "**taxi**" is used to very generally define a kind of service more than a specific vehicle design, and I will discuss the **taxi** vehicles (which must be available in at least two configurations) in greater detail shortly.

The primary boom is 224 meters (735 ft) in length. With respect to artificial gravity we need to control the rotation rate as well as the center of mass, and we need to keep our despun axis fixed on the center of mass. We accomplish that by using tethered countermasses off of either end of the cycling space station. The tethers are adjustable and can be as long as 4,300 meters, varying between 3.7 to 4.3 km (2.3 to 2.7 mi). Since there will be considerable activity within this kind of station, we feel it is necessary to put active counterbalances on the ends of the tethers to dampen induced oscillations.

The artist's concept identified above as a **Cycler** (Fig. 3D-13, pg A-5) is a *single-dumbbell* configuration, and it was specified in Figure 3D-12 as a spaceport version of the **Cycler** concept. It is similar to the tethered configurations illustrated among the advanced concepts shown in Figure 3D-14 (next page). The **single-dumbbell** spaceports are a more modest concept that we use at Earth or at Mars as a transportation node. It requires a crew of only ten people. This type of station has a mass of about 450 metric tons, and we have a tethered countermass on the opposite end. The transportation systems are located on the despun axis, and one can also see propellant tanks in the illustration. A pressurized facility also is included in the configuration to process and service **taxi** vehicles.

The Nuclear Electric Cargo Vehicle

Major specifications for the unmanned nuclear-electric cargo vehicle are given in Figure 3D-15, and Figure 3D-16 (page A-5, Color Display Section) is an artist's concept showing the cargo

445

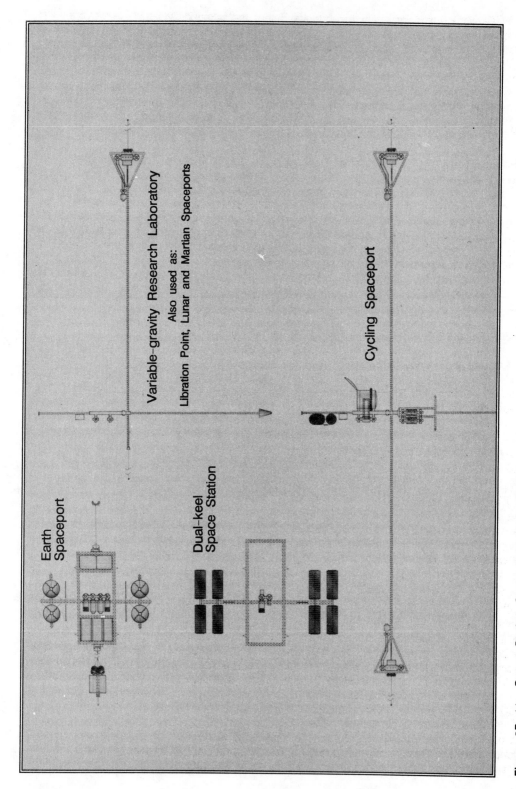

Figure 3D–14: Space Station Concepts and Applications

Figure 3D-15: Nuclear Electric Cargo Vehicle Configuration Requirements/Characteristics

vehicle near Phobos and approaching Mars. Our concept has an open-truss design measuring about 130 meters by 30 meters, providing a large area to which different kinds of equipment can be attached. Figure 3D-17 (next page) illustrates its structural elements and attachments in somewhat greater detail. The vehicles equipped with aerobrakes, for example, are Mars MEM (manned excursion module) landers. The cargo vehicle itself is powered by a 5-MW nuclear reactor located near the forward section, and the large conical shape in the illustration is the radiator for waste heat from the reactor.

Our performance specification for the vehicle is 180 metric tons delivered to low Mars orbit (LMO). The vehicle's initial mass, departing from LEO, is 340 metric tons; 300 metric tons when departing from L1. The propulsion system requires 200 metric tons of argon propellant, and transit times will be 600 days from LEO or 350 days from L1.

Secondary Elements: Taxi Vehicles

A number of secondary elements are used in support of the Cyclers and spaceports. Important among these, for example, are the taxi vehicles. There are other secondary elements that this presentation opportunity does not afford space to discuss, including the Mars surface propellant

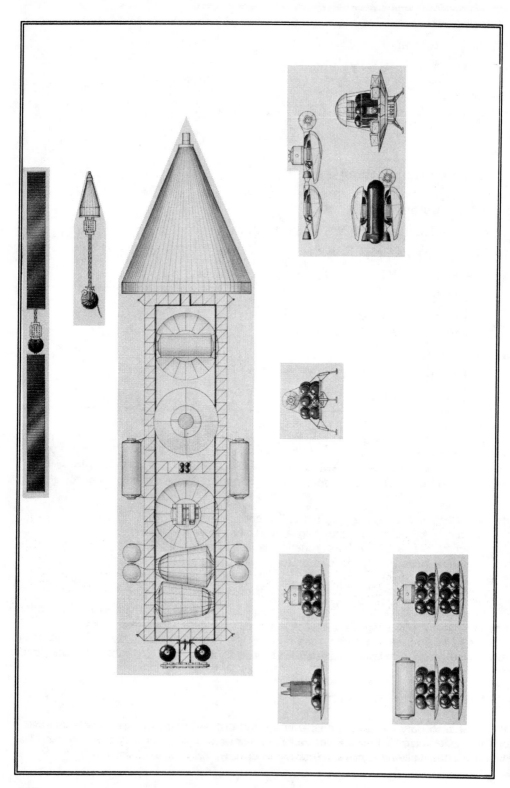

Figure 3D-17: Cargo and Transfer/Taxi Vehicles

plant briefly defined in Figure 3D-18. Among them are atmospheric reduction plants as well as drills that make it possible to access the martian permafrost and process it to produce H_2O, and then H_2 and O_2 using electrolysis.

Figure 3D-18 also lists some of the characteristics of the two types of taxi vehicles associated with our conceptual transportation system. We feel that a taxi vehicle might be an advanced version of the aerobraking, hydrogen-oxygen (cryogenic) orbit transfer vehicle (OTV) like the one illustrated in Figure 3D-19 (page A-6, Color Display Section). This particular vehicle has a circular cross section when viewed from the top; if we end up with a configuration similar to it, one could say that NASA has actually finally produced a flying saucer.

Mars Surface Propellant Plant

■ Atmospheric Reduction Plant:

  · LO_2: 300 MT/Year
  · Mass: 80 MT
  · Power: Nuclear, 750 kW

Taxis

■ Aerobraking, H_2/O_2, Orbit Transfer Vehicles (OTV's)

■ Get-Off Taxi: Aerobrake in Planetary Atmosphere

■ Get-On Taxi:

  · Earth -- 80 MT Propellants (2-stage)
  · Mars -- 200 MT Propellants (2-stage plus auxiliary tanks)

■ Crew: 4-8

Figure 3D-18: Secondary Elements

The "Get-Off" taxi vehicle is essentially like the OTV described above (and in Fig. 3D-19, pg A-6), i.e., an aerobraking vehicle that utilizes a planet's atmosphere to reduce its energy to a point where it is "captured" by the planet -- Mars or Earth[3]. The Get-On taxi vehicles will require chemical propellant. Get-On flight activity at Earth will be much easier, energetically, requiring 80 metric tons of propellants in two stages. To Get-On at Mars, because of differences in the relative trajectory geometries between Mars and the Cyclers (the Cycler orbital ellipse crosses the Mars circular orbit at very high angles), the job of simply turning a taxi's ΔV onto the correct orbit involves a significant performance penalty. In fact, 200 metric tons of propellants will be needed at Mars to match the velocity of the passing Cycler, and it will require two stages and the use of auxiliary tanks. The crew for both types of taxi vehicles is four to eight people.

---

3 One should note that the taxi terms "Get-Off" and "Get-On" reflect departure from (Get-Off) or ascent to (Get-On) a Cycler spaceport. There is a tendency to relate these terms to a planetary body, i.e., to "Get-Off" (take off from) or "Get-On" (land on) Mars, the opposite of what they actually imply. {Ed.}

CLOSING REMARKS: MOVING FORWARD

While conceptual work like this is interesting and very attractive to consider, we also need to concern ourselves with the issues and questions that must be addressed in the near-term. These include both the human issues and the development of required technologies, which are discussed at greater length by others at this conference (Jim French [NMC-3E], Arnauld Nicogossian [NMC-3F], Penelope Boston [NMC-3G], and Chris McKay [NMC-3H]). For example, we clearly need to understand how NSS-1 will impact future mission planning and design. I will close with a chart that reflects some of the characteristics of the kind of interplanetary space base I have tried to mentally image through my presentation.

Technology Issues

I have essentially discussed many of the technologies we will need and which Bill Snoddy's presentation reviewed in some detail. Figure 3D-20 lists those that I have tended to emphasize.

---

[1] Extraterrestrial Material Processing

[2] Large Space Nuclear Reactors

[3] Magnetoplasma Dynamic Thrusters for Cargo OTV

[4] Closed Ecological Life Support Systems (CELSS):
- Chemical Physical (initially)
- Biologically (later)

[5] Space Radiation:
- Environment Characterization
- Materials Interaction (cross sections)
- Acceptable Human Doses (cell damage?)

[6] Large-Scale Autonomous Space Systems

[7] Angular Momentum and Center of Rotation Control
   Using Tethers With Active Counterbalance Masses

Figure 3D-20: Technology Issues

Of course, one of the key technologies supporting the type of architecture I have discussed is **extraterrestrial materials processing**, and we currently have very little activity going on within NASA in that respect. Among the other technologies in need of critical research are the large space-transport nuclear reactors, the **magnetoplasma dynamic** (MPD) thrusters needed for the nuclear-electric unmanned cargo ship, and the development of life-support systems with chemical/physical closure (perhaps with the objective of achieving biological closure later). This issue is discussed in greater detail by Dr. Boston.

We should also begin a serious study aimed at the development of large scale, autonomous space systems. The kinds of large interplanetary systems that I and others have discussed at this conference will essentially be on their own, relative to Earth. That is, the kinds of interactive control we are accustomed to seeing exercised by mission control people on Earth, wherein hundreds of consoles are manned for the purpose of monitoring spacecraft and personnel systems in order to predict trends and then recommend or even initiate corrective action, will simply no longer be realistic.

For example, it would do little good to transmit the message: "Guys, you've got to do something with that water boiler, it's getting too hot very fast!" when, in fact, the one-way communication time (because of distance) may be 30 minutes. In the hour (plus evaluation time on Earth) that will have transpired before the warning is received on the station, the water boiler will no doubt have exploded[4]. One must remember, too, that the station will sometimes be occulted as it goes behind Mars or the Sun, relative to Earth. For these reasons, then, the kind of monitoring and responsive action that have typically involved mission control on Earth during near-space mission activity must now somehow be incorporated into the manned Mars vehicle itself (essentially "implanting" mission control in its system).

Space radiation is also a problem that must be the subject of serious study in the near term, particularly from the standpoint of environment characterization. We also need to obtain additional data on some of the cross sections for the cosmic particles which describe how they react with some of our materials. Also there are still some questions to be answered on acceptable human doses, and we need to better understand cell damage done by some of the highly energetic particles. We also need to pursue angular-momentum management, as well as center-of-rotation control associated with space stations that use tethers and active counterbalance masses.

One should remember that it was relatively easy for us to deal with most of these issues because we had only to satisfy the requirements of our simplified modeling process. All we had to do was assume that something could be done and that it would be represented by a given mass, and then go on about our business. When concepts are worked out in this manner, the *one* thing we can count on is that a lot of **practical** issues will ultimately surface and "eat our lunch!"

Human Issues

Some of the human issues with which we must become more involved include the physiological concerns discussed by Dr. Nicogossian, the psychological problems/concerns other Mars Conference participants have addressed, and a variety of social issues not often considered. The physiological and health-maintenance issues associated with long-term space residency have, quite literally, life and death implications. Psychological issues, on the other hand, are related more to the psychology of long-term space residency and isolation, but once again represent serious consequences because of the unalterable course of events associated with confinement to an artificial environment in deep space. That is, once the mission gets underway, its crew

---

4 An excellent example of the communication-delay factor was experienced during the Viking landings on Mars, both of which occurred when Mars was approaching conjunction and one-way communication time was 20 minutes or more (a range of roughly 200 million miles). The elapsed time from the point at which each lander began its final aerodynamic descent in the martian atmosphere to the instant of touchdown on the martian surface was less than ten minutes. Both landers were actually safely on the surface taking pictures before mission controllers knew for certain that terminal descent had even begun, and it would therefore have been impossible for them to actively participate in that critical process had there been any reason to do so. {Ed.}

members will be totally dependent on each other and their technological systems for the duration of that mission (up to three years), no matter what happens.

Psychological Analogy -- To help put the psychological problem in perspective, Michael Duke [NMC-3A, Session·3 Chairman] related to me his experience at a conference in northeastern Alaska. It was held at Prudhoe Bay, an extremely remote site on the coast of the Arctic Ocean. The crew at Prudhoe Bay must be rotated every two weeks because of problems with isolation and confinement. That situation certainly does not reflect the extent of the isolation and confinement one will experience on a Mars ship, and we certainly are not going to be able to rotate our personnel every few weeks; indeed, it is really more on the order of every three or four years. Clearly, the psychological issue is a problem as serious as any we will have to address relative to the manned exploration of Mars, and it cannot be taken lightly.

Social Order -- There are social issues we need to concern ourselves with, as well. Mission and personal activity will not be scripted for these deep space populations as it has been aboard near-Earth space vehicles. For example, they will have opportunities for--and may reasonably demand--a degree of self-government. In addition, recreation must also be provided to help maintain morale as well as health, and probably bears very importantly on the physiological and psychological issues just reviewed. We certainly wouldn't want our autonomous, self-reliant, self-governed crews to rebel and go somewhere else; we would prefer that the first residents of Mars be friendly!

NASA Space Station Implications

A sketch of the NASA Block·1 Space Station is presented in Figure 3D-21 (and page A-7, Color Display Section). As has been noted by others, the impacts of and implications for NASA's LEO space station (referred to only herein as NSS-1, representing the Block·1 phase, to avoid confusion with Cycler and other advanced space station concepts) will be extremely important in terms of mission planning and design for Mars missions. We will need to pursue elements of technology, research, and operations through opportunities afforded on NSS-1. With respect to the development of technologies, for example, we will want a propulsion test bed among its facilities. In addition, a space station like NSS-1 will provide necessary opportunities to address a variety of other problems, including the physiological/psychological issues I have touched upon. There will also be some operational impacts associated with the station's evolution, from the standpoint of significant mass throughput and assembly required for routine Mars visits.

Life Support Systems -- We may also need to perform *closed ecological life support system* (CELSS) research on NSS-1. In terms of life support systems development, depending on applicable decisions relative to early manned Mars missions, we may even want to augment plans for system closure on NSS-1 by adding biological closure to make it complete. There is debate going on even now, concerned with whether the NSS-1 life support system should be fully or only partially closed at the outset. If we are going to have an early mission to Mars, driven by some sort of public decree, then we must actively involve ourselves in that decision, trying to achieve full life support closure as early as possible. This will be extremely important in the development of expert knowledge and experience for application to the design of the autonomous systems we will need for going to Mars.

Concepts Representative of Advanced Planning

Advanced LEO Space Station -- Figure 3D-22 (page A-6, Color Display Section) is an artist's concept of an advanced LEO space station configuration we conceptualized, and it is based on the amount of traffic that we think will be moving through a LEO space station of this kind on its way to a spaceport located at a Earth-moon libration point. The illustration shows, for example, the bays needed to dock and process the orbit transfer vehicles (OTV's), and the Shuttle-2 is seen coming up to rendezvous with the station -- perhaps transporting a rotation crew. We have included enough storage for propellants on this advanced station to satisfy the scenario I have described.

Mars Outpost -- Figure 3D-23 (page A-6, Color Display Section) is another fairly nice artistic representation of an aggressive Mars base[5]. The characteristics of the illustrated base are that it represents a mature, permanently manned outpost on Mars, and that it is responsible for delivering approximately 100 metric tons of resource material at every **Cycler** opportunity (occurs, as previously noted, on 2.14-year cycles). The steady-state population of such a base would be approximately 20, and the Mars surface tour-of-duty would average about four years. Cost estimates for something of this magnitude, including development and operations, is approximately two times the Apollo program, spread out over 25 years. ■

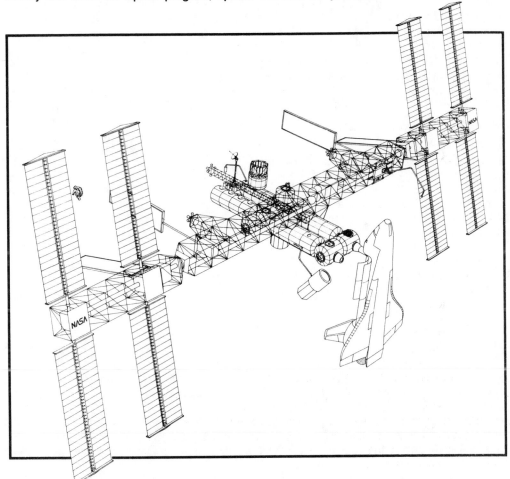

Figure 3D-21: U.S. NASA Space Station (see also pg A-7, Color Display Section)

---

5 The beautifully produced, artistically rendered engineering concepts used in this presentation, most of which have been reproduced in full color in the color section at the back of the document, were created by Pat Rawlings and Mark Dowman of Eagle Engineering, Houston, TX. {Ed.}

REFERENCES

This space is otherwise available for notes.

~~~~~~~~

KEY TECHNOLOGIES FOR EXPEDITIONS TO MARS
[NMC-3E]

Mr. James R. French, Jr.
Vice President, Engineering
American Rocket Company

A round-trip manned Mars landing mission, while difficult, is within the realm of today's technology. Massive vehicles would be required for departing Earth, however, and new technologies are needed to help reduce this mass. High technology concepts like fusion propulsion and antimatter propulsion may one day greatly reduce the requirements, but these technologies are still far in the future and may prove impractical. However, better understood and more practical advanced technologies offer promising improvements in mass performance, and some also enhance certain aspects of self-reliance for manned missions. Among these are the use of aerodynamic drag to aid or replace high energy propulsion to achieve aerocapture into closed orbits about a planet, and aerobraking to modify those orbits. These techniques reduce the need for propulsion-related technology and its mass. In addition, in-situ propellant production (ISPP) could provide fuel for both exploration vehicles on the surface and Earth-return vehicles, in addition to generating other consumables for human use. The use of aerocapture and ISPP together can reduce mission mass (approaching Mars) by a factor of 6 or 8 for a given useful payload mass. And, since the amplification factor projected back to Earth's surface is 60 or 70, the savings in launch tonnage becomes very significant. In the case of multiple missions or support for a permanent outpost on Mars, transport costs will be a major factor -- enhancing the importance of these technologies.

INTRODUCTION AND OPENING COMMENTS

In keeping with my character for this Mars Conference, I will start with some mild sarcasm. Barney Roberts [NMC-3D] used a cartoon illustration ("Hayburner" analogy, Fig. 3D-1, pg 436) to suggest how little the western border of the United States would have been pushed westward (only to the Appalachians) had we explored our country the way we are exploring space. In fact, had we explored the United States the way we have explored space during the last ten or twelve years, we would still be sitting on our rumps in the Atlantic surf near Plymouth Rock doing high-tech studies of wagons that never got built!

The subject of my Session 3 presentation is: *key* technologies for missions to Mars. As one might imagine, there is a long list of technologies that might be considered "key technologies" with respect to the manned exploration of Mars, and those that make up a given list essentially depend on the technical prejudices of the individual(s) responsible for it. If I were to do nothing more than present the entire list, my presentation space would be used up by that alone. I have chosen, therefore, to discuss in some depth two particular technologies toward which I have developed a degree of partiality and with which I have had a significant involvement over the years: 1) aerodynamic capture, and 2) extraterrestrial propellant production through in-situ resource exploitation.

MISSION ENHANCEMENT TECHNOLOGIES

The first of these technologies, which includes aerodynamic capture into orbit as well as some related concepts involving aerodynamic influences, is generally referred to as *aerobraking* or *aerocapture*. The second technology has been referred to as *in-situ propellant production* (ISPP), the production of propellants using the natural resources of the planet (or satellite). My knowledge of these technologies grew out of work at JPL with respect to sample return missions, and they are NOT the ones I consider to be "enabling" in importance; that is, a manned Mars mission *can* be performed without them. However, I think they are extremely crucial if we are going to do this kind of mission well and do it on a continuing productive basis.

Scale of Mars Mission Technologies

While quite difficult in its complexity, a round-trip manned Mars landing mission is within the realm of today's technology. Clearly, though, massive vehicles would be required. Basically, this implies that if we do the kind of manned Mars mission we have typically considered, wherein everything needed (all of the human consumables, fuel, et-cetera) is provided in the initial manifest at the outset on Earth, the Mars vehicles departing Earth orbit will have to be huge. Consequently, the transportation costs will likely eat us alive, especially if we plan to do it more than once.

There are, of course, some high-technology concepts on the horizon that can help us deal with that problem, including things like fusion propulsion or even antimatter propulsion. The latter is not just a *Star Trek* fantasy, but may actually be practical (with perhaps a more realistic concept of the velocities it might produce). These propulsion technologies may one day significantly reduce the transportation burden imposed by vehicle mass, but they are well in the future and are not really practical in terms of current consideration. If we sit around and wait for somebody to come up with a transportation system that makes it easy, we may never get off that Atlantic beach I alluded to previously. These advanced technologies, then, should perhaps be considered more as being *almost* in hand, and that they may eventually be "doable" in the sense that they could ultimately--through proper utilization--help make the manned exploration of Mars routine and a permanent base there a reality.

THE USE OF AERODYNAMIC DRAG

In the meantime, new technologies have already been demonstrated and are still evolving with respect to the use of aerodynamic drag. These technologies could greatly enhance current capabilities, and they require only engineering for the specific application in order to achieve significant improvements.

Aerocapture

Aerocapture, defined in Figure 3E-1, is perhaps one of the most productive techniques we have studied. It involves the use of controlled aerodynamic drag to slow a spacecraft and divert it from its flyby trajectory. A Mars flyby trajectory is a hyperbolic approach trajectory that

458

essentially targets the spacecraft to shoot past Mars and out into space. Aerocapture makes it possible to turn the spacecraft's trajectory into a closed orbit around the planet without the need for a major propulsive maneuver. By significantly reducing or even eliminating the requirement for orbit-insertion propulsion, one greatly reduces the mass of the vehicle required to get a given payload into orbit about the planet.

One of the important advantages of this is that one can achieve, essentially for free, a substantial plane change using aerocapture. This capability becomes possible because the drag of the spacecraft itself is being used to wear away the approach energy. The trajectory is controlled by modulating the lift of the spacecraft, usually by banking the airframe. On the other hand, it is also possible to essentially "S-turn" along a straight trajectory if a plane change is not desired. If a plane change *is* desired, one simply leans the bank angle one way or the other and bends the plane around. This is perhaps comparable to the manner in which a space shuttle orbiter is flown back from Earth orbit when it must land someplace that is not under its flight path. The Apollo spacecraft could perform the same sort of maneuver, but only to a very limited degree.

DEFINITION:

■ The use of **Aerodynamic Braking** to slow an arriving spacecraft from approach velocity to that required to achieve capture into a Closed Orbit about a planet.

ADVANTAGES

■ Greatly Reduced Mass Penalty for Achieving Low Circular Orbit

■ Substantial Change of Orbit Plane "for Free"

■ Reduces Payload Sensitivity to Mission Opportunity

DISADVANTAGES

■ Tight Approach Guidance Requirements (tradeoff with L/D)

■ Packaging for Aerodynamic Shape Complications

■ Complex Guidance and Control Problem

■ Flight at Relatively Low Altitude at Mars

REQUIREMENT

■ Moderate to High Lift/Drag Shape

■ Predictive Guidance and Control Loop

■ On-Board Approach Guidance (optical and/or Mars beacon)

Figure 3E-1: Overview of Aerocapture and Associated Technology

459

An additional advantage is that aerocapture substantially reduces sensitivity to mission opportunity. Because the flight energy is being taken out nonpropulsively upon arrival at Mars, it doesn't matter a great deal if the current opportunity is worse than another one in terms of planetary geometry.

Importance of L/D Shaping -- In spite of its advantages, aerocapture is not a cut and dried solution; i.e., "there ain't no free lunch!" One must pay for aerocapture through complexity, in that it imposes very tight navigation requirements. For example, the spacecraft must hit the "entry corridor" very precisely, and that corridor tends to be fairly narrow. The higher the lift-over-drag (L/D) of the vehicle, the wider the entry corridor becomes. I must respectfully take issue with some of the configurations suggested at this conference on the basis that they are very blunt bodies, and I am not at all certain one could successfully achieve aerocapture with such designs. In order to have adequate control authority, one needs a fairly high L/D; it is highly debatable--based on our analysis--that control can be achieved if the L/D is less than roughly 0.4 to 0.6, which is not obtainable for a very blunt body. What one would like is a shape on the order of the "bent biconic" shape that I and Doug Blanchard described in our Session·2 presentations concerned with Mars sample return missions. Such a vehicle, depending on its trim and angle-of-attack, would reflect a L/D in the range of 0.6 up to about 1.5, and these numbers satisfy the requirement we feel would be necessary for aerodynamic capture.

There are several aerodynamic shapes suitable to aerocapture: the bent-biconic shape already mentioned, the raked-cone approach similar to the concept Barney Roberts discussed, and then, if one really wants to get into high L/D numbers, there are the winged vehicles. The space shuttle orbiter itself is an example of a shape that would do a nice job of aerocapture at Mars.

As implied, then, a space vehicle capable of performing successful aerocapture requires a reasonably good L/D shape as well as a precise guidance and control system. The latter is dictated by the fact that critical energy is being bled off during the pass through the atmosphere, and the spacecraft *must* be capable of ejecting itself from the atmosphere at the correct point. The guidance and control system is crucial because it must be able to determine very precisely when enough energy has been bled off so that the maneuver needed to lift the spacecraft back out of the atmosphere and into the desired orbit can be conducted. In other words, the spacecraft must have a sufficiently predictive guidance system to let it know when enough energy has been removed and when the time is right to--as it were--pull back on the stick and pop itself out of the atmosphere. In this sense, aerocapture is a relatively complex procedure. More importantly, perhaps, the capability to perform it has to be on board the spacecraft -- there is no way one could control it from Earth.

A final important point worth noting is that when the aerocapture vehicle exits the atmosphere, it is in a slightly elliptical orbit. Its apoapsis will be at the desired orbit altitude, but its periapsis will still be in the atmosphere. A circularization burn of the propulsion unit (typical ΔV = 100 m/s) at the first apoapsis pass is therefore mandatory to prevent entry on the following periapsis pass.

Aerocapture Entry Performance -- As I indicated earlier, the nature of the entry corridor represents a critical problem in itself. If the spacecraft enters atmosphere that is too thin, it will exit the atmosphere before it has slowed down enough to achieve its closed orbit (or at least the orbit desired). In the opposite case, I do not believe I have to elucidate the problem of being too low, unless of course we want to preempt our Japanese colleagues and their proposed penetrator mission. As I noted earlier, there are some very tall mountains on Mars, and I don't believe the crew would appreciate having to look up rather than down at them -- while still in their "orbital" vehicle.

460

■ Aerocapture Can Deliver 80-85% as Useful Payload to Circular Low Mars Orbit (LMO)

   • The Remainder Being Airframe, Thermal Protection, Propulsion Hardware, and Expended Propellant

■ Propulsion Insertion Delivers Less Than 50% as Useful Payload to Circular LMO

ENTRY ENVIRONMENT

■ Maximum Entry Airspeed at Mars About 6 km/s Compared to 8 km/s from Low Earth Orbit (LEO)

■ Nonoxidizing Atmosphere

AERODYNAMIC SHAPES

■ Bent Biconic

■ Raked Cone

■ Winged

Figure 3E-2: Overview of Aerocapture Performance

Figure 3E-2 (above) reflects the performance contribution made possible using aerocapture. For a given mass delivered on the approach trajectory to Mars, one can put 80-85 percent of it into orbit as useful payload. In comparison, using propulsion (with reasonable storable propellants), the best one can achieve is perhaps 50 percent. Further, if we have to perform a plane change or some other major propulsive maneuver in addition to orbit insertion, the ratio is much worse. Clearly, then, in this respect there is a distinct advantage in aerocapture.

Aerocapture at Mars -- Because of Mars' small size, it is worth noting that for the entry environment encountered, even from a hyperbolic flyby trajectory (i.e., entering at the maximum velocity one would ordinarily see on any such mission), velocity is still slower by perhaps roughly 2 km/s than we see when simply coming home from low Earth orbit (LEO) here at Earth. The fact that the martian atmosphere is nonoxidizing helps, but some of the advantage of lower velocity is lost because the atmosphere is mostly carbon dioxide (tends to radiate strongly in the infrared[1]).

There are some additional technical issues that have to be dealt with even though the physics of aerocapture at Mars are possible (achieving it is essentially a matter of figuring out the right

---

[1] The nonoxidizing nature of the martian atmosphere may allow more options for the selection of heatshield material than is possible in Earth's case. This is due to the fact that in Earth's atmosphere oxygen can react with hot heatshield material during entry. The lower velocity at Mars helps as well, in that it reduces aerodynamic heating due to both radiation from the shock front and convection from the heated atmosphere. However, the predominance of CO_2 partially counteracts this benefit by increasing the radiative heating over what it would be in an Earth-like atmosphere at the same velocity. This effect is small but probably should not be ignored.

implementation). One of the disadvantages with respect to Mars is that the atmosphere is quite thin, which necessitates having the spacecraft--for some cases--fly at a fairly low altitude. Indeed, by "fairly low" in this context, I mean possibly as low as 10 km (approximately 33,000 ft) above the mean surface elevation. As a suggestion of scale, the Tharsis shield volcanoes have summits that are more than 80,000 feet above their basal elevations. If one is standing on Earth watching an aircraft in flight at that altitude, one can perceive that it is safely up where airliners commonly operate -- certainly well above Earth's surface and mountain peaks. If, however, one is here on Earth considering a spacecraft flying at that altitude above the martian surface, 10 km is virtually indistinguishable from the surface itself. It is all a matter of perspective.

Aerobraking

The technique we call *aerobraking* is perhaps inappropriately named. Anytime one uses the atmosphere to slow down (which also is true of **aerocapture**), aerobraking is being applied. We are, in a sense, applying a term we otherwise use somewhat generically as a name for a specific technique. To apply the aerobraking technique once in orbit around the planet, in what might be a highly elliptical but nonetheless closed orbit, the periapsis (low point of the orbit) is lowered into the upper atmosphere. Then, during repetitive passes through the atmosphere, energy is gradually removed and the orbit is circularized. This, then, is the technique that we refer to more specifically as aerobraking, using essentially the same kind of aerodynamic drag to achieve it that was used to achieve aerocapture. The difference is that drag induced by the aerocapture technique is used only once to achieve capture, while it is used repetitively in the aerobraking technique to modify the orbit. The aerobraking technique is defined and outlined in Figure 3E-3.

Upper Atmosphere -- If plenty of time is available to perform it, the aerobraking technique can be conducted in the fairly thin upper part of a planet's atmosphere. This has the benefit of keeping the heating rates low and requires only a very simple drag brake -- one represented by a fairly large area but which is structurally uncomplicated. It might be little more than a sheet of foil made of a simple high-temperature material, e.g., inconel foil or something similar, shaped to achieve aerodynamic drag in the very tenuous upper atmosphere. "High temperature" in this context implies a range of 1000-1800 ° F. We have, for example, studied an aerobraking vehicle designed for Venus that has two heatshields, a large deployable shield and a smaller fixed shield, and most of the vehicle's structure is then sandwiched between the two shields. Such a design is peculiar to both the spacecraft and the mission for which it evolved, and one could as easily have just one big shield out in front for the application I have described.

The orbit can be circularized quite easily using upper atmosphere aerobraking, and at minimal design cost. But it does require a relatively large amount of time (a period of weeks at Mars or months at larger planets like Earth and Venus). Aerobraking is also quite "ground control intensive" because it necessitates paying close attention to what is going on every time the spacecraft completes an atmospheric pass to determine if its orbital geometry is as predicted. If it is too high, it simply means that we wasted a pass through the atmosphere. However, allowing it to get too low could be catastrophic, because the high-altitude aerobrake can't withstand the aerodynamic heating or structural loads of a low pass in denser atmosphere.

As the orbit geometry approaches being circular, the aerobraking process becomes exceedingly critical. Once a spacecraft orbit is fully circular at periapsis altitude, it will not leave the atmosphere again (without propulsion). I call that the "Skylab syndrome," because the space vehicle will then be on its way toward the ground -- whether or not we want it that way. It is therefore very important and critical that we quit aerobraking *before* the orbit becomes

DEFINITION: ■ Use of Controlled Orbital Decay during multiple passes through the upper atmosphere to circularize a loose elliptical orbit following propulsive insertion.
ADVANTAGES ■ Minimum Impact on Spacecraft Design ■ Approach Guidance Same as for Any Propulsive Insertion ■ Low Heating Rates Allow Simple Drag Brake, Minimum Thermal Protection
DISADVANTAGES ■ Time Consuming (weeks) ■ Ground Control Intensive ■ Control of Final Passes (near circular) Very Critical
REQUIREMENTS ■ Large Area Drag Brake (inconel foil or similar) ■ Constant Maneuvering of Spacecraft to Maintain Attitude

Figure 3E-3: Overview of Aerobraking

absolutely circular. Adding to the criticality of the situation is the fact that the periods between atmospheric passes become increasingly shorter as the orbit approaches circular geometry, and are *very* short toward the end of the process. Indeed, prudence dictates quitting early and paying a little bit of "propulsive penalty" to avoid reaching that point. As noted elsewhere, some amount of propulsion capability is required in conjunction with these aerodynamic techniques to facilitate orbit trims, circularization, and emergency correction. The earlier the aerobraking ceases (i.e., a higher orbital eccentricity), the larger the propulsion requirement. This topic requires some analysis and "risk definition" to define the capability.

Aerobraking Performance -- In terms of performance, aerobraking produces significant benefits, as noted in Figure 3E-4. When starting out in a 24-hour elliptical orbit, this type of aerobraking vehicle can again deliver about 85 percent as useful payload down to a nearly circular low Mars orbit (LMO). This compares with about 65 percent useful payload if circularization is achieved using propulsion. Some propulsion is required on the aerobraker to facilitate orbit trims and the final circularization, but those requirements are modest. Propulsion will also be required to enter the initial elliptical orbit. As a result, aerobraking generally does not offer better performance than aerocapture, in terms of mass in low orbit versus approach mass, but both methods are clearly far superior to an all-propulsion mission profile. One great advantage of aerobraking over aerocapture (shared with the all-propulsion mode) is flexibility in spacecraft design. This contrasts with the aerocapture requirement to package the spacecraft in a protective aerodynamic shell to contend with the more severe aerodynamic environment it will encounter.

In order to effect aerobraking as described, it is *very* important to keep the spacecraft up at the bottom of the free-molecular-flow regime (the very top of the atmospheric transition region). Because the wake does not close rapidly behind the aeroshield in this tenuous atmosphere, one does not have to worry about aerodynamic heating of the spacecraft and its payload (even though aerodynamic heating might induce a temperature of up to 1000°F (538°C) on the aeroshield itself).

FOR A GIVEN MASS IN 24-HOUR ELLIPTICAL ORBIT AT MARS

- Aerobraking Can Deliver 85% as Useful Payload to a Circular Low Mars Orbit (LMO)

  • The Remainder Being Drag Brake and Propulsion

- Propulsive Circularization Can Deliver 65% as Useful Payload to Circular LMO

ENVIRONMENT

- As Long as Atmosphere Passes Remain in Free-Molecular-Flow Regime or Upper Transition Region, There is No Significant Heating to Payload

- Drag Brake Temperature May Reach about 1000°F

AERODYNAMIC SHAPES

- Flat or Cone-Shaped Drag Brake
- Spacecraft on Lee Side

Figure 3E-4: Overview of Aerobraking Performance

Aeromaneuvering

Referring back to our generic concept of what **aerobraking** represents, *aeromaneuvering* simply reflects what can be done with it by maximizing its potential through the use of the higher L/D vehicle shapes. Aeromaneuvering offers the advantages and problems reflected in Figure 3E-5. The problem with low L/D vehicles, in simple terms, is that they don't glide very well. Low L/D vehicles do not afford good control for correcting trajectory errors while being guided to a precise landing. We have done very well with them on Earth with capsule spacecraft like Mercury (which afforded very little control at all), Gemini and Apollo, but that is because we had tracking stations on Earth and could control the deorbit operations with extreme precision. We won't have that advantage on Mars -- at least in the early stages.

A great deal of the martian surface is available to a landing spacecraft that has made a direct entry from a hyperbolic approach. By manipulating L/D shapes and setting up approach trajectories, one can essentially land anywhere on the planet. This, of course, is not true for low L/D

DEFINITION:

■ Use of **Medium** and **High Lift/Drag** vehicles during entry for landing, to control landing accuracy and choice of landing sites.

ADVANTAGE

■ Allows Landing Anywhere Rather Than Just Area Beneath Orbit
■ Steer Out Errors for Higher Accuracy

DISADVANTAGES

■ Complex Guidance
■ Slender Aerodynamic Shapes

PERFORMANCE

■ For Entry At or Above Escape Velocity, Using Vehicle With L/D of 1.5, Virtually Entire Planet Available
■ Several Hundred Kilometers Cross-Range Available From Low Orbit
■ For Slightly Higher L/D Vehicle, Error in Primary Landing Point is Map Error Only

Figure 3E-5: Overview of Aeromaneuvering

(blunt-bodied) shapes using semi-ballistic approaches. These vehicles are committed, beyond a given approach point, to land within a relatively small footprint with limited control flexibility.

In spite of their advantages, however, higher L/D shapes must pay a penalty in terms of constraints imposed by the slender configurations and the increased complexity of the necessary guidance system. In exchange, one can get several hundred kilometers of cross-range capability (even from LMO). And, when using the L/D shapes with the highest reasonable lift values, it is possible to create a situation where the only significant error in landing accuracy is a result of errors in the map being used to plan the landing, i.e., the precision of the location of the landing site on our maps versus where it is in reality. Speaking as a pilot, I know how airports tend to move around relative to where they are supposed to be according to my FAA sectional charts, and that will probably be a greater problem on Mars (with all due respect to the Mars mapping work being done by the USGS, which I am confident is superior to that reflected in my FAA charts).

When a high-L/D vehicle has slowed down to a predetermined velocity during the final phase of its landing sequence (around Mach 2 for that shape), a drogue chute is opened. At around Mach 1, the main parachute is opened and the shroud lines can then be manipulated to help orient the vehicle into a tail-down configuration (somewhat the inverse of what we did on Gemini, wherein the vehicle was reoriented from vertical to horizontal attitude prior to touchdown). Once its orientation is correct, terminal descent engines can be used to achieve a precise, powered touchdown.

465

In·situ Propellant Production (ISPP)

In·situ propellant production (ISPP) associated with manned Mars missions, as defined in Figure 3E-6, simply implies that we will tap available resources on or in the vicinity of Mars (including Phobos and Deimos) for the materials needed to make propellants (fuels and/or oxidizers) to operate a variety of space and surface vehicles. And, while it is certainly possible to plan and conduct a Mars mission without ISPP, the payback for developing and utilizing this technology as early as possible is so great that it is hard to imagine a truly committed and productive program moving forward without it.

DEFINITION:

- Use of **Martian Resources** to generate the propellant necessary for launching the landing spacecraft on the return trip and to power exploration vehicles on Mars.

ADVANTAGES

- Greatly Reduced Mass Departing Earth Due to Sending Lander Vehicle With Nearly Empty Tanks (just enough to land)
  - Probable Mass Reduction Factor: 3 or 4
- Reduced Mass Simplifies Entry/Landing

DISADVANTAGES

- Reliability Concerns
- Limited Propellant Choice Based on Resources

REQUIREMENTS

- Means of Gathering/Processing Feedstock
- Production Cell
- Liquefaction/Storage Capability
- Compact Source of Hundreds of kW of Electricity and Heat

Figure 3E-6: In·situ Propellant Production (ISPP)

Non-ISPP Approach -- The best way I can make an argument for ISPP technology is through an analogy using a transatlantic airline. If the airline service between New York and London elected to fuel its 747 for the entire round trip in New York, essentially taking off with all of the fuel needed to complete the trip in order to avoid refueling in London, it will not be very profitable because the entire payload capacity will be eaten up by the mass of the fuel. In fact, I'm not even sure that a 747's payload capacity would accommodate enough fuel to complete that particular round trip. This is exactly the way we typically propose to explore space, and it makes no sense to me to do it that way when there are resources available at the other end to facilitate refueling for the return journey.

By incorporating ISPP into the mission plan, not only is the otherwise necessary mass departing Earth greatly reduced (because propellant for the return flight is no longer part of the payload at departure), but the entry and landing problem at Mars is made a good deal simpler. One of the driving parameters in atmospheric aerodynamic braking is the so-called ballistic coefficient of the vehicle. The ballistic coefficient includes the drag coefficient and must also take into account the amount of mass per unit area of the vehicle. The more propellant mass carried by the vehicle, the bigger it has to be to get the ballistic coefficient low enough to make slowing down in the thin martian atmosphere possible. One can even postulate a vehicle that behaves like a brick, in that it will drop right on through the atmosphere without slowing down much at all and hitting the ground essentially at entry velocity. That has obviously been done a lot by meteorites. In this sense, then, reducing the mass of a vehicle used to enter the atmosphere and land on Mars clearly makes the whole problem a lot simpler.

Problems and Options -- There are certainly some disadvantages associated with ISPP, as well, some of which come to mind quite naturally. For example, the mission is clearly quite dependant on the reliability of the ISPP system -- it has *got* to work! In addition, the propellant choices are somewhat limited based on the available resources, particularly for the early missions. A capability must be provided to get at the resources; in other words, making it possible to acquire "feedstock." The methods of production and storage must be both well understood and facilitated (by necessary hardware). And, finally, a means of generating power must be provided; it has to be reasonably compact and reliable, and it also must be capable of producing several hundred kilowatts of electricity as well as thermal energy.

Figure 3E-7 outlines the options that are most obvious for Mars. I should note, however, that they are not the only options possible; one can postulate making all sorts of things out of the known martian resources. My outline essentially confines itself to those that most readily suggest themselves, such as liquid oxygen (LO_2) and liquid carbon monoxide (LCO), a propellant combination that has never been used on Earth because there is no particular reason or need.

Liquid Oxygen/Liquid Carbon Monoxide -- The performance aspect of the LO_2/LCO combination is not particularly good compared to other propellants generally available to us, so there is no driver for its development here on Earth. Still, there is no reason why it should not work reasonably well, and it would certainly be the most readily available propellant option in the case of Mars.

Indeed, availability is its singular advantage on Mars, in that it can be produced entirely from the martian atmosphere (which is predominately carbon dioxide -- CO_2). And, while the martian atmosphere is not very dense, it is everywhere and therefore very accessible. To produce oxygen (O_2) and carbon monoxide (CO), one simply compresses a volume of the atmospheric CO_2. This is not necessarily easy, but it is perhaps a lot easier than strip-mining permafrost or drilling wells to obtain water ice. The O_2 and CO products are relatively easy to liquify, once made, and are therefore similarly easy to store (a mature technology on Earth).

Rocket engine performance with LO_2 and LCO, while modest, is quite satisfactory on Mars. In fact, given the relative energies involved, the LO_2/LCO combination on Mars is just about as good as LO_2 and LH_2 (liquid hydrogen) on Earth. On that basis, then, its performance is not trivial when one considers it from the viewpoint of Mars applications. More importantly, perhaps, the production technology for this particular purpose is already understood reasonably well, although not widely used here on Earth.

Liquid Oxygen/Liquid Methane -- The next choice would be LO_2 and liquid methane (LCH_4). The most critical component is hydrogen, and it would be quite a coup if we could produce this combination. One can synthesize methane from CO_2 and H_2O, and it is made quite readily

---

Liquid Oxygen/Liquid Carbon Monoxide (LO$_2$/LCO)

■ Atmosphere (CO$_2$) is Only Required Feedstock-Ubiquitous, Easily Obtained

■ Relatively Easy to Liquify and Store

■ Rocket Engine Performance Modest but Adequate for Mars

■ Produced in Thermal Decomposition Cell

Liquid Oxygen/Liquid Methane (LO$_2$/LCH$_4$)

■ Use CO$_2$ From Atmosphere

■ Requires Water

  • Availability on Mars Uncertain Except at North Pole
  • Bringing Water From Earth Carries Significant Mass Penalty

■ Excellent Rocket Performance

■ Relatively Easy to Liquify and Store

■ Several Well Known Chemical Processes Can Do the Job

Liquid Oxygen/Liquid Hydrogen (LO$_2$/LH$_2$)

■ Requires Water (see comments for liquid methane above)

■ Best Rocket Engine Performance

■ Hydrogen Very Difficult to Liquify and Store

Figure 3E-7: Options for In-situ Propellant Production (ISPP)

in association with industrial processes here on Earth. It would actually easier to make LO$_2$/LCH$_4$ than LO$_2$/LCO -- but for one exception in the case of Mars. The process requires copious amounts of water (industrial quantities in multi-ton lots), and water is going to be hard to find--much less produce in large, efficient throughput quantities--on Mars.

Water *is* available on Mars, and there is probably even quite a lot of it in some places. However, we know for sure that it is present at only one location -- in the North Pole's ice cap. We have strong evidence suggesting that it also probably exists beneath the surface in many other areas (as permafrost deposits or even as frozen reservoirs), but those deposits must first be found and verified, and would not be readily available to early missions. One option that has been suggested is to bring water from Earth to use in combination with Mars' CO$_2$. Unfortunately, that amount of water would constitute 45 percent of the total mass required, and it is not clear that the higher performance afforded makes up for the mass penalty. It may be applicable for some cases, but tradeoff analyses are required.

Rocket engine performance with LO$_2$/LCH$_4$ is excellent, providing approximately 25 percent higher performance than LO$_2$/LCO (exhaust velocity of 3340 m/s versus 2650 m/s). Again, liquid methane is relatively easy to liquify and store. It is also a pretty good refrigerant, lending itself quite nicely to the process by helping to keep the whole system cold.

Liquid Oxygen/Liquid Hydrogen -- Of course, the fuel we all think most about is LO_2 and liquid hydrogen (LH_2), and we have already used it extensively throughout the history of our space program. While it is clearly a high performance option, it clearly multiplies the problem associated with its requirement for large quantities of water. Once again, it would require a large water resource that afforded relatively easy and efficient access. The atmosphere does contain water, of course, but it doesn't represent a sufficient volume for this purpose.

There is no question about the fact that LO_2/LH_2 produces excellent rocket engine performance. However, LH_2 is extremely difficult to store. Indeed, it is virtually impossible to store it without sustaining some losses. In reality, then, it really does not seem like a viable choice -- at least during the early stages of Mars exploration.

Propellant Preference -- If water were readily and efficiently available, my choice would be LO_2/LCH_4. However, since we don't know for sure that water will be available or sufficiently easy to obtain, my feeling has always been that the initial Mars surface ISPP choice should be LO_2/LCO, because it requires no water and can be produced from atmospheric CO_2 alone. Then, when the day comes wherein we do find adequate quantities of martian water, and can easily make it available on a continuing basis, we can certainly move on to the more preferred fuel (LO_2/LCH_4) in order to achieve the higher performance it represents.

ISPP Productivity and Options

The worth of the ISPP concept hinges on the mass of the propellant manufacturing system versus that of the propellant generated. In this respect, a long period of time between the arrival at and departure from Mars helps. A plant system with a throughput of a few hundred pounds per day can easily generate the mass of propellant required. Even allowing for redundancy, the plant mass would be only a few tons (e.g., three or four) versus about 100 tons of propellant produced -- clearly a good trade. However, the choice of power source is crucial. Such a production plant would require a few hundred kilowatts of electric power. Solar arrays are impractical, and the only viable source is a small nuclear reactor. By using martian soil as radiation shielding, the mass of a suitable reactor system can be held to only two or three tons. An additional advantage is that propellant manufacturing (mostly a thermal process) can efficiently utilize heat from the electrical generation process that is normally wasted, thus greatly reducing reactor/generator system size and mass.

There are, of course, many other possibilities that could bear upon the ISPP technologies I have discussed, including the possible exploitation of potential resources on Phobos or the development of advanced interplanetary propulsion technologies that would not require transporting great masses of propellant. While these technologies have been discussed in some detail by others in these proceedings, it is appropriate to my own discussion to at least quickly review them to broaden the perspective of my topic. My quick review is included in Figure 3E-8.

Barney Roberts discussed the possibility of ISPP on Phobos, and it could certainly be very advantageous. What we must continue to bear in mind, as has been frequently pointed out, is that while some data suggest that Phobos is a carbonaceous chondrite (which are known to have water bound up in them), we do not know that either is true in the case of Phobos. Indeed, we aren't likely to know for sure until we have been there. The Soviets may find out, but even their findings may not be unequivocal. For example, if there is a thick layer of regolith on the surface and the water is located down near or at the satellite's core, they may not find it, either.

Also, as Carl Sagan has pointed out, it might be necessary to strip-mine large quantities of Phobos to get at any water one might find there, and that would be extremely difficult in itself. Even if one overlooks the fact that the process could eat up the whole satellite in a few years, the question of how one performs that amount of work in a virtual zero-g environment is not a trivial matter.

Propulsion Options for Future Consideration

Other kinds of propulsion certainly need to be considered, and two are reviewed in Figure 3E-8. These types of systems have a variety of problems associated with them and are not practical for current consideration, but they offer advantages that make them worthy of continuing study.

ISPP at Phobos

■ Advantages -- Provide Propellant to Escape Mars Orbit

■ Concerns -- Are Suitable Resources Available?
    Can They Be Readily Obtained?

Nuclear Rocket Engine

■ Advantages -- Higher Performance Than Any Chemical System

■ Concerns -- Radiation from Reactor,
    National Hysteria Concerning Nuclear Systems

Low-Thrust Options, Electric (ion) and Solar Sail

■ Advantages -- Very High Performance, Hence Lower Mass

■ Concerns -- Very Long Flight Times May Be Unsuitable for Humans
    (Optional Cargo Vehicle Applications)

Figure 3E-8: Propulsion and ISPP Options

Advanced Propulsion: Nuclear Rocket Engine -- Nuclear rocket engines could offer some significant performance advantages and would be superior to the conventional cryo-chemical systems now in use or proposed for the near future. The primary technical problem with nuclear engines is that they impose the need to provide radiation shielding for the crew. In a sense, crew members would spend the entire trip hiding from the reactor behind shadow shields. This is essentially a technical problem, however, and it can be solved. Such may not be the case with the second problem associated with nuclear engines. The worst problem associated with nuclear engines is social in origin, in that I seriously doubt that we could develop a nuclear rocket engine of adequate size in the United States today. The orchestrated hysteria we see, concerning anything that has the word "nuclear" associated with it, would I think preclude even starting such a development program. That is very unfortunate, but it seems to be a fact that we cannot overcome[2].

---

[2] Although the prevailing national attitude about the use of nuclear energy in any form is one of concern and fear (with reasonable justification in view of its potential risk for humans), it is still quite possibly the most reliable and efficient form of energy available in the context of power generation capacities needed by large deep-space systems. Indeed, such technologies may be crucial to objectives involving long-duration, long-stay manned missions, wherein alternatives to nuclear systems aren't practical or could put life at greater risk through increased chance of deficiency or failure. The development of reliable and safe nuclear support technologies is therefore important as both a resource for potential future capabilities and as a basis of comparison for alternative systems. The availability of these technologies will help define the capabilities needed to conduct the aggressive mission activities or resource exploitation operations discussed at the NASA Mars Conference. {Ed.}

Low-Thrust Electric and Solar Sail -- Low-thrust propulsion, particularly as represented by electric (ion) or solar-sail technologies, are quite advantageous for many deep-space missions. However, we find that they are not especially attractive for Mars missions because most of the needed velocity change must take place within planetary "gravity wells" where these kinds of systems are least efficient[3].

With low-thrust systems, for example, a space vehicle spends much of its flight time spiraling down to orbit around a planet or spiraling back out (within the planet's gravity well). The result of this was demonstrated during one particular Mars sample return mission study we conducted, wherein our calculations for electric ion propulsion showed that we had to leave Mars before we got there in order to get back. This somehow seemed to be quite impractical!

While the very long flight times are not desirable for human crews, it is worth continuing the study of these low-thrust propulsion systems for their potential in other applications. For example, they may have potential as unmanned cargo vehicles whose missions are not time-critical, in which case they may actually offer some substantial advantages.

CONCLUDING THOUGHTS

In summary, then, the use of aerocapture and ISPP together could reduce the mass required for a given payload approaching Mars by a factor of six to eight. Secondly, if one projects these mass savings all the way back to the launch pad on Earth, the amplified mass factor increases to 60 or 70. While the results of these calculations are highly variable at best, depending on who does them and for what purpose, the important point is that the mass reduction is significant no matter how one figures it. This clearly demonstrates the importance of being able to apply these kinds of technologies with respect to Mars, especially when planning to go there more than once.

As just suggested, I should explain that these are *my* figures. One will find that mine do not agree with those given by Barney Roberts in his presentation. This is not necessarily a case of figures lying or liar's figuring; it is simply a matter of how one approaches and utilizes the appropriate data. Barney's system, for example, represents a massive piece of machinery that remains in orbit around Mars and a relatively small piece of machinery that goes to the surface, while my scenario is based on putting most of the mass on the surface. If one does it my way, manufacturing propellant on the surface of Mars becomes much more important -- even more so than doing it in orbit (Phobos). Therefore, it is important to understand that these comments are not intended to try to prove something one way or another, or to suggest that one approach is right and the other wrong. They are instead intended only to point out that conclusions like these are strongly dependent on the type of mission one is considering and how one records the pertinent factors. Anytime one sees these sorts of numbers, it is important to understand the assumptions upon which they were based.

---

3 In general terms, a "gravity well" is that distance from a given planet at which a spacecraft essentially breaks free of--or becomes influenced by--the planet's gravitation while either departing or approaching. The depth of a gravity well is determined by the force of a planet's gravitational attraction relative to "free space" (represented by a value of zero). The Earth's gravity well, for example, is approximately 4000 miles in depth. In comparison, Mars' gravity well is only half as deep and the moon's gravity well is only 180 miles in depth. Because enough propulsive energy must be provided to lift consumable mass out of the gravity well from which it is to be taken, it is clearly prudent to acquire as much resource material as possible in-situ on Mars or from shallow gravity wells like the moon, the two martian satellites, and perhaps even (some day) from asteroids available in interplanetary space). {Ed.}

If we "plan" to go to Mars only one time--essentially to plant the flag, stick out our chests, take a few pictures, and then come home--it probably doesn't matter how that mission is done. But, if we want to go, return, and then establish a continuing presence, reflecting the kind of continuity most of us want to see, then we'd better figure out how to do it efficiently and well -- and as early as possible. In the case of multiple missions or a permanent outpost, transport costs will be a major factor of the support and maintenance required. In that eventuality, the key technologies I have discussed will be crucial to the achievement of the objectives we have defined at this conference, and that is really what these concepts are all about. ∎

REFERENCES

This space is otherwise available for notes.

~~~~~~~~

HUMAN FACTORS FOR MARS MISSIONS
[NMC-3F]

Dr. Arnauld E. Nicogossian, M.D.[1]
Director: Life Sciences Division, NASA

An expedition to Mars would be a mission of unprecedented distance and scope. Indeed, such a mission--with reasonable exploration time on the surface and perhaps an excursion to one of the martian satellites--would represent a possible round-trip mission time of two to three years. Before humans can safely undertake such a mission, however, medical implications and consequences peculiar to long-duration space habitation and work must be exhaustively studied to facilitate the development of responsive systems and methodologies to cope with its unique problems. There are many issues and unknowns to be resolved or understood, but such matters have always been an integral part of science research, including space medicine, serving as both challenge and reward. A set of three factors must be considered in dealing with these issues: (1) human, (2) system (spacecraft/habitat technology), and (3) environment. The latter must be studied in terms of both its external (space) and internal (artificially generated) effects and impacts. The environment will be confining and inescapable, perhaps complicated by long periods of low- or zero-g conditions, such that a variety of physiological and psychological problems could evolve if proper systems, responses and solutions are not well developed in advance. Experience aboard orbiting space stations (e.g., Skylab, Salyut and Mir) has already proven itself to be vital and necessary to this process.

MANNED MISSIONS TO MARS

Whether or not indigenous life is eventually discovered on Mars, it is likely that living creatures of a more familiar sort will visit that planet in the not too distant future. The question of life on Mars remains a fascinating dilemma for the closing years of the 20th Century, and it continues to capture the imagination of the science community and the public -- internationally. As conceptually depicted in Figure 3F-1, it now seems virtually certain that humans from Earth will soon be exploring the surface of Mars. Whether the initial expedition will be conducted by an American crew on a NASA spaceship, a Soviet crew on a Soviet spaceship, or an international crew on a spaceship developed through international cooperation is yet to be determined, but the technological capability is within reach and the conceptual planning is well under way.

---

[1] Dr. Nicogossian has been at NASA Headquarters since 1976, serving until 1983 as manager of medical operations. {Ed.}

Figure 3F-1: Depiction of Individual at Work in Martian Surface Environment

Mission Planning: Factors

In planning for any manned mission, a set of three factors must be taken into account: the *human,* the *system,* and the *environment.* In turn, each of these primary factors evolves out of, or with, a subset of requirements or characteristics with which each must be considered, as illustrated in Figure 3F-2. The *system* (space vehicle/habitat) is characterized by design, function, and performance. The *human* is required to manage and operate the *system,* i.e., to make pertinent judgments and to act as a backup for the automated systems, which--we have learned through experience--are virtually never *absolutely* perfect or *absolutely* reliable.

The third factor, *environment,* has a dual designation. The first is the *external environment* (relative to the space system), which can be space itself or the atmospheric environment on another planetary body. As one can imagine, the external environment is remarkably inhospitable to human life. The second of these is the *internal environment* (within the space vehicle/habitat), which is designed to sustain human life. For the physiologist, this combined environment imposes combined stresses, and human crew members must be selected on the basis of the system's function and the environment in which they will operate. For missions of extended duration, crew members must be thoroughly trained, and sufficient capabilities must be developed to protect them from the medical risks inherent in space flight.

Only when these three factors are integrated in the planning stages, tested for feasibility (e.g., in pre-mission comparable environments, such as on the NASA Earth-orbiting space station), and implemented in the mission systems to be used can a manned mission to Mars have a reasonable chance of success.

476

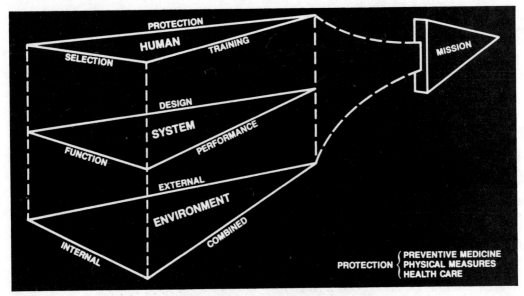

Figure 3F-2: Interactive Factors Involved in Planning Manned Missions

Space Flight: Inescapable Confinement, Tension, Boredom

The stress and risks associated with the space flight environment begin at launch. At that point, the cabin becomes a closed environment vulnerable to a number of potentially toxic materials that could be released slowly into the atmosphere. For example, nitrogen tetroxide was released into the **Apollo 3** cabin during reentry from its Earth-orbital flight. We were fortunate in that the crew did not experience the bends as a result, but they were very ill and had to spend some recovery time in intensive care. Unlike workers in occupational environments on Earth or even in Earth orbit, crew members on a Mars flight will be unable to remove themselves from this artificial environment in the event of a malfunction in the life support system. Therefore, the consequences of *any* toxicological event, e.g., the accidental introduction of a toxic material into that closed system, could be disastrous. This is a crucial point to remember when considering the nine- or ten-month transit time (each way) required for a Mars mission.

A journey of many months in the closed, necessarily constrained environment of an interplanetary spaceship will have a variety of psychological ramifications. Tensions and difficulties can arise even when crew members are highly motivated and well trained (participants in a Mars mission would, undoubtedly, be so motivated). We know from experience that it is important to facilitate a functional, interactive relationship between the *human* factor and the *system,* as I suggested during my introductory comments, so that the crew has a good opportunity to overcome technical problems that arise. One might recall the **Apollo 13** mission, for example, during which the spacecraft had to continue on its orbital trajectory to and around the moon before returning to Earth following the explosion in the command module system. Innovative use of the lunar module helped get the crew safely home. This, on a small scale, is perhaps analogous to the circumstances that could arise during a Mars mission. Another example of unanticipated technical difficulty was experienced with the initial habitation of **Skylab** when the crews had to deal with the problem of low power, due to the loss (during launch) of one of the lab's large solar panels, and subsequent excessive solar heating. The crew had a very difficult time with both the degraded environment and the hard work involved in solving those initial problems, and they in fact rebelled somewhat against the "harassment" of ground crews trying to monitor their conditions and progress while doing so. One must, of course, be very proud of how the entire Skylab team, and particularly the mission crews, solved those very critical problems.

One of the psychological stresses involved in a three-year voyage can be, quite simply, boredom. One way to alleviate much of this boredom would be to allow crew members to undertake certain kinds of work during the flight. For example, they could conduct solar and interplanetary-medium investigations while in transit, or they could assemble and prepare some of the components that will be needed for the mission at Mars (orbital investigations, Mars or Phobos surface experiments, et-cetera). Similarly, providing the crew with onboard facilities for scientific work (such as sample processing) and additional data analysis during the return leg of the voyage is also an important consideration, such that crew members are kept busy with important and meaningful projects.

Much of what we know about behavior and performance in isolated environments comes from our experience with Antarctic expeditions, submarines, remote work assignments in isolated and harsh environments, Soviet long-duration space missions, and, to some extent, from America's **Apollo** and **Skylab** programs. Problems encountered include: anxieties resulting from peer pressure or group acceptance, and uncertainties about information from outside sources (ground control, family, et-cetera). Sensory deprivation is also a very important aspect of isolation. The Soviets, for example, have used tapes on their space stations to provide familiar Earth sounds, such as rain, wind, and singing birds. These sounds are a welcome break from the monotony of the hum of spacecraft systems, and one can imagine that modern video tape capabilities could enhance the value of this concept significantly.

An example of how stress can be induced by unexpected problems in mission event scheduling was experienced during **Skylab 4** operations -- the last of the long-duration **Skylab** missions. Crews on earlier **Skylab** missions had been unable to conduct desired Earth observations due to extensive cloud obscuration. As a result, considerable pressure was generated to make up for the lost opportunities during the **Skylab 4** mission, and its program of scheduled activities was both adjusted and intensified to reflect this sensitivity. The crew had little time to train and prepare for the new work, and a variety of conflicts between ground personnel and the mission crew developed as pressure to satisfy the unexpected objectives stressed the crew's capacity.

The Soviets also have experienced some problems that reflect how crew stress can evolve. They have addressed such problems with varying degrees of success, including some solutions that may then have become--to an extent--part of a larger problem. In one case, the father of a crew flight engineer died during a mission, and the mission commander--who was given the news via private uplink communication--elected to withhold the news from his flight engineer to avoid possible impact to his morale and work. How other (future) crew members view an experience of this kind is representative of a problem that may be experienced during very long missions, which is the stress that can develop into a kind of paranoia in some individuals when they begin to believe they are NOT being told something they should know. The Soviets have tried to address the problem of crew morale in a variety of ways, as I suggested earlier, one of which was to facilitate two-way audio/visual (television) communication between crew members and their families. At first the concept worked quite well, but the family communication sessions became stressed as the missions progressed. Crew members began to complain that they were being told that everything was okay at home when in fact (the crew member believed) something was wrong. This is a difficult form of stress to deal with and alleviate, and it is the kind of stress that could become quite serious over the long duration of a Mars mission.

On the Surface: Radiation and Medical Concerns

Upon landing on the martian surface, crew members will have to readapt to planetary gravity (the gravity of Mars in this case, one-third that of Earth's). Considerable consideration must therefore be given to the length of stay-time on the martian surface, to the amount of work to be done, and, in case of injury or illness, to the medical facilities and procedures that will have to be used in a location many millions of miles from an equipped hospital. In addition, crew members will be living in the confinement of an enclosed habitat which may be covered with martian regolith as shielding to protect its occupants from radiation. Chris McKay's Mars Conference presentation [NMC-3H] describes these aspects of human activity and facilities on the surface of Mars in more detail.

Surface excursions will certainly be exciting, but the duration of exposure to the ambient radiation will most likely have to be limited. Cumulative radiation dose is a serious consideration for future missions, especially planetary missions. On Earth, we are protected by the Van Allen belts and our atmosphere which trap or filter radiation. Biologists face uncertainties as to how the body behaves when absorbing high- and low-energy particles, but we know that absorption is not uniform. Because the body is composed of water and solids which vary throughout the body and are located at differing depths, different layers of protection exist for various organs[2]. Therefore, the radiation dose absorbed varies for each organ. Studies have shown that exposure to low-level, high-energy radiation--over time--results in decreases in performance and reproductive capacity. Radiation effects are more harmful toward the end of a life span, but women of childbearing age are more susceptible and at greater risk than males of equivalent age.

As reflected in Figure 3F-3, the dose potential for galactic cosmic radiation is approximately 25-36 millirads per day, although the dose received on the martian surface also will depend on the location and nature of the landing site. Should the crew land in a low basin or valley, the greater density of the carbon dioxide atmosphere above them will afford somewhat better protection than at high elevations on the planet. The probable daily radiation dose at the martian

---

Galactic Cosmic Radiation: . . . . 25-36 mrad/day
- Mars Surface Dose . . . . 12.5-18 mrad/day
- Total Dose (30 Months) . . . . ≈20-32 rads

Solar Flares
- 11-Year Solar Cycle
  - 1 to 2 Major Events/Solar Cycle . . 5,000 rad (lethal)
  - 2 to 5/Solar Cycle . . . . 500-1000 rad
  - 20-30/Solar Cycle . . . . 50-100 rad

"SAFE HAVEN" Shielding to Reduce Dose Below 400-rad Total

Figure 3F-3: Radiation Estimates for Mars Mission (cosmic, Mars surface, solar flares)

---

2 While "rad" is the more commonly used and recognized term for the measurement of radiation, identical doses of different radiations in rad can result in differing degrees of biological impact on humans. A factor called "relative biological effectiveness" (RBE) is equated with the rad value to provide a different measurement -- the rem. A rem value represents the biological effects of a dose equivalent to 1 rad of X-ray exposure in humans. {Ed.}

surface ranges from 12.5 to 18 millirads, and the total dose received over a period of 30 months is then about 20-32 rads. It *must* be noted, however, that solar flares pose a serious problem and could significantly increase these figures. For example, one or two major events could expose a crew to up to 5,000 rads -- a lethal dose. Therefore, solar flares must be predicted and proper shelters or shielding must be designed and constructed to provide a "safe haven" and keep the total dose below 400 rad.

Implications for Long Missions

Space missions have thus far provided vast amounts of data about the changes induced in body processes by weightlessness. Much of what remains to be determined centers around the points of man's physiological limits, e.g., how much calcium loss can be tolerated and will muscle mass deterioration stop at some point?

As graphically illustrated in Figure 3F-4, it has already been shown that acclimatization to the space environment occurs after the first month on a mission. At this point, it can be seen that motion sickness, fluid and electrolyte loss, red cell loss, and cardiac deconditioning have stabilized. But serious factors and questions remain. If, in the future, someone becomes ill or must be treated during space flight, the combination of adaptive changes which occur in the first 1.5 months will create difficulties for medical treatment, i.e., the changes that have taken place may require treatment that differs in important ways from that which would otherwise be prescribed. One must remember that the loss of life that has occurred in the space program has thus far been the result of accidents rather than medical problems, and we have no experience with potentially serious human illness in space.

During the early stages of the Apollo program when crew members occupied small, very confined cabins, they did not consume a normal volume of food. It is believed that unreported

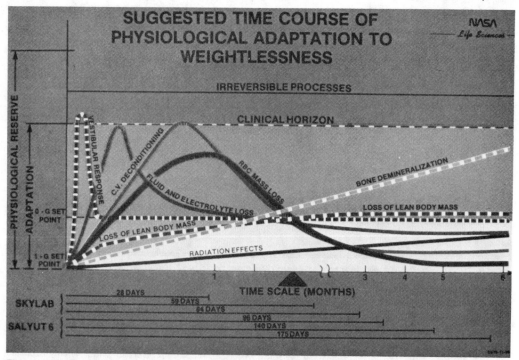

Figure 3F-4: Predicted Adaptation to Weightlessness (based on Skylab and Salyut-6)

motion sickness may have been the reason. It was decided at that time, at least for short duration flights, that crew members did not require as many calories as previously believed. The Skylab program, for example, demonstrated that both energy requirements and food consumption decreased during the first month. However, during the second and third months, this trend reversed itself. In final analysis, not only is metabolism *not* depressed (as in bed rest studies), but actually seems to increase. It is probable, then, that the decreased food consumption during the first month was due to motion sickness and acclimatization to weightlessness.

Even when food consumption increases during flight, however, there is a continuous loss of lean body mass (muscle) and calcium, as well as the nitrogen matrix for both. This is illustrated by Soviet experience during the Salyut-6 mission (185 days), during which cosmonauts consumed more calories and actually gained weight -- while still exhibiting decreased lean body mass. The muscle mass, which is needed to perform work, is probably replaced by fat. In addition to muscle atrophy, the calcium loss is continuous. One important means of calcium loss involves the kidneys, and this process seems to stabilize in flight only after about three months. The increased loss of calcium does not plateau at that point, however, but continues to rise. This indicates that the body is not absorbing calcium and/or is rejecting calcium for reasons yet unclear. It has been shown that even when the dietary intake of calcium and vitamin D are increased, we still are unable to protect crews against calcium loss and subsequent bone loss.

Bone loss is one of the most serious problems for two reasons. First, bones can become brittle with time. This brittleness may lead to fracture upon return to Earth after a long duration mission. Perhaps more importantly, the possibility of increased fracture potential while performing work on the martian surface (after a 10-month transit period in a zero-g environment) is a serious consideration. A second reason for concern is that continued loss of calcium through the kidneys may produce stones that could be incapacitating and difficult to treat (requiring surgery in some cases).

Another alarming aspect of these problems is that calcium and bone mineral levels continue to decrease during postflight (return to Earth). Loss of bone mineral from the vertebrae (a weight-bearing bone) is of great concern, since possible paralysis could result if a vertebral bone is broken. For one case study, vertebral bone loss was measured on a few individuals after two space missions. A slight decrease in vertebral bone mass--with a tendency toward further bone mineral loss--was exhibited in postflight, even after six months of recovery time. However, these are preliminary results for only three individuals, and additional studies are clearly required to confirm these findings and understand the process.

The musculoskeletal area has also been of great concern to the Soviets, who have flown very long missions in Earth orbit. To counteract bone demineralization and muscle atrophy, Soviet space scientists have pursued certain countermeasures. For example, the Soviet "penguin suit" places an axial load on the musculoskeletal system, requiring the wearer to work against the load to maintain an upright posture. Cosmonauts wear the suit for eight hours each day, during which they must continuously fight the pull of the suit. In addition, cosmonauts perform a regimen of onboard exercises, divided into two sessions on three-day cycles. Currently, cosmonauts perform about 2.5 to 3 hours of exercise per day. However, even using the treadmill and bicycle ergometer on this kind of schedule, crew members have typically returned in a deconditioned state (although more recent experience with returning crew members from the *Mir* space station suggests that improvements are being achieved).

The cardiovascular system also presents problems which must be addressed. On Earth, gravity creates a hydrostatic gradient which tends to draw body fluids toward the feet. Continuous muscle contractions in the legs, one-way valves in the veins, and the pumping of the heart all assist in fluid circulation and prevent fluids from permanently pooling in the feet. In space flight, mechanisms that force fluids headward operate unopposed, and the subsequent migration of fluids to the upper body causes tissues of the head and neck to swell and blood volume handled by the heart to increase. The tendency toward lower heart rates in flight indicates that fluid shift may result initially in enlargement of the heart to accommodate its increased workload. In response to what the body then interprets as a volume overload, fluid losses are triggered to compensate for the apparent increases in volume and pressure in the upper body. In this manner, the cardiovascular system quickly adapts to weightlessness.

Return to Earth then reverses the adaptive process. The heart beats faster to maintain cardiac output. The size of the heart is decreased upon return, and its electrical and mechanical activities indicate depressed function (a lower filling volume). Orthostatic tolerance (the ability to stand or function in an upright position) is diminished during immediate postflight. This is due to the increased pooling of blood in the lower extremities. It is also a result of the lower circulating fluid volume, which results in decreased flow of blood to the brain -- hence dizziness and possible fainting. Salt water loading four hours before landing has been used as a countermeasure against orthostatic intolerance, but four-day missions seem to be the optimum duration for using this countermeasure. For a 10-day mission, salt water loading alone might be insufficient. Therefore, there are factors involved other than hydrostatic pressure and fluid volume (possibly associated with muscle change, nervous control, and/or the resetting of baroreceptors). Surprisingly, individuals who show better orthostatic tolerance in postflight are those who have a smaller heart volume to begin with. Therefore, it might not be a requirement to be extremely fit or very young to endure space flight well.

Accompanying the fluid shifts and the resulting decrease in plasma volume is the loss of red blood cell mass (cells which carry oxygen and fuel to the body). There is also a loss of some immunity in space flight. Crew members on long-duration Soviet missions have shown some postflight allergic reactions upon return to Earth. These, however, eventually diminish. Of great concern here is the question of the body's susceptibility to infections or allergies upon reaching another environment or returning to Earth.

Partial prevention of physiological adaptation to weightlessness, through various countermeasures geared to the musculoskeletal and cardiovascular systems, seems a natural approach to successful readaptation to 1-g (Earth gravity). However, one must wonder how effective these countermeasures would be for a flight to Mars and back, wherein the conditions will be experienced continuously for up to several years rather than for only a few weeks or months. How difficult will readjustment to Earth's gravity be (or even the Mars gravity) after a zero-g voyage? If physiological countermeasures are not sufficient, some level of artificial gravity may be necessary during transit flights, as defined and discussed by Barney Roberts [NMC-3D].

The manner in which artificial gravity might be produced is also an issue in need of consideration. For example, a feasible approach to producing artificial gravity for a mission to Mars would be to spin or rotate the spaceship. However, providing artificial gravity in this manner may have certain undesirable effects on the vestibular system. These are mostly due to complex coriolis forces which may be difficult to deal with. A crew member can become disoriented in a rotating environment. For example, if one were to walk laterally to the rotation, two differing forces would be acting upon the vestibular system at once, such that the effect of the coriolis force would be worsened.

Artificial gravity environments presently being considered typically produce fairly low gravities, such that the gravities achieved must be understood in terms of how the crew will get around and function within those gravity environments. That is, the gravity environment may be variable or it may be close to margins at which conditions can change dramatically with a modest change in gravity. For example, if the gravitation produced falls below ≈0.1-g, crew members

482

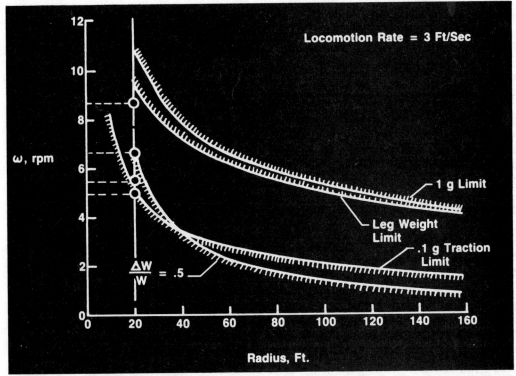

Figure 3F-5: RPM/Radius Boundaries at 3 ft/s Locomotion Rate

would lose traction needed for walking, as reflected in the locomotion data plotted in Figure 3F-5 (above). Much early research needs to be done on the NASA Space Station if we are to commit either to a tether function or rotating vehicle to produce artificial gravity. We also need to perform critical experiments in space to determine what level of gravity (duration and fraction of) will be needed to maintain the proper condition of body fitness.

Consumables

Up to the present, all NASA manned space missions have been short enough in duration to make it practical to provide life support using onboard consumables -- air, water, and food. However, for a manned flight to Mars, it may be necessary to regenerate at least some part of the food supply--as well as air and water--within the spacecraft. Penelope Boston's Mars Conference presentation [NMC-3G] describes a variety of concepts being studied with respect to possible food production on a spacecraft and on Mars.

In general, a six-member crew would require about 1,000 pounds of food material per person per year. In a closed ecological life support system, plants are grown which produce food, waste, mass, carbon dioxide, and oxygen. The crew would also be a component of the bioregenerative system. We believe that such a system is feasible if backed with modest food supplies in case of a system malfunction. It may be more advantageous to transport this system to the martian surface for use there, ultimately leaving it in dormant form when returning to Earth. In this case, the spaceship would also be provisioned with sufficient supplies for the journey home. The provision of fresh food can have a much more favorable psychological impact on the crew than consuming prepackaged, dehydrated or dried foods.

Currently, a plant growth chamber with a volume of 60 cubic meters has been set up at the Kennedy Space Center, in association with Disney World's EPCOT (which has been experimenting quite successfully with a variety of exotic plant growth/modified gravity schemes in its Land Pavilion public showplace). It is being used to develop large scale plant growth methods and solve operational problems involved in raising diverse crops in a restricted volume. Figure 3F-6 is an artist's conception of a crew module aboard an interplanetary spaceship in which food stock plants are being grown in circular chambers. The light source is in the center of the chamber in this concept, such that plants are rooted at the perimeter of the circle and their growth is toward the light source at the circular chamber's center line. This is just one of a variety of concepts being studied with respect to food production that could ultimately be used aboard space stations or in association with missions to Mars.

CONCLUSION

A manned mission to Mars will challenge the human capacity to cope with extreme environments and solve new problems associated with them. Never before has medicine been called upon to certify that an individual is healthy enough to perform for up to three years following the examination, much less under extraterrestrial conditions. We will need life support systems capable of preventing the accumulation of toxic chemicals. Psychological issues may gain greater importance. Radiation exposures could exceed Apollo mission doses by orders of magnitude.

The United States and Soviet manned programs have shown that medical issues include: cardio-vascular and musculoskeletal deconditioning, vestibular changes, immunological changes, and nutrition deficiency. All of these issues must be better understood and resolved if we are (1) to have successful flights to and from Mars in terms of human health and condition, (2) achieve productive exploration objectives on the martian surface, and (3) achieve satisfactory and safe readaptation upon return to Earth. ∎

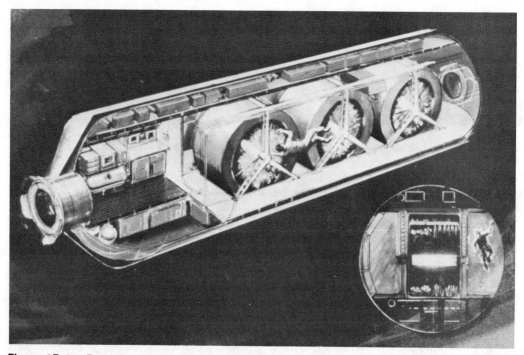

Figure 3F-6: Food Production Space Module

REFERENCES

This space is otherwise available for notes.

~~~~~~~~

MARS MISSION LIFE SUPPORT
[NMC-3G]

Dr. Penelope J. Boston
National Center for Atmospheric Research

Life support represents the most critical technology needed for a successful and safe manned mission to Mars. It involves a complex mix of biological and engineering systems upon which the lives of those making the journey will depend for virtually every aspect of their survival. It must provide the air they breathe, the water they drink, and--over time--a gradually increasing proportion of the food they consume. Developing such life support systems, which are fully closed and depend on recycling for much of what is needed, will require extensive experience in space aboard the NASA Space Station, beginning with Block·1, before the problems are well enough understood to allow us to develop reliable solutions. System constraints, the gravity environment, and the availability of extraterrestrial resources help define differences between systems designed for space and for Mars. Micro- or zero-gravity conditions will prevail in space and life support systems will necessarily be constrained to minimize mass and volume requirements. It will also be important to reduce leak and waste rates to their lowest possible values. The situation is very different on Mars. With a gravity almost 40% of Earth's, and in·situ resources available for producing a wide range of needed materials, the Mars environment makes possible a life support system that can be comprehensive in scope and which can support a more adaptive, dynamic, and aggressive effort. Indeed, life support systems for the martian surface can be thought of as "managed ecosystems" which could ultimately involve atmospheric management for food production as well as gas exchange between plant and human systems.

INTRODUCTION

Because the subject of "life support" is so broad, I will be addressing a number of issues associated with it. To do so with some degree of efficiency, I think the best way will be to discuss some of the highlights among those activities I have observed both inside and outside of NASA while involved in the matter of life support technology.

The primary question we have to address is: how will we realize our objective to establish a base on Mars beginning with what we know right now? That is, how do we get from where we are now (technologically) to a point where we can actually put humans on the surface of Mars for a reasonable period of time, such that they can safely perform the science and other activities we have in mind? The extent of that technology is perhaps suggested in Figure 3G-1,

Figure 3G-1: Conceptual Illustration of Mature Mars Base

an artist's conceptualization of such a base by Carter Emmart. Life support is clearly among the most critical elements needed. Those of us concerned with it (and biology in general) realize that we probably have the most "catching up" to do in the area of developing life support systems needed to conduct a manned Mars mission. This is true in part because biological systems are not as straightforward as physical (hardware) systems. First, they are quite complicated and difficult to develop. In addition, there has been a noticeable lag in effort in this direction. Indeed, if we are serious about manned Mars missions, we really must push research in this direction -- now!

In targeting the subject matter for my **NASA Mars Conference** presentation, I decided that it would be productive to briefly discuss some of the differences between life support criteria for a spacecraft environment and those representative of a surface system like the kind that would be needed on the moon or Mars. I have summarized the spacecraft-versus-surface life support criteria in Figure 3G-2. We feel that surface systems like the one illustrated conceptually above (Fig. 3G-1) would ultimately be needed on extraterrestrial surfaces like Mars or the moon. The objective in each case, of course, is to keep people alive, which requires fundamental capabilities common to both. But there are a number of qualitative and quantitative differences in the environments represented in these systems that distinguishes them from each other in terms of how the crew members would interact with and utilize each.

SPACECRAFT	SURFACE
High Degree of Closure Of Water/Breathable Gas	Water and Gas Available On Mars
Low Mass, Low Volume	Low Mass, Large Volume Augmentable
Food Transported	Food Grown In·situ
Solid Waste Stored	Solid Waste Recycled
Microgravity Environment	Fractional Gravity (3/8 g)

Figure 3G-2: Life Support Criteria, Spacecraft vs. Surface Environments

LIFE SUPPORT: FLIGHT VERSUS SURFACE SYSTEMS

The NASA Space Station (Block·1 and beyond) represents a critical opportunity to help resolve some of the essential concerns that reflect the differences between spacecraft and surface-based life support systems, such as gravity. A list of essential research recommendations related to the human factor and life support requirements for a manned Mars mission, in terms of work that can be done aboard a space station, is presented in Figure 3G-3. With respect to issues associated with life support, we expect that many of the solutions needed to make spacecraft flight tenable and endurable (and to get humans to the surface of Mars in good condition) will come out of the development and utilization of the space station. To deal with the issue of long-duration weightlessness, for example, it is very important that we develop ways to amelior-ate its effects on humans. This can be effectively accomplished through research and experience aboard the space station. Figure 3G-4 is a conceptualization of one of the many possible configurations envisioned for the space station we hope will be built in the near future.

In a spacecraft, one of the major problems one comes up against is the fact that the crew is confined in a small, closed environment with limited resources. No matter how spacious the

---

Research on Biological Components of Recycling
Systems is Essential

Develop Long-duration Physical and Chemical Recycling Systems

Study a Wide Variety of Organisms for Long-duration
Adaptation to Space Environment

Study Physiology and Psychological Response of Humans
To the Space Station Environment Over Long Periods of Time

Assess Human Nutritional Needs in "Working" Space Environment

Develop Medical Procedures Applicable to Zero-Gravity or
Fractional-Gravity Environments

Develop a Corps of Career Space Workers

Figure 3G-3: NASA Space Station Research Recommendations

489

Figure 3G-4: Conceptual Illustration of NASA Space Station (not current, see Fig. 3D-21 [NMC-3D], pages 453 and A-7)

spacecraft, the environment is highly circumscribed. Therefore, the objective is to constrain the spaceship's life support as much as possible in order to achieve low mass and volume in those systems that perform life support functions. Because we would certainly want to minimize the system's leak rate, as well as the total mass taken along, we would "close" the water and breathable gas loops as tight as we can. We would also want to recycle consumables to the maximum extent possible.

Another of the conditions one must contend with in a space vehicle is the microgravity (micro-g) environment that would be experienced on a spaceship while near a planetary body (like Earth, perhaps while awaiting departure from the space station), or the zero-gravity (zero-g) environment experienced while in interplanetary transit (assuming artificial gravity is not being generated in some manner during the flight). In either case, weightlessness and its associated problems will be experienced by the crew and will have to be considered in the design of the life support system. Life support systems designed for the surface of Mars, on the other hand, are given a significant advantage by that planet's gravity which is significantly greater even than the low-g environment experienced on the moon.

Food provisioning and the possibility of food production also must be addressed relative to each of these environments. I think the consensus of opinion among those thinking about the life support problem is that, even for very long missions (on the order of several years with round trip transit time between Earth and Mars), most of the food the crew will need will have to

490

be transported (at least for early missions). However, as Dr. Nicogossian points out in his presentation [NMC-3F], there is a considerable psychological advantage to be gained from having some methodology for producing fresh food on board; it would be a good research project and its product would be greatly appreciated by the crew. It might also provide some insurance against unforeseen food shortages. Therefore, the fact that virtually all of the food required (anticipated needs) will be included in the mission manifest is not meant to rule out at least some amount of food production during early missions. Presumably, though, and particularly early in an evolving program, the bulk of the calories people will need will be included as part of the initial manifest mass.

Flight Considerations

How we eventually deal with the *very*-low gravity environments will depend on many factors, one of which is represented by the physiological considerations addressed by Dr. Nicogossian. But other mission factors are a matter of operational expediency. For example, one would want to determine if the convenience and health considerations of providing artificial gravity are worth the effort (and increased cost) involved in overcoming the engineering difficulties represented in the development of such a system. Such tradeoffs certainly cannot be resolved at this time since we do not have enough information to make the necessary decisions. As reflected in several NASA Mars Conference presentations [NMC-3B/3C/3D], many people and organizations--both in and outside NASA--are working on this problem, so perhaps the necessary answer will emerge clearly once the space station is in orbit and significant long-duration, micro-g research can commence.

In determining how to close the major elements of a life support system (i.e., water, breathable atmosphere, waste), one must remember that waste in particular is not so easy to define when resources are very limited and cannot be replenished. Recycling therefore becomes a critical process and solid waste can be a very valuable resource. On a long mission, for example, it is likely to accumulate to quite a sizable bulk of material. And, since we are going to be short of certain essential commodities on Mars, at least initially, we may want to store solid waste so that particularly important constituents (like hydrogen) that will be difficult to derive from the martian environment might be recovered.

Advantage of Working with In-situ Resources on Mars

When considering systems for use on the surface of a planet like Mars, one can utilize the planet's resources to provide many things that cannot be transported in a spacecraft. For example, available in the martian environment are gaseous elements (atmospheric constituents), water (in the atmosphere and as ice or permafrost that could be reasonably abundant and accessible on/under the surface), and workable surface material (regolith). These materials are a great boon even though we may have to expend a considerable amount of energy to extract what is needed from the atmosphere or the surface. Because this resource exists, it makes food production for the life support system on Mars more "thinkable" than in a spacecraft. And eventually, as we "bootstrap" our way from an initial exploratory mission to that of what one might consider a more permanent, on-going Mars research base, it seems reasonable to assume that the people will want to rely more and more on food produced and processed on Mars. In fact, the large distance and trip times to Mars make this process essentially inescapable if we want to sustain our presence on the planet for more than just one mission.

Another important factor to be considered with respect to both the spacecraft and Mars environments, but which differs significantly on the planet, is gravity. Gravity on Mars, while only about three-eighths that of Earth's (i.e., 100 pounds in Earth weight would weigh about 38 pounds on Mars), is significantly greater than the micro-g or low-g environments experienced in low Earth orbit (LEO) or on the lunar surface[1]. In other words, things that would tend to float around in a spacecraft that is not spun up in some manner to create artificial gravity would easily stay put on the surface of Mars. This is very important because it means we would have to do less modification work on some of the components of a life support system initially developed in a ground-based (Earth) gravity environment.

Life Support Technology for Human Needs on Mars

I believe we know fairly well what a crew would need on Mars. I recall a picture from a book I have had in my personal library since I was a child (Walt Disney's *Tomorrowland Exploration of Mars*, c1959). It was a fascinating little book that the Disney people put out to make children enthusiastic about the idea of going to Mars, and it certainly worked on me! It was based on scientific ideas that Wernher von Braun and his group developed for advanced Mars missions, and little has changed--conceptually--in the interim.

Figure 3G-5 lists those aspects of the problem in which I feel important work must now be done, and it seems clear that we should include them among those issues to be resolved in the near future. The key factors associated advantageously with life support on Mars, are that (1) there are martian resources available for long-term exploitation and that (2) there is a significant planetary surface gravity. These two factors give us the option for a continuing human presence that would otherwise be far more difficult in space, and allows continuity from mission to mission. As the reservoirs of resource materials are found and understood, essentially increasing resource applicability, the economic base on the planet will increase as well. As the missions progress, life support will become easier and easier to accommodate (less critical and less dependent on deliverable resources). Of course, this also makes it possible to enhance the physical scope of the base by gradually increasing its size and capacity as resource availability or system needs dictate.

- Mars Environment and Resources
- Long-duration Physical and Chemical Recycling Systems
- Closed Life Support Systems
- Contamination Control in Air, Water, Growing Systems
- High-Reliability Automation and Control Systems
- Search for Plants Preadapted to Mars and Space Applications
- Human Nutritional Needs in Space and On Mars

Figure 3G-5: Critical Research Needs

---

[1] The moon's gravity is roughly one-sixth (0.166 g) that of Earth's. At lower gravities approaching only one-tenth (0.1 g) of Earth's, normal human mobility--i.e., walking about upright relative to the normal force of gravity--becomes increasingly difficult and finally impossible (essentially human weightlessness as would be experienced in both micro-g and zero-g environments [ref., NMC-3F]). The term "low gravity" is not often well defined, except that it is generally used in a context that associates it with the gravity margin just above the gravity at which human locomotion is lost, such as on the moon. The gravity of Mars is perhaps comparable to that achieved by some of the artificial gravity concepts discussed at the Mars Conference if it is assumed that 1 g (Earth's gravity) is not necessary during interplanetary flight [Snoddy, NMC-3C; Roberts, NMC-3D]. {Ed}

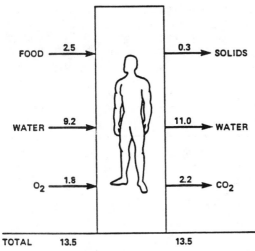

FOOD	2.5 →		0.3 → SOLIDS
WATER	9.2 →		11.0 → WATER
O_2	1.8 →		2.2 → CO_2
TOTAL	13.5		13.5

Figure 3G-6:

Human Requirements -- Food, Water, Oxygen (Person/Day)

Figure 3G-6 gives human needs in terms of mass throughput of consumables--food, water, and oxygen (pounds/day)--such that one can roughly predict total requirements for a Mars base. These numbers vary from author to author, but human consumable requirements on Mars are anticipated to be essentially the same as on Earth. We will need food, water, and breathable air, and we will have to make a variety of products that can be recycled. We have known these things for a long time, so it is not surprising that the life-support systems suggested by conceptual illustrations of bases presently envisioned for Mars, like the one depicted in Figure 3G-7, closely resemble those that were first dreamed up twenty or thirty years ago.

Figure 3G-7:   Concept of Advanced Mars Base

Key Concepts

How might we bring conceptual life support systems like these to reality on Mars? The list presented in Figure 3G-8 lists a number of key concepts for consideration, although I'm sure I have missed some that may be the pet ideas of others working on the life support problem. These, however, are the ones that I believe to be the most important.

KEY CONCEPTS

[1] Managed Ecosystem Approach

[2] Greenhouses

[3] Respiratory Gas Exchange With Human Habitats

[4] Specifically Tailored Environments for Plants
(e.g., gas and light)

[5] Microbial Processing of Food
and Microbial Protein Production

[6] Recycling of Organic Wastes

[7] Food Caches in Case of System Failure

Figure 3G-8: Life Support on Mars

[1] *Managed Ecosystem Approach* -- One of the key factors that seems to have evolved is the importance of achieving a successfully managed ecosystem. Indeed, the pendulum of conceptual thinking has now swung away from systems that are entirely engineered and in which everything is controlled automatically and invisibly. Concurrent with that former school of thought, but at the other end of the spectrum of theories, there was a belief that we could simply put organisms in a box and let them sort out their differences, essentially assuming that everything would recycle fine all by itself and with little or no human managerial interaction. We have now learned that the system must be subject to close management in order to understand in precise terms what is happening.

Reliable automation is essential, however, if the inhabitants are to have an opportunity to do more than just operate their life support system. Obviously, then, both ends of the spectrum are extreme, and what we need is some reasonable blending of these concepts so that we have as many organisms and as much diversity in the system as possible. It will take this kind of concept integration to develop the degree of system resiliency needed to secure it against failure, and to do so without handicapping its productive potential.

[2] *Greenhouses* -- One inescapable fact is that if we are going to grow plants for food on Mars, we must have something like a greenhouse (with a pressurized atmosphere and controlled environment) in which to grow them. Whether it is essentially exposed on the surface of Mars or somehow buried in the regolith, this represents a whole new and different question that I will discuss shortly.

[3] *Respiratory Gas Exchange with Human Habitats* -- It would be very productive and beneficial to have gas exchange between the human habitats and the plant growing facilities. In simple terms it would work like this: through photosynthesis (which requires solar energy), plants take in carbon dioxide (CO_2), produce food (the essential byproduct of plant growth), and give off oxygen; humans consume the food, breathe the oxygen, and produce CO_2; the plants then use that CO_2 in photosynthesis to begin the cycle anew. This gas exchange, then, is one aspect of the recycling process I mentioned earlier. However, it is doubtful that this process (particularly in a small system) could ever sufficiently sustain what goes on in these areas to be relied upon solely and without supplementation, but at least some exchange could be achieved in this way.

[4] *Tailored Environments for Plants* -- One potentially productive thing we can do with plant environments that we cannot do with human environments is manipulate their atmospheres, essentially altering what the plants are breathing. There has been some work in the academic community on how we can manipulate levels of carbon dioxide, levels of humidity, and perhaps even atmospheric pressure in order to produce as much edible food as possible while minimizing the inedible bio-mass (waste product).

[5] *Microbial Processing* -- There is clearly a need to develop plants and systems that produce as much edible foodstuff as possible, relative to the left-over inedible waste product. Microbial processing represents one possible solution. NASA originally looked at this concept during the 1960's and determined that there were a number of problems associated with it. For one thing, most of the material produced using a microbial process seemed to make people ill, and that alone seemed enough to rule it out. Since that time, however, there has been a tremendous amount of productive work done in the realm of microbial processing, which has come to be known as *single-cell protein production*.

The incentive for developing edible material produced by a microbial process during the 1960's had nothing to do with space habitation. The work was being conducted during the period that preceded the energy crisis, when researchers were looking for new ways to feed people who did not have access to enough protein. Microbial processing was looked at as a possible solution. As the technology was being developed, however, oil--which was one of the principle substrates being fed to research organisms--became very expensive, and the bottom dropped out of that work. Nevertheless, the technology exists and could be resurrected if we wished to begin to apply it to a martian situation.

[6] *Recycling* -- Inedible bio-mass is one of the big problems associated with producing food. For example, if we grow a greenhouse full of corn, there is a whole lot of plant mass left after the corn has been harvested and eaten. That left-over mass must be recycled in some manner, which may prove to be too expensive in terms of energy requirement. It might be advantageous to find ways to use the waste mass directly, such as utilizing cellulose to strengthen building materials -- perhaps analogous to using straw in adobe construction. However, it would seem that at least part of the solution must be to develop food plants in which most of the plant mass is edible, and I will return to this subject in another context.

[7] *System Insurance* -- Although I have listed this item last, I should point out that crucial consideration must be given to the need for "insurance" margins within the food production system ahead of virtually all other considerations, i.e., establishing food caches as protection against the possibility of a system failure. An insurance procedure should be incorporated into the system with the initial mission, and it should then be maintained even in the later stages when extensive food production becomes possible. A manned base should always insure itself against unforeseen emergencies in which the food resource might be significantly or totally destroyed.

Food Production in the Martian Radiation Environment

There is one aspect of life support technology that is already getting considerable attention, the development of food production methodologies. How do we go about producing food on Mars? One of the big questions, which I alluded to earlier, is whether or not food crops should be grown--in a suitable habitat, of course--on the surface or underground. This depends on a number of things, the most important of which is the radiation environment at the surface of Mars. We do not yet know how this environment will affect growing plants. The one thing we *do* know is that plants on the whole are more resistant to radiation than we humans (ionizing radiation in particular), including--to some extent for some kinds of plants--ultraviolet radiation. Figure 3G-9 is a painting by artist Mike Carroll that illustrates the idea of burying at least part of a habitat to provide radiation shielding.

A fair amount of radiation work involving affects on plants was done several decades ago, but that work was conducted at a time when our ability to study plants (and the effects of radiation) was not as advanced as it is now. One thing we need to understand is the resistance of plants to ionizing radiation. This would help us determine whether they can be grown on the surface of Mars in a transparent greenhouse or if they must be partly or completely buried like the human habitats to provide radiation protection. Of course, lighting requirements are critically dependent on the answer to this question; if the greenhouse is on the surface, supplemental light will be required only during periods of reduced solar radiation, such as during the winter season or martian dust storms[2].

If, on the other hand, we must bury the whole growing apparatus, we will have to provide artificial light. The kinds of artificial lighting now used in greenhouses is very energy-expensive,

Figure 3G-9: Concept of Buried Habitat (or greenhouse) on Mars

---

[2] Although the amount of sunlight that reaches Mars' surface is only about 60% of that which reaches Earth's, the reduction may not be a problem with respect to plant growth. Plants on Earth are over-saturated with visible light (i.e., they get more than they need). Depending on the species, they reach maximum photosynthetic rates at between 0.009 and 0.6 m^{-2} min^{-1}. For Mars' mid-latitudes ($\pm45°$) during the summer, visible light values range around 2 to 3.5 kcal m^{-2} min^{-1}. Some edible plant species (like corn) do not "light-saturate" and gradually increase photosynthesis past some value; these might therefore need supplementary light on Mars. Conversely, plants that do well in reduced-light conditions (e.g., shaded under nontropical tree canopies or in regions where it is often cloudy) may be more suitable to martian growing conditions.

so other systems are being studied. For example, researchers are looking at ways in which we might trap surface sunlight, thereafter splitting it into its component wavelengths and using only those wavelengths most appropriate to photosynthesis. If this concept can be made to work efficiently, the protection from surface conditions afforded to workers in buried facilities will tip the balance in favor of such structures.

How the greenhouses are constructed is also dependent on whether they are to be above ground or buried. The nice thing about having lightweight, easily constructed greenhouses on the surface is that it would be easier to get them there in terms of launch penalty (mass). If they are going to be buried, however, they will have to be structurally strong enough to bear the weight of a lot of regolith without collapsing. That changes how much effort will be needed to develope and transport them to begin with, and may also significantly increase the amount of effort needed to set them up on Mars[3].

Selection of Potential Food Organisms

The choice as to what kinds of food organisms (and their associated production systems) should be selected for initial use on Mars seems to boil down to two: plants or microbes. The reason animals are not on this list, at least at present, is that they represent a poor conversion ratio in terms of the amount of vegetation they need relative to the amount of protein they produce. The only hope in this respect is based in *aquaculture*, which I will discuss a bit later. Another unfortunate thing about animals is that _we_ are animals[4]. Among the factors we must therefore be concerned about is the possibility of pathogen transfer. This is somewhat less true relative to plant pathogens, and it is probable that microbes intended to be grown as food for humans would be rigorously selected to pose no pathogenic risk. Microbial production will have to be studied further to determine whether or not it can lead to a process that will be viable as a foodstuff for people without making them ill.

One aspect of plant growth technology I alluded to earlier is our increasing ability to monitor plants, and Figure 3G-10 (next page) reflects some work being conducted at **Phytoresources** (College Station, Texas). The plant pictured looks as though it is hooked up to have an EKG. Indeed, that analogy closely suggests what is actually being done. Phytoresources researchers have been developing techniques to monitor plants by hooking them up to many different kinds of sensors.

The objective of the work at Phytoresources is to be able to observe--in real time--how the plant responds to various growing conditions. In one aspect of this work, isotopically-doped carbon is traced to see how the carbon dioxide in the air is partitioned amongst the various parts of the plant. This is important when trying to maximize the production of the edible parts of a plant. That is, one would want most of the mass productivity to be in the edible portions rather than in the waste product. The benefit of this technique is that one can glean many different kinds of information about the plant through one simple series of experiments. The concept represents a very new way to study plant response and will help significantly when we actually begin to develop plants for the martian environment.

---

[3] As noted by Chris McKay [NMC-3H], it may be practical to produce building materials for structures on Mars, according to work by Benton Clark and others based on **Viking** inorganic surface chemistry data. {Ed.}

[4] Reasonable numbers of small land animals (e.g., rabbits or other small creatures) that may be suited to relatively easy transport and husbandry on Mars) would be competitive with humans for significant volumes of breathable air and water, both of which may be difficult to produce in sufficiently large enough quantities (at least initially) to satisfy the compound requirement. {Ed.}

Figure 3G-10:

*Plant Response
Research
(Phytoresources)*

Potential of Hydroponic/Aeroponic Plant Growth

There are many ways one can go about growing plants once the selection process has determined which plants will be used. *Hydroponic* technology has gotten considerable attention in recent years and has been quite successful commercially. It is a process in which plants are grown in a sterile medium, such as certain kinds of sand or gravel and occasionally a low-mass medium like vermiculite -- but *not* soil[5]. The medium is periodically flooded with a nutrient solution and the plants take up the nutrients through their root system in the normal way. This is a convenient way to grow things in greenhouses because many plants can be grown close together, and it is also easy to precisely control what the plants are fed because the nutrient solution can be tailored to specific applications.

---

[5] To a biologist or a soil scientist, Dr. Boston points out, "soil" is a complex, eroded mixture of rock and mineral fragments that includes decomposing organic detritus, bacteria, fungi, and other organisms that reside in Earth's surface material. Therefore, while the term is unfortunately often used in discussions of surface material on Mars (and even the moon), there is--as far as we know--no "soil" (as just defined) on the red planet. {Ed.}

An outgrowth of hydroponic farming is a growing technique known as *aeroponics*, which has been used with varying degrees of success. It has produced good results for certain kinds of favored plants (e.g., tomatoes). Aeroponics is characterized by hanging the plants on support wiring with their roots exposed to the air. The roots are then periodically sprayed using a nutrient solution not unlike that used in hydroponics, but with greater frequency to prevent the roots from drying out.

In the Land Pavilion at Disney World's **EPCOT (Environmental Planning Community of Tomorrow) Center** in Florida, both hydroponics and aeroponics are demonstrated for visitors. Their aeroponic system continuously moves the plants on a conveyer system that carries them through an area where the nutrient solution is sprayed in a fine mist on the roots. Aeroponics is considered a prospect for plant growth in low-g or zero-g environments (such as on an interplanetary spaceship); plants would be grown in a drum with their roots near the drum's outer margin where the sprayers are located. The light source would be mounted in the center of the drum and the plants would then essentially grow "up" toward it. This system also facilitates drum rotation to create (using inertia produced by centrifugal force) a degree of artificial gravity for the plants.

Alternative Concepts

Other growing concepts are also under study. Conventional soil horticulture has certainly not been ruled out, and for certain kinds of plants it might be essential. There is some work being conducted to study how applicable conventional soil horticulture might be in the confined area of a greenhouse[6]. Microbial production, as previously discussed, also needs to be looked at further to determine whether or not it can be processed as a viable foodstuff for people without making them ill. And, as I suggested earlier, **aquaculture** is a technique that could be considered as a Mars base matures and water production improves.

One advantage of aquaculture is that organisms that live in a fluid medium on Earth would be less affected by the lower gravitational field on Mars. Also, aquatic organisms would be easier to feed and maintain than terrestrial animals, and most have a very high conversion efficiency ratio (i.e., they are more productive as a food crop). One notable aquatic food organism being studied is *tilapia,* a species of fish from Africa. These fish have evolved to survive African dry periods during which natural water impoundments in which they are trapped (filled during wet-season flooding) are drying up, normally ending tilapia's short life cycle. As a result, tilapia can get by on very little oxygen and remarkably high concentrations can survive in a confined environment. In the case of tilapia, then, its advantage for possible propagation on Mars is that high densities can be propagated in a tank. Moreover, their conversion efficiency remains high in spite of the stress, and many pounds of fish can be produced per feed stock amount.

Figure 3G-11 is an illustration created by Carter Emmart for a *Case for Mars* presentation. It essentially illustrates most of the techniques I have been discussing; one can even see a tilapia jumping out of our aquaculture pool, which should be easier for it to do in the lower gravity of Mars. Rows of conventionally grown plants are in the foreground and aeroponically grown plants can be seen hanging toward the back-right where sprayers presumably are located. In this concept a shielding fabric can be drawn over the top of the facility when needed. But what this illustration perhaps best suggests is that no one technique will optimally satisfy all system requirements. Indeed, we might want a mix of them to grow various kinds of foodstuffs, selecting whichever is the most appropriate for each particular plant type or growth environment.

---

[6] A workshop entitled "Lunar derived soils for the growth of higher plants" was held at NASA-JSC in Houston, June 1-2, 1987, to consider this problem for possible moon bases.

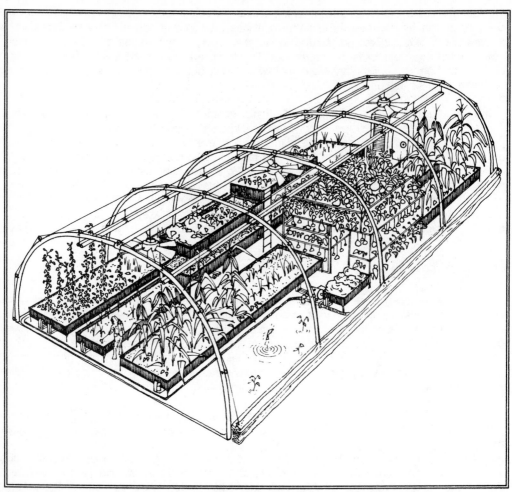

Figure 3G-11: Concept for Advanced Food Production Facility (including aquaculture) on Mars

CELSS -- Controlled Environment Life Support System

As mentioned in other Mars Conference presentations, one of the interesting projects now considering these issues is called the *CELSS (Controlled Environment Life Support System) Breadboard Project* currently being conducted at NASA-KSC in Florida. Researchers are looking at hydroponic techniques for use in a closed, recycling environment. Figure 3G-12 shows how they grow the seedling sprouts in trays. The work is focusing on wheat at this point because it is believed that wheat might be a very productive initial choice.

The researchers have devised a promising alternate method of doing it which may prove to be fairly productive. As shown in the photo presented in Figure 3G-13, the plants are essentially contained in little compartments in a pipe. This allows the roots to be protected and yet wetted down with a nutrient medium when and as required. If one were going to use this system in a micro-g or zero-g environment, the plants would be properly "rooted" and would not be floating around when weightless. The technique also allows nutrient solution to flow through the pipes without significant loss. Because the solution is confined to the pipe system to facilitate recovery, it can be pumped through the pipes as an additional benefit in zero-g where the natural flow of fluids is not possible[7].

Figure 3G-12: Hydroponically Grown Wheat Seedlings (in trays)

Figure 3G-13:

Pipe-Contained Hydroponically Grown Plants

---

7 One rewarding aspect (for many home gardeners) is that hydroponics--including pipe hydroponics--does not involve such high technology and cost that it must be confined to costly and elaborate research laboratories. Indeed, many home gardeners who had previously embraced the concept of hydroponic gardening and enjoyed considerable success with the more conventional techniques are now converting to pipe hydroponics for some kinds of plants. The opportunity for "weekend gardeners" to move into this kind of technology is due largely to the availability of readily obtained and economical PVC pipe. PVC pipe is available in a remarkable variety of products and is quite easy to work with, such that it encourages experimentation and improvement. The experience of those most familiar with its benefits indicates that the concept is not only efficient and productive, but may be more responsive when reacting to certain plant problems (e.g., fertilizer requirements, nutrient adjustment, systemic pest/disease control, et cetera). It is quite possible that many improvements in this kind of food production may be contributed by the general population if researchers take the time to observe and consider what's going on outside their own laboratories. {Ed.}

Another interesting project is being conducted at the Disney EPCOT Center in Florida with the help of some people at NASA-JSC. They are in the process of simulating Mars and lunar surface materials to learn how they might be used as a plant growth medium. We know more about the lunar soils, of course, because we have actual samples acquired during the Apollo program. At present, the properties and composition of the martian regolith are known only as determined from Viking lander data and other remote sensing systems.

The essential composition of the martian surface material was analyzed and measured by the Viking landers for elements heavier than magnesium. The Viking data suggest that the material is an iron-rich, clay-like material[8]. However, there is disagreement and uncertainty about the light elements important to life, as well as about other chemicals, compounds or minerals that may not be readily detectable. Therefore, it is not entirely clear what to include in a Mars soil simulation. Nevertheless, the JSC and EPCOT researchers are giving it a try, and the results of their first attempt to grow higher plants in the simulated material are shown in Figure 3G-14. As we get more data (and possibly some returned samples) from Mars via unmanned missions, we can improve such simulations of the chemical composition and properties of the martian regolith. Then, as that knowledge evolves, we will be able to refine our concepts about how to use the material as a substrate for growing plants.

Figure 3G-14: Plants Grown in Soil Developed from Simulated Mars Surface Material

---

8 Dr. Boston is referring to the inorganic chemistry investigations successfully conducted in both Viking landers using X-Ray Fluorescence Spectrometers (XRFS). A variety of different kinds of samples were analyzed, some acquired directly at the surface (in the form of either fine material or as chunks of duricrust) and some from deeper holes or under rocks. Results at both landing sites were remarkably similar, but minor differences were detected to represent the different kinds of samples analyzed. Although unlike any known Earth soil, the martian surface material is thought to be something like nontronite (an iron-rich clay) on Earth and it may be derived from rocks that originally contained a very high content of iron and magnesium. In what is generally an order of decreasing abundance, the elements detected were: silicon, iron, sulfur, calcium, aluminum, titanium, magnesium, cesium, and potassium. It is also possible that barium and bromine may have been detected, but their signatures are quite difficult to see or distinguish in the XRFS spectra and are therefore considered uncertain. Sulfur proved to be about 100 time more abundant than in Earth soils while potassium was less abundant by at least five times. The instrument was incapable of detecting certain elements known or believed to be present, including carbon dioxide, water, sodium, and possibly nitrogen. Among the oxides, silicon oxide was by far the most prominent (nearly 45%) followed by iron oxide (better than 18%). The team leader was Priestly Toulmin III, and included Benton C. Clark (deputy team leader), Alex K. Baird, Klaus Keil, and Harry J. Rose. {Ed.}

Figure 3G-15: Highly Advanced Food Production Facility for Moon or Mars

Getting Started: New or Needed Work

Figure 3G-15 (above) is a concept by Robert McCall that appeared in the report of the **National Commission on Space** (discussed in these proceedings by Dr. Thomas Paine [NMC-3I]). A quick study of the illustration's background reveals that it does not represent Mars and instead depicts a mature base on the moon. However, it is appropriate because all one must do to make it valid for Mars is paint the sky pink and the rocks red -- the rest remains essentially unchanged. Moreover, it illustrates what one might call the "high-priced spread" version of a complex, ongoing, mature, closed-ecological life support system. This is the ultimate goal.

The big problem with the conceptual systems, of course, is that we do not (theoretically) yet understand how to do them. Indeed, we have very little theoretical knowledge about how closed systems work -- including global environments like Earth. We need to develop that kind of theoretical background before we send astronauts out to live in such systems. We obviously can't afford to wait for a crisis to arise at the surface on Mars to develop that understanding, so we have a tremendous amount of work to do before we actually attempt to build it.

Current Work -- In contrast with the picture of a very mature system (Fig. 3G-15), Figure 3G-16 (page A-7, Color Display Section) indicates where we are now. This is a photo of the chamber developed for the NASA-KSC project mentioned earlier. Researchers are in the mid-development phase of that program. To date, plants have been grown in the chamber only under partial closure conditions. The gas control system has been updated in light of results achieved during the first growing series, and another series of growth experiments was started late in 1987. The illustration presented in Figure 3G-17 (page A-7, Color Display Section) reflects some of the methods they are using, including the use of the same trays illustrated earlier. It also shows the apparatus used to grow seedling plants in a higher humidity; the plants are kept in a closed environment until they are ready to be transplanted.

503

The first choice for a food plant is ultra high density wheat, which is shown at various stages of growth in the color photo (Fig. 3G-17). The overall approach of the NASA-KSC project is a step-by-step procedure in which plants are grown in the chamber first, and then, after perhaps several years, effort is made to begin to close the various elements of the system one by one. First, the researchers want to be able to get the plants to grow without difficulty. They are presently working to close the water and air systems, and they are also working on things like the removal of trace contaminants. Eventually, they hope to close the waste system as well.

Automation -- Figure 3G-18 is another conceptualization by Mike Carroll. The illustration provides an expansive perspective of a well developed, advanced Mars base. I like this illustration because the advanced nature of the base suggests an aspect of the system I think needs to be singled out as particularly important. We will be going to Mars for many reasons, although I'm sure a list of what they might be will differ from person to person. However, all of the reasons have one thing in common, they will have a large number of tasks associated with them once a martian base is established. That base is going to be a very busy place! Therefore, I think one of the primary goals associated with developing life support systems for the surface of Mars is to create a system in which the people are not distracted from their work by having to fuss and bother with the system itself. The goal, then, is to develop a life support system that will be as unobtrusive as possible.

Figure 3G-18:

Conceptual Perspective
of Well Established
Mars Research Center

504

There are certain strategies one can use to try to optimize the system's unobtrusive character. Ideally, I think the people at that base will want to spend no more time thinking about their life support system than we spend thinking about our home air conditioners. It should be in the background. Of course, food production is a different matter and is clearly a more labor-intensive activity. But even in the realm of food production, I think the goal should be to incorporate automation as much as possible in the process of growing and recycling food and in controlling the gas balance.

Perhaps the rule should be that if it is reasonably possible to do something automatically, automation should have developmental priority. However, I am not suggesting the extensive use of robots. I am referring to the development of very sophisticated systems that can actually monitor and control the state of the system itself while keeping the people involved aware of what it is doing (particularly in the case of a significant change in conditions or operational activity). This is a technology that needs a great deal of work and which we really are not yet adequately addressing through the planning and research activities now underway.

Use of In-situ Raw Materials -- Another area of interest is that of developing our ability to use the raw materials of Mars, and in this respect I am particularly thinking of the surface material. I believe it is imperative that we begin to think about the unconsolidated regolith on Mars in terms of its probable role as the primary component of the medium in which we will grow our food crop on Mars. The system will require the development of a large volume of soil and inert substrate, and we are certainly not going to want to transport it from Earth. Once we understand the composition of the available surface material, we will be in a position to develop ways to process it, removing any toxic components or damaging salts it might contain and balancing its pH for the plants we want to grow. Every effort should be made to develop this technology in advance so that it can be applied quickly and productively on Mars.

Gas (atmosphere) Management -- I have already suggested the importance of the gas balance between plants, humans, and any other organisms in the system. Part of that process involves working with specific gas environments for the purpose of manipulating plant growth or development to achieve maximum productivity. For example, we may want to grow one kind of plant in one kind of atmospheric mixture and another kind of plant in different mixture. This kind of technology would require considerable flexibility in gas management.

Enhancement of Food Products -- Finally, we should give some consideration not only to reclaiming or utilizing the inedible waste material that is a byproduct of any kind of plant growth, but to look at options for enhancing the edible product itself. For example, many food enhancement techniques have evolved in human cultures on Earth for perhaps thousands of years, such that foods have had their edibility and nutritional value improved through traditional, low-technology microbial processing. This is reflected in the production of such common things as bread or beer, but it is particularly well represented in the case of oriental food items like soy, tempeh, and tofu. While such enhancements essentially make these foods more palatable, they may also increase their nutritional value by improving some aspect of a food's composition -- such as its vitamin content.

CONCLUSION: WHAT DO WE DO NOW?

To get things moving from where we are right now, we must begin, as I have, by getting up and speaking out about those things that must be done in the near future. And, because we are counting on the space station to provide many of the opportunities needed to resolve problems associated with manned missions, those in whose hands the fate of the space station rests must be made aware of how important it is to the future of interplanetary human exploration.

I believe there are many things we need to be moving on now or as soon as possible. We need to take advantage of opportunities afforded by unmanned missions to learn as much as we can about the martian environment and the planet's resource potential, but we must also begin *now* to look beyond unmanned missions to the crucial technologies we will need for the manned missions. For example, the whole theoretical concept of closure (associated with life support and ecological systems) represents a critical technology in which much work must yet be accomplished, and we need to develop the long-duration physical and chemical recycling systems we presently don't have. Our long-duration experience with manned missions is very weak, and we have not been forced to cope with its problems or develop our understanding of them to a mature state. The product of this work will become increasingly necessary and crucial as concepts for the manned exploration of Mars evolve. This research should reflect our concern for microbial contamination control in the air, water, and growing systems as well. While I have not discussed this aspect of the technology for the Mars Conference, it must be recognized as a significant problem.

The need to develop high-reliability automation and control systems (as I described them earlier) also addresses the necessity for life support systems that will operate in the background as much as possible. They should afford worry-free reliability through self-monitoring and control as well as a great sense of security. We must allow our explorers to do the science and exploration they go to Mars to do rather than spend their time worrying about the life support system itself. Indeed, one can readily imagine a work or living environment not unlike that depicted in Figure 3G-19, one that minimizes concern about--or interaction with--the life support system and which distracts its occupants as little as possible from their work.

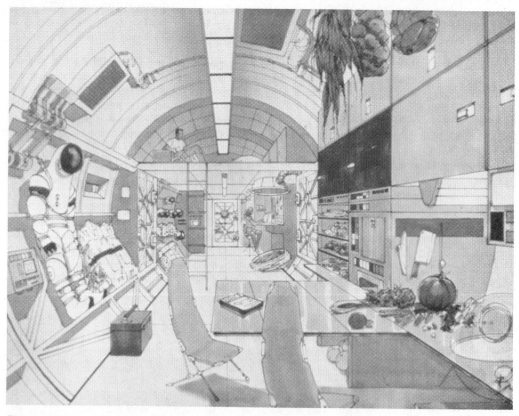

Figure 3G-19: Life Support Without Concern or Distraction

We also need to search for and develop plants that can be preadapted to Mars and space environments. Most experiments have so far considered only common domestic plant varieties like corn and wheat. In western culture, we eat only something on the order of sixty species of plants when, in fact, there are perhaps many hundreds of potentially edible species in the world -- many of which might be well suited to martian growing conditions. By considering plants that are eaten in other cultures, we may hit upon something that will be a breakthrough for the plant environment we envision for our base on Mars.

Finally, as Dr. Nicogossian suggests in his presentation, a significant amount of work must be done to determine what human nutritional needs will be like in the space environments we are contemplating. We don't really know what humans are going to require in terms of calories, and special nutrient supplements may be needed during those long space voyages. Since we have no experience with very long missions, we do not know how nutritional needs will evolve or change as people spend long periods of time in transit between the two planets -- punctuated by a period of difficult mission activity at Mars.

In closing, then, the main point I wish to drive home is that a tremendous amount of work must be started now to prepare for human missions to Mars. This development effort cannot wait until we actually have an announced plan and time schedule for doing so -- it has to start now. An encouraging point to recognize, however, is that relevant commercial and research enterprises now underway should result in a variety of off-the-shelf technologies that will be directly applicable to the human exploration of space and Mars. By carefully assessing these activities, the space science community can more effectively address itself to those areas of research that are unique to space and planetary exploration currently being neglected by other segments of science and engineering. In this way we can stretch our research budget, our *people power,* and our imaginations -- while ensuring that we will have all the capabilities needed to send humans safely and productively to Mars. ■

REFERENCES

This space is otherwise available for notes.

~~~~~~~~

LIVING AND WORKING ON MARS
[NMC-3H]

Dr. Christopher P. McKay
NASA Ames Research Center

When it becomes feasible to live and work on Mars, martian resources will be a critical part of the sustaining process. It will be necessary to drink water extracted from the martian environment, to make breathable air and fuel components from the martian atmosphere, and to shield or construct facilities using martian dirt. Carbon monoxide and oxygen can be used to fuel rocket engines (for ascent or Earth-return) as well as small, diesel-like internal combustion engines powering surface vehicles. As the human presence on Mars advances from the rudimentary first expedition to the establishment of a mature scientific community, living and working on Mars will evolve from spacecraft-afforded facilities to larger, more specialized structures. Some of these will be transported to Mars and assembled there as life support habitats, while others might be inflated and pressurized with compressed martian atmosphere to support activities for which pressure alone will be enough to overcome much of the difficulty of working in Mars' environment. Martian dirt might at first be used only to provide radiation shielding, but it would later be utilized (in association with atmospheric constituents) to develop a growing medium for plants as well as the production of fertilizers and construction materials. The ultimate use of martian resources may be the "terraforming" of that planet's global environment into an Earth-like biosphere for a population whose ancestors were born on a distant blue planet few of the "martians" will ever visit.

OPENING COMMENTS & INTRODUCTION

I am going to start by simply saying: when you go to Mars, *DO drink the water!* That statement essentially encapsulates what I am going to discuss, the utilization of resources that are on Mars right now. This involves using them to support humans through the production of critical consumables, like water and breathable air, using them to produce transportation and rocket fuel, using them to build structures, and using them to essentially achieve a permanent human presence on the red planet. The title of my presentation, *Living and Working on Mars,* is supposed to evoke images of people working at the planet's surface on a routine basis; comparable, perhaps, to living and working in California! Indeed, I believe that the activity conducted on Mars would in time become quite routine.

I am going to address two questions I think to be quite important and which have been discussed by others at this conference and elsewhere. The first is, quite simply, why Mars? What is so unique about Mars that it should inspire this NASA Mars Conference to be a conference on Mars instead of being concerned with Venus or Mercury or Jupiter? And because that is a relatively easy question to answer, I shall do so now. The most important answer one can give to the question: "Why Mars?" is that there are essential resources on Mars that can provide materials humans need to support both their existence and their purpose, and this significantly broadens the planet's potential for scientific exploration as well as human habitation.

The second question I want to address is that of the long-term potential for Mars, beginning with the establishment of a research center at which people are present on a continuing basis to conduct scientific and "community" support activities. What are these *martians* going to be interested in doing on "their" planet? While we may be earthlings in the act of going out to Mars, it is quite probable that we will have descendants born and reared there as true martians, as suggested by Ray Bradbury's enchanting classic -- *The Martian Chronicles*. What sorts of things will be most important to them at this stage? More importantly, perhaps, how will they create an environment in which their activity can be conducted? I think the matter of available resources, as well as the ability of future martians to live off the land by utilizing them, will play a key role in how these questions will ultimately be answered.

MARTIAN RESOURCES AND THEIR POTENTIAL APPLICATIONS

The picture presented in Figure 3H-1 is one of my favorite **Viking** lander photos. It clearly reveals frost on the surface in the vicinity of the lander[1]. Although the visibility of the frost was enhanced for this picture by the imaging process, it is also visible in color images (see Fig. 3H-1 in the Color Display Section, pg A-8). I included it to make two points: (1) that we know a lot about Mars as a result of the **Viking** mission, which gave us the wealth of information reviewed and updated in the proceedings of this conference, and (2) that there really is water "in them thar hills!" The frost in this picture (together with both lander meteorology and supporting orbiter data provided at the time it was taken) clearly seems to demonstrate the presence of water. Water is crucial to life, and we will need significant quantities to support long-term survival on Mars. There is ample evidence for water on Mars, and we will be able to use it for a variety of purposes once we understand its distribution and accessibility.

The Water Resource

Figure 3H-2 is a sketch illustrating where and in what forms water most likely exists on Mars, while the polar ice pictured in Figure 3H-3 offers visible support for the fact that a significant reservoir of water ice is locked up in the polar caps (with certainty at the north pole, and possibly--but with no direct evidence as in the case of the north pole--under a semi-permanent thin CO_2 hood at the south pole), such that martian polar ice represents one source of water about which we have some good evidence. We also believe, and have very strong evidence for support, that there is permafrost underneath or mixed with much of the martian surface regolith, particularly in the mid and polar latitudes. This means that there is probably a significant

---

[1] The picture was taken by Viking Lander 2 (located high in the northern hemisphere) as winter drew to a close in **Utopia Planitia**. Temperatures were essentially cold enough during the winter at that latitude to condense CO_2, at least in the atmosphere and possibly for brief periods at the surface. It is believed that a kind of precipitation occurred (possibly as a very fine snow) in which CO_2 condensed on tiny particles of water ice in the atmosphere and carried them to the surface. Although very cold, the temperature was not cold enough to hold the CO_2 at the surface for an extended period of time. It is believed to have sublimed back into the atmosphere, leaving behind a thin residual veneer of water "frost" that lasted into the martian spring. Studies of VL-2 images taken during the coldest periods (associated with two martian winters) have shown that a thin, almost transparent veneer initially covered nearly all visible surfaces. Only later as temperatures warmed slightly did the water frost reveal itself, particularly in shadows -- much as frost on Earth will remain in shadows after the Sun has caused it to "burn off" in areas receiving direct sunlight. {Ed.}

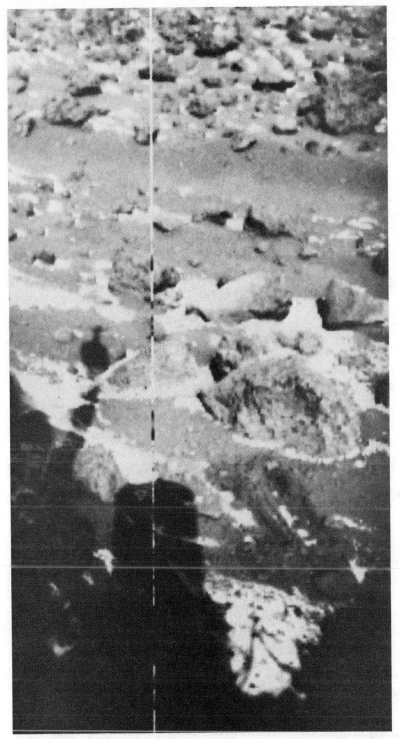

Figure 3H-1: Residual Water Frost as Imaged in Utopia Planitia by Viking Lander 2 (also Fig. 3H-1 in Color Display Section, pg A-8)

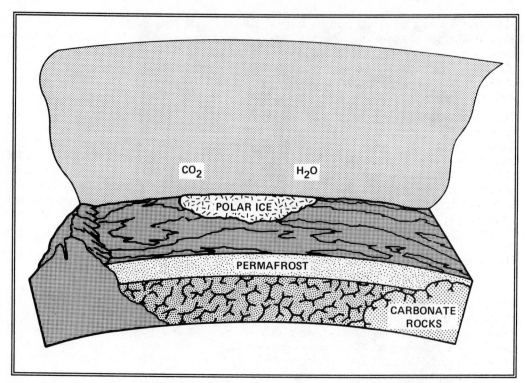

Figure 3H-2: Known and Probable Locations of Water Reserves on Mars

Figure 3H-3: Northern Ice Cap Determined to be Water Ice (residual/permanent summer state)

water resource tied up in the planet's surface material that will be available to future expeditions. Permafrost is admittedly difficult to deal with, but it is there nonetheless. During future unmanned missions it will be important to improve our understanding of the probable distribution of these hidden but potentially significant water reserves. The layered character and coloration of the residual (summer) ice cap at the north pole (opposite page) can be seen in a different version of Figure 3H-3 in the Color Display Section on page A-8.

There is a significant amount of water vapor in the atmosphere of Mars even though that atmosphere is very dry compared to Earth's. Indeed, the atmosphere is actually saturated in some places, e.g., in a broad ring around the north polar cap during the summer season in that hemisphere, which is why water frost is a real possibility in that region. While it is not a lot of water[2], a rough calculation of water consumption for a Mars base (using a baseline suggested by our Antarctic operation which consumes thirty thousand gallons a day during the summer) indicates that the martian atmospheric water would last--without replenishment--a few thousand years. There is clearly plenty of water in the martian atmosphere, no matter how dry it might be compared to Earth's.

The Availability of Oxygen

Along with water, oxygen represents a vital human need. In looking for the needed oxygen, one should first note that the martian atmosphere is ninety-six percent (96%) carbon dioxide (CO_2). If one considers only the O_2 component of the CO_2, it is clear that there is plenty of oxygen available. One must also remember that a significant amount of CO_2 that may once have been in the atmosphere could now be (as has again been reported in these proceedings) chemically bound up in carbonate rocks under the surface as an additional resource (Fig. 3H-2).

ADVANCED RESOURCE UTILIZATION ON MARS

We have clearly demonstrated, as I have already indicated, that there are potentially adequate resources of water and oxygen on Mars, particularly in the atmosphere, and that it is in some sense readily available -- even to the first expedition. As Jim French [NMC-3E] and others have already explained, that same resource can also be used to make fuel (liquid O_2 as the oxidizer and liquid CO as the propellant -- LO_2/LCO) for use in Mars surface vehicles as well as in ascent or Earth-return vehicles[3].

The technology for doing so, in fact, is not beyond the technology base for some of the unmanned precursor missions being considered, such as sample return, wherein the spacecraft could use the atmospheric resource to make small quantities of fuel for the return flight. This concept might also satisfy the requirements that Penny Boston has defined [NMC-3G], i.e., the need to develop technology that does not require constant human interaction. Extraction technology is the kind that one can simply set up and turn on to begin making water, oxygen, and fuel from the CO_2 in the martian atmosphere.

---

[2] Maximum water vapor concentrations approached 100 precipitable microns adjacent to the north polar region during the summer months, but was well below that (sometimes approaching 0) elsewhere in the martian atmosphere during the same season (consistently so in the winter hemisphere). In comparative terms, even saturated martian atmosphere is many times drier than the atmosphere above the driest regions on Earth. {Ed}

[3] More potent propellants may become possible as martian resources are found and exploited. If water becomes significantly more plentiful than it is likely to be initially, the propellants that could be made include (as Jim French [NMC-3E] has discussed) liquid methane (LCH_4) or even liquid hydrogen (LH_2). Both require large amounts of water to provide the needed hydrogen. {Ed.}

Tapping the Atmosphere for Manned Missions

To study this concept with respect to support for an early manned mission, we determined that a mission profile for the initial expedition to Mars might involve about four people who would stay on the surface for perhaps thirty days. This kind of mission would certainly benefit by getting needed oxygen from the atmosphere and might also benefit from making at least some of its water on Mars as well. Then, as one considers larger missions, it becomes increasingly more cost effective (and ultimately quite necessary) to make most of the water, breathable air (O_2, N_2 and Ar), and increasing amounts of fuel from the martian atmosphere.

The martian atmosphere is ninety-six percent (96%) CO_2, as already noted. The challenge is how to compress it and separate its components to make the products we need. Oxygen can be removed through a variety of processes. The one currently favored involves heating the CO_2 to a high enough temperature to cause it to thermally decompose into carbon monoxide (CO) and oxygen. One then simply lets the oxygen defuse through a membrane that is impermeable to the other gases that evolve, essentially creating a reservoir of oxygen for collection and storage. The CO product has potential, as well, and can be used as a fuel propellant. Energy costs depend on the system but average about three kilowatt hours per milligram of oxygen.

Note that approximately four percent (4%) of the martian atmosphere is *not* CO_2, and this is both an interesting and important point. One must remember that the air we breathe does not consist of oxygen alone, and that--for several reasons, including fire safety--we need to have a biologically inert gas mixed with it. We do not actually consume this inert gas when we breathe it; it serves only as a buffer with the needed oxygen, departing our bodies unchanged by the respiration process. Nitrogen (N_2) and argon (Ar) are two gases that can function in this capacity, and both are present in the martian atmosphere (although only in very low pro-portional abundances compared to nitrogen's role in Earth's atmosphere[4]). Nitrogen makes up about 80 percent of Earth's breathable air and therefore serves as the buffer gas for human respiration under normal circumstances. Carbon dioxide, the most abundant atmospheric con-stituent on Mars and the byproduct of human respiration on Earth, is not a suitable buffer gas because it is toxic (for humans) at high concentrations.

We have done some calculations to determine the amount of energy that would be required to get the needed nitrogen and argon out of the martian atmosphere. Because the stored volume of buffer gas will be reduced only as "lost gas" (i.e., due to escape when expedition members enter and leave a life support habitat), again because its form and volume will not be changed by human respiration within the life support environment, it won't have to be replenished to the extent that oxygen will. We find that it takes about ten kilowatt hours to produce a kilo-gram of buffer gas (nitrogen and argon), which we believe will be a reasonable energy cost at later stages of a research base development.

The amount of water in the martian atmosphere is highly variable. As I've already indicated, it is present everywhere at some times of the year, and in some regions the atmosphere can be saturated (seasonally). In truth, we still do not fully understand the water process in Mars' atmosphere, and we know even less about the water we believe is hidden under its surface. This is one of the aspects of Mars that *Mars Observer* has been designed to help us pin down. If we are going to go to Mars and plan on using its water, we must first understand this impor-tant resource as well as possible.

---

[4] Based on **Viking** data acquired during both entries and by both landers on the surface, nitrogen repre-sents only about 2.5% and argon less than 2% of Mars' atmosphere. Most of the argon, amounting to about 1.5% of the atmosphere, was determined to be the isotope ^{40}Ar -- supplemented by trace amounts of ^{36}Ar. {Ed.}

Rationale for Using Martian Resources

There are several reasons for utilizing in-situ martian resources rather than transporting them from Earth. For example, why drink martian water? First, it is cheaper; it would not reflect the cost of having been part of the original launch mass and of being transported all the way from Earth to the martian surface. Secondly, more water exists in the martian atmosphere (about one cubic kilometer) than we could reasonably bring from Earth. And because it is in the atmosphere, it is generally available anywhere on the planet (although extraction rates may vary with latitude and season). For extended or permanent operations, larger than immediately needed quantities would probably have to be extracted and stored during seasonal periods when water vapor is more plentiful. In addition, there is a large amount of carbon dioxide in the atmosphere for the other uses I have mentioned, such that it will serve as a reliable resource for a variety of critical materials. And finally there is the surface material itself, which, if initially used only in its most fundamental application, could at least afford shielding and protection for partially buried facilities. I will discuss potential uses of Mars' surface material in more detail later in the presentation.

A strategy based on utilizing martian resources is more responsive to local demand. If, for example, one wanted to take an extra shower, one would simply make more water; the water would not have to be ordered a year or two in advance from Earth. More importantly, the on-site capability represents a survival measure available to the crew. If something happens that allows a major part of the water supply to leak out of its storage facility, "phoning in" an order to Earth for more water is not a very satisfactory solution because it will take a year or two to fill that order. One must have the capability to gear up and replace lost water immediately, using some sort of extraction or production mechanism, to deal with possible accidents or unforeseen circumstances.

Indeed, insuring the water supply is an important--even vital--factor, much as Dr. Nicogossian and Penny Boston have suggested in these proceedings with respect to the food supply [NMC-3F/3G]. There will always be the possibility for critical accidents of this nature that have not been anticipated in the planned day-to-day routine. I have been involved in field research in the Antarctic, and that experience has taught me that the planning associated with long-duration remote operations rarely if ever works out as routinely as one expected it to. Usually, we have simply forgotten something unimportant like toilet paper, and the more significant oversights could be corrected within a reasonable period of time by flying in what was needed. But an oversight concerning something as critical as water replenishment on Mars could have disastrous consequences. It should also be reemphasized, following up on Dr. Nicogossian's comments, that the ability to respond to life-threatening emergencies enhances psychological security for the crew members; they could rest much easier on Mars knowing that the loss of a water supply could be dealt with effectively without the trauma of a life and death struggle for survival. It is therefore good psychological wisdom to include a water replenishment capability in the system at the outset of technology development.

In review, then, the essential atmospheric products associated with Mars resource utilization include oxygen, buffer gas, water, and fuel (propellant and oxidizer). The latter of these, in particular, has been discussed rather extensively in these proceedings because it is potentially a very important technology. ISPP (in-situ propellant production) clearly projects itself beyond its early, more limited potential with respect to mission efficiency and survival and into the realm of long-term productivity. I have highlighted the production of oxygen, buffer gas, and water because I think they are clearly the most critical. These kinds of resource exploitation capabilities are, therefore, the ones I believe we should be investigating and developing now because they may have important applicability on the very first manned mission; most certainly on the second or the third and all that follow. We should remember, too, that some of this technology could have applicability relative to earlier unmanned robotic missions, particularly through fuel production. We have also considered how other resources might be exploited, including some rather simple applications involving surface material in a habitat–shielding capacity and the utilization of martian wind for power generation. Considered together, then, all of these issues suggest the kinds of technical developments that will be important to human survival on Mars through the use of martian resources.

LOOKING FARTHER AHEAD

There are other ways in which martian resources can be utilized, some of which are relatively easy to use to advantage. As I've already suggested, for example, one can simply pile "dirt" on top of a habitat to provide solar radiation shielding[5]. As we know, the martian atmosphere does not effectively filter radiation; the UV flux at the surface is rather severe and represents a significant danger. The surface material could also be used in the same manner to improve, as an additional measure of protection, radiation shielding associated with a radioactive power plant taken to the martian surface[6]. One can also imagine wind turbines capable of generating power on Mars, and we have actually already looked at this at least in preliminary detail. The first reaction to this idea is usually that, because the atmospheric density at Mars' surface is a hundred times less than on Earth, martian wind will probably be incapable of driving turbines to the extent needed for significant power generation. But, in fact, studies suggest that it may be feasible to build units that could produce on the order of twenty kilowatts of power.

It follows, as well, that we will have to work with surface material in other ways, such as in the construction of launch and landing facilities for example. In this case a site would first have to be cleared and then developed to some extent, and this would involve moving martian dirt and rocks around in perhaps significant amounts to get it out of the way or utilize it in some constructive manner.

Early Long-Term Base

Figure 3H-4 is an illustration prepared by Carter Emmart that illustrates some of the things I have discussed. One can see, for example, a simple drag-line device for scooping up dirt (right), and surface material has been piled on top of a habitat (center) to provide radiation shielding. This kind of shielding, as explained earlier, would provide protection from the high-energy solar environment for the astronauts (or **astroscientists**) working on Mars for extended periods of time. An atmospheric gas extractor is also visible in the illustration (left of buried habitat),

---

[5] There is a continuing degree of confusion with respect to terms used to refer to the martian surface material. Although "**soil**" is often used, most probably out of habit and because it is both very convenient and very direct in the sense of its most simplistic implication, Dr. Boston has pointed out [NMC–3G] that true soil (as it is understood on Earth) is almost certainly scientifically infeasible on Mars. "**Regolith**" is similarly a confusing term to use with respect to planetary surfaces (like those of Earth or Mars) that have been formed and significantly modified by processes other than those generally associated with the production of regolith on the surfaces of moons or other bodies without atmospheres. In these cases, impact histories and debris encounters (e.g., initial and secondary impact dust/fragments) are directly and perhaps singularly important to the development of surface layers. Therefore, Dr. McKay may well have used the best term yet for the martian surface material (in lieu of referring to it as soil or regolith), calling it--quite simply--dirt! {Ed.}

[6] A nuclear power system would incorporate fully adequate shielding, of course, but the use of in·situ material to insure a broader margin of protection, which could be achieved quite effectively and economically in the manner described, would afford a benefit in psychological security and perhaps in reduced launch mass. {Ed.}

Figure 3H-4: Intermediate Mars Base (gas extraction, dirt-shielded habitat, airshells)

and it essentially sucks in martian air, compresses it, extracts the CO_2, and sends the CO_2 off to be thermally decomposed into oxygen and carbon monoxide. Since O and CO react exothermically (i.e., burn together) and can therefore be used as engine fuel (oxidizer and propellant, respectively), the CO and some of the oxygen is transported (suggested by the depiction of the surface rover pulling a storage tank) or piped to rocket facilities like those in the distance (both the oxidizer and the propellant would most probably be stored and used in their liquid state in their capacity as a fuel). The fuel produced could also be used in fuel cells or to power a small rover engine on the surface[7]. Of course, the first priority for oxygen use is as a life support gas (O_2) in the human habitat where it is needed to make a breathable atmosphere. The facilities that appear to be transparent are inflatable airshells.

The gas extractor can also compress and cool the volume of martian air that can be processed during each extraction cycle in order to condense out the small but precious amount of water vapor present in it. This water can then be used to top off the water supply in the closed system, compensating for leakage losses and normal use. In addition to providing oxygen for the habitat's life support atmosphere, the extraction process would also provide buffer gas (argon

---

[7] Research has shown that it would be possible to develop small engines for use on Mars that will run on a combination of carbon monoxide and oxygen. These would be diesel-like, internal combustion engines in which the burning of the fuel components would re-form CO_2 as their exhaust product. A more extensive discussion of fuel propellants that might be possible to produce on Mars is presented in NMC-3E by Jim French. {Ed.}

and nitrogen) for it as well, to make up for the small losses I explained earlier. It would not make good sense to go to Mars and spend the entire time inside the habitat simply to prevent the loss of minor life support gas components that could not be replenished by the extraction system. To review, a little of it is lost each time people move into and out of the habitat.

Airshells: Versatile and Simple Non-Habitat Facilities

Another productive and uncomplicated idea uses the fact that Mars has an atmosphere in a more direct, simple, and quite interesting way. Using a small compressor to compress martian atmosphere, expedition personnel would inflate a balloon-like facility presently known as an *airshell*. Because uninflated airshells would be relatively light and compact, they would be highly transportable and could be set up when and where such facilities might be needed. Two airshells are visible in Figure 3H-4 on the near and far sides of the partially buried primary habitat. A similar idea has been used for entertainment purposes at fairs and circuses for years, wherein blowers are used to inflate "balloon tents" or huge air bags with relatively low pressures. While these are not airtight (closed) and do not achieve the significant increases in interior pressure that would be needed on Mars, the technology is nonetheless similar and would be usefully economical in a more appropriate closed configuration for Mars.

The airshell concept utilizes completely unprocessed martian air and simply compresses it to bring the confined pressure up to about one atmosphere (a pressure comparable to Earth's surface pressure and in which humans can be comfortable). The amount of compression would require very little energy and the confined atmosphere could be kept warm fairly economically since the exterior martian atmosphere is so thin that it cannot carry away heat as efficiently as does Earth's. With the airshell properly inflated, inhabitants could dispense with their cumbersome, pressurized space suits and, in more comfortable attire, essentially walk around easily inside using nothing but self-contained breathing systems (very much like SCUBA systems) to provide breathable air. This rather simple and economical concept might be very useful, for example, in creating a maintenance facility for repairing and servicing rovers. They would also be useful as fairly large-volume facilities needed to perform other work that is not really compatible with the small size of a permanent habitat. For example, Penny Boston used another Carter Emmart drawing to depict a food production operation which happens to be in an airshell [Fig. 3G-11, pg 500, NMC-3G], and it illustrates some of the details I've described which aren't clearly visible in Figure 3H-4.

Moving Toward a Mature Research Base

Clearly, the fact that Mars has an atmosphere is proving to be a boon to our concept for an initial settlement and as a basis for all that follows. Starting with the first day on the surface we will be taking needed materials from the atmosphere and we will be using surface material. As the base gets bigger and more elaborate, one can imagine the introduction of more sophisticated technologies. For example, one can readily imagine the development of greenhouses and/or other types of food production systems relatively early in that evolution, as discussed by Penny Boston in her Mars Conference presentation.

Growth of Base Capacity and Capabilities -- This implies shifting gears, so to speak, from the small initial group of explorers (which might number four to ten people who would make and use only essential supplies like oxygen and water) to a more mature base that might have a greenhouse operation and a much more elaborate life support system to accommodate a larger number of people. These later "settlers" would be growing plants and living in more advanced structures. As Dr. Boston indicates, it then becomes feasible to begin using oxygen produced by plants in the greenhouse operation to supplement that extracted from the martian atmosphere, which in turn might allow more of the atmosphere to be used to make fuel components (oxidizer and propellant) and thereby improve transport potential on the surface or between the surface and orbiting vehicles. At this stage, then, the program would be absolutely tied to and dependant upon martian resources, but it would have the capacity to do much more -- including the production or construction of those things the base needs to accommodate expansion.

Using Martian Dirt -- It would also be possible at this more advanced stage to quite seriously begin considering new ways to use the martian surface material and to develop methods to process it for more advanced purposes. For example, Dr. Boston explained how information provided by the **Viking** inorganic chemistry team, which has helped us to develop a good idea of the fundamental composition of the martian surface material, is being used as a basis for the development of a possible growing medium (simulation soil) for a Mars food crop.

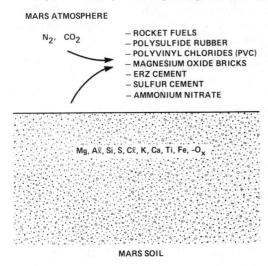

MARS ATMOSPHERE

N_2, CO_2

– ROCKET FUELS
– POLYSULFIDE RUBBER
– POLYVINYL CHLORIDES (PVC)
– MAGNESIUM OXIDE BRICKS
– ERZ CEMENT
– SULFUR CEMENT
– AMMONIUM NITRATE

Mg, Aℓ, Si, S, Cℓ, K, Ca, Ti, Fe, $-O_x$

MARS SOIL

Figure 3H-5 presents a list of product compounds suggested and studied by Dr. Benton Clark, a chemist at Martin Marietta's Denver Aerospace company[8]. These compounds utilize both the inorganic elements detected and measured on Mars and the planet's known atmospheric components. The list demonstrates that one can make a host of useful things from the chemicals found in the martian atmosphere and surface material, including fertilizers for use in the greenhouse, and they suggest that the establishment of a rudimentary industrial processing base would be well justified.

Figure 3H-5:

Possible Products With Mars Materials

One of the novel aspects of Ben's work makes use of manufactured (in·situ) hydrogen peroxide as a key resource on Mars. Hydrogen peroxide (a controversial material with respect to the Mars biology issue) can be made with water extracted using a relatively simple chemical process. Once a good reservoir of hydrogen peroxide has been produced, one would essentially have--in what amounts to a single storehouse--something that can provide rocket fuel, fuel cell fuel, heating power, water to drink, and oxygen to breathe.

Moving the martian surface material around will not be easy. Such material is quite heavy, even with the advantage afforded by Mars' lower gravity, and developing ways to process it and make useful products from it will take a significant degree of experience, study, effort, and time. Clearly, it is not something that we can expect to undertake on the first mission. But subsequently, within the context of a more mature base, it should become reasonable and cost effective to make things like bricks, cements, and possibly even blasting powder. The latter could make possible the initial stage of a resource mining industry and allow personnel to get at some of the important elements we believe are present at greater depth in the surface.

---

8 Ben Clark was deputy team leader of the **Viking** inorganic chemistry team (Priestly Toulmin III was team leader and principal investigator). The team performed direct mineralogical analyses on martian surface samples that were funneled into an X·ray fluorescence spectrometer located inside each lander (see Footnote 8 in NMC-3G, pg 502, for additional information and composition data). The understanding of the surface mineralogy and properties was supported through work conducted by the **Viking** magnetic properties team (Robert B. Hargraves, PI/team leader) and the **Viking** physical properties team (Richard W. Shorthill, PI/team leader). {Ed.}

LIFE ON A NEW MARS FOR THE "MARTIANS" OF THE FUTURE

Figure 3H-6 is an artist's sketch that addresses the long-term questions concerned with the sorts of things that may one day be considered following the evolution to a mature base. At that very advanced stage, the Mars resource budget will have to be considered on a **global scale**. For example, how much water does Mars have? ...not *just* in the atmosphere but the total for the whole planet; and, what can the inhabitants ultimately hope to do with that water? The answer to these questions may lead to the most important application of Mars' resources, the beginning of a global transformation to Earth-like conditions and the development of a biosphere. Indeed, the environment illustrated (Fig. 3H-6) is one that might exist on a Mars that has been significantly altered by human activity. The term frequently used for this very futuristic process is *terraforming*, and it represents the transformation of a hostile, alien environment into one comparable to Earth's. If this can be achieved using only the resources available on Mars, particularly the now-frozen water resource we are sure is there, we will have created an environment on the red planet that is amenable to Earth life[9].

That would be, I think, the ultimate utilization of martian resources. It would represent the application of those resources to support and propagate what will ultimately become martian life -- a kind of **genesis** on Mars. And it must again be emphasized that every indication we now have suggests that Mars is the best candidate for the development of planetary ecosystems among all the other planets and objects in our solar system. It is precisely because it has an atmosphere, along with other resources we can use, that Mars may well be the first planet in our solar system where another home for life as we know it can be established. ■

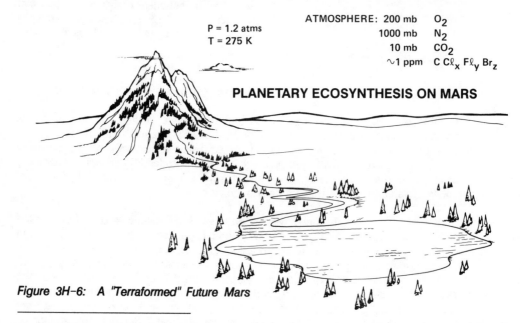

P = 1.2 atms
T = 275 K

ATMOSPHERE: 200 mb O_2
1000 mb N_2
10 mb CO_2
$\sim$1 ppm $C\ C\ell_x\ F\ell_y\ Br_z$

PLANETARY ECOSYNTHESIS ON MARS

Figure 3H-6: A "Terraformed" Future Mars

---

[9] **Terraforming** is not a process that can be performed within a time span that can be appreciated by an individual, nor are there any known mechanisms that can magically accelerate it in the Star Trek **genesis** fashion. The concepts so far studied could take from hundreds to perhaps one-hundred thousand years, with little tangible benefit visible from Earth's vantage point during the early stages. However, as Dr. McKay points out in Terraforming: Making an Earth of Mars (The Planetary Report, Vol. VII, No. 6, Nov/Dec 1987, pp 26-27): "If humans established themselves on Mars, self-sufficiency would be imperative to the long-term health of the settlement. Over the years a distinct group of martians would certainly develop. To them the benefits of terraforming Mars would be quite tangible -- the survival of their civilization." {Ed.}

REFERENCES

This space is otherwise available for notes.

~~~~~~~~

OVERVIEW:
Report of the National Commission on Space[1]
[NMC-3I]

Dr. Thomas O. Paine
Chairman, National Commission on Space

The following synopsis reflects statements of purpose for the NATIONAL COMMISSION ON SPACE, with the essential thrusts of its recommendations, as extracted from opening pages of the report:

Having been appointed by the President of the United States and charged by the Congress to formulate a bold agenda to carry America's civilian space enterprise into the 21st century;

and having met together throughout the better part of a year to obtain testimony from experts and from a cross-section of citizens across the country;

and having projected the next 50 years of the space age and deliberated on America's goals for the next 20 years;

and having prepared thereby to place before the nation a rationale and a program to assure continuing American leadership in space;

we, the members of the NATIONAL COMMISSION ON SPACE, now propose these space goals for 21st Century America....

The purpose of the comprehensive civilian space program proposed for the United States through the recommendations of the NATIONAL COMMISSION ON SPACE is: "To lead the exploration and development of the space frontier -- advancing science, technology and enterprise, and building institutions beyond Earth orbit from the highlands of the moon to the plains of Mars." Three primary thrusts are proposed to achieve it: (1) Advancing our understanding of our planet, our solar system, and the universe; (2) Exploring, prospecting, and settling the solar system; and (3) Stimulating space enterprises for the direct benefit of the people on Earth. Long-range commitments needed to make it possible to achieve the program's goals economically include two additional thrusts: (1) Advancing technology across a broad spectrum to assure timely availability of critical capabilities; and (2) Creating and operating systems and institutions to provide low-cost access to the space frontier.

---

[1] The final report of the NATIONAL COMMISSION ON SPACE was presented to Congressional House and Senate committees, and to President Reagan, on July 22, 1986, only one day prior to Dr. Paine's NASA Mars Conference presentation. {Ed.}

OPENING COMMENTS AND INTRODUCTION

Several years ago, after I gave a talk to the **Mars Underground** in Boulder, Colorado, it occurred to me that the time had come to give an award to those persons who had done the most to make a case for opening the Mars frontier. I didn't believe it was necessary to form a committee to identify the recipient(s) at that point because I was convinced that the award clearly had to go to Carol Stoker (NASA-ARC) and Tom Meyer (Boulder Center for Service & Policy) for initiating both the Mars Underground and **The Case for Mars** conferences[2]. To that end, I had a Mars banner created to serve as the award and presented it to them.

It then occurred to the three of us that the Mars banner could serve as an excellent annual award, so I am using the opportunity afforded by the **NASA Mars Conference** to announce that the Mars flag will now be awarded annually to the person or persons recognized for having done the most during the preceding period to advance the human psyche--and, later, humanity itself--outward to Mars. The original Mars Flag will simply be passed from the previous winner to the new winner each year, and we will mark the annual award by the same kind of presentation I made when presenting it to Carol and Tom. The banner will be proposed for inclusion in the cargo manifest for America's first manned mission to Mars, to recognize the achievement of our goal to establish a human presence on the red planet and to represent its continuance. I should perhaps add that whoever actually takes it there (the crew of the first expedition) will automatically be the winner that year!

Background: Pursuing the Mars Mystique

I thought it might be fitting to conclude the **NASA Mars Conference** by reminding ourselves of the public's tremendous interest in Mars, first as a mysterious legend and then more recently in fascinating reality. Mars is a fantastic world, and the attendance at this conference--with its broad participation--reflects the appreciation of Mars' great future. My late friend, Wernher von Braun, certainly felt that the chief object of his life's work was to open the trail to Mars. Although he was enormously elated and gratified by our missions to the moon, he regarded lunar exploration as a precursor to humanity's ultimate destination -- resource-rich Mars. Working closely with Wernher, I too became convinced that Mars represents the next "giant leap for mankind."

We first took a serious look at a national program to mount a manned expedition to Mars during the Apollo Program when it became reasonably clear that we would achieve our goal of landing on the moon. President Nixon then organized a high-level study to determine what goals should be established for NASA beyond Apollo. An interagency committee identified as the **Space Task Group**[3] (STG) was appointed by the President in February of 1969 to address this complex issue. It was headed by then Vice President Spiro Agnew, to whom the **National Space Council** (later abolished) reported. The Secretary of State, the Secretary of Defense, and the NASA Administrator were members of the STG. As NASA Administrator at the time, I was determined to do everything I could to win the group's support for moving our civilian space program in

---

[2] Carol Stoker served as co-chairperson for the presentations represented in the proceedings of Session 3 of the NASA Mars Conference. Dr. Stoker is a National Research Council Fellow at NASA's Ames Research Center. {Ed.}

[3] The 1969 Presidential Space Task Group (STG) should not be confused with the operating unit of the same name based at the Langley Research Center (Hampton, VA) during NASA's earliest days. The historic Langley group, headed by Robert Gilruth in planning Project Mercury, ultimately seeded new space programs at other centers, and its personnel and expertise provided the nucleus for NASA's Johnson Space Center. {Ed.}

the directions I wanted NASA to go[4]. Our relatively small staff developed and laid out a comprehensive long-range NASA plan, and we presented it to the President with a series of options. We proposed a bold civilian space program that would have resulted in continuing lunar surface activity, an orbital base (space station), and the first manned expeditions to Mars (for the 1980's). We also proposed a robotic *Grand Tour* of the outer planets for the 1980's. As I pointed out to President Nixon at the time, the giant outer planets would then be so aligned that we could use gravity-assist to "sling" a spacecraft from one planet to the next, from Jupiter all the way out to Neptune. "The last time such a favorable opportunity existed," I told him, "Thomas Jefferson was President -- and he blew it!"

Good News · Bad News

President Nixon was unmoved by my eloquence, but he did see the advantages of the outer planets mission. I blush to confess that we had structured the **Grand Tour** concept so that, if the President didn't want the "grandest" plan, we could scale the mission down to a more acceptable budget. And that, of course, is what happened; the **Grand Tour** concept served as the basis of the highly successful **Voyager** Program[5]. Although we didn't get all of the "bells and whistles" we wanted, NASA still got a magnificent mission with the two **Voyager** spacecraft. From scientists to citizens all, Americans have enjoyed exciting encounters with Jupiter, Saturn, and Uranus, as well as their incredible satellites, and eagerly look forward to 1989 (August) when **Voyager** 2 will provide us with our first close-up observations of Neptune and it's largest moon, Triton[6].

---

[4] The 1969 STG included Robert Seamans, Secretary of the Air Force, and Lee DuBridge, the President's Science Advisor. The final report was made in September, 1969, and included a three-level funding program that strongly favored the objectives previously put forth by Dr. Paine, Wernher von Braun, and George E. Mueller. However, the most ambitious objectives, including plans for manned Mars missions by the end of the century, were deleted when the first two levels of the plan were discarded. Only the third was ultimately endorsed, which essentially concluded with no post-Apollo major space goals; Apollo-Soyuz, Skylab, and the development of the space shuttle were the principal products. (From: **Managing NASA in the Apollo Era**, NASA SP-4102, 1982.) {Ed.}

[5] The name **Voyager** has been used twice as a planetary project name and is familiar to those who were associated with planning the mission that ultimately became the Viking Project. In that context, it was first used as the name of an ambitious mission to send unmanned landers to Mars. Although the Mars-**Voyager** mission was cancelled for economic reasons, it was later resurrected in a scaled down program and renamed **Viking**. The "Grand Tour" program began as MJS (Mariner Jupiter/Saturn) before being more appropriately renamed **Voyager** (encounters with Uranus and Neptune were not reflected in its original name). Another **Voyager** gained fame more recently by flying non-stop around the world, adding still more luster to the history of that name. {Ed.}

[6] Neptune, presently the outer-most known planet in the solar system (until 1999 when Pluto's elliptical orbit will again carry that small planet beyond Neptune's), is as yet known to have only two moons. The largest, Triton, is believed to be comparable in size to Ganymede (the largest of Jupiter's four Galilean satellites) and Titan (Saturn's largest satellite). More importantly, Triton, like Titan, is known to have an atmosphere. Methane appears to be a major constituent of Triton's atmosphere, and an absorption has been detected in the satellite's spectrum that suggests the possible presence of liquid nitrogen on its surface. This data also suggests that Triton's atmosphere may be significantly more transparent than Titan's. If this model of Triton is verified, the differences between these two cold planetary satellites in deep space will pose something of a puzzle in terms of differing chemistry. [see **A Look at the Outer Solar System**, by Dr. Tobias Owen, **The Planetary Report**, Vol.VII/No.6 (Nov/Dec 1987), pp 16-18. Dr. Owen is a former member of the **Viking** science team (atmospheric analyses associated with the entry and lander GCMS investigations), and he is on science teams representing both the **Voyager** and Galileo missions. In addition, he is the American chairman of the NASA-ESA Joint Science Study Team concerned with the proposed **Cassini** mission to Saturn.] {Ed.}

So, the good news is: we have an exciting planetary encounter ahead of us (Neptune and Triton). The bad news is that the last mission launched by the United States was **Pioneer Venus** (orbiter and probes) in 1978. Both of the **Voyager** spacecraft were launched in 1977 (**Voyager 2** in August and **Voyager 1** in September). The sad prospects are that, aside from **Mars Observer** and **Galileo**, NASA won't be launching any new planetary missions for a long time. The most recent deep space missions were conducted by the Japanese, the Europeans, and the Soviets when their spacecraft rendezvoused with Comet Halley -- an exciting undertaking in which America was not represented. I think the proceedings of this conference are therefore particularly important at this stage, because it helps us renew our resolve to lead in exploring the rest of the planets as well as Mars. We should never again allow such a long hiatus in U.S. planetary missions.

THE NATIONAL COMMISSION ON SPACE

The **National Commission** on Space completed its work and published its final report on the eve of this conference, representing in its recommendations a highly significant step toward a new NASA. We presented the report to the House Committee the morning of July 22, 1986, to the Senate Committee that afternoon, and later to President Reagan. Having done so, we have discharged the responsibility given to the Commission in 1984 to take a long-range look at America's future and national goals in space.

21st Century America

The way in which we were asked to undertake our task had quite a bearing on the nature of our final report. The Commission was not simply asked to extrapolate the U.S. and other space programs out into the next century. Instead, we were told to consider what 21st Century America would be like, and then to propose a long-range U.S. civilian space program that would meet the broader goals of American society over the next 20 years. This was a very thoughtful way for Congress to ask the question, but it presented us with what was indeed a challenging task; we had to develop a sound concept of what 21st Century America would be like before we could develop our recommendations.

At first we hoped to "plagiarize" a description of 21st Century America from other work, but we could find no convincing model. So we rolled up our sleeves and developed our own projection of where the nation is going, basing it on an analysis of American economic history since 1800. We did this through an effort to understand the advancing American technological and economic character in 50-year increments. We examined the last 50 years (1935 to 1985) in detail, but also focused attention on prior 50-year periods (particularly 1885 to 1935).

In reviewing the basic environmental factors that caused the major social changes during those 50-year periods, both in our own country and also around the world, we were struck by the impact of technological advances on medical, industrial, and agricultural productivity, noting that they led to spectacular long-term economic growth and rising living standards. Moreover, at no time in history did we find more opportunity for advancement than we see today. This helped us to understand and project the character of 21st Century America so that we could determine the national goals in space that would be most appropriate for future Americans.

Public Opinion

To help develop sound conclusions on public interest in space for our report, we were instructed by Congress to travel across the length and breadth of the country and hold open forums for citizens from all walks of life. We visited Seattle, Honolulu, Los Angeles, San Francisco, Boston,

Florida, and communities throughout the midwest. From this we got a fascinating insight into the broad, grass-roots support that exists everywhere we visited; Americans want and expect U.S. leadership on the space frontier. They told us so over our microphones in no uncertain terms.

The oldest space enthusiast to address us was a gentleman born in 1894 whose name will be familiar to many: Dr. Hermann Oberth. Over the past 90 years Dr. Oberth has had much to do with the historic development of rocket technology and it's application to planetary travel. He was involved with the founding of the German Rocket Society even before World War I, and is one of the great pioneers of rocket propulsion technology[7]. We also heard from young people (aged 9 to 12) as well as many college students. Overall, we took testimony from a very broad spectrum of Americans who took time off from other pursuits to appear before us.

One of the most interesting citizens to testify came to our first open forum in Los Angeles, the famed author Louis L'Amour. He was very interested in the analogy he sees between the development of our own western frontier and the anticipated movement out into space ("the final frontier") during the next century -- beginning with the colonization of the moon and Mars. Although he feels there will be many differences, he also foresees many similarities. His testimony was particulary valuable; if anybody understands the present mood of Americans it must be Louis L'Amour, surely one of the most widely published authors in America today (as the number of his books available at virtually every airport magazine stand clearly suggests).

In addition to our written "Gutenberg" report, we put together a 30-minute video tape that has been widely distributed. Copies have been given to all of the participants in NASA's **Teacher in Space** Program, for example, and we've also made the tape broadly available to the television industry: the **Voice of America**, public television (PBS), and other electronic media in the U.S. and abroad. When we officially presented our final report to President Reagan in the Oval Office, we gave him not only the standard **Bantam Book** version but a nice leather-bound volume (which we are confident will be able to withstand the heat, smog, glare, and seismic upheavals of California) for his future presidential library. We also provided the President with a copy of the video tape and informed him that Louis L'Amour's testimony made him one of our TV stars (Mr. L'Amour is one of the President's favorite writers). The President expressed great delight at this and indicated his intention to view the tape at the first opportunity[8].

Changing Mood Suggests Commission Rationale

Now America faces a major problem: what goals should we set for our civilian space program. It seems remarkable to me that after so many years with little consideration of long-range

---

[7] Hermann Oberth, as an Austro-Hungarian mathematician, published his highly successful: **The Rocket Into Interplanetary Space** in 1923, and **Way to Space Travel** in 1929. The first paper effectively proposed, without knowledge of similar work already performed by Robert Goddard (USA) or Konstantin Tsiolkovski (USSR), how rockets could achieve escape velocity. The second addressed the concept of ion propulsion 30 years in advance of serious work on such technology. He later worked for his former assistant, Wernher von Braun, at the German Rocket Development Center at Peenemünde. He lived for a time in the U.S. and made valuable contributions to the American space program during its infancy. {Ed.}

[8] As noted elsewhere in these proceedings, the final report of the **National Commission on Space** was released nationally as **A Bantam Book** under the title: Pioneering the Space Frontier (May, 1986). Sadly, author Louis L'Amour died in June, 1988, as these proceedings were being delivered to AAS for publishing. {Ed.}

goals, this Mars Conference publicly presents--on the same agenda--representatives of both NASA and the Mars Underground mutually focusing on such goals. Hurrah! The Mars Underground seems to be surfacing at last -- perhaps even becoming semi-respectable. It may now finally be possible for speakers from NASA installations to eliminate (from the introductions to their Mars papers) those lengthy disclaimers assuring everyone that the opinions expressed do *not* represent the thoughts of: the OMB ... the NASA Administrator ... the Deputy NASA Administrator ... the Associate Administrator ... the speaker's boss, and, in fact, may not even represent the official views of the speaker him/herself! Listening to authors at a **Case for Mars** conference several years ago, I became quite a connoisseur of disclaimers. Remembering what it was like then gives me a feeling of relief that the long-held but restrained anticipation of Mars' potential for mankind is at last compatible with NASA's concept of the solar system.

The Commission: Creation, Rationale, Product

When I testified before the Congressional Committee considering the creation of the **National Commission on Space**, and was asked whether I thought the proposed commission should be directed to do a long-range planning study for Congress, I said I did not think it was necessary and that if the committee wanted to know where NASA should be going, it should simply ask the NASA Administrator. James Beggs, who was sitting beside me and was then NASA Adminis-trator, chimed in with: "That's right, I've got file cabinets bulging with plans for the future of NASA." I knew he did, of course, because I'd filled half of those files myself! Despite my testimony, however, Congress elected to create the Commission. And then, in one of those tricks-of-fate that convince me that God is a humorist, President Reagan appointed me as its chairman.

I was wrong and Congress was right -- we *did* need an independent Commission! When we called on NASA to define their current thinking on future expeditions to Mars, with respect to "living off the land" in closed-ecology life support environments [see NMC-3G/3H by P. Boston and C. McKay], we were gratified on one hand but dispirited on the other. Many of the viewgraphs they presented in their response were the very same ones I'd authorized while Administrator sixteen years earlier. It's nice to see one's work last, but....! The thumb prints that appeared on several of them were undoubtedly those of my old friend, the late Wernher von Braun, still trying to help us move forward. That 1970 material is now clearly outdated, considering what has been presented in these proceedings, but isn't it encouraging to see that NASA can now generate fascinating new ideas and present them at a meeting like the **NASA Mars Conference!?** That is evidence of U.S. planning and technology moving forward over a broad front, thereby giving us the kind of knowledge base we're going to need to support moving out to Mars. It tells me that the idea of a Mars outpost may indeed be coming of age.

Value and Nature of Report

I hope that everyone who reviews these proceedings--but has not yet read the final report of the **National Commission on Space**--will read it soon, taking time to consider the balanced pro-gram we've recommended. Our recommendations include a vigorous science program, low-cost access to orbit, and a bold program for exploring the planets. Early missions represent a pros-pecting phase that gradually leads to eventual settlement using the techniques discussed in the proceedings of this conference. Moreover, the comprehensive nature of the long-range plan includes a broad space applications program that can produce short-term economic returns. Finally, our long-range goals are consistent with the continuing growth we foresee for the American economy, and we're confident that they can be achieved within a reasonable, sustained NASA budget.

As we approach the 500th anniversary of the discovery of the new world, it seems to me that re-investing 0.5 percent of our growing gross national product in space exploration is a very small price to pay in return for an even greater new frontier. This expenditure seems all the more reasonable when one considers that Americans are the beneficiaries of all of the pioneering generations who preceded us in developing the new world. Why shouldn't we enjoy the challenges our ancestors faced? Why shouldn't our children? From this perspective, I think that the financial requirements of the program our Commission has laid out are indeed reasonable. The question is really one of national vision, resolution, and leadership.

We hope very much that the positive message voiced during our nation-wide hearings will give Washington leaders the inspiration and the courage to set America's goals far beyond low Earth orbit. We were specifically directed by Congress not to look at NASA's STS or space station programs, but to concentrate our attention on space goals that would challenge the nation over the next twenty years. And that's what our final report presents: bold space goals for 21st Century America. It received a fine reception in the Congress, which initiated and provided the money for our work. The book represents 1.2 million dollars worth of time and effort--staff, people, travel, conferences, publication costs--over the course of the year. Our budget was 1.5 million, however, so we came in substantially under budget as well as on time! We are proud of that achievement, which we feel represents a promising first step on the high road to Mars.

A Question of Commitment and Leadership

The real question is whether--in Washington today--the vision and statesmanship exists to institute the sustained, long-range programs required to establish the new technology base America will need, and then whether our nation can maintain a purposeful year-after-year commitment to capitalize on that base and move steadily outward in the solar system. The answer to this complex question will determine whether the United States can achieve its far-reaching objectives in the decades to come and maintain the leadership it has held in the past.

Are the American people and their representatives ready to support U.S. leadership on the space frontier? I'm sure that if we asked the question the other way: Are Americans content to give up that leadership? the answer would be a resounding *"NO!"* But, when we face the necessity of providing sustained support over the next twenty years, which is what will be needed to open both the moon and Mars to human activity, the wavering begins. Politicians agree that bold exploration is a fine policy *after* 1990, but until then (they tell us) we've got serious immediate problems to solve and that "now" is not (and rarely is) a good time to address a long-range commitment. This is not how leadership is achieved; this is how it is surrendered.

I submit that "now" *IS* the time to establish a national commitment, and that it should set NASA's sights beyond low Earth orbit: to initiate a bold science program, to rebuild our space technology base, and to develop new transportation systems for low cost access to space. Now, in the wake of the Challenger tragedy, while NASA is pausing to regroup and rebuild; this is the right time to set a bold future course for America's civilian space program. We need to look at how the rest of the world is accelerating space activities and then find the determination to maintain America's leadership. Indeed, our Commission report recommends a pioneering thrust that can be conducted in concert with our allies, perhaps in partnership with the Soviet Union, if either or both should prove to be mutually beneficial to our goals. Such a program would allow the Soviet Union to examine its own Mars plans relative to ours, so that the national leaders of our two countries might develop new opportunities to increase harmony and cooperation in space as an alternative to the atmosphere of suspicion and confrontation in which we find ourselves here on Earth.

The Kennedy Legacy: A Model for Leadership

Twenty-five years ago, President John F. Kennedy motivated and mobilized America to demonstrate its full capacity for leadership in sailing the new ocean of space by committing NASA to the national goal of landing American astronauts on the moon. He knew there were other national problems to solve, but he foresaw the immense value of going to the moon and accepted the challenge "...not because it is easy, but because it is hard." He added that in "accomplishing that difficult task, we will learn all the things we need to be pre-eminent." President Kennedy's Apollo-commitment speech, delivered to a cheering joint session of Congress just four days before his 44th birthday in 1961, is a superb example of the strong presidential leadership we need. And indeed, Apollo **did** give America pre-eminence in space that lasted into the 1980's. Unfortunately, there has been no comparable presidential statement of national goals in space since those stated so eloquently by President Kennedy were achieved, and the opportunity afforded by the remarkable achievement his commitment made possible has since been shamefully abandoned to the corrosive effects of time and neglect. Decisions to build space shuttle (the space transportation system) and to develop a space station were hardware decisions, not thoughtful determinations of what America should accomplish in space over the long term.

Now the time has come for the nation to move forward once again and to set appropriate goals in space. The report of the **National Commission on Space**, with it's emphasis on bold science objectives and on moving out to the moon and Mars, is the kind of civilian space agenda our country should adopt. We must set our sights on Mars, not as an Apollo-like, short-term "space spectacular," but with a sustained, long-term commitment to achieving a permanent presence on that planet that can truly expand the future of mankind. Such a goal will "pull through" the new technology base our nation needs for the future, from robotics and artificial intelligence to aerospace planes and closed-ecology life support systems. A commitment to such a national goal would also allow us to more precisely plan our space station program and set the development of future orbital systems on an appropriate schedule. It would also facilitate the evolution of the space station to the kind of future orbiting spaceport needed to support America's deep space missions. A clearly stated long-range goal of establishing a martian base (outpost) would allow America to convincingly demonstrate to other nations where NASA is going so that they, in turn, can determine the degree to which they would like to participate and contribute. Finally, young people considering careers in space would be given the confidence they need to feel secure in selecting appropriate study programs to further their professional careers.

PRINCIPAL LONG-RANGE RECOMMENDATIONS

In particular, I think that the Commission's recommendations to develop a network of transportation systems and bases throughout the inner solar system provide NASA's manned space flight program with a much needed long-range rationale. Indeed, these goals address the essential role of humanity in space and optimize our basic human characteristics -- the urge to explore and to extend mankind beyond the boundaries of current existence. The solar system could hardly have been designed better, with "Mother Earth" as the magnificent womb for the birth of life and rise of intelligence whose true destiny lies elsewhere. Our species has finally evolved to the point where it can rationally develop technology to free it from the comfortable but confining cradle of our own planet. Civilization on Earth is now nearly 200 years into the industrial revolution: that subtle, exponential advance in society's productive (if not always sensible) use of resources and labor through its continuing progress in science, technology, and economic organization. Moreover, we have now taken our first steps outward into space and our explorers have set foot on another world -- Earth's moon. Scientists and astronauts have already opened the new frontier there (the Apollo Program), for the moon can serve as a stepping stone to worlds that lie beyond. In this way the Malthusian limits on the human race can be removed[9]; our future will be what we make it.

We've already sent our robotic precursors throughout the inner solar system and we're in the last stages of our reconnaissance of the outer planets. We understand a great deal about conditions on Venus with respect to that planet's unsuitability for even extended robotic activity. We also understand a great deal about Mercury, the moon, Mars (and its satellites), the satellites of the outer planets (where only those of Neptune and Pluto still tantalize us with sparse, uncertain clues about their characteristics), and the receptivity that the environments of these distant worlds represents for long-range robotic and human activities. Now we're ready to tackle the problems of "living off the land" on other worlds through next-generation technology.

In Full Agreement

One of the great pages in our report, as far as I'm concerned, is the one that contains the fifteen signatures of our commissioners (page 22). Fifteen Americans, coming from very different walks of life, all agreed that NASA should adopt bold, long-range goals consistent with 21st Century America's "new society." With their signatures, they have stated that we need a renaissance in space sciences and technologies, and that we need a sustained national commitment to extend the human frontier throughout the inner solar system. As Wernher von Braun once pointed out: "The 21st Century will be known as the *extraterrestrial century*." Our report concludes that America should be a leader in this great pioneering movement. The United States is the wealthiest nation on Earth and we have the most advanced technology. More importantly, perhaps, we have the pioneering heritage that made our country what it is today, and it is for this reason as much as any other that I believe America will insist that our nation lead the world out onto the space frontier.

CONCLUSION

When I handed our Commission's final report to President Reagan, he said to me: "I know what you want me to say, and I'm going to say it. We *will* carry out the recommendations of the Commission." The President's acting science advisor, Dr. Richard Johnson, was given the responsibility of preparing an administration response to our report, and Congress concurrently began preparing legislation instructing the NASA Administrator to study our report and to propose the steps that NASA would need to take to move toward achieving our recommended long-range goals. In response, Jim Fletcher has assigned hundreds of NASA's top people to the task of rethinking long-range goals in space.

---

9 Malthusian theory, (Thomas Malthus, 1766-1834), holds that the expansion of humanity (population growth) will generally outstrip production capacity. It suggests that as Earth populations grow ever larger, unchecked by needed changes in social attitude, the planet's resources and production capacities will gradually be overwhelmed by the mass of humanity they must support. This process would have the effect of increasing the focus of humanity on stresses induced with respect to subsistence and survival, thereby reducing opportunities for advancement in other areas of social improvement and technology. Malthusian theory suggests that Earth will achieve a maximum population at which both technical progress and further population growth will essentially be neutralized by global economic poverty and resource exhaustion. While many of the projections made in conjunction with the theory have been suppressed by technology factors Malthus could not have foreseen, as in the cases of food production and industrial/resource capacities, or the capability of politically circumscribed and motivated populations to wage wars of mass destruction on each other, the human problems he identified and the results he believed they would produce appear to be increasingly in evidence today. {Ed.}

If space professionals and enthusiasts who feel as we do get behind our political leaders and let them know that our nation should adopt a program like that proposed by the **National Commission on Space**, the U.S. civilian space program will once again lead the world. The United States can find the money to do it, can put in place the necessary management, and can mobilize the technological, industrial and academic resources needed. We must once again organize those magnificent academia-industry-government teams that achieved such great success for America over the past 50 years whenever they've been challenged to achieve great national goals and have been given sustained support. A new generation is ready to show that a young America still has the pioneering spirit that built our nation.

The American public is, in my view, far ahead of its leaders when it comes to civilian space goals. The Commission has given America a rational plan. I don't think anybody would be more pleased to see NASA moving again than President Reagan. So, *let's get going!* ■(10)

Comments by author Louis L'Amour
for **Pioneering the Space Frontier**

*Everything that has happened in the world,
from the beginning of time up to this moment,
is preliminary. That's the threshold.
Now the real world begins.
We start moving out into space.*

*There is no end to the possibilities.
No end to the exploration we can do.
And I think that we're the people who can do it.*

*There's always going to be those people
who look out there and say
I wonder.*

And then they're going to find out.

This will definitely be the great age of pioneering.

Louis L'Amour died June 10, 1988
His knowledge of America's pioneering spirit
will live on in his words and in his books
as an inspiration for all that lies ahead

---

10 Since the completion of the **National Commission on Space** report, a variety of studies have been initiated in association with some of the Commission's recommendations -- although significant decisions have not yet been made with respect to major goals or programs. In addition, the NASA Administrator formed a task group "to define potential U.S. space initiatives, and to evaluate them in light of the current space program and the nation's desire to regain and retain space leadership." The group was chaired by former astronaut Dr. Sally K. Ride, and its report, titled: <u>NASA Leadership and America's Future in Space</u>, was completed and submitted in August, 1987. {Ed.}

REFERENCES

This space is otherwise available for notes.

~~~~~~~~

APPENDIX A

NASA MARS CONFERENCE

Color Display Section

This section of the NASA MARS CONFERENCE proceedings presents the color illustrations referenced by individual authors within the context of a number of the Mars Conference presentations. As noted in the contents, each page may be folded out face up, such that all of its color illustrations will be visible to the reader while reading in the front of the book. Constraints imposed by cost considerations and AAS convention prevented their use within the text, but it is hoped that this presentation technique will make the illustrations convenient and useful to readers of the Mars Conference proceedings.

Mars Observer -- 1992

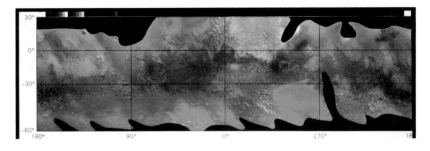

1B-8, Ref Pg 51

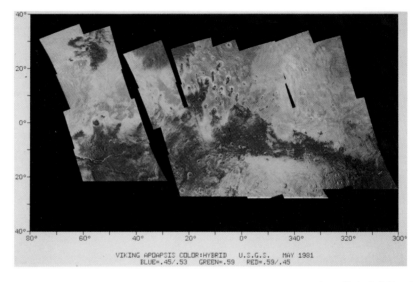

VIKING APOAPSIS COLOR:HYBRID   U.S.G.S.   MAY 1981
BLUE=.45/.53   GREEN=.59   RED=.59/.45

1B-9, Ref Pg 51

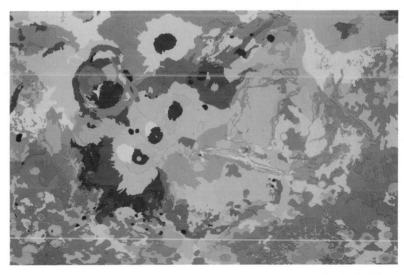

1B-10, Ref Pg 51

Figure
Reprocessed Viking
Image of South Pola

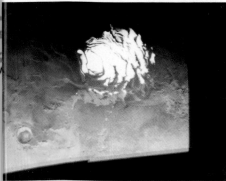

Figure, Ref Pg 52

Sinusoidal Project
of South Pola

, Ref Pg 52

Figure

Reprocessed, E
Viking Orbiter Image Data (

Figure

Reprocessed, E
Viking Orbiter Image Data (

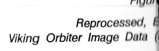
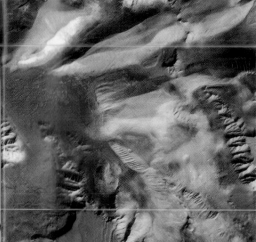

Figure
Reprocessed, E
Viking Orbiter Image Data (, Ref Pg 52

1B-14, Ref Pg 52

1B-15, Ref Pg 52

2G-2, Ref Pg 326

2G-23, Ref Pg 346

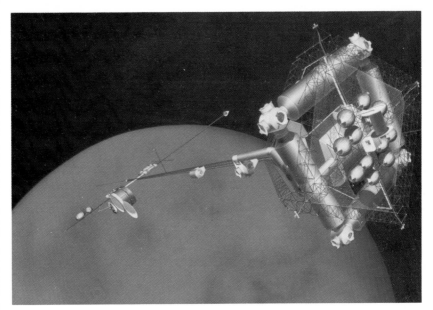

3D-13, Ref Pg 444

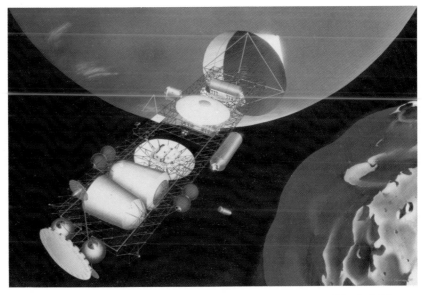

3D-16, Ref Pg 445

3D-23, Ref Pg 453

3G-17, Ref Pg 503

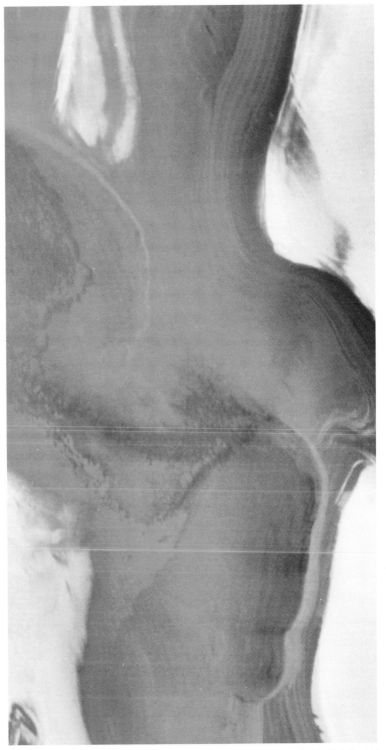

3H-3, Ref Pg 514

APPENDIX B

NASA MARS CONFERENCE
Presentation Illustrations

A·# page numbers are for illustrations presented in the Color Display Section appendix. Page numbers in parentheses are those on which color illustrations are referenced and discussed.

NASA Mars Conference figure index continued on next page

NASA Mars Conference figure index continued on next page

APPENDIX B

NASA MARS CONFERENCE
Presentation Illustrations

*A·# page numbers are for illustrations presented in the Color
Display Section appendix. Page numbers in parentheses are those on
which color illustrations are referenced and discussed.*

FIGURE CAPTION   PAGE

NASA Mars Conference figure index continued on next page

NASA Mars Conference figure index continued on next page

NASA Mars Conference figure index continued on next page

NASA Mars Conference figure index continued on next page

NASA Mars Conference figure index continued on next page

NASA Mars Conference figure index continued on next page

NASA Mars Conference figure index continued on next page

NASA Mars Conference figure index continued on next page

NASA Mars Conference figure index continued on next page

NASA Mars Conference figure index continued on next page

NASA Mars Conference figure index continued on next page

NASA Mars Conference figure index continued on next page

NASA Mars Conference figure index continued on next page

NASA Mars Conference figure index concluded on next page

■■■

NUMERICAL INDEX

AUTHOR INDEX[*]

---

[*] For each author the AAS number is given. The page numbers refer to Volume 71, Science and Technology Series.

AMERICAN *ASTRONAUTICAL* SOCIETY

AVAILABLE IN TWO VOLUMES IS AN INDEX TO ALL AMERICAN ASTRONAUTICAL SOCIETY PAPERS AND ARTICLES 1954-1985

This index is a **numerical/ chronological index** *(which also serves as* **citation index***) and an* **author index.** *(A subject index volume will be forthcoming in 1987.)*

It covers all articles that appear in the following:

Advances in the Astronautical Sciences (1957-August 1986)

Science and Technology Series (1964 - September 1986)

AAS History Series (1977 - 1986)

AAS Microfiche Series (1968 - August 1986)

Journal of the Astronautical Sciences (1954 - March 1986)

Astronautical Sciences Review (1959 - 1962)

If you are in aerospace you will want this excellent reference tool which covers the first 30 years of the Space Age.

Numerical/Chronological/Author Index in two volumes, Library Binding (both volumes) $95; Soft Cover (both Volumes) $80; Volume 1 (1954-1978) Library Binding $40; Soft Cover $30; Volume II (1979-1985/86) Library Binding $60; Soft Cover $45. Order from Univelt, Inc., P.O. Box 28130, San Diego, California 92128.